Linear Algebra

Vector Spaces and Linear Transformations

Pure and Applied
UNDERGRADUATE TEXTS • 57

Linear Algebra

Vector Spaces and Linear Transformations

Meighan I. Dillon

AMERICAN MATHEMATICAL SOCIETY

Providence, Rhode Island USA

EDITORIAL COMMITTEE

Giuliana Davidoff   Tara S. Holm
Steven J. Miller   Maria Cristina Pereyra
Gerald B. Folland (Chair)

Cover credit: GeorgePeters/DigitalVision Vectors via Getty Images.

2020 *Mathematics Subject Classification.* Primary 15-01.

For additional information and updates on this book, visit
www.ams.org/bookpages/amstext-57

Library of Congress Cataloging-in-Publication Data

Cataloging-in-Publication Data has been applied for by the AMS.
See http://www.loc.gov/publish/cip/.
DOI: https://doi.org/10.1090/amstext/57

Copying and reprinting. Individual readers of this publication, and nonprofit libraries acting for them, are permitted to make fair use of the material, such as to copy select pages for use in teaching or research. Permission is granted to quote brief passages from this publication in reviews, provided the customary acknowledgment of the source is given.

Republication, systematic copying, or multiple reproduction of any material in this publication is permitted only under license from the American Mathematical Society. Requests for permission to reuse portions of AMS publication content are handled by the Copyright Clearance Center. For more information, please visit www.ams.org/publications/pubpermissions.

Send requests for translation rights and licensed reprints to reprint-permission@ams.org.

© 2023 by the American Mathematical Society. All rights reserved.
The American Mathematical Society retains all rights
except those granted to the United States Government.
Printed in the United States of America.

∞ The paper used in this book is acid-free and falls within the guidelines
established to ensure permanence and durability.
Visit the AMS home page at https://www.ams.org/

10 9 8 7 6 5 4 3 2 1   28 27 26 25 24 23

In memory of John Loustau

Contents

List of Figures	xi
Preface	xiii
How To Use This Book	xvii
Notation and Terminology	xxi
To the Student	xxiii
Introduction	1
Chapter 1. Vector Spaces	3
1.1. Fields	3
1.2. Vector Spaces	8
1.3. Spanning and Linear Independence	14
1.4. Bases	18
1.5. Polynomials	24
1.6. $\mathbb{R}$ and $\mathbb{C}$ in Linear Algebra	28
Chapter 2. Linear Transformations and Subspaces	31
2.1. Linear Transformations	32
2.2. Cosets and Quotient Spaces	35
2.3. Affine Sets and Mappings	41
2.4. Isomorphism and the Rank Theorem	44
2.5. Sums, Products, and Projections	47
Chapter 3. Matrices and Coordinates	55
3.1. Matrices	55
3.2. Coordinate Vectors	61

3.3.	Change of Basis	66
3.4.	Vector Spaces of Linear Transformations	69
3.5.	Equivalences	73

Chapter 4. Systems of Linear Equations — 77
Introduction — 77
- 4.1. The Solution Set — 78
- 4.2. Elementary Matrices — 83
- 4.3. Reduced Row Echelon Form — 86
- 4.4. Row Equivalence — 89
- 4.5. An Early Use of the Determinant — 93
- 4.6. LU-Factorization — 98

Chapter 5. Introductions — 107
- 5.1. Dual Spaces — 107
- 5.2. Transposition and Duality — 112
- 5.3. Bilinear Forms, Their Matrices, and Duality — 116
- 5.4. Linear Operators and Direct Sums — 121
- 5.5. Groups of Matrices — 126
- 5.6. Self-Adjoint and Unitary Matrices — 130

Chapter 6. The Determinant Is a Multilinear Mapping — 133
- 6.1. Multilinear Mappings — 133
- 6.2. Alternating Multilinear Mappings — 136
- 6.3. Permutations, Part I — 139
- 6.4. Permutations, Part II — 145
- 6.5. The Determinant — 149
- 6.6. Properties of the Determinant — 152

Chapter 7. Inner Product Spaces — 157
- 7.1. The Dot Product: Under the Hood — 157
- 7.2. Inner Products — 162
- 7.3. Length and Angle — 164
- 7.4. Orthonormal Sets — 168
- 7.5. Orthogonal Complements — 174
- 7.6. Inner Product Spaces of Functions — 176
- 7.7. Unitary Transformations — 182
- 7.8. The Adjoint of an Operator — 187
- 7.9. A Fundamental Theorem — 191

Chapter 8.	The Life of a Linear Operator	199
8.1.	Factoring Polynomials	199
8.2.	The Minimal Polynomial	201
8.3.	Eigenvalues	206
8.4.	The Characteristic Polynomial	212
8.5.	Diagonalizability	216
8.6.	Self-Adjoint Matrices Are Diagonalizable	221
8.7.	Rotations and Translations	223
Chapter 9.	Similarity	231
9.1.	Triangularization	231
9.2.	The Primary Decomposition	234
9.3.	Nilpotent Operators, Part I	242
9.4.	Nilpotent Operators, Part II	245
9.5.	Jordan Canonical Form	248
Chapter 10.	$GL_n(\mathbb{F})$ and Friends	255
10.1.	More about Groups	256
10.2.	Homomorphisms and Normal Subgroups	259
10.3.	The Quaternions	264
10.4.	The Special Linear Group	270
10.5.	The Projective Group	274
10.6.	The Orthogonal Group	281
10.7.	The Unitary Group	285
10.8.	The Symplectic Group	290
Appendix A.	Background Review	297
A.1.	Logic and Proof	297
A.2.	Sets	301
A.3.	Well-Definedness	306
A.4.	Counting	308
A.5.	Equivalence Relations	312
A.6.	Mappings	315
A.7.	Binary Operations	318
Appendix B.	$\mathbb{R}^2$ and $\mathbb{R}^3$	321
B.1.	Vectors	321
B.2.	The Real Plane	324
B.3.	The Complex Numbers and $\mathbb{R}^2$	326
B.4.	Real 3-Space	328

B.5.	The Dot Product	332
B.6.	The Cross-Product	334
Appendix C.	More Set Theory	341
C.1.	Partially Ordered Sets	341
C.2.	Zorn's Lemma	343
Appendix D.	Infinite Dimension	351
Bibliography		357
Index		359

List of Figures

2.1.	The cosets of $W = \mathrm{Span}\{(1,1)\} \subseteq \mathbb{R}^2$ comprise a parallel class of lines in the xy-plane.	36
2.2.	Three parallel lines in $\mathbb{F}_3{}^2$.	42
2.3.	$L[V] \cong V/\mathrm{Ker}\, L$ and $\tilde{L} \circ f = L$.	45
3.1.	$[\![L]\!]_\mathcal{B}^\mathcal{C} \circ []_\mathcal{B} = []_\mathcal{C} \circ L$.	63
3.2.	L is the orthogonal reflection through ℓ in $\mathbb{R}^2$.	67
4.1.	The plane in $\mathbb{R}^3$ given by $x + y + z = 1$.	80
4.2.	The curves described by equations in (4.2).	81
4.3.	Add products along the arrows pointing down; subtract products along the arrows pointing up.	98
7.1.	The dot product links coordinates of points in $\mathbb{R}^2$ to the Pythagorean Theorem.	158
7.2.	The Law of Cosines holds in any inner product space.	168
7.3.	Orthogonal complements in the domain and codomain of a matrix transformation $\mathbb{R}^n \to \mathbb{R}^m$.	191
7.4.	Orthogonal complements in the domain and codomain of a matrix transformation $\mathbb{C}^n \to \mathbb{C}^m$.	193
8.1.	L is the projection onto the line $x + y = 0$ along the line $x + 4y = 0$.	207
8.2.	The shear on $\mathbb{R}^2$ that fixes $\mathbf{e}_1$ and maps $\mathbf{e}_2$ to $(2,1)$.	208
8.3.	A nonorthogonal reflection on $\mathbb{R}^2$ determined by ℓ.	209
10.1.	Four symmetries of an equilateral triangle.	258
10.2.	Multiplication rule for quaternions.	266

10.3.	A negatively oriented parallelogram in $\mathbb{R}^2$.	271
10.4.	A shear on $\mathbb{R}^2$ in the x-direction.	272
10.5.	$\int_a^b f(x)\,dx = \int_a^c f(x)\,dx + \int_c^b f(x)\,dx.$	274
10.6.	The sides of a projective triangle are lines, not line segments.	277
10.7.	The complete quadrangle determined by vertex set $\{P_1, P_2, P_3, P_4\}$.	278
10.8.	The Fano plane can be modeled using $\mathbb{F}_2^3$.	279
10.9.	The angle between $\mathbf{v}$ in $\mathbb{R}^3$ and the z-axis is φ. The angle between the orthogonal projection of $\mathbf{v}$ onto the xy-plane and the x-axis is θ.	284
A.1.	A Venn diagram.	304
A.2.	Coterminal angles in $\mathbb{R}^2$.	315
B.1.	Identify a point in $\mathbb{R}^2$ with an arrow.	322
B.2.	A vector has many different representations as arrows in the xy-plane.	322
B.3.	Adding vectors by the Parallelogram Rule.	323
B.4.	Proving the Law of Cosines using $\mathbb{R}^2$.	325
B.5.	Polar coordinates for $\mathbf{v}$ in $\mathbb{R}^2$ come from basic trigonometry.	328
B.6.	The standard coordinatization of $\mathbb{R}^3$.	329
B.7.	A parametric description of a line in $\mathbb{R}^3$ depends on a point and a direction vector.	330
B.8.	The projection of $\mathbf{v}$ onto $\mathbf{w}$ in $\mathbb{R}^n$.	333
B.9.	The force due to gravity on the block is indicated by the arrow.	334
B.10.	The areas of the quadrilateral and the rectangle are identical.	336
B.11.	If $\mathbf{v}$ and $\mathbf{w}$ determine the base of the parallelepiped given by $\mathbf{v}, \mathbf{w}, \mathbf{u}$, then the height of the parallelepiped is $\|\mathbf{u}\|\|\cos\varphi\|$, where φ is the angle between $\mathbf{v} \times \mathbf{w}$ and $\mathbf{u}$.	336
B.12.	Cross-products in a right-handed coordinate system for $\mathbb{R}^3$.	337
B.13.	To construct a right-handed coordinate system, start by choosing $\mathbf{u}_2$ to the left of $\mathbf{u}_1$.	338
D.1.	$(\mathcal{B}_1, L_1) \leq (\mathcal{B}_2, L_2)$ means $L_1 = L_2$ on $\mathcal{B}$ except on $\mathcal{B}_2 \setminus \mathcal{B}_1$.	353

Preface

As a springboard for so many technical applications, linear algebra is now taught to a large and varied audience of undergraduate students. The courses we offer these students straddle two worlds. One revolves around solutions to systems of linear equations and matrix manipulation. The second revolves around the algebraic framework that supports the first. Courses — or segments of courses — designed to introduce students in the early years of college to linear algebra are often squarely planted in the world of systems of linear equations and matrices. Whether their first experience with the subject was from a dedicated linear algebra course that de-emphasized proof, from a calculus course with a linear algebra component, or from a strong high school background in algebra, many students can benefit from a second exposure to the subject. The power of linear algebra derives from its algebraic underpinnings. A basic understanding of those underpinnings better positions students to use the tools linear algebra offers.

This book was written for undergraduate students who already know how to work with matrices and how to use them to solve systems of linear equations. Students in the intended audience have the curiosity and ambition to take on the algebra underlying, and emanating from, a study of systems of linear equations and matrix manipulation. Typically, this audience is a mix of students studying or intending to study mathematics, science, engineering, or computer science. The approach here is proof-based with an emphasis on algebraic foundations. As important as applications and software are, our aim is not to explore those, but to set out the mathematics that supports the tools students will need for applications, inside and outside mathematics, with and without the help of machine computing.

This may be the terminal mathematics course for many students. That makes it the perfect setting for opening their eyes to the theories of groups, affine geometry, fields, and algebras. We introduce these ideas early in the course and refer to them throughout the book. Once aware of them, students will see them in the applications and motivating examples they are likely to meet. These are some of the elements that make linear algebra such a workhorse.

Aside from offering a general grounding in the algebra associated to vector spaces and linear transformations, the text has two more specific goals. One is to undertake the journey necessary to answer a question I have heard students ask repeatedly: What is a determinant? The second is to prove that every vector space has a basis.

Many of us would agree that "[d]eterminants are difficult, non-intuitive, and often defined without motivation" [4]. Several years ago, while teaching an introductory linear algebra course — and avoiding determinants — I was confronted by a student who said, "We've learned about determinants in other courses: Why can't we use them in here?" At that moment I realized that this simple question had an answer only a mathematician could love. I resolved then to find an introduction to the determinant that would justify its use in calculations but that was more satisfying than a definition via cofactor expansion. In the meantime, Danny Otero's note [19] about Cramer's use of determinants crossed my desk. This gave me what I was looking for. It placed the determinant in an interesting historical context that would be meaningful to any student who ever met an $n \times n$ system of linear equations. Students could have their coveted determinants early in the game, and we could go on from there to pursue other ways of looking at the determinant.

That every vector space has an algebraic basis is a fundamental theorem that is routinely given short shrift, or overlooked entirely, in the undergraduate curriculum. Once the shock of having to use Zorn's Lemma wears off, though, it is not difficult to prove. Related, but more delicate, is proof that the dimension of a vector space is well-defined. We prove both results here, the latter in an appendix.

These three topics — determinants, bases, dimension — inform the development of the course.

The first four chapters of the book roughly follow the arc of a traditional linear algebra course taken by college sophomores, but in more detail, more generality, and with more supporting theory. Chapter 5 introduces some of the ideas that make linear algebra so rich: dual spaces, for instance, and special types of matrices that take us into the classical groups.

Chapter 6 treats enough multilinear algebra to provide context to the definition of the determinant in terms of the nth exterior power of an n-dimensional vector space. Chapter 7 develops real and complex inner product spaces, positioning us for least-squares approximation and Gil Strang's *Fundamental Theorem of Linear Algebra* [21]. In Chapter 8, we look at how to study a linear operator via the polynomial algebra it generates. This leads naturally to the idea of the minimal polynomial and, thence, to eigenvectors and the finite-dimensional spectral theorem for self-adjoint operators. In Chapter 9, we work through Jordan canonical form. By this point, the general linear group and several of its notable subgroups have made appearances. We take them on more earnestly in Chapter 10, where we discuss enough introductory group theory to define normal subgroups. With normal subgroups in hand, we define the "special" versions of the general linear group, orthogonal group, and unitary group as kernels of the determinant mapping. Normal subgroups give us access to the projective group, thus to projective spaces and more connections between linear algebra and geometry. The quaternions make an appearance in Chapter 10, as does the symplectic group.

Because the expectation is that students using this book have at least a passing familiarity with linear algebra, words like *vector space, scalar, basis,* and *dimension* enter the conversation before they are formally introduced. Once the formal definitions appear, so do discussion and exercises that assume a reasonable technical command of the terminology and associated results. Students needing a refresher with concrete examples should work through Appendix B, which supplies a treatment of elementary properties of $\mathbb{R}^2$ and $\mathbb{R}^3$, including the meaning of the word *vector* as it arises in high school physics and mathematics classes. The Euclidean inner product, the cross-product, and the complex numbers also come under discussion in Appendix B.

A brief explanation for some stylistic choices may be helpful. Generally, a theorem is a result that captures a property essential to the development here or essential to understanding an object at hand. A lemma either paves the way towards a theorem or lacks the heft of a theorem. A corollary follows nearly immediately either from the statement or the proof of a previous result. Definitions are usually, but not always, framed apart from the narrative flow. Definitions that seem to be universal among mathematicians working today are in bold. Definitions of devices or notation that are convenient but perhaps inessential may appear in italics as part of the narrative flow. The first mention of a term that may be unfamiliar to students is sometimes set in italics, as well.

The word *canonical* comes up frequently in the text. Its first appearance is in connection to the mapping $V \to V/W$ where V is a vector space and W is the kernel of a linear transformation. In my experience, people who apply linear algebra use the word canonical freely and the dictionary definition does not capture the way mathematicians use it. Students should know what it means and how it is used.

My choice for denoting the conjugate transpose of a matrix using a dagger instead of an asterisk is an attempt to obviate the confusion students may have in distinguishing the adjoint of a matrix from the dual of a linear transformation. Likewise, my choice to define the Hermitian inner product on $\mathbb{C}^n$ by $\mathbf{v}^\dagger \mathbf{w}$ instead of $\mathbf{v}^T \overline{\mathbf{w}}$ is an effort to remove any barrier to a view of the connection between Hermitian forms and self-adjoint matrices.

Finally, it may be worth noting that where some authors would prefer to introduce the space of $n \times n$ matrices over a field $\mathbb{F}$ as a ring, I chose to introduce it as an $\mathbb{F}$-algebra. An $\mathbb{F}$-algebra is a vector space with a second binary operation, while a ring belongs to an entirely new class of algebraic objects. For the sake of what could be budding mathematics students, I wanted the subject matter here to seem of a piece.

There is much more here than anyone can cover in a one-semester course and the range in difficulty of the material is nontrivial. An instructor using this book will have to make choices. Below I have provided a guide to using the text in a course.

A project like this book takes many years, much support, and a great deal of patience on the part of people who surround the author. I am grateful to Kennesaw State University, which granted me a sabbatical to do the initial work on the book, and to Sean Ellermeyer, who chaired the Department of Mathematics at Kennesaw State in my final years there, while I was getting the book off the ground. My husband, Steve Edwards, offered invaluable support as the cook of the house, as a fellow mathematician and teacher, and as my best friend. His patience seems to know no bounds. The moral support of my adult children, Annabel Grace Edwards and Miles Dillon Edwards, has been a steady, reliable source of solace through the years. As a fellow mathematician and teacher, Miles offered material support for the book, specifically in the form of a proof that an infinite-dimensional vector space has a defined dimension. Eriko Hironaka, at the American Mathematical Society (AMS), was a tremendous help in the preparation of the manuscript. She worked with me for years, with an apparently unshakeable belief in the project. I offer her heartfelt thanks. Finally, I am indebted to the committee members at AMS who offered amendments to my manuscript. I believe their suggestions made this a better book.

I am dedicating this work to John Loustau, whose 79th birthday is today as I write this. John died while I was preparing the manuscript for the book, but his guiding hand shows in everything I do as a teacher and mathematician. He was my teacher at Hunter College, my mentor through graduate school and my career, and finally, one of my dearest friends. Without his influence, it is unlikely I would have become a mathematician.

Meighan I. Dillon
Roswell, GA
July 4, 2022

How To Use This Book

Early in their undergraduate studies, students may be shy around or averse to the complex numbers. To use this text successfully, one must embrace the complex numbers. In light of this, we included a section on complex arithmetic in Appendix B. We visit and revisit properties of $\mathbb{C}$ several times in the narrative, starting in Chapter 1. By the end of any course based on this book, students should be much more comfortable toggling between $\mathbb{R}$ and $\mathbb{C}$ and understanding what to take and what to leave behind when they do so.

Like the complex numbers, spaces of functions have a recurring role in this book. We include a dedicated discussion of complex functions in Section 7.6. While it is fundamental to mathematics that every vector space has a basis — at least for those of us who accept the Axiom of Choice — it is equally fundamental that, in general, you cannot access an algebraic basis for an infinite-dimensional vector space. The text serves regular reminders of both of those concepts.

The Appendices. Students using this book will come from a variety of backgrounds. With the exception of Appendix D, the appendices were written with an eye towards bridging gaps between what students may know and what they will be expected to manage as they navigate a course based on the material here. To take full advantage of the appendices, students and instructors should be familiar with what they can find in them before starting Chapter 1. Depending on the backgrounds of the students, the appendices should be consulted as needed. For example, an instructor who detects student mystification at the first mention of equivalence relations in Chapter 2 may elect to pause the course for a ten-minute lecture based on Section A.5 or to assign readings and problems from that section.

Appendix A. This appendix should help students who have not had a course dedicated to logic, sets, counting, equivalence relations, mappings, and proofs. It includes a discussion of well-definedness, which can be a stumbling block on a first encounter with cosets. An introduction to modular arithmetic on $\mathbb{Z}$ is also included.

Appendix B. This appendix is devoted to $\mathbb{R}^2$ and $\mathbb{R}^3$ as models. For instance, $\mathbb{R}^2$ is a model for high school physics problems, the complex plane, and analytic geometry. One of our objectives in this appendix is to consolidate the results that most students will have seen regarding the dot product and the cross-product. We also include a discussion of the affine geometry of $\mathbb{R}^2$ and $\mathbb{R}^3$.

Appendix C. Proof of the existence of a basis for every vector space, regardless of dimension, requires Zorn's Lemma, which we discuss in Section C.2. Section C.2, in turn, requires knowledge of partially ordered sets, which we discuss in Section C.1.

We also discuss cardinality in this appendix and include a proof that the real numbers are uncountable. This is a fact that we refer to informally throughout the course. Many students will be familiar with it but may not know how to prove it.

Appendix D. This appendix contains proof that the cardinality of a basis is an invariant of any vector space. Instructors who are inclined to include it in the course may wish to outline the proof in class.

The Body of the Text. This book was written so that, as much as possible, results are framed to apply to vector spaces of any dimension, over any field. There was also an effort to include proofs that are often left out in linear algebra textbooks, for instance that matrix multiplication is associative. No instructor should set out to go through all of these in class. Judicious choices will serve both the audience and course goals.

Chapter 1. A vector space is an introductory generalization of a field so we start the course with fields. Throughout the course, we refer to fields with p elements, for p prime, and to the characteristic of a field, but an instructor need not belabor these points. They can be dispatched via examples.

An important example that we refer to throughout the course is the vector space of mappings from any set into a vector space.

Theorem 1.31 implies that every vector space has a basis. At that point, there is a bifurcation in the text: We present a proof of the theorem in the finite-dimensional case and a proof of the general case. Instructors will have to decide whether to work through the proof in the finite-dimensional setting only, work through the proof in the general setting only, or use proof for the finite-dimensional case as a warm-up to proof in the general case. (I recommend the third option.)

The last two sections of the chapter should not be dismissed. The section on polynomials introduces $\mathbb{F}$-algebras. References to algebras recur throughout the course. The section on $\mathbb{R}$ and $\mathbb{C}$ in linear algebra addresses the idea that a complex vector space is also a real vector space, again, something we mention several times throughout the course.

Chapter 2. The emphasis in this chapter is on relationships between linear transformations and subspaces. We treat quotient spaces here, as well as internal and external direct sums, projections, and the affine objects in a vector space. The Rank Theorem is one of the goals of the chapter.

Chapter 3. This chapter is where we introduce coordinatization. The treatment of matrices emphasizes how matrices represent and determine linear transformations. One goal of the chapter is to establish that matrices are equivalent if and only if they have the same dimensions and rank.

Chapter 4. In this chapter, we work with systems of linear equations in the context of vector spaces and linear transformations, showing that the affine sets in $\mathbb{F}^n$ are solutions to $m \times n$ systems of linear equations over $\mathbb{F}$. We treat row reduction here as the result of premultiplication by elementary matrices. A goal of this chapter is proof that the reduced row echelon form of a matrix exists and is unique.

Chapter 5. This is where we introduce several new ideas, among them dual spaces, block matrices, and the conjugate transpose of a matrix. A goal for this chapter is to show that the dual of a matrix transformation is the transpose of the matrix. This is also where we introduce bilinear forms and discuss the connection between dual spaces and nondegenerate bilinear forms. Our treatment of block matrices in this chapter is largely restricted to diagonal block matrices. Block matrices do not have a strong presence in the text until we discuss symplectic matrices in Chapter 10.

Chapter 6. We introduce the symmetric group on n letters, $\mathcal{S}_n$, in this chapter and work up enough material to prove that a permutation in $\mathcal{S}_n$ can be classified as even or odd. This is also where we first discuss group cosets, though normal subgroups do not come up until Chapter 10. As we have mentioned, one goal of Chapter 6 is to see the determinant as an alternating n-linear form on $\mathbb{F}^n$, for an arbitary field $\mathbb{F}$. With this result, properties of the determinant enter the story effortlessly.

Chapter 7. We discuss inner products on real and complex vector spaces here, proving the Cauchy-Schwarz inequality and the Law of Cosines. In Section 7.6, we discuss working with complex-valued functions. One goal of the chapter is to treat the material in [21] about least-squares approximation and the fundamental subspaces associated to a matrix transformation. We generalize treatment of the latter to include matrices over $\mathbb{C}$. We also introduce unitary/orthogonal transformations here.

Chapter 8. The polynomial algebra generated by a linear operator is our subject here. We treat diagonalization in this chapter.

Chapter 9. The goal of this chapter is the Jordan form of a matrix.

Chapter 10. This is an introductory chapter on groups of matrices. Our approach is to see each of these groups as preserving some structure on an underlying vector space. We discuss the quaternions here and make the connection between $\mathbb{H}$ and rotations in $\mathbb{R}^3$. We also introduce projective planes based on vector spaces.

Notation and Terminology

The following sets of numbers come up frequently in our studies: the integers, $\mathbb{Z}$; the real numbers, $\mathbb{R}$; the rational numbers, $\mathbb{Q}$; the complex numbers, $\mathbb{C}$. If S is a collection of numbers that contains real numbers, we use S_+ to designate the positive numbers in S and we use $S_{\geq 0}$ to designate the nonnegative numbers in S.

We use the words *mapping* and *function* interchangeably.

If A and B are sets and there is a mapping from A to B, we may indicate that by writing $A \to B$ without naming the mapping.

When referring to mappings, we use *injective* and *one-to-one* interchageably, and we use *surjective* and *onto* interchangeably. *Bijective* means one-to-one and onto.

We draw on calculus for examples in the course. Though unnecessary for the study of linear algebra, calculus will be familiar to most students at this point in their studies. The set of continuous real-valued functions on $D \subseteq \mathbb{R}$ is $C(D)$, while $C'(D)$ (pronounced "C one functions on D") is the set of functions on D with continuous first derivatives. We add more primes if we want functions on D with continuous higher-order derivatives. The set of functions on D with derivatives of every order is $C^\infty(D)$.

We use $\mathscr{C}([a,b])$ to indicate the set of continuous complex-valued functions defined on the interval $[a,b] \subseteq \mathbb{R} \subseteq \mathbb{C}$.

The notation $:=$ indicates that the object on the left is defined by the expression on the right. For instance, $f(x) := y$ means $f(x)$ is defined to be y.

When we list elements in a set using brackets, for instance $\{\mathbf{v}_1, \ldots, \mathbf{v}_n\}$, the implication is that the $\mathbf{v}_i$s are distinct.

Geometric terminology arises naturally in the study of linear algebra in several different contexts. The words *point*, *line*, and *plane* refer to geometric objects.

Orthogonal, perpendicular, and *at right angles* are all synonymous descriptions for the relative positions of two objects, usually two lines, a line and a plane, or two planes.

Points that lie on one line are *collinear*. Points that do not lie on one line are *noncollinear*. Objects that lie in one plane are *coplanar*. Objects that do not lie in one plane are *noncoplanar*.

The word *intersection* and the locutions *lies on* and *passes through* may be used to describe what we call incidence relations among geometric objects. *Incidence geometry* is the study of points, lines, planes, and possibly higher-dimensional objects and their incidence relations. "Does this line lie in this plane? Does this line pass through this point? Are these three points collinear? Are these two lines coplanar?" These are questions about incidence. "Are these angles congruent? Are these lines orthogonal? Are these triangles similar?" These are questions about geometry but not about incidence.

The words *equation, solution, unknown, variable,* and *parameter* are words we use to talk about algebra. "The point A lies on the line ℓ which intersects the line m at the point B," is an example of a geometric description of something. "$(2, -2/3)$ is a solution to the equation $2x - 3y = 6$. The solution to the system

$$\begin{aligned} 2x - 3y &= 6 \\ x + y &= 3 \end{aligned}$$

is $(3, 0)$." This is an algebraic description, possibly of the same configuration.

We make several references to *axioms*. Another word for *axiom* is *assumption*. One of our assumptions is the Axiom of Choice.

To the Student

Mathematics is a hands-on pursuit. Study it with pen or pencil at the ready, the concepts on simmer in the back of your mind at all times.

Learning linear algebra means learning definitions and how to apply them, first and foremost. The calculations are mostly addition and multiplication, but to see your way through the subject, do them by hand, as much as possible. Very few problems in this book require machinery.

When we want to understand a mathematical fact, the first place we look is to its proof. Study proofs to better understand theorems. Study proofs to learn how to write proofs.

When studying mathematics, remember that it is about ideas. The calculations are done in service to the ideas. When you start to see mathematics this way, a door opens.

Introduction

Let A be an $m \times n$ matrix over $\mathbb{R}$.

If $\mathbf{b}$ is a vector in $\mathbb{R}^m$ and $\mathbf{x}$ is an unknown in $\mathbb{R}^n$, then $A\mathbf{x} = \mathbf{b}$ is a generalization of an equation of the form $ax = b$, where a and b are real numbers and x is an unknown in $\mathbb{R}$. While $ax = b$ is a linear equation in one unknown, $A\mathbf{x} = \mathbf{b}$ represents a system of m linear equations in n unknowns.

As long as $a \neq 0$, the solution to $ax = b$ is
$$x = a^{-1}b.$$
As long as A is a square matrix with nonzero determinant, the solution to $A\mathbf{x} = \mathbf{b}$ is
$$\mathbf{x} = A^{-1}\mathbf{b}.$$
In general, though, A will not be square, and if it is, its determinant may be zero. Depending on the circumstances, the solution to $A\mathbf{x} = \mathbf{b}$ may be one point, many points, or it may not exist at all.

What is it about A and $\mathbf{b}$ that determines whether $A\mathbf{x} = \mathbf{b}$ has a solution? If it has a solution, what — if anything — does that solution have in common with the solution to $A\mathbf{x} = \mathbf{c}$ when $\mathbf{c}$ in $\mathbb{R}^m$ is different from $\mathbf{b}$? If $A\mathbf{x} = \mathbf{b}$ does not have a solution, is there $\mathbf{y}$ in $\mathbb{R}^n$ so that $A\mathbf{y}$ is as close as possible to $\mathbf{b}$ in $\mathbb{R}^m$? The answers to these questions come out of the theories of vector spaces and linear transformations. These are the objects of our study.

Chapter 1

Vector Spaces

A vector space is a generalization of $\mathbb{R}^n$. That its scalars are real numbers makes $\mathbb{R}^n$ a real vector space. That every vector $(a_1, \ldots, a_n)$ in $\mathbb{R}^n$ can be specified with n independent coordinates a_i makes $\mathbb{R}^n$ an n-dimensional vector space.

Here we study vector spaces that are not necessarily real and not necessarily finite-dimensional. Our first order of business, then, is to consider sets of scalars that may not be real numbers.

1.1. Fields

The scalars associated to a vector space comprise a *field*.

Definition 1.1. Let $\mathbb{F}$ be a set with two associative, commutative, binary operations: addition, which we denote $\oplus$, and multiplication, which we denote $\odot$.[1] When all the following axioms hold, $(\mathbb{F}, \oplus, \odot)$ is a **field**.

(a) There is 0 in $\mathbb{F}$ such that $0 \oplus a = a$ for all a in $\mathbb{F}$.
(b) For each a in $\mathbb{F}$ there is $-a$ in $\mathbb{F}$ so that $a \oplus (-a) = 0$.
(c) There is $1 \neq 0$ in $\mathbb{F}$ so that $a \odot 1 = a$ for all a in $\mathbb{F}$.
(d) For each nonzero a in $\mathbb{F}$ there is a^{-1} in $\mathbb{F}$ so that $a \odot a^{-1} = 1$.
(e) For all a, b, and c in $\mathbb{F}$, $(a \oplus b) \odot c = (a \odot c) \oplus (b \odot c)$.

When $(\mathbb{F}, \oplus, \odot)$ is a field, we also say $\mathbb{F}$ is a **field under** $\oplus$ **and** $\odot$.

A **subfield** of a field $(\mathbb{F}, \oplus, \odot)$ is a subset of $\mathbb{F}$ that is also a field under $\oplus$ and $\odot$.

When $(\mathbb{F}, \oplus, \odot)$ is a field and $\oplus$ and $\odot$ are understood, we may simply say that $\mathbb{F}$ is a field.

The element guaranteed in Definition 1.1(a) is the **additive identity** in $\mathbb{F}$ and is called *zero*.

---
[1] We discuss binary operations in Section A.7.

When a is in a field, $-a$ is the **additive inverse** of a and is read "negative a." The concepts of positive as greater than zero and of negative as less than zero are meaningless in most fields, though. That these concepts are meaningful in $\mathbb{R}$ and $\mathbb{Q}$, in the sense of being consistent with the field axioms, makes $\mathbb{R}$ and $\mathbb{Q}$ *ordered fields*.

The element guaranteed in Definition 1.1(c) is the **multiplicative identity** in $\mathbb{F}$ and is called *one*.

When a is a nonzero element in a field, a^{-1} is its **multiplicative inverse** and may also be written $1/a$.

By Lemma A.34, the additive identity, the multiplicative identity, the additive inverse of a given element, and the multiplicative inverse of a given element in a field are all unique.

Definition 1.1(e) is called the *distributive law*.

The field we know best is $\mathbb{R}$. It is a subfield of the field of complex numbers, $\mathbb{C}$. The rational numbers, $\mathbb{Q}$, form a subfield of both $\mathbb{R}$ and $\mathbb{C}$. The integers, $\mathbb{Z}$, do not comprise a field because, except for 1 and -1, its nonzero elements do not have multiplicative inverses in $\mathbb{Z}$.

The next theorem cites features of the arithmetic in any field. Since every field contains a multiplicative identity, 1, every field contains its additive inverse, -1.

Theorem 1.2. Let $(\mathbb{F}, \oplus, \odot)$ be a field. Let a be arbitrary in $\mathbb{F}$.

(a) If $a \oplus b = b$ for any b in $\mathbb{F}$, then $a = 0$.

(b) $0 \odot a = 0$.

(c) If $a \oplus b = 0$ for any b in $\mathbb{F}$, then $a = -b$.

(d) If $a \odot b = b$ for any nonzero b in $\mathbb{F}$, then $a = 1$.

(e) $-1 \odot a = -a$.

(f) If $a \odot b = 0$ for some b in $\mathbb{F}$, then either $a = 0$ or $b = 0$.

Proof. Let everything be as hypothesized.

(a) If $a \oplus b = b$ for some b in $\mathbb{F}$, then
$$0 = b \oplus -b = (a \oplus b) \oplus -b = a \oplus (b \oplus -b) = a.$$

(b) The distributive law gives us
$$(0 \odot a) \oplus a = (0 \odot a) \oplus (1 \odot a) = (0 \oplus 1) \odot a = 1 \odot a = a.$$
This shows that $(0 \odot a) \oplus a = a$. By part (a), $0 \odot a = 0$.

(c) Suppose $a \oplus b = 0$ for some b in $\mathbb{F}$. We then have
$$-b = 0 \oplus -b = (a \oplus b) \oplus -b = a \oplus (b \oplus -b) = a \oplus 0 = a.$$

(d) Suppose there is nonzero b in $\mathbb{F}$ such that $a \odot b = b$. Since $\odot$ is associative, we have
$$1 = b \odot b^{-1} = (a \odot b) \odot b^{-1} = a \odot (b \odot b^{-1}) = a \odot 1 = a.$$

1.1. Fields

(e) Applying the distributive law and part (b) we have
$$(-1 \odot a) \oplus a = (-1 \odot a) \oplus (1 \odot a) = (-1 \oplus 1) \odot a = 0 \odot a = 0.$$
By part (c), $-1 \odot a = -a$.

(f) Suppose there is nonzero b in $\mathbb{F}$ so that $a \odot b = 0$. Applying part (b), we get
$$0 = 0 \odot b^{-1} = (a \odot b) \odot b^{-1} = a \odot (b \odot b^{-1}) = a \odot 1 = a.$$
$\square$

Theorem 1.2(f) says that a field has no nonzero **zero divisors**.

Our next example is one in a family of fields, each denoted $(\mathbb{F}_p, \oplus_p, \odot_p)$, where p is a prime number.

Example 1.3. Let $\mathbb{F}_3$ be the set of symbols $\{0, 1, 2\}$ under binary operations $\oplus_3$ and $\odot_3$ given by the following *Cayley tables*.

(1.1)

$\oplus_3$	0	1	2
0	0	1	2
1	1	2	0
2	2	0	1

(1.2)

$\odot_3$	0	1	2
0	0	0	0
1	0	1	2
2	0	2	1

The symmetry in the tables tells us that both binary operations are commutative. The additive identity is 0 and the multiplicative identity is 1. Every element has an additive inverse: $-0 = 0$, $-1 = 2$, $-2 = 1$. Every nonzero element has a multiplicative inverse: $1^{-1} = 1$, $2^{-1} = 2$.

We leave it as an exercise to check associativity of addition and multiplication, as well as the distributive law. This will verify that $(\mathbb{F}_3, \oplus_3, \odot_3)$ is a field.

The field associated to a prime number p, $(\mathbb{F}_p, \oplus_p, \odot_p)$, is called **the field with p elements**. The set underlying $(\mathbb{F}_p, \oplus_p, \odot_p)$ is usually denoted
$$\{0, 1, 2, \ldots, p-1\}.$$
The operations on $\mathbb{F}_p$ are **addition modulo p** and **multiplication modulo p**, or addition and multiplication $\bmod p$ for short. A calculation $\bmod p$ yields the remainder on division by p after we apply the usual arithmetic, that is, treating the elements in $\mathbb{F}_p$ as integers. For instance, $4 \oplus_7 6 = 3$, and $3 \odot_5 4 = 2$.

Whenever we write $\mathbb{F}_p$ in the sequel, p is prime unless we indicate otherwise.

If $\mathbb{F}$ is any field and
$$a \oplus \cdots \oplus a \neq 0$$
for all nonzero a in $\mathbb{F}$, then $\mathbb{F}$ has **characteristic zero**. In this case, we write char $\mathbb{F} = 0$.

Notice that in $\mathbb{F}_3$, $1 \oplus_3 1 \oplus_3 1 = 0$. This is enough to establish that there are fields with nonzero characteristic.

Proof of the following is outlined in the exercises.

Theorem 1.4. If $\mathbb{F}$ is a field with nonzero characteristic, then there is a prime number p such that
$$\overbrace{a \oplus \cdots \oplus a}^{p \text{ summands}} = 0$$
for all a in $\mathbb{F}$.

When p is the prime number guaranteed by Theorem 1.4, $\mathbb{F}$ has **characteristic** p. In this case, we write char $\mathbb{F} = p$.

Theorem 1.5. Any field containing $\mathbb{Q}$ as a subfield has characteristic zero.

Proof. Let $\mathbb{F}$ be any field for which $\mathbb{Q}$ is a subfield. Lemma 1.2(d) implies that the multiplicative identity 1 in $\mathbb{F}$ is the same as the 1 in $\mathbb{Q}$. If there is a positive integer n such that
$$\overbrace{1 \oplus \cdots \oplus 1}^{n \text{ summands}} = 0$$
in $\mathbb{F}$, it follows that the same must hold in $\mathbb{Q}$, which is absurd.

The distributive law guarantees that if a is arbitary in $\mathbb{F}$, then
$$\overbrace{a \oplus \cdots \oplus a}^{n \text{ summands}} = a \odot \left(\overbrace{1 \oplus \cdots \oplus 1}^{n \text{ summands}} \right).$$
Since there are no nonzero zero divisors in $\mathbb{F}$, we see that if $a \neq 0$, then for n in $\mathbb{Z}_+$,
$$\overbrace{a \oplus \cdots \oplus a}^{n \text{ summands}} \neq 0.$$
This is enough to prove that char $\mathbb{F} = 0$. $\square$

It is safe, for the purposes of this course, to think of a field with characteristic zero as $\mathbb{Q}$, $\mathbb{R}$, or $\mathbb{C}$ and to think of a field with nonzero characteristic as $(\mathbb{F}_p, \oplus_p, \odot_p)$. Sometimes we insist that char $\mathbb{F} \neq 2$. In those cases, it is because we want to be able to divide by 2 in $\mathbb{F}$ or, equivalently, because we need $1 \neq -1$ in $\mathbb{F}$.

Exercises 1.1.

1.1.1. What is the minimum number of elements in a field?

1.1.2. Write out the Cayley tables for $\mathbb{F}_2$.

1.1.3. Show that $-0 = 0$ in any field.

1.1.4. Show that $-1 \neq 0$ in any field.

1.1.5. Show that $-(-a) = a$ for all elements a in a field.

1.1.6. Let a be a nonzero element in a field. Show that $(a^{-1})^{-1} = a$.

1.1.7. Consider $<$ as defined on $\mathbb{R}$ and suppose we attempt to extend the definition of $<$ to all of $\mathbb{C}$. To get started, we insist that $0 < i$. What familiar properties of $<$ fail? Answer the same question if we insist that $i < 0$.

1.1. Fields

1.1.8. Verify that addition and multiplication mod 3 are associative and that the distributive law holds in $\mathbb{F}_3$.

1.1.9. What is -6 in $\mathbb{F}_7$? What is 3^{-1} in $\mathbb{F}_7$?

1.1.10. Is the set of 2×2 matrices over $\mathbb{R}$ a field?

1.1.11. Does any subset of the integers form a field?

1.1.12. Show that $\{a + ib \mid a, b \in \mathbb{Q}\}$ is a subfield of $\mathbb{C}$.

1.1.13. Show that $\{a + b\sqrt{2} \mid a, b \in \mathbb{Q}\}$ is a subfield of $\mathbb{R}$.

1.1.14. Let $(\mathbb{F}, \oplus, \odot)$ be a field and let n in $\mathbb{Z}_+$. Given a in $\mathbb{F}$, we define na by
$$na := \overbrace{a \oplus \cdots \oplus a}^{n \text{ summands}}.$$
  (a) Show that $na = a \odot (n1)$.
  (b) Show that $na = 0$ for some nonzero a in $\mathbb{F}$ if and only if $nb = 0$ for all b in $\mathbb{F}$.

1.1.15. Using the notation from the previous problem, prove that for positive integers m and n, $(mn)1 = m(n1)$ in any field.

1.1.16. This problem outlines a proof of Theorem 1.4, referencing the notation and results from the previous two exercises.

Let $(\mathbb{F}, \oplus, \odot)$ be a field. Suppose $\operatorname{char} \mathbb{F} \neq 0$. Let n be minimal in $\mathbb{Z}_+$ so that $n1 = 0$.
  (a) Argue that $n \geq 2$.
  (b) By the Fundamental Theorem of Arithmetic, every integer greater than 1 has a prime divisor.[2] Let p be a prime divisor of n so that $n = pq$ for some q in $\mathbb{Z}_+$. By Exercise 1.1.15, $n1 = (pq)1 = p(q1)$. Argue that $q1 \neq 0$.
  (c) Argue that $p1 = 0$, thus, that $pa = 0$ for all a in $\mathbb{F}$.
  (d) Conclude that $n = p$.

1.1.17. Let $\mathbb{F}$ be a field. Show that $-1 = 1$ in $\mathbb{F}$ if and only if $\operatorname{char} \mathbb{F} = 2$.

1.1.18. Show that $\operatorname{char} \mathbb{F}_p = p$.

1.1.19. Show that if $\mathbb{F}_p$ is a subfield of $\mathbb{F}$, then $\operatorname{char} \mathbb{F} = p$.

1.1.20. There is a field with four elements, $\mathbb{F}_4 = \{0, 1, a, b\}$, under $\oplus$ and $\odot$ given by the following Cayley tables.

(1.3)

$\oplus$	0	1	a	b
0	0	1	a	b
1	1	0	b	a
a	a	b	0	1
b	b	a	1	0

(1.4)

$\odot$	1	a	b
1	1	a	b
a	a	b	1
b	b	1	a

---

[2] The Fundamental Theorem of Arithmetic is the subject of a brief discussion in Section A.1.

(a) Verify that $a \odot (a \odot b) = (a \odot a) \odot b$ and that $b \odot (a \odot b) = (b \odot a) \odot b$.
(b) Verify that $a \odot (1 \oplus b) = (a \odot 1) \oplus (a \odot b)$.
(c) Prove that $\operatorname{char} \mathbb{F}_4 = 2$.

1.2. Vector Spaces

A vector space is a nonempty set closed under addition and closed under scaling by the elements in some underlying field.[3] The following definition specifies the full set of axioms that a vector space must satisfy.

Definition 1.6. Let V be a set with an associative, commutative, binary operation, $+$, called addition. Let $(\mathbb{F}, \oplus, \odot)$ be a field with a mapping

$$\cdot : \mathbb{F} \times V \to V.$$

We say that $(V, \mathbb{F}, +, \cdot)$ is a **vector space**, or that V is a **vector space over** $\mathbb{F}$ **under** $+$ **and** $\cdot$, provided the following axioms hold.

(a) There is $\mathbf{0}$ in V such that $\mathbf{v} + \mathbf{0} = \mathbf{v}$ for all $\mathbf{v}$ in V.
(b) For each $\mathbf{v}$ in V there is $-\mathbf{v}$ in V such that $\mathbf{v} + (-\mathbf{v}) = \mathbf{0}$.
(c) $c \cdot (\mathbf{v} + \mathbf{w}) = c \cdot \mathbf{v} + c \cdot \mathbf{w}$, for all c in $\mathbb{F}$ and all $\mathbf{v}, \mathbf{w}$ in V.
(d) $(a \oplus b) \cdot \mathbf{v} = a \cdot \mathbf{v} + b \cdot \mathbf{v}$, for all a, b in $\mathbb{F}$ and $\mathbf{v}$ in V.
(e) $(a \odot b) \cdot \mathbf{v} = a \cdot (b \cdot \mathbf{v})$, for all a, b in $\mathbb{F}$ and $\mathbf{v}$ in V.
(f) $1 \cdot \mathbf{v} = \mathbf{v}$, for all $\mathbf{v}$ in V.

When $(V, \mathbb{F}, +, \cdot)$ is a vector space, $+$ is called **vector addition** and $\cdot$ is called **scaling**.

A **subspace** of a vector space $(V, \mathbb{F}, +, \cdot)$ is a subset of V that is also a vector space over $\mathbb{F}$ under $+$ and $\cdot$.

When $(V, \mathbb{F}, +, \cdot)$ is a vector space, we usually just say that V is a vector space. When V is a vector space over $\mathbb{F}$, we refer to $\mathbb{F}$ as the **underlying field**. To emphasize the underlying field, we may say that V is an $\mathbb{F}$ **vector space** or a vector space **over** $\mathbb{F}$. Below we will see that one set may be viewed as a vector space over different fields.

The element guaranteed in Definition 1.6(a) is called the **zero vector** in V. It may be denoted $\mathbf{0}_V$, especially when there is more than one vector space under consideration.

When $\mathbf{v}$ is in a vector space, $-\mathbf{v}$ is its **additive inverse** and is called *negative* $\mathbf{v}$.

By Lemma A.34, the zero vector is the unique additive identity in a vector space and $-\mathbf{v}$ is the unique additive inverse of an element $\mathbf{v}$ in a vector space.

A **vector** is an element in any vector space.[4] A **scalar** is an element in the field underlying a vector space.

---

[3] What we call "scaling" is usually called "scalar multiplication." The modification in terminology is an effort to sidestep confusion between scalar multiplication and the scalar product on $\mathbb{R}^n$.
[4] The vectors used in physics are elements in vector spaces. See Section B.1.

1.2. Vector Spaces

If **v** is a vector and c is a scalar in the underlying field, then instead of $c \cdot \mathbf{v}$, we usually write $c\mathbf{v}$, that is, we indicate scaling by *juxtaposition*. In this case, we say that $c\mathbf{v}$ is a **scalar multiple** of **v**.

Notice that the axioms of Definition 1.6(c)–(f) specify how the field operations and the vector space operations work together.

Linear structure refers to the properties detailed in Definition 1.6. Sometimes vector spaces are called *linear spaces*.

A **trivial vector space** has exactly one element, **0**. Any vector space with more than one element is **nontrivial**. When we discuss vector spaces, the word "nonzero" applies both to vectors different from the zero vector and to scalars different from zero in the underlying field. To avoid problems with this, remain aware of context.

Example 1.7. If $\mathbb{F}$ is any field, the set of *ordered n-tuples* over $\mathbb{F}$,

$$\mathbb{F}^n = \{(a_1, \ldots, a_n) \mid a_i \in \mathbb{F}\},$$

is a vector space under *componentwise addition and scaling*. This means that if $\mathbf{v} = (a_1, \ldots, a_n)$ and $\mathbf{w} = (b_1, \ldots, b_n)$ belong to $\mathbb{F}^n$, then $\mathbf{v} = \mathbf{w}$ only if $a_i = b_i$ for all $i \in \{1, \ldots, n\}$, that

$$\mathbf{v} + \mathbf{w} := (a_1 + b_1, \ldots, a_n + b_n),$$

and that for c in $\mathbb{F}$,

$$c\mathbf{v} := (ca_1, \ldots, ca_n).$$

In words, the ith component of $\mathbf{v} + \mathbf{w}$ is $a_i + b_i$ and the ith component of $c\mathbf{v}$ is ca_i.

That $\mathbb{F}^n$ satisfies the axioms of Definition 1.6 is a direct consequence of the fact that $\mathbb{F}$ is a field.

We often denote the zero vector in $\mathbb{F}^n$ by $\mathbf{0}_n = (0, \ldots, 0)$.

Any nontrivial vector space V has at least two subspaces: $\{\mathbf{0}\}$ and V itself. A **proper subspace** of V is any subspace different from V.

Example 1.8. Consider $\mathbb{F}_2^2 = \{(0,0), (1,0), (0,1), (1,1)\}$. The only scalars in the field are 0 and 1 so the only scalar multiples of each vector are $\mathbf{0} = (0,0)$ and the vector itself. Each vector is its own additive inverse since in $\mathbb{F}_2$, $-1 = 1$. The proper subspaces of $\mathbb{F}_2^2$ are $\{(0,0), (1,0)\}$, $\{(0,0), (0,1)\}$, and $\{(0,0), (1,1)\}$.

When $\mathbb{F}$ is a field, we can identify it with $\mathbb{F}^1$. This is enough to prove the following.

Lemma 1.9. *Every field is a vector space over itself.*

Going forward, we use + and juxtaposition to denote addition and multiplication in a field $\mathbb{F}$. While common practice, this means that when $(V, \mathbb{F}, +, \cdot)$ is a vector space, we use the same symbol for addition in V and addition in $\mathbb{F}$, and we use juxtaposition both for multiplication in $\mathbb{F}$ and for scaling in V. Again: Remain aware of context!

We can combine addition and scaling into one operation. Let V be a vector space with $S \subseteq V$. If $\mathbf{v}_1, \ldots, \mathbf{v}_n$ belong to S and $c_1, \ldots, c_n$ are scalars in the underlying field, then

(1.5)
$$c_1 \mathbf{v}_1 + \cdots + c_n \mathbf{v}_n$$

is a **linear combination** of vectors in S. Each $c_i \mathbf{v}_i$ is a **term** of the linear combination in (1.5) and c_i is a **coefficient** or **weight**.

A **trivial linear combination** is one for which all coefficients are zero. The expression in (1.5) is a **nontrivial linear combination** provided the $\mathbf{v}_i$s are distinct and c_i is nonzero for some i. Typically, we avoid working with linear combinations of the form (1.5) for which the $\mathbf{v}_i$s are not distinct vectors.

We can characterize a vector space as a nonempty set closed under linear combination over some field.

A linear combination is a sum, and in algebra, *a sum is an operation on finitely many elements*. This may not be the case in other settings. We revisit this idea several times as we go through the course.

The axioms of Definition 1.6 guarantee that a linear combination of linear combinations is a linear combination, as long as all the vectors come from the same vector space.

The following is a compilation of properties of vector arithmetic common to all vector spaces. Note the parallels to Theorem 1.2.

Theorem 1.10. Let V be a vector space over a field $\mathbb{F}$. Let $\mathbf{v}$ be arbitrary in V.

(a) If $\mathbf{v} + \mathbf{w} = \mathbf{w}$ for any $\mathbf{w}$ in V, then $\mathbf{v} = \mathbf{0}$.

(b) $0\mathbf{v} = \mathbf{0}$.

(c) If $\mathbf{v} + \mathbf{w} = \mathbf{0}$ for any $\mathbf{w}$ in V, then $\mathbf{v} = -\mathbf{w}$.

(d) $(-1)\mathbf{v} = -\mathbf{v}$.

Proof. Let V and $\mathbf{v}$ be as hypothesized.

(a) If $\mathbf{v} + \mathbf{w} = \mathbf{w}$, for some $\mathbf{w}$ in V, then
$$\mathbf{0} = \mathbf{w} + -\mathbf{w} = (\mathbf{v} + \mathbf{w}) + -\mathbf{w} = \mathbf{v} + (\mathbf{w} + -\mathbf{w}) = \mathbf{v} + \mathbf{0} = \mathbf{v}.$$

(b) Since
$$0\mathbf{v} + \mathbf{v} = 0\mathbf{v} + 1\mathbf{v} = (0 + 1)\mathbf{v} = \mathbf{v},$$
it follows by part (a) that $0\mathbf{v} = \mathbf{0}$.

Proofs of the last two statements are exercises. $\square$

The next two lemmas are about behaviors of the zero vector and the scalar zero. Proof of the first is an exercise.

Lemma 1.11. Let V be a vector space over a field $\mathbb{F}$. If c is in $\mathbb{F}$, then $c\mathbf{0} = \mathbf{0}$.

Lemma 1.12. Let V be a vector space over a field $\mathbb{F}$. Suppose c belongs to $\mathbb{F}$ and $\mathbf{v}$ belongs to V. If $c\mathbf{v} = \mathbf{0}$, then $c = 0$ or $\mathbf{v} = \mathbf{0}$.

1.2. Vector Spaces

Proof. Let V, $\mathbb{F}$, c, and $\mathbf{v}$ be as hypothesized. Suppose $c\mathbf{v} = \mathbf{0}$ and that $c \neq 0$. Since c^{-1} is in $\mathbb{F}$, we have
$$\mathbf{v} = 1\mathbf{v} = (c^{-1}c)\mathbf{v} = c^{-1}(c\mathbf{v}) = c^{-1}\mathbf{0}.$$
By Lemma 1.11, $c^{-1}\mathbf{0} = \mathbf{0}$, so $\mathbf{v} = \mathbf{0}$. □

Fix a field $\mathbb{F}$ and positive integers m and n. The set of $m \times n$ matrices over $\mathbb{F}$, $M_{m,n}(\mathbb{F})$, is a vector space under componentwise addition and scaling. We use $M_n(\mathbb{F})$ to designate the set of $n \times n$ (square) matrices over $\mathbb{F}$.

Example 1.13. $M_2(\mathbb{F}_2)$ is the vector space of 2×2 matrices over the field with two elements. Since a 2×2 matrix has four entries, $M_2(\mathbb{F}_2)$ has $2^4 = 16$ elements.

We identify $M_{n,1}(\mathbb{F})$, the space of column vectors with n entries from $\mathbb{F}$, with $\mathbb{F}^n$. This means we may denote an element in $\mathbb{F}^n$ as an n-tuple or as a column vector:
$$(a_1, \ldots, a_n) = \begin{bmatrix} a_1 \\ \vdots \\ a_n \end{bmatrix}.$$

We use $\mathbb{F}_n$, instead of $M_{1,n}(\mathbb{F})$, to designate the space of row vectors with n entries from $\mathbb{F}$:
$$\mathbb{F}_n = \left\{ \begin{bmatrix} a_1 & \cdots & a_n \end{bmatrix} \mid a_i \in \mathbb{F} \right\}.$$

Note that $(a_1, \ldots, a_n)$ and $\begin{bmatrix} a_1 & \cdots & a_n \end{bmatrix}$ belong to different vector spaces.

The distinction between $\mathbb{F}^n$ and $\mathbb{F}_n$ seems purely cosmetic right now, but an interesting relationship between the two materializes once we have more ideas in the mix.

Let A and B be arbitrary sets. We use $\mathcal{F}(A, B)$ to designate the set of all functions $A \to B$ and we use $\mathcal{F}(A)$ to designate the set of all functions $A \to A$.[5]

Suppose that A is an arbitrary set and that V is a vector space over a field $\mathbb{F}$. If f and g belong to $\mathcal{F}(A, V)$, we define $f + g$ in $\mathcal{F}(A, V)$ by where it maps an element in A. Since addition is defined in V, we can say
$$(f+g)(a) := f(a) + g(a), \forall a \in A.$$
Likewise, if c is in $\mathbb{F}$, we can define cf in $\mathcal{F}(A, V)$ by
$$(cf)(a) := c(f(a)), \forall a \in A.$$

Theorem 1.14. *If A is an arbitrary set and V is a vector space over $\mathbb{F}$, then $\mathcal{F}(A, V)$ is a vector space over $\mathbb{F}$.*

Proof. Evidently, $\mathcal{F}(A, V)$ is closed under addition and scaling. That $+$ in $\mathcal{F}(A, V)$ is commutative and associative is a consequence of the fact that addition in V is commutative and associative. The identity element in $\mathcal{F}(A, V)$ is O given by $O(a) = \mathbf{0}_V$ for all a in A. For any f in $\mathcal{F}(A, V)$, the mapping $-f = (-1)f$ is the additive inverse of f. The axioms of Definition 1.6(c)–(f) all hold because of properties of addition and mulitplication in V. □

---

[5]The set of all functions $A \to B$ is often denoted B^A.

Many examples in this course derive from $\mathcal{F}(V,W)$ where V and W are vector spaces over a field $\mathbb{F}$.

When $\mathbb{F}$ is small, we can write down all the elements in $\mathcal{F}(\mathbb{F})$.

Example 1.15. Let $V = \mathcal{F}(\mathbb{F}_2)$. A mapping in $\mathcal{F}(\mathbb{F}_2)$ sends each element in $\mathbb{F}_2$ to itself or to the other element in $\mathbb{F}_2$. There are four different ways to do this so $\mathcal{F}(\mathbb{F}_2)$ has four elements. There are two constant mappings: the zero mapping, $\mathbf{0}(a) = 0$ for a in $\mathbb{F}_2$, and $\kappa(a) = 1$ for all a in $\mathbb{F}_2$. The identity mapping is $\iota(a) = a$, for all a in $\mathbb{F}_2$. Finally, there is τ, which swaps 0 and 1: $\tau(0) = 1$ and $\tau(1) = 0$. Scaling an element in $\mathcal{F}(\mathbb{F}_2)$ by 1 gives us the element back again. Scaling an element by 0 gives us $\mathbf{0}$. There is no trick to sorting out addition in $\mathcal{F}(\mathbb{F}_2)$. For example,

$$(\iota + \tau)(0) = \iota(0) \oplus_2 \tau(0) = 0 \oplus_2 1 = 1$$

and

$$(\iota + \tau)(1) = \iota(1) \oplus_2 \tau(1) = 1 \oplus_2 0 = 1$$

so $\iota + \tau = \kappa$. We leave it as an exercise to check the other sums. A Cayley table shows all of them.

+	$\mathbf{0}$	κ	ι	τ
$\mathbf{0}$	$\mathbf{0}$	κ	ι	τ
κ	κ	$\mathbf{0}$	τ	ι
ι	ι	τ	$\mathbf{0}$	κ
τ	τ	ι	κ	$\mathbf{0}$

Let V be a vector space over $\widetilde{\mathbb{F}}$ and suppose $\mathbb{F} \subseteq \widetilde{\mathbb{F}}$ is a subfield. (Think of $\widetilde{\mathbb{F}} = \mathbb{C}$ and $\mathbb{F} = \mathbb{R}$.) It is certainly the case that

$$c\mathbf{v} + \mathbf{w} \in V$$

whenever $\mathbf{v}$ and $\mathbf{w}$ are in V and c in $\mathbb{F}$. Theorem 1.2 guarantees that the additive and multiplicative identities in $\mathbb{F}$ are identical to those in $\widetilde{\mathbb{F}}$ so none of the axioms in Definition 1.6 are affected when we shift the scalar field from $\widetilde{\mathbb{F}}$ to its subfield, $\mathbb{F}$. This is enough to prove the following theorem.

Theorem 1.16. If V is a vector space over $\mathbb{F}$, then V is a vector space over any subfield of $\mathbb{F}$.

Theorem 1.16 guarantees that every complex vector space is also a real vector space and that every real vector space is also a rational vector space.

Example 1.17. Since $\mathbb{C}$ is a vector space over itself, it is also a vector space over $\mathbb{R}$. When we think of z in $\mathbb{C}$ as having the form $z = a + ib$, for a, b in $\mathbb{R}$, we are, in a sense, treating $\mathbb{C}$ explicitly as a vector space over $\mathbb{R}$. (See Section B.3 for more details about the complex numbers.)

The complex numbers arise repeatedly in our study of vector spaces. Going forward, assume that $a, b, x,$ and y are real numbers whenever we write $z = a + ib$, or $z = x + iy$, for z in $\mathbb{C}$.

1.2. Vector Spaces

Exercises 1.2.

1.2.1. Determine whether each of the following is a vector space. Give reasons for your answers.
  (a) $\mathbb{R}$ over $\mathbb{Q}$
  (b) $\mathbb{Q}$ over $\mathbb{R}$
  (c) $\mathbb{R}$ over $\mathbb{Z}$
  (d) $\mathbb{Z}$ over $\mathbb{R}$
  (e) $\mathbb{Z}$ over $\mathbb{Q}$

1.2.2. Finish the proof of Theorem 1.10.

1.2.3. Prove Lemma 1.11.

1.2.4. Verify the sums in the Cayley table in Example 1.15.

1.2.5. List the elements in $M_2(\mathbb{F}_2)$.

1.2.6. The **cardinality of a finite set** is the number of elements it contains.[6] If $S = \{a_1, \ldots, a_n\}$, we write $|S| = n$. What is $|\mathcal{F}(\mathbb{F}_3)|$?

1.2.7. What theorems from calculus guarantee that $C(D)$, the set of continuous functions on an open interval $D \subseteq \mathbb{R}$, is a vector space?

1.2.8. Let $(V, \mathbb{F}, +, \cdot)$ be a vector space. Show that a subset $W \subseteq V$ is a subspace of V if and only if W is nonempty and $c\mathbf{w} + \mathbf{x}$ is in W for any c in $\mathbb{F}$, and $\mathbf{w}$ and $\mathbf{x}$ in W.

1.2.9. Prove that a subset of a vector space is a subspace if and only if it is nonempty and closed under linear combinations.

1.2.10. Prove that if we consider $\mathbb{C}$ as a vector space over $\mathbb{R}$, then the real numbers form a subspace of $\mathbb{C}$.

1.2.11. Is $\mathbb{R}$ a subspace of $\mathbb{C}$ if we consider $\mathbb{C}$ as a complex vector space?

1.2.12. Let V be a vector space over $\mathbb{F}$ and let $\mathbf{v}$ be in V. Verify that
$$W = \{t\mathbf{v} \mid t \in \mathbb{F}\}$$
is a subspace of V.

1.2.13. Let V be a vector space with subspace W_1. Suppose W_2 is a subspace of W_1. Show that W_2 is a subspace of V. (Informally: A subspace of a subspace is a subspace.)

1.2.14. Let V be a vector space with subspaces $\{W_i\}_{i \in I}$ for some index set I.
  (a) Show that $\bigcap_{i \in I} W_i$ is a subspace of V. (Informally: The arbitrary intersection of subspaces is a subspace.)
  (b) Is the union of any two sets in $\{W_i\}_{i \in I}$ a subspace of V? Justify your answer with a proof or a counterexample.

1.2.15. Let A be an $m \times n$ matrix with entries in some field, $\mathbb{F}$. Use what you already know about matrices to show that the solution set to $A\mathbf{x} = \mathbf{0}$ is a subspace of $\mathbb{F}^n$.

1.2.16. Let V be an arbitrary vector space over a field $\mathbb{F}$ and let $\mathbf{v}$ be nonzero in V. Let $\ell = \{\mathbf{v}_0 + t\mathbf{v} \mid t \in \mathbb{F}\}$, for some nonzero $\mathbf{v}_0$ in V. Define addition in

---

[6] Section C.2 has a more extensive discussion of cardinality.

ℓ by
$$(\mathbf{v}_0 + s\mathbf{v}) + (\mathbf{v}_0 + t\mathbf{v}) := \mathbf{v}_0 + (s+t)\mathbf{v}$$
and define scaling in ℓ by
$$c(\mathbf{v}_0 + t\mathbf{v}) := \mathbf{v}_0 + ct\mathbf{v}.$$

(a) Is ℓ a vector space? If so, what is the zero vector?

(b) Is ℓ a subspace of V? Justify your answer.

1.3. Spanning and Linear Independence

Let V be a vector space over $\mathbb{F}$ and let $\mathbf{v}$ belong to V. Another way to say that $W = \{t\mathbf{v} \mid t \in \mathbb{F}\}$ is a subspace of V is to say that W is the *span* of the set $\{\mathbf{v}\} \subseteq V$.

Definition 1.18. Let V be a vector space with $S \subseteq V$.

When $S = \emptyset$, the **span** of S is $\{\mathbf{0}\}$. When S is nonempty, its span is the set of all vectors in V that can be written as linear combinations of vectors in S.

We designate the span of a set S by $\operatorname{Span} S$. If $\operatorname{Span} S = W$, then S **spans** W, and S is a **spanning set** for W.

Example 1.19. Consider $S = \{(1,1,0),(1,1,1)\} \subseteq \mathbb{F}_2^3$. A linear combination of vectors from S is determined by two coefficients from $\mathbb{F}_2$: one for $(1,1,0)$ and one for $(1,1,1)$. The four resulting linear combinations, in this case, are equal to four distinct vectors in $\mathbb{F}_2^3$. The elements in $\operatorname{Span} S$ are then

$$0(1,1,0) + 0(1,1,1) = (0,0,0),$$
$$1(1,1,0) + 0(1,1,1) = (1,1,0),$$
$$0(1,1,0) + 1(1,1,1) = (1,1,1),$$
$$1(1,1,0) + 1(1,1,1) = (0,0,1).$$

When we refer in the sequel to an abstract set of vectors, assume that all the vectors in the set belong to a single vector space.

Evidently, if S is a set of vectors and $T \subseteq S$, then $\operatorname{Span} T \subseteq \operatorname{Span} S$. In particular, $\mathbf{0}$ is an element in $\operatorname{Span} S$ and $S \subseteq \operatorname{Span} S$.

If W is a subspace of a vector space V, then V is the *ambient space*.

Proof of the following is left as an exercise.

Lemma 1.20. *The span of a set of vectors is a subspace of the ambient vector space.*

Linear independence is the property of a set of vectors S guaranteeing that each element in $\operatorname{Span} S$ can be expressed in only one way as a linear combination of distinct elements in S. It is important to understand that we distinguish linear combinations of distinct vectors by their terms, rather than what the terms add up to or how they are ordered. For instance, if $S = \{(1,1),(1,0),(0,1)\} \subseteq \mathbb{R}^2$, then since

$$(1,1) + (1,0) + (0,1)$$

and

$$2(1,0) + 2(0,1)$$

1.3. Spanning and Linear Independence

have different coefficients, they are different nontrivial linear combinations of the vectors in S. (We usually omit terms with a coefficient of zero.) We express the fact that
$$(1,1) + (1,0) + (0,1) = 2(1,0) + 2(0,1)$$
by saying there is a *dependence relation* on S, or that S is *linearly dependent*.

Definition 1.21. A set of vectors S is **linearly dependent** provided there is a vector in Span S that can be written as two different linear combinations of distinct elements in S. When S is not linearly dependent, it is **linearly independent**.

Any equation proving that a set of vectors S is linearly dependent is called a **dependence relation** on S.

Example 1.22. Let $S = \{\cos 2t, \cos^2 t, \sin^2 t\} \subseteq \mathcal{F}(\mathbb{R})$. Since

(1.6) $$\cos 2t = \cos^2 t - \sin^2 t,$$

(1.6) is a dependence relation on S.

The following lemma is often used as the definition of linear dependence. The proof is an exercise.

Lemma 1.23. A set of vectors, S, is linearly dependent if and only if $\mathbf{0}$ can be written as a nontrivial linear combination of distinct vectors in S.

The next example details an approach to testing the linear independence of a set of vectors in a function space.

Example 1.24. Consider $S = \{\sin t, \cos t, e^t\} \subseteq C(\mathbb{R})$. To test the linear independence of S, we apply Lemma 1.23, analyzing the coefficients when we express $\mathbf{0}$ as a linear combination of elements in S. If we find that the coefficients must all be zero, we will have shown that S is linearly independent.

Note that the zero vector in $C(\mathbb{R})$ is the constant function $f(t) = 0$ for all t in $\mathbb{R}$.

Suppose there is some choice of c_i in $\mathbb{R}$ so that
$$f(t) = c_1 \sin t + c_2 \cos t + c_3 e^t = 0 \, \forall t \in \mathbb{R}.$$
We then have
$$f(0) = c_1 \sin 0 + c_2 \cos 0 + c_3 e^0 = c_2 + c_3 = 0,$$
so $c_3 = -c_2$. This gives us
$$f(\pi/2) = c_1 - c_2 e^{\pi/2} = 0,$$
so $c_1 = c_2 e^{\pi/2}$. Finally,
$$f(\pi) = -c_2 - c_2 e^\pi = -c_2(1 + e^\pi) = 0.$$
Since $1 + e^\pi \neq 0$, $c_2 = 0$, which implies that c_1 and c_3 are zero as well. Since a linear combination of elements in S equal to the zero mapping must be trivial, S is linearly independent.

Since there are no vectors in it, the empty set is linearly independent. A set containing a single nonzero vector is linearly independent. If a set contains two nonzero vectors, it is linearly independent if and only if one vector is not a scalar multiple of the other. For example $\{(1,2,-1),(0,1,3)\} \subseteq \mathbb{R}^3$ is linearly independent.

It should be evident that any set of vectors with a linearly dependent subset is itself linearly dependent. The next theorem is a stronger statement. Its proof is an exercise.

Theorem 1.25. A set of vectors is linearly independent if and only if all of its finite subsets are linearly independent.

Proof of the next theorem requires a routine application of the definitions so we leave it, too, as an exercise. The results in the statement of the theorem come up repeatedly in the sequel, sometimes in the form of the contrapositive.[7]

Theorem 1.26. (a) A set of vectors, S, is linearly dependent if and only if there is $\mathbf{v}$ in S so that
$$\mathrm{Span}(S \setminus \{\mathbf{v}\}) = \mathrm{Span}\, S.$$
(b) A set of vectors, S, is linearly dependent if and only if every vector in $\mathrm{Span}\, S$ can be expressed using at least two different linear combinations of distinct elements in S.

The next theorem pinpoints a connection between linear independence and spanning.

Theorem 1.27. Let S be a linearly independent set of vectors in a vector space, V, and let $\mathbf{v}$ be a vector in V. The set $S \cup \{\mathbf{v}\}$ is linearly independent if and only if $\mathbf{v}$ is not in the span of S.

Proof. Let S, V, and $\mathbf{v}$ be as hypothesized. Suppose $\mathbf{v}$ is not in $\mathrm{Span}\, S$ and that we have distinct $\mathbf{v}_1, \ldots, \mathbf{v}_n$ in S and scalars c and c_i such that

(1.7) $$\mathbf{0} = c\mathbf{v} + c_1\mathbf{v}_1 + \cdots + c_n\mathbf{v}_n.$$

If c were nonzero, then
$$\mathbf{v} = -\frac{c_1}{c}\mathbf{v}_1 - \cdots - \frac{c_n}{c}\mathbf{v}_n,$$
which is impossible since $\mathbf{v}$ is not in $\mathrm{Span}\, S$. We conclude that in any equation of the form given in (1.7), c must be zero. Since S is linearly independent, $c = 0$ implies $c_i = 0$ for $i = 1, \ldots, n$. It follows by Lemma 1.23 that $S \cup \{\mathbf{v}\}$ is linearly independent. This proves the result in one direction. Proof of the other direction is an exercise. □

Theorem 1.27 often appears in the form of its contrapositive. We state it as a corollary.

Corollary 1.28. Let S be a linearly independent set of vectors in a vector space, V, and let $\mathbf{v}$ be a vector in V. The set $S \cup \{\mathbf{v}\}$ is linearly dependent if and only if $\mathbf{v}$ is in $\mathrm{Span}\, S$.

---
[7]See Section A.1 for the definition, and a discussion of, the contrapositive of a logical statement.

Exercises 1.3.

1.3.1. Verify that if S is a subset of a vector space and $T \subseteq \operatorname{Span} S$, then $\operatorname{Span} T \subseteq \operatorname{Span} S$.

1.3.2. Prove Lemma 1.20.

1.3.3. Prove that every subspace of a vector space is the span of some set.

1.3.4. Let V be a vector space with subset S. The span of S is often defined as the minimal subspace of V containing S. In this problem, we take two different perspectives to argue that this is a true description of $\operatorname{Span} S$.
   (a) Show that if W is any subspace of V containing S, then $\operatorname{Span} S \subseteq W$.
   (b) Show that $\operatorname{Span} S$ is the intersection of all subspaces of V containing S.

1.3.5. Give a geometric description of $\operatorname{Span}\{(1, 1, 1)\} \subseteq \mathbb{R}^3$.

1.3.6. Let $S = \{(1, 1, 2), (2, 2, 1)\} \subseteq \mathbb{F}_3{}^3$. List the elements in $\operatorname{Span} S$.

1.3.7. Let $S = \{(1, -1, 2), (2, 0, 1)\} \subseteq \mathbb{R}^3$ and let $W = \operatorname{Span} S$. Find another two-vector spanning set for W that contains neither vector in S. Justify your answer.

1.3.8. Show that in $\mathbb{F}_3{}^2$, $\operatorname{Span}\{(1, 2)\} = \operatorname{Span}\{(2, 1)\}$.

1.3.9. Use Definition 1.21 to prove that $\{\mathbf{0}\}$ is linearly dependent in any vector space.

1.3.10. Use Definition 1.21 to prove that if S is a subset of a vector space that contains a linearly dependent set, then S itself is linearly dependent.

1.3.11. Prove Lemma 1.23.

1.3.12. Prove Theorem 1.25.

1.3.13. Prove Theorem 1.26.

1.3.14. Write down the contrapositive to each statement in Theorem 1.26.

1.3.15. Let $S = \{(1, 1, 1), (1, 0, 1), (0, 1, 0)\} \subseteq \mathbb{F}^3$ for some field, $\mathbb{F}$. Show that S is linearly dependent.

1.3.16. Let $S = \{(1, 1, 1), (1, 0, 1), (0, 1, 1)\} \subseteq \mathbb{F}^3$ for some field, $\mathbb{F}$. Show that S is linearly independent.

1.3.17. Let $S = \{(1, 0), (0, 1), (1, 1)\} \subseteq \mathbb{F}_5{}^2$. Find all nontrivial linear combinations of vectors in S equal to the zero vector.

1.3.18. Let S be a linearly dependent set over a field of characteristic zero. Show that for every $\mathbf{v}$ in $\operatorname{Span} S$, there are infinitely many nonidentical linear combinations of distinct elements in S equal to $\mathbf{v}$.

1.3.19. Find an example of a linearly dependent set of vectors S that contains a vector $\mathbf{v}$ not in $\operatorname{Span}(S \setminus \{\mathbf{v}\})$.

1.3.20. Finish the proof of Theorem 1.27.

1.3.21. Suppose $S = \{\mathbf{v}, \mathbf{w}, \mathbf{x}\}$ is a linearly independent set of vectors. Under what circumstances is $\{\mathbf{v} + \mathbf{x}, \mathbf{v} + \mathbf{w}, \mathbf{x} + \mathbf{w}\}$ linearly independent?

1.4. Bases

The **Kronecker delta** is the symbol defined by

$$(1.8) \qquad \delta(i,j) := \begin{cases} 1 \text{ if } i = j, \\ 0 \text{ if } i \ne j. \end{cases}$$

This is used widely in mathematics and may also be denoted δ_{ij}.

Fix a field $\mathbb{F}$ and define $\mathbf{e}_j$ in $\mathbb{F}^n$ by

$$\mathbf{e}_j = (\delta(1,j), \delta(2,j), \ldots, \delta(n,j)).$$

For example, if $n = 3$, $\mathbf{e}_1 = (1,0,0)$, $\mathbf{e}_2 = (0,1,0)$, and $\mathbf{e}_3 = (0,0,1)$. Let

$$\mathcal{E} = \{\mathbf{e}_1, \ldots, \mathbf{e}_n\}.$$

Without much effort, we can see that $\mathcal{E}$ is a linearly independent spanning set for $\mathbb{F}^n$.

Definition 1.29. A **basis** is a linearly independent spanning set for a vector space.

We call $\mathcal{E}$ the **standard basis** or the **usual basis** for $\mathbb{F}^n$. It comes up frequently in the sequel. To emphasize that we care about how the $\mathbf{e}_i$s are ordered, we may also call $\mathcal{E}$ the **standard ordered basis** for $\mathbb{F}^n$.

The next result is a compilation of equivalent formulations for the definition of basis. All of these arise frequently, both in our studies here and in applications. Since proofs follow readily from the definitions, we leave them to the exercises.

Theorem 1.30. Let V be a vector space.

(a) A set $\mathcal{B} \subseteq V$ is a basis for V if and only if each element in V can be written in exactly one way as a linear combination of distinct elements in $\mathcal{B}$.

(b) A set $\mathcal{B} \subseteq V$ is a basis for V if and only if it is a minimal spanning set for V, that is, $\mathcal{B}$ spans V and for each $\mathbf{v}$ in $\mathcal{B}$, $\mathrm{Span}(\mathcal{B} \setminus \{\mathbf{v}\})$ is a proper subspace of V.

(c) A set $\mathcal{B} \subseteq V$ is a basis for V if and only if it is a maximal linearly independent set, that is, $\mathcal{B}$ is linearly independent and for any $\mathbf{v}$ in $V \setminus \mathcal{B}$, $\mathcal{B} \cup \{\mathbf{v}\}$ is linearly dependent.

A vector space with a finite spanning set is **finite-dimensional**. A vector space that is not finite-dimensional is **infinite-dimensional**.

We state the next theorem here because of its importance. In Subsection 1.4.1, we prove it in the finite-dimensional case. In Subsection 1.4.2, we prove it in the general case. Unlike the proof of the general case, the proof for the finite-dimensional case is *constructive*: It suggests methods we can apply to expand a linearly independent set into a basis and to identify a basis in a given spanning set. This makes it worthwhile even for readers who plan to study Subsection 1.4.2.

Theorem 1.31. (a) Any linearly independent set in a vector space is a subset of a basis for the space.

(b) Any spanning set for a vector space has a subset that is a basis for the space.

1.4. Bases

Since the empty set is linearly independent and a vector space spans itself, every vector space has both linearly independent subsets and spanning sets. Theorem 1.31 thus implies the following corollary.

Corollary 1.32. Every vector space has a basis.

1.4.1. Bases for Finite-Dimensional Spaces.
We start with the finite-dimensional version of Theorem 1.31(b).

Lemma 1.33. A finite spanning set for a vector space has a subset that is a basis for the space.

Proof. Let S be a finite spanning set for a vector space, V. We prove the lemma by induction on $n = |S|$.

If $n = 0$, S is the empty set, so $V = \{\mathbf{0}\}$. Since the empty set is a linearly independent spanning set for the trivial vector space, this is enough to prove the lemma in the base case.

Suppose now that $n > 0$. If S is not a basis, it is linearly dependent. Theorem 1.26 then guarantees that there is $\mathbf{v}$ in S so that $S \setminus \{\mathbf{v}\}$ spans V. Since $|S \setminus \{\mathbf{v}\}| = n - 1$, the result follows by the induction hypothesis. $\square$

The next result follows Lemma 1.33 immediately, by our definition of "finite-dimensional."

Theorem 1.34. Every finite-dimensional vector space has a basis.

The next theorem allows us to define "dimension" in the finite-dimensional setting.

Theorem 1.35. Let V be a finite-dimensional vector space. If $S \subseteq V$ is linearly independent and $S' \subseteq V$ is a spanning set for V, then $|S| \leq |S'|$.

Proof. Let V, S, and S' be as hypothesized. Since we must show that $|S| \leq |S'|$, we lose no generality in assuming that S' is finite. Say $S' = \{\mathbf{v}_i\}_{i=1}^n$.

Given $\mathbf{w}_1$ in S there are scalars c_i so that

(1.9) $$\mathbf{w}_1 = c_1 \mathbf{v}_1 + \cdots + c_n \mathbf{v}_n.$$

As an element in a linearly independent set, $\mathbf{w}_1$ is nonzero so there must be a nonzero coefficient among the c_is in (1.9). Reindexing the elements in S' if necessary, we may assume $c_1 \neq 0$. This allows us to write

$$\mathbf{v}_1 = \frac{1}{c_1}(\mathbf{w}_1 - c_2 \mathbf{v}_2 - \cdots - c_n \mathbf{v}_n),$$

showing that $\mathbf{v}_1$ is in $\mathrm{Span}\{\mathbf{w}_1, \mathbf{v}_2, \ldots, \mathbf{v}_n\}$, thus, that

$$\{\mathbf{v}_1, \ldots, \mathbf{v}_n\} \subseteq \mathrm{Span}\{\mathbf{w}_1, \mathbf{v}_2, \ldots, \mathbf{v}_n\}.$$

It follows that $\{\mathbf{w}_1, \mathbf{v}_2, \ldots, \mathbf{v}_n\}$ spans V.

If we repeat this process k times, where $k < \min\{n, |S|\}$, we arrive at a spanning set for V of the form

$$\{\mathbf{w}_1, \ldots, \mathbf{w}_k, \mathbf{v}_{k+1}, \ldots, \mathbf{v}_n\},$$

where each $\mathbf{w}_i$ belongs to S.

Suppose $|S| > n$. We can then replace all n elements in S' with the elements in S'', some proper subset of S, to get a spanning set for V. Since S is linearly independent, so is $S'' \cup \{\mathbf{w}\}$ for any $\mathbf{w}$ in $S \setminus S''$. By Theorem 1.27, though, $S'' \cup \{\mathbf{w}\}$ must be linearly dependent. The contradiction implies that $|S| \leq |S'|$. □

Let V be finite-dimensional with bases $\mathcal{B}_1$ and $\mathcal{B}_2$. Since $\mathcal{B}_1$ is linearly independent and $\mathcal{B}_2$ is a spanning set for V, Theorem 1.35 says $|\mathcal{B}_1| \leq |\mathcal{B}_2|$. By the same argument, $|\mathcal{B}_2| \leq |\mathcal{B}_1|$. We conclude that $|\mathcal{B}_1| = |\mathcal{B}_2|$. This gives us a definition of dimension that applies in the present setting.

Definition 1.36. The **dimension** of a finite-dimensional vector space is the number of elements in a basis for the space.

When V is a vector space over $\mathbb{F}$ and a basis for V contains n elements, we write $\dim V = n$. When we want to emphasize the underlying field, we write $\dim_\mathbb{F} V = n$.

Since $\mathcal{E}$ contains n elements, $\dim \mathbb{F}^n = n$.

Example 1.37. We can view $\mathbb{C}$ as a vector space over itself. Since any nonzero element in $\mathbb{C}$ spans all of $\mathbb{C}$, $\dim_\mathbb{C} \mathbb{C} = 1$. When we view $\mathbb{C}$ as a vector space over $\mathbb{R}$, it is easy to see that $\{1, i\}$ is a basis. It follows that $\dim_\mathbb{R} \mathbb{C} = 2$.

Corollary 1.38. Let V be a vector space with $\dim V = n$. If S is a subset of V containing exactly n elements, then S is linearly independent if and only if it is a spanning set for V.

Proof. Suppose V and S are as hypothesized. If S is linearly independent but does not span V, take $\mathbf{v}$ in $V \setminus \operatorname{Span} S$. Theorem 1.27 guarantees that $S' = S \cup \{\mathbf{v}\}$ is linearly independent. Since $|S'| = n+1$, we have a violation of Theorem 1.35, forcing us to conclude that S spans V. This proves the corollary in one direction. The rest of the proof is an exercise. □

The next theorem completes our proof of the finite-dimensional version of Theorem 1.31.

Theorem 1.39. Any linearly independent set in a finite-dimensional vector space is a subset of a basis for the space.

Proof. Let S be a linearly independent set in an n-dimensional vector space, V. Theorem 1.35 guarantees that S is finite so say $|S| = k$. We prove the result by induction on $n - k$. By Corollary 1.38, the theorem is true in the base case, $n - k = 0$.

Let $n - k = t > 0$. In this case, there is $\mathbf{v}$ in V that is not in $\operatorname{Span} S$. By Theorem 1.27, $S \cup \{\mathbf{v}\}$ is linearly independent. Since $|S \cup \{\mathbf{v}\}| = k+1$ and $n-(k+1) < t$, the induction hypothesis ensures that there is a basis for V containing $S \cup \{\mathbf{v}\}$, thus, containing S. □

If you study them, the proofs of Lemma 1.33 and Theorem 1.39 suggest methods for constructing a basis for a finite-dimensional vector space. You will have a chance to apply those methods in the exercises.

1.4. Bases

1.4.2. Proof of Theorem 1.31. We rely on relations and on Zorn's Lemma to prove Theorem 1.31 so this is a good place to review the associated terminology. (Section A.5 and Section C.2 have more details.)

Let S be a set with a relation $\leq$. If (a,b) belongs to $\leq$, we write $a \leq b$.

A relation $\leq$ is an *ordering* on S provided it is

(1) *transitive*, meaning that if a, b, c in S satisfy $a \leq b$ and $b \leq c$, then $a \leq c$; and

(2) *antisymmetric*, meaning that if a, b in S satisfy $a \leq b$ and $b \leq a$, then $a = b$.

In this case, $(S, \leq)$ is a *partially ordered set* or a *poset*.

Example 1.40. Let S be the collection of all subsets of $\{a, b\}$. We have
$$S = \{\emptyset, \{a\}, \{b\}, \{a, b\}\}.$$
We see that $(S, \subseteq)$ is a poset. For instance, $\emptyset \subseteq \{b\}$ and $\{a\} \subseteq \{a, b\}$.

We leave it as an exercise to show that the collection of all subsets of any set is a poset under the relation $\subseteq$.

Let $(S, \leq)$ be a poset. We say that $\leq$ is a *total ordering* on S provided $a \leq b$ or $b \leq a$ for every a, b in S. In this case, $(S, \leq)$ is a *totally ordered set*. Since $\{a\}$ is not a subset of $\{b\}$ and $\{b\}$ is not a subset of $\{a\}$, the set in Example 1.40 is not totally ordered under $\subseteq$.

A *chain* in a poset $(S, \leq)$ is a totally ordered subset of S. Considering again the poset $(S, \subseteq)$ in Example 1.40, we can identify $\{\emptyset, \{a\}, \{a, b\}\}$ as a chain. Every poset has chains. Any subset of a chain is also a chain. If $C \subseteq S$ is a chain and there is b in S such that $a \leq b$ for all a in C, then b is an *upper bound* for C in S. For example, considering the chain $\{\emptyset, \{a\}\}$ associated to Example 1.40, we see that both $\{a\}$ and $\{a, b\}$ are upper bounds. An upper bound for a chain is an element in the ambient poset; it need not be in the chain itself.

Zorn's Lemma says that if every chain in a poset $(S, \leq)$ has an upper bound in S, then S has a *maximal element*, that is, m in S such that if $m \leq a$ for any a in S, then $a = m$. The poset in Example 1.40 has maximal element $\{a, b\}$.

We now prove Theorem 1.31, restating each part of the theorem for reference.

Theorem 1.31(a). *Any linearly independent set in a vector space is a subset of a basis for the space.*

Proof. Let V be a vector space and let S be a linearly independent subset of V. Let $\mathcal{S}$ be the collection of linearly independent sets in V that contain S as a subset. Since S belongs to $\mathcal{S}$, $\mathcal{S}$ is nonempty and $(\mathcal{S}, \subseteq)$ is a poset.

Let $\mathcal{G}$ be a chain in $\mathcal{S}$ and let S_1 and S_2 be elements in $\mathcal{G}$: S_1 and S_2 are linearly independent subsets of V, each containing S as a subset. As elements in a chain, either $S_1 \subseteq S_2$ or $S_2 \subseteq S_1$.

Let **G** be the union of the sets in $\mathcal{G}$: **G** is a set of vectors, each of which belongs to a linearly independent set in a chain of linearly independent sets, each of which contains S as a subset.

We claim that **G** is linearly independent.

Suppose there are distinct $\mathbf{v}_1, \ldots, \mathbf{v}_k$ in $\mathbf{G}$ so that for some scalars c_i,
$$c_1\mathbf{v}_1 + \cdots + c_k\mathbf{v}_k = \mathbf{0}.$$
As an element in $\mathbf{G}$, each $\mathbf{v}_i$ belongs to some S_i in $\mathcal{G}$. Since $\mathcal{G}$ is a chain, we may assume that $S_1 \subseteq S_2 \subseteq \cdots \subseteq S_k$: For $i = 1, \ldots, k$, $\mathbf{v}_i$ belongs to S_k. Since S_k is linearly independent, $c_i = 0$ for $i = 1, \ldots, k$. This is enough to establish that $\mathbf{G}$ is itself linearly independent, thus, that $\mathbf{G}$ belongs to $\mathcal{S}$, the collection of all linearly independent subsets of V containing S as a subset.

Since any element in $\mathcal{G}$ is a subset of $\mathbf{G}$, $\mathbf{G}$ is an upper bound for $\mathcal{G}$. As every chain in $\mathcal{S}$ thus has an upper bound, Zorn's Lemma guarantees the existence of a maximal element, $\mathcal{B}$, in $\mathcal{S}$. As a maximal linearly independent subset of V, $\mathcal{B}$ is a basis for V. Since $S \subseteq \mathcal{B}$, the proof of the theorem is complete. □

Theorem 1.31(b). *Any spanning set for a vector space has a subset that is a basis for the space.*

Proof. Let S be a spanning set for a vector space, V. Let $\mathcal{S}$ be the collection of linearly independent subsets of S. Now proceed as in the proof of Theorem 1.31(a). □

Since Theorem 1.31 holds in every vector space, we have established that every vector space has a basis.

Knowing that an infinite-dimensional vector space has a basis is often the best we can do. There are no algorithms for constructing bases in general. One approach to working with linear structure in the general setting is to use linearly independent sets that do not span the space in the algebraic sense but that give us a foundation for defining infinite series that converge to the elements in the space. In this case, the word *sum* may be extended to apply to convergent series, and the words *span* and *basis* may be extended similarly. The locutions *Hamel basis*, *linear basis*, and *algebraic basis* may then be used to refer to what we call a basis in linear algebra.

The cardinality of a basis for any vector space is fixed, whether or not the space is finite-dimensional. Proof of this is more subtle than proof of the existence of a basis and is best approached after a review of linear transformations. We provide the details in Appendix D. Since we use this result going forward, we state it here.

Theorem 1.41. Let V be a vector space. Every basis for V has the same cardinality, that is, if $\mathcal{B}_1$ and $\mathcal{B}_2$ are bases for V, there is a bijection $\mathcal{B}_1 \to \mathcal{B}_2$.

Definition 1.42. The **dimension** of a vector space is the cardinality of any basis for the space.

As in the finite-dimensional case, we designate the dimension of an arbitrary vector space V over a field $\mathbb{F}$ by $\dim V$. To emphasize the field, we write $\dim_{\mathbb{F}} V$.

Exercises 1.4.

1.4.1. Prove Theorem 1.30.

1.4.2. Let $\mathbb{F}$ be a field, considered as a vector space over itself. Describe all bases for $\mathbb{F}$.

1.4.3. Verify the claims made in Example 1.37.

Finite-Dimensional Spaces

1.4.4. Describe every 0-dimensional vector space.

1.4.5. Describe every 1-dimensional vector space.

1.4.6. Let $V = \mathcal{F}(\mathbb{F}_2)$. What is $\dim V$?

1.4.7. Let V be a finite-dimensional vector space. Prove that if W is a subspace of V, then $\dim W \leq \dim V$.

1.4.8. Show that if $\dim V = n$, then for each nonnegative integer $k < n$, there is a subspace of V, W, with $\dim W = k$.

1.4.9. Show that if $\dim V = n$, then the only subspace of V with dimension n is V itself.

1.4.10. Find a basis for the subspace of $\mathbb{R}^4$ spanned by
$$\{(1,0,1,1), (1,0,1,0), (-1,0,-1,1)\}.$$

1.4.11. Expand the following set of vectors into a basis for $\mathbb{R}^4$:
$$\{(1,1,1,1), (0,-1,1,0)\}.$$

1.4.12. Expand the following set of vectors into a basis for $M_2(\mathbb{R})$:
$$\left\{ \begin{bmatrix} 1 & 1 \\ 0 & 1 \end{bmatrix}, \begin{bmatrix} 0 & 1 \\ 1 & 0 \end{bmatrix} \right\}.$$

1.4.13. Finish the proof of Corollary 1.38.

Infinite-Dimensional Vector Spaces

1.4.14. Let $\mathcal{S}$ be any nonempty collection of subsets in a vector space, V. Show that $\subseteq$ is a partial ordering on $\mathcal{S}$.

1.4.15. Show that chains exist in every poset.

1.4.16. There are several different sets that arise in the proof of Theorem 1.31(a). For each set discussed in the proof, determine whether a single element is a vector in V or a set of vectors in V.

1.4.17. Let $V = C(\mathbb{R})$ and let $S = \{\sin t, \cos t\}$. The problems here refer to sets that arise in the proof of Theorem 1.31(a).
   (a) Let $\mathcal{S}$ be the collection of linearly independent sets in V that contain S. Write down three elements of $\mathcal{S}$ that do not form a chain.
   (b) Write down the union of the sets you found in the previous exercise.
   (c) Write down three elements of $\mathcal{S}$ that form a chain.
   (d) Write down an upper bound for the chain you found in the previous exercise.

1.4.18. Finish the proof of Theorem 1.31(b).

1.4.19. Prove or disprove the statement in Exercise 1.4.7 after removing the assumption that V is finite-dimensional.

1.4.20. State the generalization of Exercise 1.4.8 so that it applies to vector spaces that are not necessarily finite-dimensional. Prove or disprove that statement.

1.4.21. Let $V = \{(a_1, \ldots, a_k, 0, 0, \ldots) \,|\, a_i \in \mathbb{R}\}$. An element in V is a real sequence with at most finitely many nonzero terms.
   (a) Show that V is a real vector space.
   (b) Argue that V is infinite-dimensional.
   (c) Let $W = \{(0, a_1, a_2, \ldots, a_k, 0, 0, \ldots) \,|\, a_i \in \mathbb{R}\}$. Prove that W is a proper subspace of V.
   (d) Prove that $\dim W = \dim V$. This shows that the result in Exercise 1.4.9 does not apply to infinite-dimensional spaces.

1.5. Polynomials

Polynomials play several different roles in linear algebra. They define elements in $\mathcal{F}(\mathbb{R})$, for example, such as $f(x) = x^2$. Strictly speaking, this is not a polynomial but a *polynomial function*. This may seem like a distinction without a difference but polynomial functions over arbitrary fields can be wildly different from their underlying polynomials. Tools we have developed to this point will facilitate our introduction to polynomials.

Fix a field $\mathbb{F}$ and symbols $t, t^2, t^3, \ldots$. Let
$$\mathbb{F}[t] := \{c_0 + c_1 t + \cdots + c_k t^k \,|\, c_i \in \mathbb{F}\}.$$

Elements in $\mathbb{F}[t]$ are **polynomials**. Since we do not actually calculate a sum when presented with a polynomial, we say that polynomials are *formal sums*. We call t an **indeterminate** or **variable**. Defining $t^0 := 1$ lets us write
$$c_0 + c_1 t + \cdots + c_k t^k = \sum_{i=0}^{k} c_i t^i.$$

Define addition in $\mathbb{F}[t]$ by adding coefficients of *like terms*,
$$\sum_{i=0}^{k} c_i t^i + \sum_{i=0}^{k} b_i t^i := \sum_{i=0}^{k} (c_i + b_i) t^i.$$

Scaling in $\mathbb{F}[t]$ is done term by term:
$$a \sum_{i=0}^{k} c_i t^i := \sum_{i=0}^{k} a c_i t^i$$

for any a in $\mathbb{F}$. The **zero polynomial** is the polynomial for which every coefficient is zero. Two polynomials are equal provided they have identical nonzero terms. Under these definitions, $\mathbb{F}[t]$ is a countably infinite-dimensional vector space with basis $\mathcal{B} = \{1, t, t^2, t^3, \ldots\}$. We call $\mathbb{F}[t]$ the space of **polynomials in t over $\mathbb{F}$**.

The **nth term** of $p(t) = \sum_{i=1}^{k} c_i t^i$ in $\mathbb{F}[t]$ is the **monomial** $c_n t^n$. The **nth coefficient** of $p(t)$ is c_n. A **nonzero term** is one with a nonzero coefficient. The **constant term** of $p(t)$ is c_0. A **constant polynomial** has the form $p(t) = c_0$.

The **degree** of $p(t) = \sum_{i=1}^{k} c_i t^i \neq 0$ is the maximum value of k in $\mathbb{Z}_{\geq 0}$ such that $c_k \neq 0$. In this case, we write $\deg p(t) = k$. The degree of the zero polynomial is defined to be $-\infty$.

1.5. Polynomials

When $\deg p(t) = k$, the **leading term** of $p(t)$ is $c_k t^k$ and c_k is the **leading coefficient** of $p(t)$. When its leading coefficient is 1, $p(t)$ is **monic**.

The prefix *poly* in the word *polynomial* refers to what may be the several terms — addends — in a polynomial. For instance, $3t^2$ is a monomial, $1-2t$ is a binomial, and $1 + 3t - 5t^2$ is a trinomial.

We can define polynomials over sets such as $\mathbb{Z}$, but we no longer have a vector space when we do that. Note, moreover, that every nonzero polynomial over a field has a scalar multiple that is monic. We leave the easy verification of that as an exercise.

Example 1.43. Here we consider $\mathbb{F}_2[t]$. The constant polynomials are 0 and 1. The degree one polynomials are t and $1 + t$. The degree two polynomials are t^2, $1 + t^2$, $t + t^2$, $1 + t + t^2$. We leave it as an exercise to write down all the degree three polynomials over $\mathbb{F}_2$.

Next, fix n in $\mathbb{Z}_{\geq 0}$. Let $\mathbb{F}[t]_n \subseteq \mathbb{F}[t]$ be the collection of polynomials of degree n or less. Certainly 0 is in $\mathbb{F}[t]_n$. Adding two polynomials in $\mathbb{F}[t]_n$ produces another polynomial of degree n or less and scaling a polynomial by a nonzero element in $\mathbb{F}$ preserves the degree of the polynomial. This proves the following.

Corollary 1.44. If $\mathbb{F}$ is a field and n is in $\mathbb{Z}_{\geq 0}$, then under polynomial addition and scaling, $\mathbb{F}[t]_n$ is a subspace of $\mathbb{F}[t]$.

Example 1.45. We can write down all the elements in
$$\mathbb{F}_2[t]_2 = \{0, 1, t, 1+t, t^2, t+t^2, 1+t^2, 1+t+t^2\}.$$
We leave it as an exercise to write out the addition table for $\mathbb{F}_2[t]_2$.

Just as a polynomial over $\mathbb{R}$ determines a real-valued function, a polynomial over an arbitrary field $\mathbb{F}$ determines a function from $\mathbb{F}$ to $\mathbb{F}$. To make this precise, we must define exponentiation in $\mathbb{F}$. For a nonzero in $\mathbb{F}$, $a^0 := 1$. If a is in $\mathbb{F}$ and k is in $\mathbb{Z}_+$, the *kth power of a* is
$$a^k := \overbrace{a \cdots a}^{k \text{ factors}}.$$

When a is nonzero and k is in $\mathbb{Z}_+$, we define
$$a^{-k} := \overbrace{a^{-1} \cdots a^{-1}}^{k \text{ factors}}.$$

It is not difficult to prove that with these definitions, the usual rules of exponentiation apply in $\mathbb{F}$.

Let $p(t) = \sum_{i=1}^{k} c_i t^i$ belong to $\mathbb{F}[t]$. Define the associated mapping $p : \mathbb{F} \to \mathbb{F}$ by
$$p(a) = \sum_{i=1}^{k} c_i a^i,$$
for a in $\mathbb{F}$. A mapping f in $\mathcal{F}(\mathbb{F})$ that arises from a polynomial this way is a *polynomial function* or *polynomial mapping* on $\mathbb{F}$.

Example 1.46. Consider $p(t) = 1 + 2t - t^2 + t^3 + 2t^4$ in $\mathbb{F}_3[t]$. Since $-1 = 2$ in $\mathbb{F}_3$, $p(t) = 1 + 2t + 2t^2 + t^3 + 2t^4 = 1 - t - t^2 - 2t^3 - t^4$, etc. We can calculate the values of the polynomial function $p : \mathbb{F}_3 \to \mathbb{F}_3$ by substituting $0, 1, 2$, respectively, into $p(t)$. Notice for instance that $2^3 = 2$ and that $2^4 = 1$ in $\mathbb{F}_3$ so that $p(0) = 1$, $p(1) = 1 + 2 + 2 + 1 + 2 = 2$, and $p(2) = 1 + 1 + 2 + 2 + 2 = 2$. It is easy to show that $p : \mathbb{F}_3 \to \mathbb{F}_3$ is the same mapping as q defined by $q(t) = 1 + t^2$.

We have seen that $\mathcal{F}(\mathbb{F}_3)$ is a vector space. The polynomial functions in $\mathcal{F}(\mathbb{F}_3)$ comprise a subspace of $\mathcal{F}(\mathbb{F}_3)$. Since there are 27 elements in $\mathcal{F}(\mathbb{F}_3)$, there can be no more than 27 elements in the subspace of polynomial functions on $\mathbb{F}_3$. We leave it as an exercise to find a basis for this subspace.

Blurring the distinction between a polynomial and the function it defines does not cause problems when we are working over fields of characteristic zero. Different polynomials over these fields define different functions. We can see from Example 1.46, though, that this is not the case for fields of arbitrary characteristic. While a polynomial is determined by its coefficients, a function is determined by its values on its domain.

There is another vector space that we naturally associate to $\mathbb{F}[t]$.

Let $\mathbb{F}$ be a field and let t be an indeterminate. A **formal power series in t over $\mathbb{F}$** is an expression of the form

$$c_0 + c_1 t + c_2 t^2 + \cdots = \sum_{i=0}^{\infty} c_i t^i,$$

where c_i in $\mathbb{F}$ are the **coefficients**. Though we usually drop the modifier, we call these formal power series to emphasize that we are treating them as expressions, not as functions and not as series that might converge.

The collection of all power series in t over $\mathbb{F}$ is designated $\mathbb{F}[[t]]$.

A polynomial can be identified as a power series for which all but finitely many terms are zero. Unless it happens to be a polynomial, a power series does not have a defined degree. Where appropriate, though, the terminology for polynomials extends to power series. For example, the 10th degree term of

$$t + 2t^2 + 3t^3 + \cdots + nt^n + \cdots$$

is $10t^{10}$. As in $\mathbb{F}[t]$, we define addition in $\mathbb{F}[[t]]$ by adding coefficients of like terms. We scale an element in $\mathbb{F}[[t]]$ by scaling each of its coefficients. It is thus evident that $\mathbb{F}[[t]]$ is a vector space, with subspaces that we can identify with $\mathbb{F}[t]$ and with $\mathbb{F}[t]_n$ for n in $\mathbb{Z}_{\geq 0}$.

Since $\mathbb{F}[t]$ can be identified as a proper subspace of $\mathbb{F}[[t]]$, a basis for $\mathbb{F}[t]$ cannot be a basis for $\mathbb{F}[[t]]$. We have seen that $\mathbb{F}[[t]]$ must have a basis. Finding one is another matter.

Multiplication of polynomials is familiar from our work with $\mathbb{R}[t]$ in calculus and precalculus. The idea of multiplication extends to polynomials over arbitrary fields. Formally, we start with $\mathcal{B} = \{1, t, t^2, \ldots\} \subseteq \mathbb{F}[t]$ and the rule $t^k t^m = t^{k+m}$ for m and k in $\mathbb{Z}_{\geq 0}$. With the help of the distributive law, we can then multiply any two elements in $\mathbb{F}[t]$. This makes $\mathbb{F}[t]$ an example of an $\mathbb{F}$-*algebra*.

1.5. Polynomials

Definition 1.47. Let $\mathcal{A}$ be a vector space over $\mathbb{F}$ with a second binary operation, **multiplication**, denoted by juxtaposition. We say that $\mathcal{A}$ is an $\mathbb{F}$-**algebra** provided the following conditions hold for all c in $\mathbb{F}$ and vectors $\mathbf{a}_i$ in $\mathcal{A}$:

(a) $c(\mathbf{a}_1\mathbf{a}_2) = (c\mathbf{a}_1)\mathbf{a}_2 = \mathbf{a}_1(c\mathbf{a}_2)$;

(b) $(\mathbf{a}_1 + \mathbf{a}_2)\mathbf{a}_3 = \mathbf{a}_1\mathbf{a}_3 + \mathbf{a}_2\mathbf{a}_3$; and

(c) $\mathbf{a}_1(\mathbf{a}_2 + \mathbf{a}_3) = \mathbf{a}_1\mathbf{a}_2 + \mathbf{a}_1\mathbf{a}_3$.

A subset of an $\mathbb{F}$-algebra that is also an $\mathbb{F}$-algebra under the same operations is an $\mathbb{F}$-**subalgebra**.

Properties (b) and (c) of Definition 1.47 comprise the distributive law for algebras.

Descriptions that we apply to an $\mathbb{F}$-algebra often refer to its multiplication. For instance, an **associative** $\mathbb{F}$-**algebra** is one for which the multiplication is associative. A **commutative** $\mathbb{F}$-**algebra** is one for which the multiplication is commutative. An $\mathbb{F}$-algebra with identity has a multiplicative identity element, ι. In an $\mathbb{F}$-algebra with identity, an element $\mathbf{a}$ with inverse has a multiplicative inverse, $\mathbf{a}^{-1}$.

When $\mathbb{F}$ is any field, $\mathbb{F}[t]$ is an associative, commutative $\mathbb{F}$-algebra with identity, 1.

Note that when $\mathcal{A}$ is an associative $\mathbb{F}$-algebra with identity, ι, and $\mathbf{a}$ in $\mathcal{A}$ has multiplicative inverse, Lemma A.34 guarantees that the identity is unique and that $\mathbf{a}^{-1}$ is the unique inverse of $\mathbf{a}$.

We may refer to an $\mathbb{F}$-algebra as an algebra, or an algebra over $\mathbb{F}$.

Many vector spaces are spaces of functions, and functions can be composed. This is reason enough to expect algebras to abound. Later we will verify that $M_n(\mathbb{F})$ is an associative $\mathbb{F}$-algebra with identity.

An algebra may have unexpected peculiarities but the zero vector is always a zero divisor.

Lemma 1.48. *If $\mathcal{A}$ is an $\mathbb{F}$-algebra, then $\mathbf{0}\mathbf{a} = \mathbf{a}\mathbf{0} = \mathbf{0}$ for all $\mathbf{a}$ in $\mathcal{A}$.*

Proof. Let $\mathcal{A}$ and $\mathbf{a}$ be as hypothesized. We have
$$\mathbf{0}\mathbf{a} = (\mathbf{a} - \mathbf{a})\mathbf{a} = \mathbf{a}\mathbf{a} - \mathbf{a}\mathbf{a} = \mathbf{0}$$
and
$$\mathbf{a}\mathbf{0} = \mathbf{a}(\mathbf{a} - \mathbf{a}) = \mathbf{a}\mathbf{a} - \mathbf{a}\mathbf{a} = \mathbf{0}. \qquad \square$$

Exercises 1.5.

1.5.1. Let $p(t)$ belong to $\mathbb{F}[t]$ where $\mathbb{F}$ is a field. Show that there is c in $\mathbb{F}$ so that $cp(t)$ is monic.

1.5.2. Referring to Example 1.45, write out the addition table for $\mathbb{F}_2[t]_2$.

1.5.3. Write down all the elements in $\mathbb{F}_3[t]_2$.

1.5.4. Let p be prime. How many elements are in $\mathbb{F}_p[t]_2$? How many elements are in $\mathbb{F}_p[t]_n$? (For help with counting, see Section A.4.)

1.5.5. We found all functions from $\mathbb{F}_2$ to $\mathbb{F}_2$ in Example 1.15. Now find all distinct polynomial functions from $\mathbb{F}_2$ to $\mathbb{F}_2$.

1.5.6. Find $p(t)$ in $\mathbb{F}_2[t]$, $p(t)$ different from 0, where p in $\mathcal{F}(\mathbb{F}_2)$ is the zero mapping.

1.5.7. Repeat the last exercise, this time substituting $\mathbb{F}_3$ for $\mathbb{F}_2$.

1.5.8. Verify the claim made in Example 1.46 that if $q(t) = 1 + t^2$, then q in $\mathcal{F}(\mathbb{F}_3)$ is the same mapping as p, where $p(t) = 1 + 2t - t^2 + t^3 + 2t^4$. Is there a degree three polynomial $r(t)$ in $\mathbb{F}_3[t]$ so that $r = p = q$?

1.5.9. Let S_n be the subset of $\mathbb{F}[[t]]$ consisting of power series of the form
$$c_n t^n + c_{n+1} t^{n+1} + \cdots,$$
c_i in $\mathbb{F}$. Is S_n a subspace of $\mathbb{F}[[t]]$?

1.5.10. Let $\mathcal{A}$ be an $\mathbb{F}$-algebra. Show that $\mathcal{W} \subseteq \mathcal{A}$ is a subalgebra of $\mathcal{A}$ if and only if $\mathcal{W}$ is a subspace of $\mathcal{A}$ that is closed under multiplication.

1.5.11. Show that $W = \{tp(t) \,|\, p(t) \in \mathbb{F}[t]\}$ is a subalgebra of $\mathbb{F}[t]$.

1.5.12. Identify the invertible elements in $\mathbb{F}[t]$.

1.5.13. Let $\mathcal{A}$ be an associative $\mathbb{F}$-algebra with identity. Show that if $\mathbf{a}_1$ and $\mathbf{a}_2$ are invertible elements in $\mathcal{A}$, then $\mathbf{a}_1 \mathbf{a}_2$ is invertible with $(\mathbf{a}_1 \mathbf{a}_2)^{-1} = \mathbf{a}_2^{-1} \mathbf{a}_1^{-1}$.

1.5.14. We mentioned above that the usual rules of arithmetic with exponents apply in any field. Here you will show that they apply in any associative algebra, under the appropriate hypotheses. Let $\mathcal{A}$ be an associative $\mathbb{F}$-algebra. For $\mathbf{a}$ in $\mathcal{A}$ and k in $\mathbb{Z}_+$, define $\mathbf{a}^k$ as we do in any field.
 (a) Show that $\mathbf{a}^m \mathbf{a}^n = \mathbf{a}^{m+n}$, for all m and n in $\mathbb{Z}_+$.
 (b) Show that $(\mathbf{a}^m)^n = \mathbf{a}^{mn}$, for m and n in $\mathbb{Z}_+$.
 (c) If $\mathcal{A}$ has identity, ι, define $\mathbf{a}^0 := \iota$, for all $\mathbf{a}$ in $\mathcal{A}$. Show that the above two rules apply for all $\mathbf{a}$ in $\mathcal{A}$, m and n in $\mathbb{Z}_{\geq 0}$.
 (d) If $\mathcal{A}$ has identity, ι, and if $\mathbf{a}$ is invertible, show that $\mathbf{a}^k$ is invertible, for all k in $\mathbb{Z}_+$. Next show that $(\mathbf{a}^k)^{-1} = (\mathbf{a}^{-1})^k$.
 (e) Show that if $\mathcal{A}$ has identity, ι, and $\mathbf{a}$ is invertible, $\mathbf{a}^m \mathbf{a}^n = \mathbf{a}^{m+n}$ for all m and n in $\mathbb{Z}$.

1.5.15. Let $W = \{p(t) \in \mathbb{F}[t] \,|\, p(0) = 0\}$.
 (a) Show that W is a subspace of $\mathbb{F}[t]$.
 (b) Find a basis for W.
 (c) Is W a subalgebra of $\mathbb{F}[t]$?

1.5.16. Is $\mathbb{F}[t]_n$ a subalgebra of $\mathbb{F}[t]$?

1.5.17. Let $p(t)$ be fixed in $\mathbb{F}[t]$. Show that the set $\{p(t)q(t) \,|\, q(t) \in \mathbb{F}[t]\}$ is a subspace of $\mathbb{F}[t]$. Is it a subalgebra?

1.6. $\mathbb{R}$ and $\mathbb{C}$ in Linear Algebra

Arbitrary fields are interesting and important in applications inside and outside of mathematics. Partly because of their roles in physics and geometry, though, the real numbers and the complex numbers have a special place in linear algebra. We use $\mathbb{R}^2$ to model the Euclidean plane and $\mathbb{R}^3$ to model the physics of motion at the human scale. Complex numbers underlie the theories of electricity, magnetism, and quantum mechanics, all of which use linear algebra one way or another.

1.6. ℝ and ℂ in Linear Algebra

We have seen that every complex vector space is a real vector space. The next theorem is more specific.

Theorem 1.49. If V is a complex vector space, then
$$\dim_{\mathbb{R}} V = 2 \dim_{\mathbb{C}} V.$$

Proof. Let V be a complex vector space with basis $\mathcal{B}$. Let $i\mathcal{B} = \{i\mathbf{b} \mid \mathbf{b} \in \mathcal{B}\}$. We claim that $\mathcal{B} \cup i\mathcal{B}$ is linearly independent over $\mathbb{R}$.

Suppose $\mathbf{b}_1, \ldots, \mathbf{b}_n, \mathbf{b}_{n+1}, \ldots, \mathbf{b}_{n+m}$ are elements in $\mathcal{B}$ and that

(1.10) $\qquad a_1 \mathbf{b}_1 + \cdots + a_n \mathbf{b}_n + ic_1 \mathbf{b}_{n+1} + \cdots + ic_m \mathbf{b}_{n+m} = \mathbf{0},$

for real coefficients a_j and c_j. Since a_j and ic_j are also complex, we can reindex the $\mathbf{b}_j$s if necessary to rewrite (1.10) in the form

$$z_1 \mathbf{b}_1 + \cdots + z_k \mathbf{b}_k = \mathbf{0},$$

where $z_j = a_j + ic_j$ and $\mathbf{b}_1, \ldots, \mathbf{b}_k$ are distinct in $\mathcal{B}$. Since $\mathcal{B}$ is linearly independent, $z_j = 0$ for all j, implying that all a_j and c_j in (1.10) are zero. This is enough to prove our claim.

A similar argument shows that any linear combination of elements in $\mathcal{B}$ over $\mathbb{C}$ can be written as a linear combination of elements in $\mathcal{B} \cup i\mathcal{B}$ over $\mathbb{R}$. This is enough to prove that $\mathcal{B} \cup i\mathcal{B}$ is a basis for V over $\mathbb{R}$. The theorem follows. □

Example 1.50. Let $\mathcal{B} = \{(1, i), (1, -1)\} \subseteq \mathbb{C}^2$. Given z_1, z_2 in $\mathbb{C}$, we have
$$z_1(1, i) + z_2(1, -1) = (z_1 + z_2, iz_1 - z_2).$$

If this linear combination is equal to the zero vector, then we can solve the following system of equations over $\mathbb{C}$ to find the coefficients:

$$\begin{array}{rcl} z_1 + z_2 & = & 0 \\ iz_1 - z_2 & = & 0. \end{array}$$

Adding the two equations we get $(1 + i)z_1 = 0$. Since $\mathbb{C}$ is a field, $z_1 = 0$, from which it follows that z_2 is zero. This establishes that $\mathcal{B}$ is linearly independent. Since $\mathbb{C}^2$ is 2-dimensional over $\mathbb{C}$, $\mathcal{B}$ must be a basis for $\mathbb{C}^2$.

Now consider $i\mathcal{B} = \{(i, -1), (i, -i)\}$. Theorem 1.49 guarantees that $\mathcal{B} \cup i\mathcal{B}$ is a basis for $\mathbb{C}^2$ as a vector space over $\mathbb{R}$. This means that we must be able to write $(1, 3)$, for instance, as a linear combination of elements in

$$\mathcal{B} \cup i\mathcal{B} = \{(1, i), (1, -1), (i, -1), (i, -i)\},$$

using coefficients in $\mathbb{R}$. If we have c_k in $\mathbb{R}$ with

$$c_1(1, i) + c_2(1, -1) + c_3(i, -1) + c_4(i, -i) = (1, 3),$$

we can find the coefficients by solving the following system of equations:

$$\begin{array}{rcl} c_1 + c_2 + ic_3 + ic_4 & = & 1 \\ ic_1 - c_2 - c_3 - ic_4 & = & 3. \end{array}$$

Rewriting the first equation we have
$$c_1 + c_2 + i(c_3 + c_4) = 1$$

implying that $c_3 + c_4 = 0$ and $c_1 + c_2 = 1$. The second equation implies that $c_1 - c_4 = 0$ and that $c_2 + c_3 = -3$. Continuing in this fashion, we find $c_1 = 2$, $c_2 = -1, c_3 = -2$, and $c_4 = 2$. In conclusion, we have
$$(1,3) = 2(1,i) - (1,-1) - 2(i,-1) + 2(i,-i).$$

Complex conjugation on $\mathbb{C}$ can be extended naturally to many complex vector spaces. Recall that if $z = a + ib$ is in $\mathbb{C}$, the *complex conjugate* of z is
$$\bar{z} = a - ib.$$
For instance, $\overline{1+3i} = 1 - 3i$. If $A = [a_{ij}]$ is a complex matrix, its complex conjugate is $\bar{A} = [b_{ij}]$, where $b_{ij} = \overline{a_{ij}}$. Likewise, if f is a mapping into $\mathbb{C}$, we can define $\bar{f}$ by $\bar{f}(x) := \overline{f(x)}$ for any x in the domain of f.

If z is a real number, $\bar{z} = z$. In fact, we can identify $\mathbb{R}$ as a subset of $\mathbb{C}$ by
$$\mathbb{R} = \{z \in \mathbb{C} \mid \bar{z} = z\}.$$

Discussion of complex conjugation is often in terms of the real and imaginary parts of complex numbers. If $z = a + ib$, then $a = \operatorname{Re} z$ in $\mathbb{R}$ is the *real part* of z, and $b = \operatorname{Im} z$ in $\mathbb{R}$ is the *imaginary part* of z. The real and imaginary parts of complex matrices and complex functions are defined likewise. If $A = [a_{ij}]$ is a complex matrix, then the ij-entry of $\operatorname{Re} A$ is $\operatorname{Re} a_{ij}$ and the ij-entry of $\operatorname{Im} A$ is $\operatorname{Im} a_{ij}$. For any complex matrix, A, we can write
$$A = \operatorname{Re} A + i \operatorname{Im} A.$$
Notice that $\operatorname{Re} A$ and $\operatorname{Im} A$ are both real matrices. Likewise, if f is a mapping into $\mathbb{C}$, then $\operatorname{Re} f$ and $\operatorname{Im} f$ are mappings into $\mathbb{R}$ defined by $(\operatorname{Re} f)(x) := \operatorname{Re}(f(x))$ and $(\operatorname{Im} f)(x) := \operatorname{Im}(f(x))$ for any x in the domain of f.

The complex conjugate of a matrix is an important tool in the study of inner product spaces. In Section 5.6, we will see how complex conjugation gets along with other matrix operations, such as addition, scaling, transposing, and multiplication.

Exercises 1.6.

1.6.1. Cite an application of linear algebra in which the underlying field is finite.

1.6.2. Finish the proof of Theorem 1.49.

1.6.3. Let $W \subseteq M_2(\mathbb{C})$ be given by
$$W = \{A \mid \bar{A} = -A\}.$$
(a) Is W a vector space over $\mathbb{C}$?
(b) Is W a vector space over $\mathbb{R}$?

1.6.4. Let $W = \{(z_1, z_1 + z_2, -z_2) \mid z_1, z_2 \in \mathbb{C}\}$.
(a) Verify that W is a complex subspace of $\mathbb{C}^3$ by writing it as the span of some set in $\mathbb{C}^3$.
(b) Find a basis for W as a real vector space.

1.6.5. Let $f : \mathbb{C} \to \mathbb{C}$ be given by $f(z) = z^2$.
(a) Find $f(2 - 3i)$.
(b) Find $\bar{f}(2 - 3i)$.
(c) Write $\bar{f}(z)$ in terms of $a = \operatorname{Re} z$ and $b = \operatorname{Im} z$.

Chapter 2

Linear Transformations and Subspaces

The role of linear transformations in linear algebra is analogous to the role of continuous functions in calculus. We study continuous functions in calculus because they respect the topology of $\mathbb{R}$ and, in so doing, help reveal the nature of $\mathbb{R}$. We study linear transformations in linear algebra because they respect the structure of a vector space and, in so doing, help reveal the nature of a vector space. Just as continuous functions are an essential tool in the study of calculus, linear transformations are central to the study of linear algebra.

One of our goals here is to clarify relationships between the subspaces of a vector space and the linear transformations defined on that space. Another is to explore constructions using subspaces of a vector space, as well as constructions of new vector spaces from old vector spaces. All of these goals are entwined. We will soon see that, in some sense, the subspaces of a vector space determine the linear transformations defined on that space.

Recall that if $f : A \to B$ is any mapping, then A is the *domain* and B is the *codomain* of f. If $S \subseteq A$, the *image of S* is

$$f[S] = \{f(s) \mid s \in S\}.$$

The *range* of f is the image of A, $f[A]$. If $C \subseteq B$, the *preimage* of C is

$$f^{-1}[C] = \{a \in A \mid f(a) \in C\}.$$

We say that f is *injective* or *1-1* when $f(a_1) = f(a_2)$ implies $a_1 = a_2$. We say that f is *surjective* or *onto* when $f[A] = B$. When f is both injective and surjective, it is bijective. [1]

---

[1] For a more detailed review of mappings, see Section A.6.

2.1. Linear Transformations

We use V and W to designate arbitrary vector spaces over a field, $\mathbb{F}$, throughout this section and in the exercises.

Definition 2.1. A **linear transformation** or **linear mapping** $L: V \to W$ is a mapping that satisfies
$$L(c\mathbf{v}_1 + \mathbf{v}_2) = cL(\mathbf{v}_1) + L(\mathbf{v}_2)$$
for all $\mathbf{v}_1, \mathbf{v}_2$ in V, c in $\mathbb{F}$.

A **linear operator** is a linear transformation $L: V \to V$.

The *identity mapping*, $\mathbf{v} \mapsto \mathbf{v}$, is a linear operator on any vector space. The *zero mapping*, $\mathbf{v} \mapsto \mathbf{0}_W$, is a linear mapping $V \to W$. A linear transformation is nontrivial if it is different from the zero mapping.

We have seen that a vector space over a field $\tilde{\mathbb{F}}$ is also a vector space over any subfield of $\tilde{\mathbb{F}}$. This means that occasions arise in which we may wish to clarify the underlying field when referring to a linear transformation. In such a case, we say that $L: V \to W$ is $\mathbb{F}$-linear. For example, the mapping $z \mapsto \bar{z}$ is $\mathbb{R}$-linear on $\mathbb{C}$.

Our next example identifies an important class of mappings that arise in linear algebra and in applications.

Example 2.2. Let $\tilde{\mathbb{F}}$ be a field with subfield $\mathbb{F}$. Let a belong to $\tilde{\mathbb{F}}$. Define
$$\xi_a : \mathbb{F}[t] \to \tilde{\mathbb{F}}$$
by $\xi_a(p(t)) = p(a)$. Because of the way we define vector addition and scaling in $\mathbb{F}[t]$, ξ_a is an $\mathbb{F}$-linear transformation. We say that ξ_a is an **evaluation** mapping and we call ξ_a **evaluation at** a.

An easy exercise reveals that the only linear operators on a 1-dimensional vector space are those of the form $\mathbf{v} \mapsto c\mathbf{v}$, for a scalar c. This interpretation of linear operators does not extend beyond 1-dimensional spaces.

Example 2.3. Define L on $\mathbb{R}^2$ by
$$L(a,b) = (a-b, a+b).$$
We leave it as an exercise to verify that L is linear. We claim, moreover, that there is no nonzero $\mathbf{v}$ in $\mathbb{R}^2$ such that $L(\mathbf{v}) = c\mathbf{v}$ for some c in $\mathbb{R}$.

Suppose $\mathbf{v} = (a,b)$ satisfies $L(\mathbf{v}) = c\mathbf{v}$, for some c in $\mathbb{R}$. Since $ca = a - b$ and $cb = a + b$, both of the following equations must hold:
$$\begin{array}{rcrc} (c-1)a & + & b & = 0 \\ -a & + & (c-1)b & = 0. \end{array}$$
Routine calculations reveal that these hold only if $\mathbf{v} = (0,0)$. This is enough to prove our claim.

The functions we work with in calculus are not usually linear, but many of the operations we perform on the functions in calculus — taking limits, derivatives, and

2.1. Linear Transformations

definite integrals, for example — are linear. If f and g are differentiable functions on some subset of $\mathbb{R}$, for instance, and c is a real number, then

$$(cf + g)' = cf' + g'$$

says that $f \mapsto f'$ is linear.

We can — and often do — define linear transformations by *linear extension*. Suppose $\mathcal{B} = \{\mathbf{b}_i\}_{i \in I}$ is a basis for V and that f maps $\mathcal{B}$ into W. Define

$$L(\mathbf{v}) := c_1 f(\mathbf{b}_1) + \cdots + c_k f(\mathbf{b}_k)$$

for $\mathbf{v} = \sum_{i=1}^{k} c_i \mathbf{b}_i$ in V. By its definition, $L : V \to W$ is a linear transformation. We say that L is the **linear extension** of f.

Example 2.4. Define $L : \mathbb{R}[t]_2 \to \mathbb{R}[t]_3$ by extending linearly from $1 \mapsto 1 + t$, $t \mapsto 1 - t + t^2$, and $t^2 \mapsto t + t^3$. We then have

$$L(a_0 + a_1 t + a_2 t^2) = a_0(1 + t) + a_1(1 - t + t^2) + a_2(t + t^3)$$
$$= (a_0 + a_1) + (a_0 - a_1 + a_2)t + a_1 t^2 + a_2 t^3.$$

The next theorem cites some of the features of linear transformations that follow directly from the definitions. We leave the proof as an exercise.

Theorem 2.5. If $L : V \to W$ is a linear transformation, then

(a) $L(\mathbf{0}_V) = \mathbf{0}_W$;
(b) $L(-\mathbf{v}) = -L(\mathbf{v})$ for all $\mathbf{v}$ in V;
(c) if X is a subspace of V, then $L[X]$ is a subspace of W;
(d) if Y is a subspace of W, then $L^{-1}[Y]$ is a subspace of V.

Every nontrivial vector space has two subspaces: $\{\mathbf{0}\}$ and the space itself. Theorem 2.5 thus gives us the means to identify two subspaces associated to any linear transformation $L : V \to W$. One is $L[V] \subseteq W$, the range of L. The second is $L^{-1}[\{\mathbf{0}_W\}] \subseteq V$.

Definition 2.6. The **kernel** of a linear transformation $L : V \to W$ is $L^{-1}[\{\mathbf{0}_W\}]$.

We denote the kernel of $L : V \to W$ by $\operatorname{Ker} L$.

Example 2.7. Using the notation we established in Example 2.2, consider the $\mathbb{R}$-linear map

$$\xi_i : \mathbb{R}[t] \to \mathbb{C}.$$

An element $p(t)$ belongs to $\operatorname{Ker} \xi_i$ if and only if $p(i) = 0$. Based on our experience with polynomials, we know this means that

$$\operatorname{Ker} \xi_i = \{(1 + t^2) q(t) \mid q(t) \in \mathbb{R}[t]\}.$$

Both the kernel and the range reveal important information about a linear transformation.

Theorem 2.8. A linear transformation is injective if and only if its kernel is trivial.

Proof. Let $L : V \to W$ be a linear transformation.

If L is injective, it is immediate that its kernel is $\{\mathbf{0}_V\}$.

Suppose $\operatorname{Ker} L = \{\mathbf{0}_V\}$. If $L(\mathbf{v}_1) = L(\mathbf{v}_2)$ for some $\mathbf{v}_i$ in V, then
$$L(\mathbf{v}_1 - \mathbf{v}_2) = L(\mathbf{v}_1) - L(\mathbf{v}_2) = \mathbf{0}_W.$$
This says that $\mathbf{v}_1 - \mathbf{v}_2 = \mathbf{0}_V$, thus, that $\mathbf{v}_1 = \mathbf{v}_2$. It follows that L is injective. $\square$

Proof of the following important corollary is an exercise.

Corollary 2.9. A linear transformation is injective if and only if the image of each linearly independent set in its domain is linearly independent in its range.

There is an analogous criterion for surjectivity of linear transformations.

Lemma 2.10. A linear transformation is surjective if and only if the image of each spanning set for its domain is a spanning set for its codomain.

Proof. Suppose $L : V \to W$ is a surjective linear transformation. Let S be a spanning set for V and let $\mathbf{w}$ belong to W. Say that $L(\mathbf{v}) = \mathbf{w}$. Writing $\mathbf{v} = c_1 \mathbf{v}_1 + \cdots + c_k \mathbf{v}_k$ for $\mathbf{v}_i$ in S and c_i in $\mathbb{F}$, we have
$$\mathbf{w} = L(\mathbf{v}) = c_1 L(\mathbf{v}_1) + \cdots + c_k L(\mathbf{v}_k).$$
This shows that every $\mathbf{w}$ in W is in $\operatorname{Span} L[S]$, proving the result in one direction.

We leave proof of the lemma in the other direction as an exercise. $\square$

Suppose $L : V \to W$ is an arbitrary linear transformation. Since $L[V]$ is a subspace of W, we can think of L has having codomain $L[V]$ when it suits us. Invoking this trick allows us to restrict our attention to linear transformations that are surjective. This suggests that injectivity is the more pressing concern when we study linear transformations.

Exercises 2.1.

2.1.1. Is complex conjugation $\mathbb{C}$-linear on $\mathbb{C}^n$? Is it $\mathbb{R}$-linear on $\mathbb{C}^n$?

2.1.2. Prove that the only linear operators on a 1-dimensional vector space are those of the form $\mathbf{v} \mapsto c\mathbf{v}$ where c is a scalar.

2.1.3. Define $L : V \to \mathbb{F}^m$ by
$$L(\mathbf{v}) = (f_1(\mathbf{v}), \ldots, f_m(\mathbf{v}))$$
for some $f_i : V \to \mathbb{F}$. Show that L is a linear transformation if and only if f_i is a linear transformation for each $i = 1, \ldots, m$.

2.1.4. Verify that the mapping L in Example 2.3 is linear. Finish the proof that $L(\mathbf{v}) = c\mathbf{v}$ only if $\mathbf{v} = (0,0)$.

2.1.5. Let $V = \mathbb{F}_3{}^2$. How many linear operators are there on V? Find three that are bijective.

2.1.6. Let $L : V \to V$ be a linear operator. Fix c in $\mathbb{F}$. Show that
$$\{\mathbf{v} \in V \mid L(\mathbf{v}) = c\mathbf{v}\}$$
is a subspace of V.

2.1.7. Prove Theorem 2.5.

2.1.8. Is the converse of Theorem 2.5(c) true? In other words, if $L: V \to W$ is a linear transformation and $X \subseteq V$ has the property that $L[X]$ is a subspace of W, must X be a subspace of V?

2.1.9. Consider the evaluation mapping given in Example 2.7. Let
$$\mathcal{C} = \{1+t^2, t+t^3, t^2+t^4, \ldots\} \subseteq \mathbb{R}[t].$$
(a) Show that $\mathcal{C}$ is linearly independent.
(b) Show that $\mathcal{C}$ spans $\operatorname{Ker} \xi_i$. Conclude that $\mathcal{C}$ is a basis for $\operatorname{Ker} \xi_i$.

2.1.10. Prove Corollary 2.9.

2.1.11. Suppose $L: V \to W$ is an injective linear transformation. Why is it immediately evident that $\operatorname{Ker} L = \{\mathbf{0}_V\}$?

2.1.12. Show that the mapping in Example 2.4 is injective.

2.1.13. Show that a linear transformation $L: V \to W$ is injective if and only if $L^{-1}[\{L(\mathbf{v})\}] = \mathbf{v}$ for any one vector $\mathbf{v}$ in V.

2.1.14. Let $L: V \to W$ be an injective linear transformation. Let $\mathcal{B}$ be a basis for V. Prove directly that $L[\mathcal{B}]$ is a basis for $L[V]$.

2.1.15. Is the kernel of the mapping in Example 2.7 a subalgebra of $\mathbb{R}[t]$?

2.1.16. Let $V = C(\mathbb{R})$. For a fixed in $\mathbb{R}$, define $\xi_a: V \to \mathbb{R}$ by $\xi_a(f) = f(a)$. What is $\dim \xi_a[V]$?

2.1.17. Here we consider $\xi_{\sqrt{2}}: \mathbb{Q}[t] \to \mathbb{R}$. Let $X = \operatorname{Ker} \xi_{\sqrt{2}}$.
(a) Show that $t^2 - 2$ is the lowest degree monic polynomial in X.
(b) Let $Y \subseteq X$ be the set of polynomials of degree three or less in X. Show that Y is a subspace of X.
(c) Show that if $p(t) = at^3 + bt^2 + ct + d$ is in Y, then $p(t) = (t^2 - 2)q(t)$, for some $q(t)$ in $\mathbb{Q}[t]$.

2.1.18. Finish the proof of Lemma 2.10.

2.1.19. Suppose $L: V \to W$ is an injective linear transformation. Define T on $L[V]$ by $T(L(\mathbf{v})) = \mathbf{v}$.
(a) Prove that T is well-defined and linear.
(b) Prove that $T \circ L$ is the identity mapping on V.
(c) Prove that $L \circ T$ is the identity mapping on $L[V]$.

2.2. Cosets and Quotient Spaces

Our work here will show that every subspace of a vector space V is the kernel of a linear transformation defined on V. We start by introducing the notion of a vector space *coset*.

Definition 2.11. Let V be a vector space with subspace W. A *W-coset* is a set of the form
$$\mathbf{v} + W = \{\mathbf{v} + \mathbf{w} \mid \mathbf{w} \in W\}$$
where $\mathbf{v}$ is any vector in V.

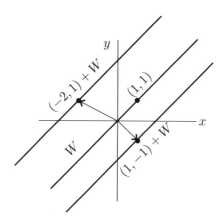

Figure 2.1. The cosets of $W = \text{Span}\{(1,1)\} \subseteq \mathbb{R}^2$ comprise a parallel class of lines in the xy-plane.

Example 2.12. Let $W = \text{Span}\{(1,1)\} \subseteq \mathbb{R}^2$. If we identify W with the line in the xy-plane given by $y = x$, how can we describe the other W-cosets in $\mathbb{R}^2$? We claim that they are the lines in $\mathbb{R}^2$ with slope equal to 1, that is, the parallel class of lines in $\mathbb{R}^2$ determined by W. To see this, take (a,b) in $\mathbb{R}^2$ and notice that

$$(a,b) + W = (a,b) + \{(t,t)\,|\,t \in \mathbb{R}\} = \{(t+a, t+b)\,|\,t \in \mathbb{R}\}.$$

This means we can identify $(a,b) + W$ with the line in the xy-plane given by $y = x + (b-a)$. On the other hand, if ℓ is any line in $\mathbb{R}^2$ with slope 1, then ℓ is given by $y = x + b$, for some b in $\mathbb{R}$. In this case, we can write

$$\ell = \{(t, t+b)\,|\,t \in \mathbb{R}\},$$

allowing us to see ℓ as the coset $(0,b) + W$.

Figure 2.1 shows W along with two other W-cosets.

Next we show that the cosets determined by a subspace form a partition on the ambient space.[2] We start with a lemma.

Lemma 2.13. Let V be a vector space with subspace W. If $\mathbf{v}$ and $\mathbf{x}$ belong to V, then the following statements are logically equivalent:

(a) $\mathbf{v}$ is in $\mathbf{x} + W$;

(b) $\mathbf{v} - \mathbf{x}$ is in W;

(c) $\mathbf{x} - \mathbf{v}$ is in W;

(d) $\mathbf{x}$ is in $\mathbf{v} + W$;

(e) $\mathbf{v} + W = \mathbf{x} + W$.

---

[2] For definitions and a discussion of partitions and equivalence relations, see Section A.5.

2.2. Cosets and Quotient Spaces

Proof. Suppose V, W, $\mathbf{v}$, and $\mathbf{x}$ are as hypothesized.

If $\mathbf{v}$ belongs to $\mathbf{x} + W$, then there is $\mathbf{w}$ in W such that
$$\mathbf{v} = \mathbf{x} + \mathbf{w}.$$
In this case, $\mathbf{v} - \mathbf{x} = \mathbf{w}$. This proves that (a) implies (b).

Suppose next that $\mathbf{v} - \mathbf{x}$ is in W. Since W is a subspace,
$$-(\mathbf{v} - \mathbf{x}) = \mathbf{x} - \mathbf{v}$$
is in W so (b) implies (c).

Suppose that $\mathbf{x} - \mathbf{v} = \mathbf{w}$ is in W. We then have
$$\mathbf{x} = \mathbf{v} + \mathbf{w} \in \mathbf{v} + W$$
proving that (c) implies (d).

Suppose next that $\mathbf{x}$ is in $\mathbf{v} + W$ with $\mathbf{x} = \mathbf{v} + \mathbf{w}_1$ for $\mathbf{w}_1$ in W. This means that for any $\mathbf{w}_2$ in W,
$$\mathbf{x} + \mathbf{w}_2 = \mathbf{v} + (\mathbf{w}_1 + \mathbf{w}_2) \in \mathbf{v} + W.$$
This shows that (d) implies $\mathbf{x} + W \subseteq \mathbf{v} + W$. Switching the roles of $\mathbf{v}$ and $\mathbf{x}$ and then repeating the same argument, we get $\mathbf{v} + W \subseteq \mathbf{x} + W$. This is enough to prove that (d) implies (e).

Finally, suppose $\mathbf{v} + W = \mathbf{x} + W$. This means that for each $\mathbf{w}_1$ in W there is $\mathbf{w}_2$ in W so that
$$\mathbf{v} + \mathbf{w}_1 = \mathbf{x} + \mathbf{w}_2.$$
We then have
$$\mathbf{v} = \mathbf{x} + (\mathbf{w}_2 - \mathbf{w}_1) \in \mathbf{x} + W.$$
This shows that (e) implies (a), thus, that all statements of the lemma are logically equivalent. □

Theorem 2.14. If V is a vector space with subspace W, then the W-cosets partition V.

Proof. Let V and W be as hypothesized. We must show that each element in V belongs to a W-coset, that each W-coset is nonempty, and that the intersection of distinct W-cosets is empty.

Since $\mathbf{0}$ is in W, any $\mathbf{v}$ in V belongs to $\mathbf{v} + W$. This proves that every $\mathbf{v}$ in V belongs to a W-coset and that each W-coset is nonempty.

That parts (a) and (e) of Lemma 2.13 are logically equivalent means precisely that the intersection of distinct W-cosets is empty. □

Let V be a vector space with subspace W and let $\mathbf{v}$ belong to V. Any element $\mathbf{x}$ in $\mathbf{v} + W$ is a **coset representative** of $\mathbf{v} + W$. A W-coset may have many different representatives. Referring to Example 2.12, for instance, the line given by $y = x + 3$ has coset representatives $(0, 3)$, $(1, 4)$, and $(-1, 2)$, simply because all of those points are on that line.

Denote the set of W-cosets in V by V/W.

Example 2.15. Let $V = \mathbb{F}_2^3$ and let
$$W = \text{Span}\{(1,1,1)\} = \{(0,0,0),(1,1,1)\} = (0,0,0) + W = (1,1,1) + W.$$
Since the W-cosets partition the eight elements in V, we can find a second W-coset by identifying an element in V that is not in W. For instance, since $(1,0,0)$ is not in W, $(1,0,0) + W$ is different from W, with
$$(1,0,0) + W = \{(1,0,0),(0,1,1)\} = (0,1,1) + W.$$
Continuing in this fashion, we have
$$(0,1,0) + W = \{(0,1,0),(1,0,1)\} = (1,0,1) + W$$
and finally
$$(0,0,1) + W = \{(0,0,1),(1,1,0)\} = (1,1,0) + W.$$
This shows that V/W is a four-element set.

The next theorem clarifies our interest in W-cosets.

Theorem 2.16. If V is a vector space with subspace W, then V/W is a vector space with addition defined by
$$(\mathbf{v}_1 + W) + (\mathbf{v}_2 + W) := (\mathbf{v}_1 + \mathbf{v}_2) + W$$
and scaling defined by
$$c(\mathbf{v} + W) := c\mathbf{v} + W.$$

Proof. Let V and W be as hypothesized. Since it contains W, V/W is nonempty.

Since we define addition and scaling in V/W in terms of coset representatives, we must show that the operations are well-defined, in other words, that the result of adding or scaling cosets depends only on the underlying cosets, not on the coset representatives.[3]

Suppose $\mathbf{v}_i + W = \mathbf{x}_i + W$ for $\mathbf{v}_i$ and $\mathbf{x}_i$ in V, $i = 1, 2$. Applying Lemma 2.13, we have $\mathbf{v}_i - \mathbf{x}_i$ belonging to W. Since W is a subspace,
$$\mathbf{v}_1 - \mathbf{w}_1 + \mathbf{v}_2 - \mathbf{w}_2 = (\mathbf{v}_1 + \mathbf{v}_2) - (\mathbf{x}_1 + \mathbf{x}_2) \in W.$$
It follows again by Lemma 2.13 that
$$\mathbf{v}_1 + \mathbf{v}_2 + W = \mathbf{x}_1 + \mathbf{x}_2 + W,$$
proving that addition is well-defined on V/W.

We leave it as an exercise to show that scaling on V/W is well-defined.

Since $(\mathbf{v} + W) + W = \mathbf{v} + W$, the zero vector in V/W is W. The additive inverse of $\mathbf{v} + W$ in V/W is $-\mathbf{v} + W$.

Because they hold in V, the remaining properties of a vector space can be applied to coset representatives of elements in V/W. Those properties thus hold in V/W. □

---
[3] Well-definedness is the subject of Section A.3.

2.2. Cosets and Quotient Spaces

Definition 2.17. Let V be a vector space with subspace W. The **quotient space** of V by W is the set V/W with addition and scaling determined by adding and scaling coset representatives.

We read V/W as $V \mod W$.

Example 2.18. Let $V = \mathbb{C}$, viewed as a real vector space. Let W be the set of real numbers in V. Given $z = a + ib$ in $\mathbb{C}$,
$$z + W = a + ib + W = ib + a + W = ib + W,$$
because $a + W = W$. It follows that $V/W = \{ib + W \mid b \in \mathbb{R}\}$: V/W can be identified with the set of pure imaginary numbers in $\mathbb{C}$.

Example 2.19. What is V/W for V and W as in Example 2.12? As a set, V/W is the collection of all lines in $\mathbb{R}^2$ with slope 1. Since there is exactly one line in $\mathbb{R}^2$ with slope 1 and a given y-intercept, we can identify each W-coset by its y-intercept: V/W can be identified with $\mathbb{R}$.

Theorem 1.31(a) implies the following. We leave the details of a proof to the exercises.

Corollary 2.20. Let V be a vector space with subspace, W. If $\dim V = n$ and $\dim W = k$, then $\dim V/W = n - k$.

We are ready now for the theorem we promised at the beginning of this section.

Theorem 2.21. Every subspace of a vector space V is the kernel of a linear transformation on V.

Proof. Let V be a vector space with subspace W. Define $L : V \to V/W$ by $L(\mathbf{v}) = \mathbf{v} + W$. Given $\mathbf{v}_1, \mathbf{v}_2$ in V and c in the underlying field, we have
$$L(c\mathbf{v}_1 + \mathbf{v}_2) = (c\mathbf{v}_1 + \mathbf{v}_2) + W = (c\mathbf{v}_1 + W) + (\mathbf{v}_2 + W) = cL(\mathbf{v}_1) + L(\mathbf{v}_2).$$
This shows that L is a linear transformation. Since $L(\mathbf{v}) = W = \mathbf{0}_{V/W}$ if and only if $\mathbf{v}$ is in W, $W = \operatorname{Ker} L$. $\square$

The mapping $L : V \to V/W$ described in the proof of Theorem 2.21 is called the **canonical mapping** $V \to V/W$. When we say that a mapping is canonical, we mean that it is defined without reference to a coordinate system, that is, a basis. This idea comes up frequently in the sequel.

Let V be a vector space and let $L : V \to X$ be a linear transformation. If W is a subspace of V and $\mathbf{v}$ is an element in V, then by the definition of $L[S]$ for any subset S in V, we have
$$L[\mathbf{v} + W] = \{L(\mathbf{v} + \mathbf{w}) \mid \mathbf{w} \in W\} = \{L(\mathbf{v}) + L(\mathbf{w}) \mid \mathbf{w} \in W\} = L(\mathbf{v}) + L[W].$$
By Theorem 2.5, $L[W]$ is a subspace of X.

With the same definitions of L, V, and W, define $\tilde{L}(\mathbf{v} + W) := L(\mathbf{v}) + L[W]$ for any $\mathbf{v}$ in V. If $\mathbf{v}_1 + W = \mathbf{v}_2 + W$, then $L(\mathbf{v}_1 - \mathbf{v}_2)$ belongs to $L[W]$. It follows that
$$\tilde{L}(\mathbf{v}_1 + W) = \tilde{L}(\mathbf{v}_2 + W),$$

thus, that $\tilde{L}$ is well-defined on V/W. Since L is linear and $\tilde{L}$ is defined by applying L to coset representatives, $\tilde{L}$ is a linear transformation $V \to X/L[W]$. We say that L *induces* the linear transformation $\tilde{L}$.

When does a linear operator on V determine a linear operator on V/W? The next definition goes toward answering the question.

Definition 2.22. Let L be a linear operator on a vector space V. A subspace $W \subseteq V$ is L-**invariant** provided $L[W] \subseteq W$.

The most obvious example of an L-invariant subspace is the kernel of a linear operator $L: V \to V$. This is just to say that

$$L[\operatorname{Ker} L] = \{\mathbf{0}_V\} \subseteq \operatorname{Ker} L.$$

We leave it as an exercise to show that $L[V]$ is also L-invariant.

Theorem 2.23. Let L be a linear operator on a vector space, V. Let W be a subspace of V. The mapping

(2.1) $$\tilde{L}(\mathbf{v} + W) := L(\mathbf{v}) + W$$

is a linear operator on V/W if and only if W is L-invariant.

Proof. Let L, V, and W be as hypothesized.

If the rule given in (2.1) is a linear operator on V/W, then it is well-defined. This means that if $\mathbf{v}_1 + W = \mathbf{v}_2 + W$, then

$$\tilde{L}(\mathbf{v}_1 + W) = \tilde{L}(\mathbf{v}_2 + W) = L(\mathbf{v}_1) + W = L(\mathbf{v}_2) + W.$$

It follows that when $\mathbf{v}_1 - \mathbf{v}_2$ is in W,

$$L(\mathbf{v}_1) - L(\mathbf{v}_2) = L(\mathbf{v}_1 - \mathbf{v}_2) \in W.$$

This is enough to establish that if $\mathbf{w}$ in W, then $L(\mathbf{w})$ is also in W, which proves that W is L-invariant.

We leave proof of the converse as an exercise. □

Quotient spaces and associated mappings remain with us for the rest of the course.

Exercises 2.2.

Unless indicated otherwise, assume throughout these exercises that V is an $\mathbb{F}$ vector space with subspace W.

2.2.1. Let $S = \mathbf{v} + W$ for some $\mathbf{v}$ in V. Show that S is a subspace of V if and only if it contains the zero vector.

2.2.2. Show that any two W-cosets in V have the same cardinality, that is, that there is a bijection between any two W-cosets.

2.2.3. Consider the line ℓ comprised of points (x, y) in $\mathbb{R}^2$ that satisfy $2x + 3y = 1$. Identify a subspace $W \subseteq \mathbb{R}^2$ so that ℓ is a W-coset.

2.2.4. Find three different coset representatives for each of the lines depicted in Figure 2.1. Sketch the associated vectors, as $(1, -1)$ is depicted in the figure.

2.2.5. Prove that
$$\mathbf{v} \stackrel{\sim}{w} \mathbf{x} \iff \mathbf{v} - \mathbf{x} \in W$$
is an equivalence relation on V.

2.2.6. Write out an addition table for V/W in Example 2.15.

2.2.7. Let S be a linearly independent subset of V/W. Let $\mathbf{v}_1, \ldots, \mathbf{v}_k$ be coset representatives for k distinct W-cosets in S. Show that $\{\mathbf{v}_1, \ldots, \mathbf{v}_k\}$ is linearly independent in V.

2.2.8. Let $V = \mathbb{R}[t]$ and let $W = \text{Span}\{t^2, t^3, \ldots\} \subseteq V$. Find a basis for V/W.

2.2.9. Prove Corollary 2.20.

2.2.10. Finish the proof of Theorem 2.16.

2.2.11. Show that there is a bijective correspondence between subspaces in V/W and subspaces in V that contain W.

2.2.12. This problem refers to Example 2.7.
   (a) Find a basis for $\mathbb{R}[t]/\text{Ker}\,\xi_i$.
   (b) Is $\mathbb{R}[t]/\text{Ker}\,\xi_i$ an algebra?

2.2.13. Prove that if L is a linear operator on V, then $L[V]$ is L-invariant.

2.2.14. Finish the proof of Theorem 2.23.

2.3. Affine Sets and Mappings

That the lines through the origin in $\mathbb{R}^2$ are its 1-dimensional subspaces affords us a satisfying connection between linear algebra and the Euclidean geometry of the real plane. While a line not containing the zero vector cannot be a subspace, it can still be accessible to the tools of linear algebra. In this section, we digress briefly to develop this idea.

Let V be a nontrivial vector space with subspace W. We established in Section 2.2 that the W-cosets are vectors in the quotient space, V/W. Here, we view the W-cosets as elements in the *affine geometry* associated to V. Specifically, the objects in the affine geometry associated to V are all cosets associated to all subspaces of V.

Definition 2.24. An **affine set** in a vector space V is a coset for any subspace in V. The **dimension of an affine set** is the dimension of the underlying subspace.

A **k-plane** is a k-dimensional affine set. A **point** is a 0-plane. A **line** is a 1-plane. A **plane** is a 2-plane.

If distinct affine sets have a nonempty intersection, we say they are *incident*. We use the same vocabulary to talk about incidence among affine sets that we use when talking about geometric objects in other settings. For instance, when a point P and a line ℓ are incident, we say that P lies on ℓ, that ℓ passes through P, that P is incident to ℓ, and that ℓ is incident to P.

Definition 2.25. The **affine geometry** associated to a vector space V is the collection of affine sets in V together with the incidence relation.

When W is the subspace underlying an affine set α, any nonzero vector in W is a *direction* or *direction vector* for α. If $\mathbf{v}$ and $\mathbf{w}$ are points in an affine set α, then $\mathbf{v} - \mathbf{w}$ is a direction vector for α.

The W-cosets we studied in Section 2.2 were the affine sets associated to one subspace W of some vector space. There we saw that distinct W-cosets are nonintersecting. We describe the associated affine objects by saying they are *parallel*: Not only are they nonintersecting, they also have the same directions. When W is a fixed subspace of a vector space, V, the W-cosets thus comprise the entire parallel class of affine objects with underlying subspace W.

Affine sets behave much as we expect points, lines, and planes to behave, based on our experience with Euclidean space. For example, since a field has at least two elements, there are at least two points on every line and at least three points in any plane in a vector space. What affine sets do not have, in general, are the features that Euclidean objects owe to the real numbers. For instance, an affine line may have finitely many points.

Figure 2.2 shows the points in $\mathbb{F}_3{}^2$, partitioned into three parallel lines. The solid black points form a line that passes through the origin, that is, a 1-dimensional subspace. The three gray points form a parallel line, and the three white points form a third line parallel to the other two.

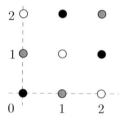

Figure 2.2. Three parallel lines in $\mathbb{F}_3{}^2$.

Theorem 2.26. *Two points determine a unique line in any vector space.*

Proof. Let $\mathbf{v}$ and $\mathbf{w}$ be distinct points in a vector space, V. Since $\mathbf{x} = \mathbf{v} - \mathbf{w}$ is nonzero, $\mathrm{Span}\{\mathbf{x}\}$ is a 1-dimensional subspace of V. Let $\ell = \mathbf{v} + \mathrm{Span}\{\mathbf{x}\}$. Since both $\mathbf{v}$ and $\mathbf{w} = \mathbf{v} - \mathbf{x}$ belong to ℓ, we can say that $\mathbf{v}$ and $\mathbf{w}$ determine the line ℓ.

If ℓ' is another line passing through $\mathbf{v}$ and $\mathbf{w}$, then for some $\mathbf{y}$ in V,
$$\ell' = \mathbf{v} + \mathrm{Span}\{\mathbf{y}\} = \mathbf{w} + \mathrm{Span}\{\mathbf{y}\}.$$
This means that $\mathbf{v} - \mathbf{w}$ is in $\mathrm{Span}\{\mathbf{y}\}$. Since $\mathbf{v} - \mathbf{w} \neq \mathbf{0}$, $\mathrm{Span}\{\mathbf{v} - \mathbf{w}\} = \mathrm{Span}\{\mathbf{y}\}$. We conclude that $\ell' = \ell$, proving that the line determined by $\mathbf{v}$ and $\mathbf{w}$ is unique. $\square$

A corollary follows immediately.

Corollary 2.27. *Distinct lines in a vector space intersect in a single point or not at all.*

2.3. Affine Sets and Mappings

When working with the affine geometry underlying a vector space, we use affine transformations and mappings instead of linear transformations.

Definition 2.28. Let V and W be vector spaces over $\mathbb{F}$. An **affine mapping** $A(L, \mathbf{w}) : V \to W$ is a mapping of the form

$$A(L, \mathbf{w}) = T_\mathbf{w} \circ L,$$

where $L : V \to W$ is a linear transformation and $T_\mathbf{w}(\mathbf{y}) = \mathbf{y} + \mathbf{w}$, for $\mathbf{y}$ in W, is **translation** by $\mathbf{w}$ in W.

An **affine transformation** is an affine mapping $A(L, \mathbf{v}) : V \to V$, where $L : V \to V$ is a bijective linear operator and $\mathbf{v}$ is in V.

While the terms *linear mapping* and *linear transformation* are interchangeable, the terms *affine mapping* and *affine transformation* are not.

Example 2.29. Define $f : \mathbb{R} \to \mathbb{R}$ by $f(x) = 2x + 1$. Though its graph is a line, f violates Theorem 2.5(a) so it is not a linear mapping. Since we can write $f = T_1 \circ L$, where $L(x) = 2x$ for x in $\mathbb{R}$, f is an affine transformation.

Example 2.30. Define $f : \mathbb{R} \to \mathbb{R}^3$ by

$$f(t) = (2t + 1, 3t - 1, t + 2).$$

This mapping gives us a parametric description of a line in $\mathbb{R}^3$:

$$x = 2t + 1, \quad y = 3t - 1, \quad z = t + 2, \quad t \in \mathbb{R}.$$

If we define $L(t) := (2t, 3t, t)$, then $L : \mathbb{R} \to \mathbb{R}^3$ is a linear transformation. Letting $\mathbf{w} = (1, -1, 2)$, we have

$$f = T_\mathbf{w} \circ L.$$

This shows that f is an affine mapping $\mathbb{R} \to \mathbb{R}^3$.

Where linear transformations map subspaces to subspaces, affine mappings map affine sets to affine sets. Affine transformations preserve dimension, so they map affine k-planes to affine k-planes. We leave the verifications of these statements as an exercise.

Affine sets arise several times in the sequel.

Exercises 2.3.

2.3.1. Is incidence a transitive relation in affine geometry? Justify your answer.

2.3.2. Show that if $\mathbf{v}$ and $\mathbf{w}$ belong to an affine set α, then $\mathbf{v} - \mathbf{w}$ is a direction vector for α.

2.3.3. What are the directions for the lines in Figure 2.2?

2.3.4. How many parallel classes of lines are there in $\mathbb{F}_3^2$?

2.3.5. Can distinct subspaces of a vector space be "parallel"? Explain.

2.3.6. Find $\ell = \text{Span}\{(1, 1)\}$ in $\mathbb{F}_5^2$. How many lines in $\mathbb{F}_5^2$ are parallel to ℓ?

2.3.7. Nonintersecting lines that are not parallel are *skew*. Show that
$$\ell_1 = \{t(1,2,0) \,|\, t \in \mathbb{R}\}$$
and
$$\ell_2 = \{(1,-1,3) + t(0,1,0) \,|\, t \in \mathbb{R}\}$$
are skew in $\mathbb{R}^3$.

2.3.8. Let $V \cong \mathbb{F}^3$ for some field $\mathbb{F}$. Show that two distinct planes in V intersect in a line or not at all.

2.3.9. Show that parallelism is transitive among the affine sets in a vector space. In other words, if α, β, and γ are k-planes in a vector space V, such that α is parallel to β and β is parallel to γ, then α is parallel to γ.

2.3.10. Give an example of nonparallel, nonintersecting planes in $\mathbb{R}^4$.

2.3.11. Prove that an affine transformation maps any affine k-plane to an affine k-plane.

2.4. Isomorphism and the Rank Theorem

When different vector spaces have identical linear structures, we say they are *isomorphic*.

Definition 2.31. A **vector space isomorphism** or **linear isomorphism** is a bijective linear transformation. A **linear automorphism** is a linear isomorphism from a vector space to itself.[4]

Example 2.32. Let $L : \mathbb{R}[t]_2 \to \mathbb{R}^3$ be defined by
$$L(a + bt + ct^2) = (a, b, c).$$
An easy exercise verifies that L is an isomorphism.

Our primary interest is in linear isomorphisms but the word *isomorphism* may refer to any bijection that preserves some kind of algebraic structure. Different types of isomorphisms preserve different types of structure. When context makes it evident that we are only concerned with linear structure, we refer to vector space isomorphisms simply as isomorphisms.

As a bijection, any isomorphism has an inverse. The results in Exercise 2.1.19 imply that if $L : V \to W$ is a linear isomorphism with inverse $L^{-1} : W \to V$, then L^{-1} is also an isomorphism, $L \circ L^{-1}$ is the identity mapping on W, and $L^{-1} \circ L$ is the identity mapping on V.

Definition 2.33. Vector spaces V and W over $\mathbb{F}$ are **isomorphic** provided there is an isomorphism $L : V \to W$. In this case we write $V \cong W$, or, when we want to emphasize $\mathbb{F}$, $V \cong_\mathbb{F} W$.

Since the identity mapping is an automorphism, $V \cong V$ for any vector space V. That the inverse of an isomorphism is an isomorphism means that isomorphism is symmetric: $V \cong W$ implies $W \cong V$. We leave it as an exercise to show that isomorphism is transitive: If V, W, and X are vector spaces over $\mathbb{F}$ with $V \cong W$

---
[4]*Iso* means equal and *morph* means shape. The prefix *auto* means self.

2.4. Isomorphism and the Rank Theorem

and $W \cong X$, then $V \cong X$. This makes vector space isomorphism an equivalence relation on the vector spaces over a field $\mathbb{F}$.

Corollary 2.9, Lemma 2.10, and the notion of linear extension come together now to establish the following important theorem.

Theorem 2.34. *A linear transformation is an isomorphism if and only if it maps a basis for the domain to a basis for the codomain. Consequently, if V and W are vector spaces over a field $\mathbb{F}$, then $V \cong W$ if and only if $\dim V = \dim W$.*

The following more general result brings us back to the notion of a quotient space. While the theorem is central to an understanding of linear transformations, its proof is a routine verification that we leave to the exercises.

This theorem has many analogs throughout mathematics.

Theorem 2.35. *If $L : V \to W$ is any linear transformation, then*
$$L[V] \cong V/\operatorname{Ker} L.$$

Figure 2.3 captures Theorem 2.35. Note that $f(\mathbf{v}) := \mathbf{v} + \operatorname{Ker} L$ is the canonical mapping $V \to V/\operatorname{Ker} L$, while $\tilde{L}$ is defined by
$$\tilde{L}(\mathbf{v} + \operatorname{Ker} L) := L(\mathbf{v}).$$
Because $\tilde{L} \circ f = L$, we say that Figure 2.3 is a *commutative diagram*.

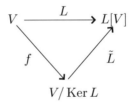

Figure 2.3. $L[V] \cong V/\operatorname{Ker} L$ and $\tilde{L} \circ f = L$.

Example 2.36. Here we consider the evaluation mapping $\xi_{\sqrt{2}}$ on $\mathbb{Q}[t]$. Since $(\sqrt{2})^k$ is an integer multiple of $\sqrt{2}$ for any integer k, the range of $\xi_{\sqrt{2}}$ is
$$\mathbb{Q}[\sqrt{2}] = \{a + b\sqrt{2} \mid a, b \in \mathbb{Q}\}.$$
Notice, for instance, that
$$5t^3 + 3t^2 + 2t - 7 \mapsto 5(\sqrt{2})^3 + 3(\sqrt{2})^2 + 2\sqrt{2} - 7 = 10\sqrt{2} + 6 + 2\sqrt{2} - 7 = 12\sqrt{2} - 1.$$
Now we can say
$$\mathbb{Q}[t]/\operatorname{Ker} \xi_{\sqrt{2}} \cong \mathbb{Q}[\sqrt{2}].$$
Exercise 1.1.13 was to show that the range of $\xi_{\sqrt{2}}$ is a subfield of $\mathbb{R}$. As a $\mathbb{Q}$ vector space, the range of $\xi_{\sqrt{2}}$ is a 2-dimensional subspace of $\mathbb{R}$ with basis $\{1, \sqrt{2}\}$.

The next theorem is immediate by Theorem 2.35 together with Corollary 2.20.

Theorem 2.37 (Rank Theorem). Let V and W be vector spaces over a field $\mathbb{F}$. Let $L : V \to W$ be a linear transformation. If V is finite-dimensional, then
$$\dim V = \dim \operatorname{Ker} L + \dim L[V].$$

The **rank** of a linear transformation $L : V \to W$ is $\dim L[V]$. The **nullity** of L is $\dim \operatorname{Ker} L$.

Example 2.38. Define $L : \mathbb{R}[t]_2 \to \mathbb{R}^3$ by $L(p(t)) = (p(0), p(1), p(0) + p(1))$. Since each component of $L(p(t))$ is a linear combination of linear mappings into $\mathbb{R}$, Exercise 2.1.3 ensures that L is linear.

Suppose $p(t) = a_0 + a_1 t + a_2 t^2$ belongs to $\operatorname{Ker} L$. Since $p(0) = a_0$, we see that $a_0 = 0$. Since $p(1) = a_0 + a_1 + a_2$, we see also that $a_1 + a_2 = 0$. This means that any element in $\operatorname{Ker} L$ has the form $at - at^2$, for some a in $\mathbb{R}$. It follows that $\{t - t^2\}$ is a basis for $\operatorname{Ker} L$ and that $\dim \operatorname{Ker} L = 1$. The Rank Theorem implies that $\dim L[\mathbb{R}[t]_2] = 2$.

Considering that $L(1) = (1, 1, 2)$, $L(t) = (0, 1, 1)$, and $L(t^2) = (0, 1, 1)$, we see that $\{(1, 1, 2), (0, 1, 1)\}$ is a basis for $L[\mathbb{R}[t]_2]$.

The following observations are immediate by the Rank Theorem.

Corollary 2.39. Let V and W be finite-dimensional vector spaces over $\mathbb{F}$. Let $L : V \to W$ be a linear transformation.

(a) L is injective if and only if $\dim V = \dim L[V]$.

(b) L is surjective if and only if $\dim W = \dim V - \dim \operatorname{Ker} L$.

We can say more when V and W are both n-dimensional over $\mathbb{F}$.

Corollary 2.40. Let V and W be n-dimensional vector spaces over $\mathbb{F}$. Let $L : V \to W$ be a linear transformation. The following are then equivalent.

(a) L is injective.

(b) L is surjective.

(c) L is an isomorphism.

Note that Corollary 2.40 is about a linear transformation between isomorphic finite-dimensional vector spaces. It does not apply when $\dim V \neq \dim W$, nor does it apply in the infinite-dimensional case.

Example 2.41. Consider $L : \mathbb{F}[t] \to \mathbb{F}[t]$ given by $p(t) \mapsto tp(t)$. It is easy to see that L is linear. It is injective because $\operatorname{Ker} L = \{0\}$. It is not surjective because 1 is not in $L[\mathbb{F}[t]]$.

Exercises 2.4.

2.4.1. Let $V = \mathbb{F}_2^3$ and let $W = \operatorname{Span}\{(1, 1, 1)\} \subseteq V$. In Example 2.15 we established that V/W contained four elements. It follows that $V/W \cong \mathbb{F}_2^2$. Identify two different isomorphisms $V/W \to \mathbb{F}_2^2$.

2.4.2. Show that the mapping in Example 2.32 is an isomorphism.

2.4.3. Show that the inverse of an isomorphism is unique.

2.5. Sums, Products, and Projections

2.4.4. Let $L : V \to W$ be a linear transformation. Show that L is an isomorphism if and only if $L[\mathcal{B}]$ is a basis for W for any one basis $\mathcal{B}$ for V.

2.4.5. Show that isomorphism is transitive, that is, if $V, W,$ and X are vector spaces over $\mathbb{F}$ with $V \cong W$ and $W \cong X$, then $V \cong X$.

2.4.6. Let $\mathbb{F}$ be a field. Find a vector space different from, but isomorphic to, $\mathbb{F}[t]$.

2.4.7. Find two different isomorphisms $M_2(\mathbb{F}) \to \mathbb{F}^4$.

2.4.8. Treating $\mathbb{C}^n$ as a real vector space, prove that the mapping given by
$$(z_1, \ldots, z_n) \mapsto (\operatorname{Re} z_1, \ldots, \operatorname{Re} z_n, \operatorname{Im} z_1, \ldots, \operatorname{Im} z_n)$$
is an isomorphism $\mathbb{C}^n \to \mathbb{R}^{2n}$.

2.4.9. Find three isomorphisms $\mathbb{C}^n \to \mathbb{R}^{2n}$ different from the one in the previous exercise.

2.4.10. Prove Theorem 2.35.

2.4.11. Let V be the real vector space of infinite sequences $(a_1, a_2, \ldots)$. Show that
$$(a_1, a_2, \ldots) \mapsto (0, a_1, a_2, \ldots)$$
is an injective linear operator on V that is not a bijection $V \to V$.

2.4.12. Show that if a linear transformation $L : \mathbb{F}^n \to \mathbb{F}^n$ has rank 1, then there is $\mathbf{x}$ in $\mathbb{F}^n$ so that $L(\mathbf{x}) = \lambda \mathbf{x}$ for some λ in $\mathbb{F}$.

2.5. Sums, Products, and Projections

Let V be a vector space with subspaces W_1 and W_2. The **sum of subspaces** W_1 and W_2 is
$$W_1 + W_2 = \{\mathbf{w}_1 + \mathbf{w}_2 \mid \mathbf{w}_1 \in W_1, \mathbf{w}_2 \in W_2\}.$$
We leave it as an exercise to verify that $W_1 + W_2$ is a subspace of V.

Example 2.42. Let $V = \mathbb{R}^3$. Let
$$W_1 = \{(t, 2t, 0) \mid \in \mathbb{R}\}.$$
Let
$$W_2 = \{(-t, t, -t) \mid t \in \mathbb{R}\}.$$
We can identify each W_i with a line in V. The sum of W_1 and W_2 is
$$W_1 + W_2 = \{(s, 2s, 0) + (-t, t, -t) \mid s, t \in \mathbb{R}\}.$$
An easy exercise shows that we can write
$$W_1 + W_2 = \{(x, y, z) \mid 2x - y - 3z = 0\}.$$
This means that we can identify $W_1 + W_2$ with a plane in $\mathbb{R}^3$.

Example 2.43. A straightforward calculation reveals that $\mathbb{R}[t]_2 = W_1 + W_2$ when $W_1 = \operatorname{Span}\{1, t\}$ and $W_2 = \operatorname{Span}\{1 + t, t^2\}$.

Definition 2.44. A vector space V is the **internal direct sum** of subspaces W_1 and W_2 in V provided $V = W_1 + W_2$ and $W_1 \cap W_2 = \{\mathbf{0}\}$. In this case, we write $V = W_1 \oplus W_2$ and we say that W_1 and W_2 are **complementary** in V or **direct complements** in V or **complementary subspaces** in V.

Evidently, the subspaces in Example 2.43 are not complementary: $W_1 \cap W_2$ is nontrivial.

The next theorem articulates the sense in which the subspaces in a direct sum generalize a basis. Its proof is an exercise.

Theorem 2.45. *A vector space V is the direct sum of subspaces W_1 and W_2 if and only if each $\mathbf{v}$ in V has a unique decomposition, $\mathbf{v} = \mathbf{w}_1 + \mathbf{w}_2$, for $\mathbf{w}_i$ in W_i.*

We can extend the definition of internal direct sum to apply to any number of — or infinitely many — subspaces. Let $\{W_j\}_{j \in J}$ be an arbitrary collection of pairwise trivially intersecting subspaces of V: $W_{j_1} \cap W_{j_2} = \{\mathbf{0}\}$ for any $j_1 \neq j_2$ in J. If each element in V is a sum

$$\mathbf{v} = \mathbf{w}_{j_1} + \cdots + \mathbf{w}_{j_k}$$

where $\mathbf{w}_{j_i}$ is in W_{j_i} for W_{j_i} in $\{W_j\}_{j \in J}$, then $V = \bigoplus_{j \in J} W_j$ is an internal direct sum.

The next theorem establishes that an internal direct sum of nontrivial subspaces corresponds to a partition on a basis for the ambient space.

Theorem 2.46. *Let V be a vector space.*

If V is an internal direct sum of nontrivial subspaces $\{W_j\}_{j \in J}$ and if $\mathcal{B}_j$ is a basis for W_j, then $\mathcal{B} = \bigcup_{j \in J} \mathcal{B}_j$ is a basis for V and $\{\mathcal{B}_j\}_{j \in J}$ is a partition on $\mathcal{B}$.

If $\mathcal{B}$ is a basis for V with partition $\{\mathcal{B}_j\}_{j \in J}$, then for $W_j = \operatorname{Span} \mathcal{B}_j$,

$$V = \bigoplus_{j \in J} W_j.$$

Proof. Let V be a vector space.

If $V = \bigoplus_{j \in J} W_j$ is an internal direct sum of nontrivial subspaces of V, let $\mathcal{B}_j$ be a basis for W_j. As a basis for a nontrivial space, each $\mathcal{B}_j$ is nonempty. Since $W_{j_1} \cap W_{j_2} = \mathbf{0}_V$, for $j_1 \neq j_2$ in J, $\mathcal{B}_{j_1} \cap \mathcal{B}_{j_2}$ is empty when $j_1 \neq j_2$. It follows that if

$$\bigcup_{j \in J} \mathcal{B}_j = \mathcal{B},$$

then $\{\mathcal{B}_j\}_{j \in J}$ is a partition on $\mathcal{B}$. We claim that $\mathcal{B}$ is a basis for V.

Since $V = \bigoplus_{j \in J} W_j$ is a direct sum, we can write any $\mathbf{v}$ in V as

$$\mathbf{v} = \mathbf{w}_{j_1} + \cdots + \mathbf{w}_{j_k}$$

for $\mathbf{w}_{j_i}$ in W_{j_i}, where $W_{j_1}, \ldots, W_{j_k}$ are distinct in $\{W_j\}_{j \in J}$. Since each $\mathbf{w}_{j_i}$ is in the span of $\mathcal{B}_{j_i}$, $\mathcal{B}$ spans V.

Suppose next that there are scalars c_j so that

(2.2) $$\mathbf{0} = c_1 \mathbf{b}_1 + \cdots + c_m \mathbf{b}_m$$

for some $S = \{\mathbf{b}_1, \ldots, \mathbf{b}_m\} \subseteq \mathcal{B}$. We can rewrite the sum on the right-hand side of (2.2) in the form

$$\mathbf{w}_{j_1} + \cdots + \mathbf{w}_{j_k}$$

where $\mathbf{w}_{j_i}$ belongs to W_{j_i} and $W_{j_1}, \ldots, W_{j_k}$ are distinct in $\{W_j\}_{j \in J}$. Since V is a direct sum, each $\mathbf{w}_{j_i}$ must be zero. Since each $\mathcal{B}_j$ is linearly independent, the

2.5. Sums, Products, and Projections

coefficients c_i in (2.2) must all be zero, as well. This is enough to prove that $\mathcal{B}$ is linearly independent, thus, a basis for V. This proves the first statement of the theorem.

We leave proof of the second statement as an exercise. $\square$

Theorem 2.46 makes it evident that there may be many ways to write a given vector space as a direct sum. Even a fixed subspace of a vector space may have many different direct complements. For example, both $W_1 = \{(t,t) \,|\, t \in \mathbb{R}\}$ and $W_2 = \{(t,0) \,|\, t \in \mathbb{R}\}$ are direct complements of $X = \{(t,-t) \,|\, t \in \mathbb{R}\}$ in $\mathbb{R}^2$. We leave it as an exercise to verify this and to find other ways to write $\mathbb{R}^2$ as an internal direct sum.

While an internal direct sum is a way of decomposing a given vector space, *direct products* and *external direct sums* are ways of constructing new vector spaces from a given collection of vector spaces defined over a fixed field. As both constructions are based on Cartesian products, we start this part of the discussion with a reminder of the definition of a Cartesian product.[5]

Definition 2.47. The **Cartesian product** of an indexed family of sets, $\{S_j\}_{j \in J}$, is the set of all mappings $f : J \to \bigcup_{j \in J} S_j$ where $f(j)$ is in S_j.

We designate the Cartesian product of sets $\{S_j\}_{j \in J}$ by $\prod_{j \in J} S_j$. We may refer to the elements in $\{S_j\}_{j \in J}$ as the **components** of $\prod_{j \in J} S_j$. If f belongs to $\prod_{j \in J} S_j$, we may say that $f(j)$ is the jth component of f.

When $J = \{1, \ldots, n\}$, we may write

$$\prod_{j \in J} S_j = S_1 \times \cdots \times S_n.$$

We usually indicate an element in $S_1 \times \cdots \times S_n$ by $(s_1, \ldots, s_n)$. In terms of Definition 2.47, $f = (s_1, \ldots, s_n)$ means $f(j) = s_j$.

Consider $\{S_j\}_{j \in \mathbb{R}}$, where each $S_j = \mathbb{R}$. Here $\bigcup_{j \in J} S_j = \mathbb{R}$, so each element in the Cartesian product is a mapping $f : \mathbb{R} \to \mathbb{R}$. In this case, $\prod_{j \in J} S_j = \mathcal{F}(\mathbb{R})$.

The Cartesian product of a family of vector spaces over a fixed field inherits linear structure from its component spaces. Let $\mathbf{V} = \prod_{j \in J} V_j$ where V_j is a vector space over $\mathbb{F}$ for all j in J. The linear structure on $\mathbf{V}$ comes out of defining addition and scaling componentwise. In particular, for any $\mathbf{v}, \mathbf{w}$ in $\mathbf{V}$, c in $\mathbb{F}$, and $j \in J$, we define $\mathbf{v} + c\mathbf{w}$ in $\mathbf{V}$ by

$$(\mathbf{v} + c\mathbf{w})(j) := \mathbf{v}(j) + c\mathbf{w}(j).$$

The zero mapping in $\mathbf{V}$ is given by

$$\mathbf{0}(j) = \mathbf{0}_{V_j}.$$

The remaining axioms from Definition 1.6 follow because each V_j is a vector space.

We have done enough now to prove the following.

Theorem 2.48. *The Cartesian product of an indexed family of vector spaces over a field $\mathbb{F}$ is a vector space over $\mathbb{F}$ under componentwise addition and scaling.*

---
[5] We also discuss Cartesian products in Section A.2.

The next definition gives us the vocabulary to distinguish between the Cartesian product of a family of arbitrary sets and the Cartesian product of a family of vector spaces over a fixed field.

Definition 2.49. Let $\{V_j\}_{j \in J}$ be an indexed family of vector spaces over a field $\mathbb{F}$. The **direct product** of $\{V_j\}_{j \in J}$ is the vector space $\mathbf{V} = \prod_{j \in J} V_j$.

We use $\mathbf{V}$ to indicate a direct product $\prod_{j \in J} V_j$ for the duration of our discussion.

There is often room for interpretation of a direct product.

Example 2.50. Let $V_j = \mathbb{R}$ for $j \in \mathbb{Z}_+$. We can interpret $\mathbf{V} = \prod_{j \in J} V_j$ as the set of formal series in $\mathbb{R}$. This parallels our interpretation of the elements in an internal direct sum as sums. For instance, if $\mathbf{v}(j) = (2j+1)/(3j^2 - 2)$, then

$$\mathbf{v} = \sum_{n=1}^{\infty} \frac{2n+1}{3n^2 - 2}.$$

Depending on the application, we might instead interpret $\mathbf{V}$ as the set of infinite sequences over $\mathbb{R}$. In that case, we think of the same element $\mathbf{v}$ as given by

$$\mathbf{v} = \left(\frac{2n+1}{3n^2 - 2}\right)_{n=1}^{\infty}.$$

In terms of linear algebra, the two interpretations are interchangeable.

While V_j is not necessarily a subspace of $\mathbf{V} = \prod_{j \in J} V_j$,

(2.3) $$W_j = \{\mathbf{v} \in \mathbf{V} \mid \mathbf{v}(i) = \mathbf{0}_{V_i} \text{ for } i \neq j\}$$

is a subspace of $\mathbf{V}$ that is canonically isomorphic to V_j. This means that we can define an isomorphism $W_j \to V_j$ without reference to a basis. We leave the verification of this as an exercise.

This brings us to another important subspace of $\mathbf{V}$.

Definition 2.51. The **external direct sum** of an indexed family of vector spaces $\{V_j\}_{j \in J}$ is the set of $\mathbf{v}$ in $\mathbf{V} = \prod_{j \in J} V_j$ such that $\mathbf{v}(j) = \mathbf{0}_{V_j}$, except for finitely many j in J.

We use $\bigoplus_{j \in J} V_j$ to designate the external direct sum of the vector spaces in $\{V_j\}_{j \in J}$.

An element in $\mathbf{V} = \prod_{j \in J} V_j$ may be nonzero for infinitely many $j \in J$. An element in $\bigoplus_{j \in J} V_j$ is nonzero for at most finitely many $j \in J$.

Example 2.52. Let t be an indeterminate and say that $V_j = \{ct^j \mid c \in \mathbb{R}\}$ for $J = \mathbb{Z}_{\geq 0}$. We can interpret $\mathbf{V} = \prod_{j \in J} V_j$ as the vector space of formal power series over $\mathbb{R}$. If

$$\mathbf{v} = \sum_{n=0}^{\infty} \frac{t^n}{(n+1)^2} = 1 + \frac{t}{4} + \frac{t^2}{9} + \cdots,$$

for instance, then $\mathbf{v}(j) = t^j/(j+1)^2$. In this case, we would interpret $\bigoplus_{j \in J} V_j$ as the space of polynomials over $\mathbb{R}$.

2.5. Sums, Products, and Projections

When J is finite, the direct product and the external direct sum of vector spaces $\{V_j\}_{j \in J}$ are identical. In this case, we use the same notation for both, which is also the notation we use for internal direct sums. Context precludes any possible confusion. When we write

$$\mathbb{R}^3 = \mathbb{R} \oplus \mathbb{R} \oplus \mathbb{R}, \tag{2.4}$$

for instance, this must be an external direct sum because $\mathbb{R}$ is not a subspace of $\mathbb{R}^3$. On the other hand,

$$\mathbb{R}^3 = \mathrm{Span}\{(1,0,0)\} \oplus \mathrm{Span}\{(0,1,0)\} \oplus \mathrm{Span}\{(0,0,1)\}$$

must be an internal direct sum.

We can define a linear operator $L : \mathbf{V} \to \mathbf{V}$ by specifying $L(\mathbf{v})(j)$ for each element j in the underlying index set. This idea applies to internal direct sums, as well.

Example 2.53. Letting $W_1 = \mathrm{Span}\{(1,1,1)\}$ and $W_2 = \mathrm{Span}\{(1,0,1),(0,1,1)\}$ in $\mathbb{R}^3$, we can see that

$$\mathbb{R}^3 = W_1 \oplus W_2.$$

If $\mathbf{v} = (1,-1,2)$, for instance, then $\mathbf{v}(1) = (-2,-2,-2)$ and $\mathbf{v}(2) = (3,1,4)$. To check this, verify that $\mathbf{v}$ is the sum $\mathbf{v}(1) + \mathbf{v}(2)$, that $\mathbf{v}(1)$ is in W_1, and that $\mathbf{v}(2)$ is in W_2.

Suppose we define L on $\mathbb{R}^3$ by

$$L(\mathbf{v})(1) = 2\mathbf{v}(1) + \mathbf{v}(2)$$

and

$$L(\mathbf{v})(2) = \mathbf{v}(1) - \mathbf{v}(2).$$

This is like defining a linear transformation by linear extension. Since $(1,1,1)(1) = (1,1,1)$ and $(1,1,1)(2) = (0,0,0)$, for instance, $L(1,1,1)(1) = (2,2,2)$ and $L(1,1,1)(2) = (1,1,1)$. Normally, we would just say

$$L(1,1,1) = (3,3,3).$$

It is easy to check that $W_2 \subseteq \mathrm{Ker}\, L$.

We next consider projection mappings. The most obvious setting for these is an internal direct sum, say $V = W_1 \oplus W_2$. Let L be the operator on V given by

$$L(\mathbf{w}_1 + \mathbf{w}_2) = \mathbf{w}_1$$

for all $\mathbf{w}_1$ in W_1 and $\mathbf{w}_2$ in W_2. This mapping, which is easily seen to be linear, is called **projection onto W_1 along W_2**. We would like to generalize this idea.

Consider $\mathbf{V} = \prod_{j \in J} V_j$. Fix $j \in J$ and define $L_j : \mathbf{V} \to \mathbf{V}$ by

$$L_j(\mathbf{v})(i) = \delta(i,j)\mathbf{v}(j)$$

for each i in J. This means $L_j[\mathbf{V}] = W_j$ as defined in (2.3) and $\mathrm{Ker}\, L_j = \prod_{i \neq j} V_i$. While $L_j(\mathbf{v})$ belongs to $\mathbf{V}$, informally we say that L_j maps $\mathbf{v}$ to its jth component, $\mathbf{v}(j)$.

A quick calculation reveals that L_j is linear. We also have

$$L_j(L_j(\mathbf{v}))(i) = \delta(i,j)L_j(\mathbf{v})(j) = \delta(i,j)\mathbf{v}(j) = L_j(\mathbf{v})(i).$$

This proves that $L_j^2 = L_j$.

Definition 2.54. A **projection** is a nontrivial linear operator L with the property
$$L \circ L = L^2 = L.$$

Our next theorem establishes the connection between projections and direct sum decompositions.

Theorem 2.55. If V is a vector space with projection $L : V \to V$, then
$$V = L[V] \oplus \operatorname{Ker} L.$$

Proof. Let V be a vector space and let $L : V \to V$ be a projection. Let $\mathbf{w}$ belong to $L[V] \cap \operatorname{Ker} L$ and say $\mathbf{w} = L(\mathbf{v})$. Since $\mathbf{w}$ is in $\operatorname{Ker} L$,
$$\mathbf{0} = L(\mathbf{w}) = L^2(\mathbf{v}) = L(\mathbf{v}) = \mathbf{w}.$$
We conclude that $L[V] \cap \operatorname{Ker} L = \{\mathbf{0}\}$.

Now let $\mathbf{v}$ be arbitrary in V and say that $L(\mathbf{v}) = \mathbf{x}$. Since we can write
$$\mathbf{v} = (\mathbf{v} - \mathbf{x}) + \mathbf{x},$$
we can prove the lemma if we show that $\mathbf{v} - \mathbf{x}$ is in $\operatorname{Ker} L$. This is immediate, since, for instance,
$$L(\mathbf{v} - \mathbf{x}) = L(L(\mathbf{x}) - \mathbf{x}) = L^2(\mathbf{x}) - L(\mathbf{x}) = L(\mathbf{x}) - L(\mathbf{x}) = \mathbf{0}. \qquad \square$$

Since the direct complement to a subspace is not unique, there may be many different projections onto a given subspace of V. We can specify a projection by its range and its kernel. In lower dimensions, we may describe the kernel as a line or plane. In this case, we may say that a projection is onto its range, and along the line or plane that is its kernel. For instance the operator L on $\mathbb{R}^2$ given by $L(a, b) = (a - b, 0)$ is the projection onto the x-axis, $\operatorname{Span}\{(1,0)\}$, along the line $\ell = \operatorname{Span}\{(1,1)\}$. This means $\operatorname{Ker} L = \ell$.

Exercises 2.5.

2.5.1. Verify that if V is a vector space with subspaces W_1 and W_2, then $W_1 + W_2$ is a subspace of V.

2.5.2. Show that in Example 2.42, $W_1 + W_2 = \{(x, y, z) \mid 2x - y - 3z = 0\}$.

2.5.3. Find $W_1 \cap W_2$ for W_1 and W_2 as given in Example 2.42.

2.5.4. Verify the claims made in Example 2.43.

2.5.5. Let V be a vector space with subspaces W_1 and W_2. Show that $V = W_1 + W_2$ if and only if the union of bases for W_1 and W_2 is a spanning set for V.

2.5.6. Suppose V is a vector space and that $V = W_1 + W_2$. When is the union of bases for W_1 and W_2 a basis for V?

2.5.7. Let $X = \{(t, -t) \mid t \in \mathbb{R}\}$, $W_1 = \{(t, t) \mid t \in \mathbb{R}\}$, and $W_2 = \{(t, 0) \mid t \in \mathbb{R}\}$.
   (a) Show that both W_1 and W_2 are complementary to X in $\mathbb{R}^2$.
   (b) Find a third subspace in $\mathbb{R}^2$ that is complementary to X.
   (c) Find two nontrivial complementary subspaces in $\mathbb{R}^2$ different from X, W_1, and W_2.

2.5.8. Let $V = \mathbb{R}[t]$ and let $W = \text{Span}\{t^2, t^3, \ldots\} \subseteq V$. Find two different subspaces of V complementary to W.

2.5.9. Prove Theorem 2.45.

2.5.10. Suppose V is a vector space with subspaces W_i such that
$$V = \bigoplus_{i=1}^{k} W_i.$$
Generalize Theorem 2.45 to apply to V, then prove the generalization.

2.5.11. Finish the proof of Theorem 2.46.

2.5.12. Find a subspace complementary to $W = \{(t, s, t) \,|\, t, s \in \mathbb{R}\}$ in $\mathbb{R}^3$.

2.5.13. Let V be finite-dimensional and suppose W_1 and W_2 are subspaces of V.
(a) Show that $\dim(W_1 + W_2) = \dim W_1 + \dim W_2 - \dim(W_1 \cap W_2)$.
(b) What is $\dim V/W_1$?

2.5.14. Let V be a vector space with subspaces W_1 and W_2.
(a) Show that $W_1 \cap W_2$ is a subspace of $W_1 + W_2$.
(b) Suppose $V = W_1 + W_2$. Let $X = W_1 \cap W_2$. Show that
$$V/X = W_1/X \oplus W_2/X.$$

2.5.15. Check the calculations in Example 2.53.

2.5.16. Let $\mathbf{V} = \prod_{j \in J} V_j$. Fix j, and define $L_j : \mathbf{V} \to \mathbf{V}$ by
$$L_j(\mathbf{v})(i) = \delta(i, j)\mathbf{v}(j)$$
for each i in J. Prove that L_j is a linear operator.

2.5.17. Define $L : \mathbb{R}^2 \to \mathbb{R}^2$ by $L(a, b) = (a + b, 0)$.
(a) Show that L is a projection.
(b) Identify the range and kernel of L.

Chapter 3

Matrices and Coordinates

While we can view a matrix as a generalization of a constant, we can view a linear transformation as a generalization of a matrix. What, then, does a matrix equation $A\mathbf{x} = \mathbf{b}$ have to do with linear transformations?

Once we understand matrix arithmetic, we can see that if A belongs to $M_{m,n}(\mathbb{F})$, then $\mathbf{v} \mapsto A\mathbf{v}$ defines a linear transformation $L : \mathbb{F}^n \to \mathbb{F}^m$. It follows from there that $A\mathbf{x} = \mathbf{b}$ has a solution if and only if $\mathbf{b}$ belongs to $L[\mathbb{F}^n]$. In this case, the solution to $A\mathbf{x} = \mathbf{b}$ is $L^{-1}[\{\mathbf{b}\}]$.

The key to these connections is matrix multiplication. We thus begin this part of our journey by pursuing an understanding of matrix multiplication, both in terms of the rows and in terms of the columns of the matrices involved.

3.1. Matrices

Recall that if $A = [a_{ij}]$ is in $M_{m,n}(\mathbb{F})$ and $B = [b_{ij}]$ is in $M_{n,r}(\mathbb{F})$, then $AB = [c_{ij}]$ where

(3.1) $$c_{ij} = a_{i1}b_{1j} + \cdots + a_{in}b_{nj} = \sum_{k=1}^{n} a_{ik}b_{kj}.$$

Example 3.1. Let $A = \begin{bmatrix} 1 & 0 & 1 \\ 1 & 1 & 1 \end{bmatrix}$ in $M_{2,3}(\mathbb{R})$, and let $B = \begin{bmatrix} 1 & 0 \\ 2 & 1 \\ 0 & 2 \end{bmatrix}$ in $M_{3,2}(\mathbb{R})$.

In this case, AB and BA are both defined with

$$AB = \begin{bmatrix} 1 & 2 \\ 3 & 3 \end{bmatrix}$$

and

$$BA = \begin{bmatrix} 1 & 0 & 1 \\ 3 & 1 & 3 \\ 2 & 2 & 2 \end{bmatrix}.$$

Evidently, matrix multiplication is not commutative. When it is defined, though, it is associative. The proof is an exercise in working with indices and sums.

Theorem 3.2. If A is in $M_{m,n}(\mathbb{F})$, B is in $M_{n,r}(\mathbb{F})$, and C is in $M_{r,t}(\mathbb{F})$, then
$$(AB)C = A(BC).$$

Proof. Let $A = [a_{ij}]$, $B = [b_{ij}]$, and $C = [c_{ij}]$ be as hypothesized.

If $AB = [d_{ij}]$, then the ij-entry of $(AB)C$ is
$$\sum_{k=1}^{r} d_{ik} c_{kj} = \sum_{k=1}^{r} \left(\sum_{\ell=1}^{n} a_{i\ell} b_{\ell k} \right) c_{kj}.$$

Because the sums are finite, we can apply the properties of arithmetic in a field to write this expression several different ways, among them
$$\sum_{\ell=1}^{n} a_{i\ell} \left(\sum_{k=1}^{r} b_{\ell k} c_{kj} \right).$$

The last step is to check that this is the ij-entry of $A(BC)$. We leave that to the reader. $\square$

Another approach to defining matrix multiplication is to start with the product $\mathbf{ab}$ for a row vector $\mathbf{a}$ and a column vector $\mathbf{b}$. Let $\mathbf{a} = \begin{bmatrix} a_1 & \cdots & a_n \end{bmatrix}$ belong to $\mathbb{F}_n$ and let $\mathbf{b} = (b_1, \ldots, b_n)$ belong to $\mathbb{F}^n$. Define
$$\mathbf{ab} := \sum_{i=1}^{n} a_i b_i.$$

If A is in $M_{m,n}(\mathbb{F})$ and B is in $M_{n,r}(\mathbb{F})$, each row of A is in $\mathbb{F}_n$ and each column of B is in $\mathbb{F}^n$. The ij-entry in AB is then the product of row i of A and column j of B.

Example 3.3. Suppose the second row of A in $M_{3,4}(\mathbb{F})$ is $\begin{bmatrix} 1 & 0 & 0 & 1 \end{bmatrix}$. For any $\mathbf{b} = (b_1, b_2, b_3, b_4)$ in $\mathbb{F}^4$, notice that
$$\begin{bmatrix} 1 & 0 & 0 & 1 \end{bmatrix} \mathbf{b} = b_1 + b_4.$$

This means that for any B in $M_{4,r}(\mathbb{F})$, the second row of AB is the sum of the first and fourth rows of B.

If $A = [a_{ij}]$ is in $M_{m,n}(\mathbb{F})$ and $\mathbf{v} = (v_1, \ldots, v_n)$ is in $\mathbb{F}^n$, then $A\mathbf{v}$ is a vector in $\mathbb{F}^m$, its ith coordinate the product of row i of A with $\mathbf{v}$, given by

(3.2) $$\sum_{j=1}^{n} a_{ij} v_j = a_{i1} v_1 + \cdots + a_{in} v_n = v_1 a_{i1} + \cdots + v_n a_{in}.$$

If $\mathbf{a}_j$ is column j of A, then the ith coordinate of $\sum_{j=1}^{n} v_j \mathbf{a}_j$ is also given by (3.2). This shows that

(3.3) $$A\mathbf{v} = v_1 \mathbf{a}_1 + \cdots + v_n \mathbf{a}_n,$$

proving the following lemma.

3.1. Matrices

Lemma 3.4. Let A belong to $M_{m,n}(\mathbb{F})$. If $\mathbf{v}$ is in $\mathbb{F}^n$, then $A\mathbf{v}$ is a linear combination of the columns of A, with coefficients given by the coordinates of $\mathbf{v}$.

We can think of a matrix as a row of column vectors or as a column of row vectors. Each of these views is useful. When we want to view A in $M_{m,n}(\mathbb{F})$ as a row of column vectors, we write
$$A = \begin{bmatrix} \mathbf{a}_1 & \cdots & \mathbf{a}_n \end{bmatrix},$$
indicating that column j of A is $\mathbf{a}_j \in \mathbb{F}^m$. When we want to view the same matrix as a column of row vectors, we write
$$A = (\mathbf{w}_1, \ldots, \mathbf{w}_m),$$
indicating that row i of A is $\mathbf{w}_i \in \mathbb{F}_n$.

Lemma 3.5. If A belongs to $M_{m,n}(\mathbb{F})$, then the mapping $\mathbf{v} \mapsto A\mathbf{v}$ is a linear transformation $\mathbb{F}^n \to \mathbb{F}^m$.

Proof. Let $A = \begin{bmatrix} \mathbf{a}_1 & \cdots & \mathbf{a}_n \end{bmatrix}$ belong to $M_{m,n}(\mathbb{F})$. Let c belong to $\mathbb{F}$. Take $\mathbf{v} = (v_1, \ldots, v_n)$ and $\mathbf{x} = (x_1, \ldots, x_n)$ in $\mathbb{F}^n$. Appealing to (3.3), we get
$$A(c\mathbf{v} + \mathbf{x}) = (cv_1 + x_1)\mathbf{a}_1 + \cdots + (cv_n + x_n)\mathbf{a}_n.$$
Because of the properties of $\mathbb{F}^m$ as a vector space,
$$(cv_j + x_j)\mathbf{a}_j = cv_j\mathbf{a}_j + x_j\mathbf{a}_j.$$
This means we have
$$A(c\mathbf{v} + \mathbf{x}) = c\sum_{j=1}^n v_j\mathbf{a}_j + \sum_{j=1}^n x_j\mathbf{a}_j.$$
Again by (3.3),
$$c\sum_{j=1}^n v_j\mathbf{a}_j + \sum_{j=1}^n x_j\mathbf{a}_j = cA\mathbf{v} + A\mathbf{x},$$
which is enough to prove the lemma. $\square$

Next is the general case.

Theorem 3.6. If A belongs to $M_{m,n}(\mathbb{F})$, then $B \mapsto AB$ is a linear transformation $M_{n,r}(\mathbb{F}) \to M_{m,r}(\mathbb{F})$.

Proof. Let A be as hypothesized. Suppose $B = \begin{bmatrix} \mathbf{b}_1 & \cdots & \mathbf{b}_r \end{bmatrix}$ belongs to $M_{n,r}(\mathbb{F})$. We leave it as an exercise to verify that column j of AB is $A\mathbf{b}_j$. This means
$$AB = \begin{bmatrix} A\mathbf{b}_1 & \cdots & A\mathbf{b}_r \end{bmatrix}.$$
Let c be a scalar and let $C = \begin{bmatrix} \mathbf{c}_1 & \cdots & \mathbf{c}_r \end{bmatrix}$ belong to $M_{n,r}(\mathbb{F})$. We have
$$cB + C = \begin{bmatrix} c\mathbf{b}_1 + \mathbf{c}_1 & \cdots & c\mathbf{b}_r + \mathbf{c}_r \end{bmatrix}$$
so that
$$A(cB + C) = \begin{bmatrix} A(c\mathbf{b}_1 + \mathbf{c}_1) & \cdots & A(c\mathbf{b}_r + \mathbf{c}_r) \end{bmatrix}.$$

By Lemma 3.5, $A(c\mathbf{b}_j + \mathbf{c}_j) = cA\mathbf{b}_j + A\mathbf{c}_j$ so we can write
$$A(cB + C) = \begin{bmatrix} cA\mathbf{b}_1 + A\mathbf{c}_1 & \cdots & cA\mathbf{b}_r + A\mathbf{c}_r \end{bmatrix} = cAB + AC,$$
proving that $B \mapsto AB$ is linear. $\square$

If A is a matrix and X is any other matrix for which AX is defined, then the mapping $X \mapsto AX$ is called *premultiplication* by A. In the exercises, you will show that when Y and A are matrices for which YA is defined, the mapping $Y \mapsto YA$ is also a linear transformation. It is called *postmultiplication* by A.

We have clarified the sense in which a matrix defines a linear transformation. Next we show that a linear transformation $\mathbb{F}^n \to \mathbb{F}^m$ can be realized as matrix multiplication.

Let $L : \mathbb{F}^n \to \mathbb{F}^m$ be an arbitrary linear transformation. If $\mathbf{v} = (v_1, \ldots, v_n)$ is in $\mathbb{F}^n$, we can write
$$\mathbf{v} = \sum_{i=1}^n v_i \mathbf{e}_i,$$
where $\mathcal{E} = \{\mathbf{e}_i\}_{i=1}^n$ is the standard ordered basis for $\mathbb{F}^n$. Linearity of L gives us
$$L(\mathbf{v}) = L(v_1 \mathbf{e}_1 + \cdots + v_n \mathbf{e}_n) = v_1 L(\mathbf{e}_1) + \cdots + v_n L(\mathbf{e}_n).$$
This means that by taking $A_L := \begin{bmatrix} L(\mathbf{e}_1) & \cdots & L(\mathbf{e}_n) \end{bmatrix}$, we have
$$L(\mathbf{v}) = A_L \mathbf{v}.$$
This proves the following theorem.

Theorem 3.7. *If $L : \mathbb{F}^n \to \mathbb{F}^m$ is a linear transformation, then for each $\mathbf{v}$ in $\mathbb{F}^n$, $L(\mathbf{v}) = A_L \mathbf{v}$.*

Definition 3.8. The **standard matrix representation** of a linear transformation $L : \mathbb{F}^n \to \mathbb{F}^m$ is
$$A_L := \begin{bmatrix} L(\mathbf{e}_1) & \cdots & L(\mathbf{e}_n) \end{bmatrix}.$$

Given A in $M_{m,n}(\mathbb{F})$, define $L_A : \mathbb{F}^n \to \mathbb{F}^m$ by
$$L_A(\mathbf{v}) := A\mathbf{v}.$$
It is easy to show that for A in $M_{m,n}(\mathbb{F})$, $A_{L_A} = A$ and that for a linear transformation $L : \mathbb{F}^n \to \mathbb{F}^m$, $L_{A_L} = L$.

Notice that if $A = \begin{bmatrix} \mathbf{a}_1 & \cdots & \mathbf{a}_n \end{bmatrix}$ is in $M_{m,n}(\mathbb{F})$, then $A\mathbf{e}_j = \mathbf{a}_j$. This means
$$L_A[\mathbb{F}^n] = \text{Span}\{\mathbf{a}_1, \ldots, \mathbf{a}_n\}.$$
The *column space* of A belonging to $M_{m,n}(\mathbb{F})$ is $L_A[\mathbb{F}^n]$. The *rank* of A is the rank of L_A. The *null space* of A is $\text{Ker } L_A$. The following definition is the one more usually cited.

Definition 3.9. The **column space** of a matrix A in $M_{m,n}(\mathbb{F})$, written $\text{Col } A$, is the span of the columns of A in $\mathbb{F}^m$. The **rank of a matrix** is the dimension of its column space.

The **null space** of a matrix A in $M_{m,n}(\mathbb{F})$ is
$$\text{Nul } A = \{\mathbf{v} \in \mathbb{F}^n \mid A\mathbf{v} = \mathbf{0}_m\}.$$
The **nullity** of A is $\dim \text{Nul } A$.

3.1. Matrices

Next we state the Rank Theorem as it applies to matrices.

Theorem 3.10 (Rank Theorem for Matrices). If A belongs to $M_{m,n}(\mathbb{F})$, then
$$n = \dim \operatorname{Col} A + \dim \operatorname{Nul} A.$$

Suppose A in $M_n(\mathbb{F})$ has the property that for each i belonging to $\{1,\ldots,n\}$, there is $\mathbf{b}_i$ in $\mathbb{F}^n$ so that $A\mathbf{b}_i = \mathbf{e}_i$. We then say that A is **invertible** with inverse
$$B = \begin{bmatrix} \mathbf{b}_1 & \cdots & \mathbf{b}_n \end{bmatrix}.$$
In this case,
$$AB = \begin{bmatrix} A\mathbf{b}_1 & \cdots & A\mathbf{b}_n \end{bmatrix} = \begin{bmatrix} \mathbf{e}_1 & \cdots & \mathbf{e}_n \end{bmatrix} = I_n.$$
We call I_n the **identity matrix** in $M_n(\mathbb{F})$.

We leave it as an exercise to verify that whenever A is in $M_{m,n}(\mathbb{F})$, $AI_n = A$ and $I_m A = A$.

Theorem 3.11. A matrix in $M_n(\mathbb{F})$ is invertible if and only if the linear operator it determines on $\mathbb{F}^n$ is an automorphism.

Proof. Fix A in $M_n(\mathbb{F})$.

If A is invertible with inverse $B = \begin{bmatrix} \mathbf{b}_1 & \cdots & \mathbf{b}_n \end{bmatrix}$, then since $L_A : \mathbb{F}^n \to \mathbb{F}^n$ maps the columns of B to $\mathcal{E}$, it is surjective. By Corollary 2.40, L_A is an automorphism on $\mathbb{F}^n$. This is the proof in one direction.

Next assume that L_A is an automorphism on $\mathbb{F}^n$. Let B be the standard matrix representation of L_A^{-1} and write
$$B = \begin{bmatrix} \mathbf{b}_1 & \cdots & \mathbf{b}_n \end{bmatrix}.$$
For each $j \in \{1,\ldots,n\}$, we have
$$\mathbf{e}_j = L_A(L_A^{-1}(\mathbf{e}_j)) = A(B\mathbf{e}_j) = A\mathbf{b}_j.$$
It follows that $AB = I_n$, so A is invertible with inverse B. This completes the proof. $\square$

The next corollary establishes that any inverse for an $n \times n$ matrix must be two-sided.

Corollary 3.12. Let A belong to $M_n(\mathbb{F})$. There is then B in $M_n(\mathbb{F})$ such that $AB = I_n$ if and only if $BA = I_n$.

Proof. Let A and $B = \begin{bmatrix} \mathbf{b}_1 & \cdots & \mathbf{b}_n \end{bmatrix}$ belong to $M_n(\mathbb{F})$.

If $AB = I_n$, then by Theorem 3.11, L_A is an automorphism with $L_A(\mathbf{b}_j) = \mathbf{e}_j$. In this case, $L_A^{-1}(\mathbf{e}_j) = \mathbf{b}_j$, from which it follows that B is the standard matrix representation of L_A^{-1}. From there we can say
$$BA\mathbf{e}_j = L_A^{-1}(L_A(\mathbf{e}_j)) = \mathbf{e}_j.$$
This is enough to verify that $BA = I_n$.

If $BA = I_n$ and we reverse the roles of A and B in the argument above, we find that $AB = I_n$. $\square$

When A in $M_n(\mathbb{F})$ is invertible, we denote its inverse by A^{-1}.

The proof of the statement in Exercise 1.5.13 applies to prove the following.

Lemma 3.13. If A and B in $M_n(\mathbb{F})$ are invertible, so is AB and
$$(AB)^{-1} = B^{-1}A^{-1}.$$

If someone hands you an invertible matrix, A in $M_n(\mathbb{F})$, how do you find A^{-1}? Column j of A^{-1} is the solution to $A\mathbf{x} = \mathbf{e}_j$ so finding A^{-1} means solving n, $n \times n$ systems of linear equations. We discuss this further in Chapter 4. In the meantime, it is worthwhile to memorize the inverse of a 2×2 matrix with nonzero determinant. If $A = \begin{bmatrix} a & b \\ c & d \end{bmatrix}$ and $\det A = ad - bc \neq 0$, then

$$(3.4) \qquad A^{-1} = \frac{1}{ad-bc}\begin{bmatrix} d & -b \\ -c & a \end{bmatrix}.$$

Exercises 3.1.

3.1.1. Is there any value of n for which matrix multiplication in $M_n(\mathbb{F})$ is commutative for any field $\mathbb{F}$?

3.1.2. Is there a field $\mathbb{F}$ for which $M_n(\mathbb{F})$ is commutative for any n in $\mathbb{Z}_+$? Justify your answer.

3.1.3. Finish the proof of Theorem 3.2.

3.1.4. Let A, B, C, D be matrices. Theorem 3.2 establishes that we do not need parentheses when we write the product ABC as long as the product is defined. Assume here that AB, BC, and CD are all defined.
(a) Argue that both ABC and BCD are defined.
(b) Argue that $(AB)(CD)$, $A(BCD)$, and $(ABC)D$ are defined.
(c) Use Theorem 3.2 to show that $(AB)(CD) = A(BCD) = (ABC)D$.

3.1.5. Use induction to show that if $A_1, \ldots, A_n$ are matrices so that the product $(A_1 \cdots A_{n-1})A_n$ is defined, then $(A_1 \cdots A_{n-1})A_n = A_1(A_2 \cdots A_n)$.

3.1.6. Verify that if A is in $M_{m,n}(\mathbb{F})$ and $B = \begin{bmatrix} \mathbf{b}_1 & \cdots & \mathbf{b}_r \end{bmatrix}$ is in $M_{n,r}(\mathbb{F})$, then
$$AB = \begin{bmatrix} A\mathbf{b}_1 & \cdots & A\mathbf{b}_r \end{bmatrix}.$$

3.1.7. Prove that if A_i are invertible in $M_n(\mathbb{F})$ for $i = 1, \ldots, k$, then $A_1 \cdots A_k$ is invertible and
$$(A_1 \cdots A_k)^{-1} = A_k^{-1} \cdots A_1^{-1}.$$

3.1.8. Let A belong to $M_{m,n}(\mathbb{F})$.
(a) Show that if $\mathbf{y}$ is in $\mathbb{F}_m$, then $\mathbf{y}A$ is a linear combination of the rows of A. Verify that $\mathbf{y} \mapsto \mathbf{y}A$ is a linear transformation $\mathbb{F}_m \to \mathbb{F}_n$.
(b) Show that $Y \mapsto YA$ is a linear transformation $M_{r,m}(\mathbb{F}) \to M_{r,n}(\mathbb{F})$.

3.1.9. Show that if $R : \mathbb{F}_m \to \mathbb{F}_n$ is a linear transformation, then R is effected by $\mathbf{y} \mapsto \mathbf{y}A$, for some A in $M_{m,n}(\mathbb{F})$.

3.1.10. Let A and B belong to $M_n(\mathbb{F})$. Here we consider the mapping $(A,B) \mapsto AB$.
   (a) Show that for A_1, A_2 in $M_n(\mathbb{F})$ and c in $\mathbb{F}$,
   $$(A_1 + cA_2, B) = (A_1, B) + c(A_2, B).$$
   This shows that $(A,B) \mapsto AB$ is *linear in its first variable*.
   (b) Show that the mapping is also linear in its second variable. This shows that $(\cdot, \cdot) : M_n(\mathbb{F}) \times M_n(\mathbb{F}) \to M_n(\mathbb{F})$ is a *bilinear mapping*.

3.1.11. Let A belong to $M_{m,n}(\mathbb{F})$.
   (a) Verify that $AI_n = A$.
   (b) If $\mathbf{e}_i$ is the ith element in the standard ordered basis for $\mathbb{F}^m$, then
   $$\mathbf{e}_i^T = \begin{bmatrix} \delta(i,1) & \cdots & \delta(i,m) \end{bmatrix}.$$
   Find $\mathbf{e}_i^T A$. Use this to verify that $I_m A = A$.

3.1.12. Let A belong to $M_{2,3}(\mathbb{R})$. Suppose that for any B in $M_{3,5}(\mathbb{R})$, the first row of AB is double the third row of B. What does this say about A?

3.1.13. Let B belong to $M_{3,5}(\mathbb{F})$. Suppose that for any A in $M_{2,3}(\mathbb{F})$, the second column of AB is the first column plus twice the third column of A. What does this say about B?

3.1.14. Suppose A is in $M_{m,n}(\mathbb{F})$ and B is in $M_{n,r}(\mathbb{F})$.
   (a) Show that if A has a row of zeroes, then AB has a row of zeroes.
   (b) Show that if B has a column of zeroes, then AB has a column of zeroes.

3.1.15. Let $A = \begin{bmatrix} 0 & 1 \\ 1 & 0 \end{bmatrix}$. Give a geometric description of the mapping on $\mathbb{R}^2$ given by $\mathbf{v} \mapsto A\mathbf{v}$.

3.1.16. Show that if we are given A in $M_{m,n}(\mathbb{F})$, then $A_{L_A} = A$ and that if we are given a linear transformation $L : \mathbb{F}^n \to \mathbb{F}^m$, then $L_{A_L} = L$.

3.1.17. Verify the formula in (3.4).

3.1.18. Let $A = \begin{bmatrix} 1 & -1 \\ 2 & -3 \end{bmatrix}$ belong to $M_2(\mathbb{R})$. Find A^{-1} or argue that it does not exist.

3.1.19. Find all rank 1 matrices in $M_2(\mathbb{F}_2)$. Find all matrices in the same space that have rank 2.

3.1.20. Find $n \times n$ matrices A, B, C so that $AB = AC$ but $B \neq C$.

3.2. Coordinate Vectors

The idea of a standard matrix representation applies only to linear mappings from $\mathbb{F}^n$ to $\mathbb{F}^m$. Coordinate vectors enable us to generalize matrix representations to linear transformations on arbitrary finite-dimensional vector spaces. In view of Theorem 2.34, we specify an arbitrary finite-dimensional vector space V over $\mathbb{F}$ by indicating that $V \cong \mathbb{F}^n$.

Suppose $V \cong \mathbb{F}^n$ and that $\mathcal{B} = \{\mathbf{b}_1, \ldots, \mathbf{b}_n\}$ is an ordered basis for V. The vector space isomorphism we get by extending linearly from $\mathbf{b}_i \mapsto \mathbf{e}_i$ is the **$\mathcal{B}$-coordinate mapping** $V \to \mathbb{F}^n$, designated by $[\ ]_\mathcal{B}$. When $\mathbf{v}$ is in V, the **$\mathcal{B}$-coordinate vector** associated to $\mathbf{v}$ is

$$[\mathbf{v}]_\mathcal{B} := [\ ]_\mathcal{B}(\mathbf{v}).$$

For example, if $p(t) = a_0 + a_1 t + \cdots + a_n t^n$ belongs to $\mathbb{F}[t]_n$ and $\mathcal{B} = \{1, t, \ldots, t^n\}$, then

$$[p(t)]_\mathcal{B} = (a_0, a_1, \ldots, a_n) \in \mathbb{F}^{n+1}.$$

Note that $[\ ]_\mathcal{E}$ is the identity mapping on $\mathbb{F}^n$.

As an isomorphism, $[\ ]_\mathcal{B}$ is invertible with

$$[\ ]_\mathcal{B}^{-1} : \mathbb{F}^n \to V.$$

We write

$$[\mathbf{v}]_\mathcal{B}^{-1} := [\ ]_\mathcal{B}^{-1}(\mathbf{v}).$$

Sometimes we can use coordinate vectors for countably infinite-dimensional vector spaces.

Example 3.14. Let $V = \mathbb{R}[t]$ and let $\mathcal{B} = \{1, t, t^2, \ldots\}$. If $p(t) = 2t - 3t^2$, then $[p(t)]_\mathcal{B} = (0, 2, -3, 0, 0, \ldots)$. If $\mathbf{v} = (2, 0, 0, -5, 1, 0, -2, 0, 0, \ldots)$, then

$$[\mathbf{v}]_\mathcal{B}^{-1} = 2 - 5t^3 + t^4 - 2t^6.$$

Coordinate vectors associated to subspaces can be more interesting.

Example 3.15. Let $W \subseteq \mathbb{R}^3$ be the span of $\mathcal{B} = \{(1, 0, 1), (0, 1, 0)\}$. Since $(2, 3, 2) = 2(1, 0, 1) + 3(0, 1, 0)$, we have

$$[(2, 3, 2)]_\mathcal{B} = (2, 3).$$

On the other hand,

$$[(-1, 4)]_\mathcal{B}^{-1} = (-1)(1, 0, 1) + 4(0, 1, 0) = (-1, 4, -1).$$

If we use $\mathcal{C} = \{(0, 1, 0), (1, 0, 1)\}$ as our ordered basis for W, then

$$[(2, 3, 2)]_\mathcal{C} = (3, 2)$$

and

$$[(-1, 4)]_\mathcal{C}^{-1} = (-1)(0, 1, 0) + 4(1, 0, 1) = (4, -1, 4).$$

Definition 3.16. Let $L : V \to W$ be a linear transformation where $V \cong \mathbb{F}^n$ and $W \cong \mathbb{F}^m$. Let $\mathcal{B} = \{\mathbf{b}_j\}_{j=1}^n$ be an ordered basis for V. Let $\mathcal{C}$ be an ordered basis for W. The **matrix representation** of L with respect to $\mathcal{B}$ and $\mathcal{C}$ is

$$[\![L]\!]_\mathcal{B}^\mathcal{C} := \begin{bmatrix} [L(\mathbf{b}_1)]_\mathcal{C} & \cdots & [L(\mathbf{b}_n)]_\mathcal{C} \end{bmatrix}.$$

When $W = V$ and $\mathcal{C} = \mathcal{B}$,

$$[\![L]\!]_\mathcal{B} := \begin{bmatrix} [L(\mathbf{b}_1)]_\mathcal{B} & \cdots & [L(\mathbf{b}_n)]_\mathcal{B} \end{bmatrix}$$

is the **matrix representation** of L with respect to $\mathcal{B}$.[1]

---

[1] Notation for a matrix representation of a linear transformation varies throughout the literature.

3.2. Coordinate Vectors

The best way to remember Definition 3.16 may be to remember that the following must hold.

Theorem 3.17. Let $V \cong \mathbb{F}^n$ and $W \cong \mathbb{F}^m$. Let $\mathcal{B}$ be an ordered basis for V and let $\mathcal{C}$ be an ordered basis for W. If $L : V \to W$ is a linear transformation, then, for each $\mathbf{v}$ in V,
$$[L(\mathbf{v})]_\mathcal{C} = [\![L]\!]_\mathcal{B}^\mathcal{C} [\mathbf{v}]_\mathcal{B}.$$

Proof. Let everything be as hypothesized, with $\mathcal{B} = \{\mathbf{b}_j\}_{j=1}^n$. If $\mathbf{v} = \sum_{j=1}^n c_j \mathbf{b}_j$, then since L is linear,
$$L(\mathbf{v}) = \sum_{j=1}^n c_j L(\mathbf{b}_j).$$

The mapping $[\ ]_\mathcal{C}$ is also linear, so
$$[L(\mathbf{v})]_\mathcal{C} = \sum_{j=1}^n c_j [L(\mathbf{b}_j)]_\mathcal{C} = [\![L]\!]_\mathcal{B}^\mathcal{C} [\mathbf{v}]_\mathcal{B},$$
since $[\mathbf{v}]_\mathcal{B} = (c_1, \ldots, c_n)$. $\square$

The commutative diagram in Figure 3.1 captures Theorem 3.17 if we take V, W, $\mathcal{B}$, $\mathcal{C}$, and L as in the statement of the theorem. Starting at the upper left in the diagram with an arbitrary element in V, we get the same element in $\mathbb{F}^m$ whether we traverse the diagram going from V to $\mathbb{F}^n$ to $\mathbb{F}^m$ or going from V to W to $\mathbb{F}^m$.

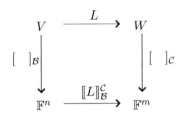

Figure 3.1. $[\![L]\!]_\mathcal{B}^\mathcal{C} \circ [\ ]_\mathcal{B} = [\ ]_\mathcal{C} \circ L.$

The next theorem follows nearly immediately from the fact that coordinate mappings are isomorphisms.

Theorem 3.18. Suppose $V \cong \mathbb{F}^n$ and $W \cong \mathbb{F}^m$. If $L : V \to W$ is linear and A is any matrix representation of L, then $\operatorname{rank} L = \operatorname{rank} A$ and $\dim \operatorname{Ker} L = \dim \operatorname{Nul} A$.

Proof. Let V, W, and L be as hypothesized. Let $\mathcal{B} = \{\mathbf{b}_j\}_{j=1}^n$ be an ordered basis for V and let $\mathcal{C}$ be an ordered basis for W. If $A = [\![L]\!]_\mathcal{B}^\mathcal{C}$, then
$$\operatorname{Col} A = \operatorname{Span}\{[L(\mathbf{b}_j)]_\mathcal{C}\}_{j=1}^n.$$

Since $[\ ]_\mathcal{C} : L[V] \to \operatorname{Col} A$ is an isomorphism, the rank of A and the rank of L must be equal.

The second statement is immediate by the Rank Theorem. $\square$

Example 3.19. Let $A = \begin{bmatrix} 1 & -1 & -2 & 1 \\ 0 & -2 & -2 & 1 \\ 3 & -3 & -6 & 3 \end{bmatrix}$ in $M_{3,4}(\mathbb{R})$. The first two columns of A form a linearly independent set in $\mathbb{R}^3$. Note, as well, that

$$(3.5) \qquad (-1)(1,0,3) + (-1,-2,-3) = (-2,-2,-6) = (-2)(1,1,3)$$

so both the third and fourth columns of A are in the span of the first two columns. It follows that rank $A = 2$.

If $L : \mathbb{R}^4 \to \mathbb{R}^3$ is given by $L(\mathbf{v}) = A\mathbf{v}$, then the Rank Theorem implies that the nullity of L, thus of A, is two. Equation (3.5) gives us dependence relations on the columns of A and implies that $(-1, 1, -1, 0)$ and $(-1, 1, 0, 2)$ belong to Nul A. Since $\{(-1,1,-1,0),(-1,1,0,2)\}$ is linearly independent, it is a basis for Nul A.

Example 3.20. Let $L : \mathbb{R}[t]_2 \to M_2(\mathbb{R})$ be given by

$$L(a_0 + a_1 t + a_2 t^2) = \begin{bmatrix} a_0 & a_1 \\ -a_2 & a_0 + a_1 \end{bmatrix}.$$

Take $\mathcal{B} = \{1, t, t^2\}$ and

$$\mathcal{C} = \left\{ \begin{bmatrix} 1 & 0 \\ 0 & 0 \end{bmatrix}, \begin{bmatrix} 0 & 1 \\ 0 & 0 \end{bmatrix}, \begin{bmatrix} 0 & 0 \\ 1 & 0 \end{bmatrix}, \begin{bmatrix} 0 & 0 \\ 0 & 1 \end{bmatrix} \right\}.$$

We have

$$L(1) = \begin{bmatrix} 1 & 0 \\ 0 & 1 \end{bmatrix}, \quad L(t) = \begin{bmatrix} 0 & 1 \\ 0 & 1 \end{bmatrix}, \quad \text{and} \quad L(t^2) = \begin{bmatrix} 0 & 0 \\ -1 & 0 \end{bmatrix}$$

so $[L(1)]_\mathcal{C} = (1,0,0,1)$, $[L(t)]_\mathcal{C} = (0,1,0,1)$, and $[L(t^2)]_\mathcal{C} = (0,0,-1,0)$. The matrix representation of L with respect to $\mathcal{B}$ and $\mathcal{C}$ is then

$$[L]_\mathcal{B}^\mathcal{C} = \begin{bmatrix} 1 & 0 & 0 \\ 0 & 1 & 0 \\ 0 & 0 & -1 \\ 1 & 1 & 0 \end{bmatrix}.$$

Checking our work, we can see that $L(1 - 2t + 3t^2) = \begin{bmatrix} 1 & -2 \\ -3 & -1 \end{bmatrix}$ while

$$[L]_\mathcal{B}^\mathcal{C}[1 - 2t + 3t^2]_\mathcal{B} = \begin{bmatrix} 1 & 0 & 0 \\ 0 & 1 & 0 \\ 0 & 0 & -1 \\ 1 & 1 & 0 \end{bmatrix} \begin{bmatrix} 1 \\ -2 \\ 3 \end{bmatrix} = \begin{bmatrix} 1 \\ -2 \\ -3 \\ -1 \end{bmatrix} = [L(1 - 2t + 3t^2)]_\mathcal{C}.$$

The corollary follows immediately from Theorem 3.18

Corollary 3.21. Let $V \cong W \cong \mathbb{F}^n$. A linear transformation $L : V \to W$ is an isomorphism if and only if any matrix representation of L is invertible.

3.2. Coordinate Vectors

Exercises 3.2.

3.2.1. The differential equation given by

(3.6) $$my'' + ky = 0$$

models the motion of an object of mass m attached to a frictionless spring that moves back and forth along a line. The spring coefficient k is a measure of the resistance of the spring to stretching/compressing. When at rest, the mass is at its equilibrium position, $y = 0$. At time $t = 0$, the mass is released after having been displaced and/or nudged. Its position at time $t > 0$ is then $y(t)$. The solution to (3.6) is the set of all functions $y(t)$ that satisfy (3.6). A solution to an initial value problem associated to (3.6) is a function in the solution set that also satisfies *initial conditions*. Initial conditions typically describe the displacement and/or nudge given to the spring at $t = 0$, so they take the form $y(0) = y_0$ and $y'(0) = v_0$. The initial position of the mass is then y_0 and its initial velocity is v_0.

(a) Show that the collection of functions satisfying (3.6) comprises a subspace of $C(\mathbb{R})$. This is the *solution space* for (3.6).

(b) A fact we will not prove is that the solution space for (3.6) is 2-dimensional. Given this information, show that if $\omega = \sqrt{k/m}$, then $\mathcal{B} = \{\sin \omega t, \cos \omega t\}$ is an ordered basis for the solution space.

(c) Find $y(t) = [(1, -2)]_\mathcal{B}^{-1}$.

(d) Let $y(t) = [(-2, 4)]_\mathcal{B}^{-1}$ and $\omega = 3$. Find $y(t)$, $y(0)$, and $y'(0)$.

(e) Let $y(t)$ be the solution to (3.6) when $\omega = 1/2$, subject to initial conditions $y(0) = 3$ and $y'(0) = -1$. Find $[y(t)]_\mathcal{B}$.

3.2.2. Let $\mathcal{B} = \{t^2 + 1, t + 1, t^2 + t\}$ and let $\mathcal{C} = \{t^2 + t + 1, t^2 - 1, t - 1\}$. Treat $\mathcal{B}$ and $\mathcal{C}$ as ordered bases for $\mathbb{R}[t]_2$.

(a) Suppose $[p(t)]_\mathcal{B} = (1, 0, 1)$. Find $[p(t)]_\mathcal{C}$.

(b) Suppose $[p(t)]_\mathcal{C} = (1, -1, -1)$. Find $[p(t)]_\mathcal{B}$.

3.2.3. Let $W = \text{Span}\, S \subseteq M_{2,3}(\mathbb{Q})$ where

(3.7) $$S = \left\{ \begin{bmatrix} 1 & 0 & 1 \\ -1 & 2 & 0 \end{bmatrix}, \begin{bmatrix} 0 & -2 & 0 \\ 0 & 0 & 1 \end{bmatrix} \right\}.$$

(a) Show that S is linearly independent.

(b) Show that

(3.8) $$A = \begin{bmatrix} 4 & 2 & 4 \\ -4 & 8 & -1 \end{bmatrix}$$

is in W.

(c) With A given in (3.8) and taking S in (3.7) to be an ordered basis for W, find $[A]_S$.

(d) Let S' be the second ordered basis for W that we can get using the set that underlies S in (3.7). Find $[A]_{S'}$, if A is given in (3.8).

(e) Find $B = [(-3, 5)]_S^{-1}$.

(f) Find $C = [(-3, 5)]_{S'}^{-1}$.

3.2.4. Referring to Example 3.15, verify that $\mathcal{C} = \{(1, 2, 1), (1, 3, 1)\}$ is an ordered basis for $W = \text{Span}\{(1, 0, 1), (0, 1, 0)\}$. Find $[(2, 3, 2)]_\mathcal{C}$ and $[(-1, 4)]_\mathcal{C}^{-1}$.

3.2.5. Define $L : \mathbb{R}[t]_2 \to M_{2,2}(\mathbb{R})$ by $L(a + bt + ct^2) = \begin{bmatrix} a+b & b-2c \\ 2a+b & -c \end{bmatrix}$. Let $\mathcal{B} = \{1, t, t^2\}$ and let
$$\mathcal{C} = \left\{ \begin{bmatrix} 1 & 0 \\ 0 & 0 \end{bmatrix}, \begin{bmatrix} 0 & 1 \\ 0 & 0 \end{bmatrix}, \begin{bmatrix} 0 & 0 \\ 1 & 0 \end{bmatrix}, \begin{bmatrix} 0 & 0 \\ 0 & 1 \end{bmatrix} \right\}.$$
(a) Find $A = [\![L]\!]_\mathcal{B}^\mathcal{C}$.
(b) What is the rank of A?

3.2.6. Let $A = \begin{bmatrix} 1 & 2 & 0 \\ 3 & -1 & 1 \end{bmatrix}$ in $M_{2,3}(\mathbb{R})$. Let $V = \mathbb{R}[t]_2$ and let W be the span of $\{\sin t, \cos t\}$ in $C(\mathbb{R})$. Let $\mathcal{B} = \{1, t, t^2\}$ and let $\mathcal{C} = \{\sin t, \cos t\}$. Define a linear transformation $L : V \to W$ so that $[\![L]\!]_\mathcal{B}^\mathcal{C} = A$.

3.2.7. Let V be any finite-dimensional vector space. Let $\mathcal{B}$ be any ordered basis for V. What is $[\![I]\!]_\mathcal{B}$?

3.3. Change of Basis

How are two matrices related when they represent the same linear transformation? Change of basis matrices give us the answer.

Definition 3.22. Let $\mathcal{B}$ and $\mathcal{C}$ be ordered bases for some $V \cong \mathbb{F}^n$. The $\mathcal{B}$ to $\mathcal{C}$ **change of basis matrix** is $[\![I]\!]_\mathcal{B}^\mathcal{C}$.

If $\mathcal{B} = \{\mathbf{b}_j\}_{j=1}^n$ and $\mathcal{C}$ are ordered bases for $V \cong \mathbb{F}^n$, then
$$[\![I]\!]_\mathcal{B}^\mathcal{C} = \begin{bmatrix} [\mathbf{b}_1]_\mathcal{C} & \cdots & [\mathbf{b}_n]_\mathcal{C} \end{bmatrix}.$$
For any $\mathbf{v}$ in V, we then have
$$[\![I]\!]_\mathcal{B}^\mathcal{C} [\mathbf{v}]_\mathcal{B} = [\mathbf{v}]_\mathcal{C}.$$

As the matrix representation of an automorphism, any change of basis matrix is invertible. Indeed, if $\mathcal{B} = \{\mathbf{b}_j\}_{j=1}^n$ and $\mathcal{C}$ are ordered bases for $V \cong \mathbb{F}^n$, then
$$([\![I]\!]_\mathcal{B}^\mathcal{C})^{-1} = [\![I]\!]_\mathcal{C}^\mathcal{B}.$$
We verify this by noting that
$$[\![I]\!]_\mathcal{C}^\mathcal{B} [\![I]\!]_\mathcal{B}^\mathcal{C} = [\![I]\!]_\mathcal{C}^\mathcal{B} \begin{bmatrix} [\mathbf{b}_1]_\mathcal{C} & \cdots & [\mathbf{b}_n]_\mathcal{C} \end{bmatrix}.$$
Since $[\![I]\!]_\mathcal{C}^\mathcal{B} [\mathbf{b}_j]_\mathcal{C} = [\mathbf{b}_j]_\mathcal{B} = \mathbf{e}_j$,
$$[\![I]\!]_\mathcal{C}^\mathcal{B} [\![I]\!]_\mathcal{B}^\mathcal{C} = I_n.$$

The next theorem gives us a formula allowing us to toggle between different matrix representations of a given linear transformation. The proof is a routine verification that we leave as an exercise.

Theorem 3.23. Let $V \cong \mathbb{F}^n$ have ordered bases $\mathcal{B}_i$, $i = 1, 2$, and let $W \cong \mathbb{F}^m$ have ordered bases $\mathcal{C}_i$, $i = 1, 2$. If $L : V \to W$ is a linear transformation, then
(3.9) $$[\![L]\!]_{\mathcal{B}_2}^{\mathcal{C}_2} = [\![I]\!]_{\mathcal{C}_1}^{\mathcal{C}_2} [\![L]\!]_{\mathcal{B}_1}^{\mathcal{C}_1} [\![I]\!]_{\mathcal{B}_2}^{\mathcal{B}_1}.$$

The next example suggests the importance of change of basis in applications.

3.3. Change of Basis

Example 3.24. Let ℓ be the line given by $y = 2x$ in the xy-plane. Let L map every point on $\ell = \text{Span}\{(1,2)\}$ to itself and every point perpendicular to ℓ to its negative. This mapping is the **orthogonal reflection** through ℓ. Taking $\mathcal{B} = \{(1,2), (-2,1)\}$ as our basis for $\mathbb{R}^2$, we can describe L as the linear extension of $(1,2) \mapsto (1,2)$, $(-2,1) \mapsto (2,-1)$. The matrix representation of L with respect to $\mathcal{B}$ is

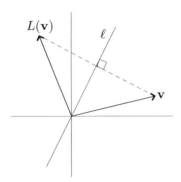

Figure 3.2. L is the orthogonal reflection through ℓ in $\mathbb{R}^2$.

$$[\![L]\!]_\mathcal{B} = \begin{bmatrix} 1 & 0 \\ 0 & -1 \end{bmatrix}.$$

We use this to find $L(a,b) = [L(a,b)]_\mathcal{E}$ for arbitrary (a,b) in $\mathbb{R}^2$.

First, we note that
$$[(a,b)]_\mathcal{B} = [\![I]\!]_\mathcal{E}^\mathcal{B}(a,b).$$

Then we have
$$[L(a,b)]_\mathcal{B} = [\![L]\!]_\mathcal{B}[\![I]\!]_\mathcal{E}^\mathcal{B}(a,b).$$

This means
$$L(a,b) = [L(a,b)]_\mathcal{E} = [\![I]\!]_\mathcal{B}^\mathcal{E}[\![L]\!]_\mathcal{B}[\![I]\!]_\mathcal{E}^\mathcal{B}(a,b).$$

In particular,
$$[\![L]\!]_\mathcal{E} = [\![I]\!]_\mathcal{B}^\mathcal{E}[\![L]\!]_\mathcal{B}[\![I]\!]_\mathcal{E}^\mathcal{B}.$$

Working out the details we find
$$[\![I]\!]_\mathcal{B}^\mathcal{E} = \begin{bmatrix} 1 & -2 \\ 2 & 1 \end{bmatrix}$$

and
$$[\![I]\!]_\mathcal{E}^\mathcal{B} = \begin{bmatrix} 1/5 & 2/5 \\ -2/5 & 1/5 \end{bmatrix}$$

so that
$$[\![L]\!]_\mathcal{E} = \begin{bmatrix} -3/5 & 4/5 \\ 4/5 & 3/5 \end{bmatrix}.$$

We double-check our work by calculating
$$\begin{bmatrix} -3/5 & 4/5 \\ 4/5 & 3/5 \end{bmatrix} \begin{bmatrix} 1 \\ 2 \end{bmatrix} = \begin{bmatrix} 1 \\ 2 \end{bmatrix} = L(1,2)$$

and
$$\begin{bmatrix} -3/5 & 4/5 \\ 4/5 & 3/5 \end{bmatrix} \begin{bmatrix} -2 \\ 1 \end{bmatrix} = \begin{bmatrix} 2 \\ -1 \end{bmatrix} = L(-2, 1).$$

In general, we can say
$$L(a, b) = \left(-\frac{3}{5}a + \frac{4}{5}b, \frac{4}{5}a + \frac{3}{5}b\right).$$

If $\mathcal{B} = \{\mathbf{b}_j\}_{j=1}^n$ is an ordered basis for $\mathbb{F}^n$, then the $\mathcal{B}$ to $\mathcal{E}$ change of basis matrix is
$$[\![I]\!]_{\mathcal{B}}^{\mathcal{E}} = \begin{bmatrix} \mathbf{b}_1 & \cdots & \mathbf{b}_n \end{bmatrix}.$$

This is enough to show that every invertible matrix is a change of basis matrix. We can then say that when A and B in $M_{m,n}(\mathbb{F})$ represent the same linear transformation, $B = PAQ$ for invertible matrices P and Q. The converse is true as well.

Theorem 3.25. *Let P in $M_m(\mathbb{F})$ and Q in $M_n(\mathbb{F})$ be invertible matrices. Suppose A and B in $M_{m,n}(\mathbb{F})$ are related by $B = PAQ$. If $V \cong \mathbb{F}^n$ and $W \cong \mathbb{F}^m$, then there is a linear transformation $L : V \to W$ for which A and B are matrix representations.*

Proof. Let P, Q, $A = [a_{ij}]$, B, V, and W be as hypothesized. Fix an ordered basis for V, $\mathcal{B}_1 = \{\mathbf{b}_j\}_{j=1}^n$, and an ordered basis for W, $\mathcal{C}_1 = \{\mathbf{c}_i\}_{i=1}^m$.

If we define $L : V \to W$ by extending linearly from
$$\mathbf{b}_j \mapsto a_{1j}\mathbf{c}_1 + \cdots + a_{mj}\mathbf{c}_m,$$
then $A = [\![L]\!]_{\mathcal{B}_1}^{\mathcal{C}_1}$.

If $Q = [q_{ij}]$, define
$$\mathbf{x}_j = q_{1j}\mathbf{b}_1 + \cdots + q_{nj}\mathbf{b}_n \in V$$
for each $j \in \{1, \ldots, n\}$. Let $\mathcal{B}_2 = \{\mathbf{x}_j\}_{j=1}^n$. Column j of Q is $[\mathbf{x}_j]_{\mathcal{B}_1}$. Since the $\mathcal{B}_1$-coordinate mapping is an isomorphism and the columns of Q comprise a basis for $\mathbb{F}^n$, $\mathcal{B}_2$ is an ordered basis for V and $Q = [\![I]\!]_{\mathcal{B}_2}^{\mathcal{B}_1}$.

The same trick gets us a second basis for W. This time, we use $P^{-1} = [p_{ij}]$ and define
$$\mathbf{y}_j = p_{1j}\mathbf{c}_1 + \cdots + p_{mj}\mathbf{c}_m \in W$$
for each $j \in \{1, \ldots, m\}$. Let $\mathcal{C}_2 = \{\mathbf{y}_j\}_{j=1}^m$. Column j of P^{-1} is $[\mathbf{y}_j]_{\mathcal{C}_1}$. By our reasoning above, $\mathcal{C}_2$ is an ordered basis for W and $P^{-1} = [\![I]\!]_{\mathcal{C}_2}^{\mathcal{C}_1}$. From there, we have $P = [\![I]\!]_{\mathcal{C}_1}^{\mathcal{C}_2}$.

Putting everything together we see that
$$B = PAQ = [\![I]\!]_{\mathcal{C}_1}^{\mathcal{C}_2} [\![L]\!]_{\mathcal{B}_1}^{\mathcal{C}_1} [\![I]\!]_{\mathcal{B}_2}^{\mathcal{B}_1} = [\![L]\!]_{\mathcal{B}_2}^{\mathcal{C}_2}$$
which proves the theorem. $\square$

Exercises 3.3.

3.3.1. Prove Theorem 3.23.

3.3.2. Let L be the orthogonal reflection in $\mathbb{R}^2$ through the line ℓ given by $3x + 2y = 0$.

(a) Find an ordered basis $\mathcal{B} = \{\mathbf{b}_1, \mathbf{b}_2\}$ for $\mathbb{R}^2$ so that $\mathbf{b}_1$ lies on ℓ and $\mathbf{b}_2$ is perpendicular to ℓ.
(b) Find the matrix representation of L with respect to $\mathcal{B}$.
(c) Find the $\mathcal{E}$ to $\mathcal{B}$ change of basis matrix and use it to find $[(1,1)]_\mathcal{B}$ and $[L(1,1)]_\mathcal{B}$.
(d) Use the $\mathcal{B}$ to $\mathcal{E}$ change of basis matrix to find $L(1,1)$.

3.3.3. Let $L : \mathbb{R}^2 \to \mathbb{R}^2$ be given by $L(x,y) = (2x - y, x + y)$.
(a) Find the standard matrix representation of L.
(b) Find the $\mathcal{E}$ to $\mathcal{B}$ change of basis matrix if $\mathcal{B} = \{(1,1),(1,-1)\}$.
(c) Find the $\mathcal{B}$ to $\mathcal{E}$ change of basis matrix.
(d) Find the matrix representation of L with respect to $\mathcal{B}$.

3.3.4. Consider ordered bases for $\mathbb{R}[t]_2$, $\mathcal{B} = \{1+t, 1-t, 1+2t+t^2\}$ and $\mathcal{C} = \{1, 1+t, 1+t+2t^2\}$.
(a) Find P, the $\mathcal{B}$ to $\mathcal{C}$ change of basis matrix.
(b) Find P^{-1}, the $\mathcal{C}$ to $\mathcal{B}$ change of basis matrix.
(c) Double-check your calculations by finding $[\mathbf{p}_j]_\mathcal{B}^{-1}$ for each column, $\mathbf{p}_j$, of the matrix you found in part (b).

3.3.5. Let $A = \begin{bmatrix} 1 & 3 & 5 & 7 \\ 2 & 4 & 6 & 8 \end{bmatrix}$. Find P and Q so that $PAQ = \begin{bmatrix} 1 & 0 & 0 & 0 \\ 0 & 1 & 0 & 0 \end{bmatrix}$.

3.4. Vector Spaces of Linear Transformations

Here we take a closer look at the connections between $M_{m,n}(\mathbb{F})$ and the set of linear transformations $V \to W$, where $V \cong \mathbb{F}^n$ and $W \cong \mathbb{F}^m$. We exploit the fact that, like $M_{m,n}(\mathbb{F})$, the set of linear transformations from V to W is a vector space. Once we see this, we consider the space of linear transformations $V \to W$, when V and W are arbitrary vector spaces over $\mathbb{F}$.

The set of linear transformations $V \to W$ is $\mathcal{L}(V,W)$. We use $\mathcal{L}(V)$ to denote the set of linear operators on V.

Theorem 3.26. If V and W are vector spaces over a field $\mathbb{F}$, then $\mathcal{L}(V,W)$ is a subspace of $\mathcal{F}(V,W)$.

Proof. Let V and W be as hypothesized. The zero mapping is in $\mathcal{L}(V,W)$, so $\mathcal{L}(V,W) \neq \emptyset$. We must show that if L and T are in $\mathcal{L}(V,W)$ and c is in $\mathbb{F}$, then $cL + T$ is linear.

Taking a in $\mathbb{F}$ and $\mathbf{v}_i$ in V, we have
$$(cL + T)(a\mathbf{v}_1 + \mathbf{v}_2) = cL(a\mathbf{v}_1 + \mathbf{v}_2) + T(a\mathbf{v}_1 + \mathbf{v}_2).$$

Since L and T are both linear and $ca = ac$, we can write
$$(cL + T)(a\mathbf{v}_1 + \mathbf{v}_2) = acL(\mathbf{v}_1) + cL(\mathbf{v}_2) + aT(\mathbf{v}_1) + T(\mathbf{v}_2).$$

From here we can say that
$$(cL + T)(a\mathbf{v}_1 + \mathbf{v}_2) = a(cL + T)(\mathbf{v}_1) + (cL + T)(\mathbf{v}_2),$$

which is what we needed to show. □

Another way to state Theorem 3.26 is to say that, under the appropriate hypotheses, a linear combination of linear transformations is a linear transformation.

Next we verify that, under the appropriate hypotheses, a linear combination of linear transformations corresponds to a linear combination of matrix representations of those transformations.

Theorem 3.27. Let $V \cong \mathbb{F}^n$ and $W \cong \mathbb{F}^m$ for some field $\mathbb{F}$. If $\mathcal{B}$ is an ordered basis for V and $\mathcal{C}$ is an ordered basis for W, then $L \mapsto [\![L]\!]_{\mathcal{B}}^{\mathcal{C}}$ is an isomorphism $\mathcal{L}(V, W) \to M_{m,n}(\mathbb{F})$.

Proof. Let everything be as hypothesized with $\mathcal{B} = \{\mathbf{b}_j\}_{j=1}^n$.

Consider L and T in $\mathcal{L}(V, W)$ and c in $\mathbb{F}$. We have

$$(cL + T)(\mathbf{b}_j) = cL(\mathbf{b}_j) + T(\mathbf{b}_j)$$

for $j = 1, \ldots, n$. Since coordinate mappings are linear,

$$[(cL + T)(\mathbf{b}_j)]_{\mathcal{C}} = [cL(\mathbf{b}_j) + T(\mathbf{b}_j)]_{\mathcal{C}} = c[L(\mathbf{b}_j)]_{\mathcal{C}} + [T(\mathbf{b}_j)]_{\mathcal{C}}.$$

This shows that column j of $[\![cL + T]\!]_{\mathcal{B}}^{\mathcal{C}}$ is the sum of columns j of $c[\![L]\!]_{\mathcal{B}}^{\mathcal{C}}$ and $[\![T]\!]_{\mathcal{B}}^{\mathcal{C}}$. It follows that $L \mapsto [\![L]\!]_{\mathcal{B}}^{\mathcal{C}}$ is linear.

If $[\![L]\!]_{\mathcal{B}}^{\mathcal{C}} = [\![T]\!]_{\mathcal{B}}^{\mathcal{C}}$ for L and T in $\mathcal{L}(V, W)$, then for each $j \in \{1, \ldots, n\}$,

$$[L(\mathbf{b}_j)]_{\mathcal{C}} = [T(\mathbf{b}_j)]_{\mathcal{C}}.$$

Since $\mathbf{w} \mapsto [\mathbf{w}]_{\mathcal{C}}$ is injective, $L(\mathbf{b}_j) = T(\mathbf{b}_j)$ for all $j \in \{1, \ldots, n\}$. Since they agree on a basis for V, L and T must be the same mapping. We conclude that $L \mapsto [\![L]\!]_{\mathcal{B}}^{\mathcal{C}}$ is injective.

Let $A = [a_{ij}] = \begin{bmatrix} \mathbf{a}_1 & \cdots & \mathbf{a}_n \end{bmatrix}$ be arbitrary in $M_{m,n}(\mathbb{F})$ and say that $\mathcal{C} = \{\mathbf{c}_j\}_{j=1}^m$. We may then define $L : V \to W$ by extending linearly from

$$L(\mathbf{b}_j) := a_{1j}\mathbf{c}_1 + \cdots + a_{mj}\mathbf{c}_m.$$

This gives us

$$[L(\mathbf{b}_j)]_{\mathcal{C}} = \mathbf{a}_j$$

and from there, $[\![L]\!]_{\mathcal{B}}^{\mathcal{C}} = A$. This proves that $L \mapsto [\![L]\!]_{\mathcal{B}}^{\mathcal{C}}$ is surjective, thus an isomorphism. $\square$

Let E_{ij} in $M_{m,n}(\mathbb{F})$ be the matrix with ij-entry equal to 1 and all other entries equal to 0. It is easy to see that $\{E_{ij}\}_{i=1,j=1}^{i=m,j=n}$ is a basis for $M_{m,n}(\mathbb{F})$ and that it contains mn elements. When we refer to it as the standard ordered basis for $M_{m,n}(\mathbb{F})$, we mean it is ordered first by i, then j, in other words,

$$\{E_{11}, E_{12}, \ldots, E_{1n}, \ldots, E_{m1}, \ldots, E_{mn}\}.$$

For instance, the standard ordered basis for $M_2(\mathbb{F})$ is

$$\left\{ \begin{bmatrix} 1 & 0 \\ 0 & 1 \end{bmatrix}, \begin{bmatrix} 0 & 1 \\ 0 & 0 \end{bmatrix}, \begin{bmatrix} 0 & 0 \\ 1 & 0 \end{bmatrix}, \begin{bmatrix} 0 & 0 \\ 0 & 1 \end{bmatrix} \right\}.$$

The following is immediate.

Corollary 3.28. If $V \cong \mathbb{F}^n$ and $W \cong \mathbb{F}^m$ for some field $\mathbb{F}$, then $\dim \mathcal{L}(V, W) = mn$.

3.4. Vector Spaces of Linear Transformations

The next lemma is another humble result with powerful implications. It says that, under the appropriate hypotheses, the composition of linear transformations is a linear transformation.

Lemma 3.29. Suppose V, W, and X are vector spaces over $\mathbb{F}$. If L is in $\mathcal{L}(V,W)$ and T is in $\mathcal{L}(W,X)$, then $T \circ L$ is in $\mathcal{L}(V,X)$.

Proof. Let everything be as hypothesized. Given a scalar c in $\mathbb{F}$ and vectors $\mathbf{v}_1, \mathbf{v}_2$ in V we have

$$(T \circ L)(c\mathbf{v}_1 + \mathbf{v}_2) = T(L(c\mathbf{v}_1 + \mathbf{v}_2)) = T(cL(\mathbf{v}_1) + L(\mathbf{v}_2))$$
$$= T(cL(\mathbf{v}_1)) + T(L(\mathbf{v}_2)) = cT(L(\mathbf{v}_1)) + T(L(\mathbf{v}_2))$$
$$= c(T \circ L)(\mathbf{v}_1) + (T \circ L)(\mathbf{v}_2). \qquad \square$$

Lemma 3.29 implies that function composition is a binary operation on $\mathcal{L}(V)$. It is routine to verify that Definition 1.47 applies to $\mathcal{L}(V)$, making it an associative $\mathbb{F}$-algebra with identity.[2]

The next result is the last detail we need to see that $M_n(\mathbb{F})$ is also an associative $\mathbb{F}$-algebra with identity.

Theorem 3.30. If A is in $M_{m,n}(\mathbb{F})$, B and C are in $M_{n,r}(\mathbb{F})$, and c is in $\mathbb{F}$, then

(a) $c(AB) = (cA)B = A(cB)$,

(b) $A(B+C) = AB + AC$, and

(c) $(B+C)A = BA + CA$.

Proof. Let everything be as hypothesized, with $A = [a_{ij}]$ and $B = [b_{ij}]$. Given c in $\mathbb{F}$, we have

$$c\left(\sum_{k=1}^n a_{ik}b_{kj}\right) = \sum_{k=1}^n (ca_{ik})b_{kj} = \sum_{k=1}^n a_{ik}(cb_{kj}).$$

This shows that the ij-entries of $c(AB)$, $(cA)B$, and $A(cB)$ are identical, which proves part (a).

The proof of part (b) is immediate by Theorem 3.6. The proof of part (c) follows likewise from Exercise 3.1.8. $\qquad \square$

The next definition gives us the device we need to compare $\mathcal{L}(V)$ and $M_n(\mathbb{F})$ as $\mathbb{F}$-algebras.

Definition 3.31. An **$\mathbb{F}$-algebra homomorphism** is a linear transformation

$$\varphi : \mathcal{A}_1 \to \mathcal{A}_2,$$

for $\mathbb{F}$-algebras $\mathcal{A}_1$ and $\mathcal{A}_2$, that satisfies

$$\varphi(\mathbf{a}_1\mathbf{a}_2) = \varphi(\mathbf{a}_1)\varphi(\mathbf{a}_2)$$

for all $\mathbf{a}_1$ and $\mathbf{a}_2$ in $\mathcal{A}_1$. An **$\mathbb{F}$-algebra isomorphism** is a vector space isomorphism that is also an $\mathbb{F}$-algebra homomorphism.

---
[2] Function composition is always associative.

Properties of algebra homomorphisms shadow properties of linear transformations.

Theorem 3.32. Let $\mathcal{A}_1$ and $\mathcal{A}_2$ be associative $\mathbb{F}$-algebras. If $\varphi : \mathcal{A}_1 \to \mathcal{A}_2$ is an $\mathbb{F}$-algebra homomorphism, then

(a) $\varphi[\mathcal{A}_1]$ is a subalgebra of $\mathcal{A}_2$;

(b) if $\mathcal{A}_1$ has identity ι, then $\varphi[\mathcal{A}_1]$ has identity $\varphi(\iota)$; and

(c) if $\mathbf{a}$ in $\mathcal{A}_1$ is invertible, then $\varphi(\mathbf{a})$ is invertible in $\varphi[\mathcal{A}_1]$ and
$$(\varphi(\mathbf{a}))^{-1} = \varphi(\mathbf{a}^{-1}).$$

Proof. Suppose $\mathcal{A}_1$, $\mathcal{A}_2$, and φ are as hypothesized. We know that $\varphi[\mathcal{A}_1]$ is a vector subspace of $\mathcal{A}_2$. Since
$$\varphi(\mathbf{a}_1 \mathbf{a}_2) = \varphi(\mathbf{a}_1)\varphi(\mathbf{a}_2),$$
for all $\mathbf{a}_1$ and $\mathbf{a}_2$ in $\mathcal{A}_1$, $\varphi[\mathcal{A}_1]$ is closed under multiplication. The other properties of Definition 1.47 follow readily. We leave the details as an exercise.

If $\mathcal{A}_1$ has identity element ι, then for all $\mathbf{a}$ in $\mathcal{A}_1$,
$$\varphi(\mathbf{a})\varphi(\iota) = \varphi(\mathbf{a}\iota) = \varphi(\mathbf{a}).$$
The same reasoning applies to show that
$$\varphi(\iota)\varphi(\mathbf{a}) = \varphi(\mathbf{a}).$$
It follows that $\varphi(\iota)$ is the identity element in $\varphi[\mathcal{A}_1]$.

If $\mathbf{a}$ in $\mathcal{A}_1$ has inverse $\mathbf{a}^{-1}$, then
$$\varphi(\iota) = \varphi(\mathbf{a}\mathbf{a}^{-1}) = \varphi(\mathbf{a})\varphi(\mathbf{a}^{-1}).$$
Likewise,
$$\varphi(\iota) = \varphi(\mathbf{a}^{-1}\mathbf{a}) = \varphi(\mathbf{a}^{-1})\varphi(\mathbf{a}).$$
It follows that $\varphi(\mathbf{a})$ is invertible in $\varphi[\mathcal{A}_1]$ with $(\varphi(\mathbf{a}))^{-1} = \varphi(\mathbf{a}^{-1})$. $\square$

The next theorem clarifies that matrix multiplication is a version of function composition.

Theorem 3.33. Let $V \cong \mathbb{F}^n$. If $\mathcal{B}$ is an ordered basis for V, then $L \mapsto [\![L]\!]_\mathcal{B}$ is an $\mathbb{F}$-algebra isomorphism $\mathcal{L}(V) \to M_n(\mathbb{F})$.

Proof. Let V and $\mathcal{B}$ be as hypothesized. Theorem 3.27 implies that $L \mapsto [\![L]\!]_\mathcal{B}$ is a vector space isomorphism $\mathcal{L}(V) \to M_n(\mathbb{F})$. It remains to show that if L and T belong to $\mathcal{L}(V)$, then
$$[\![L \circ T]\!]_\mathcal{B} = [\![L]\!]_\mathcal{B} [\![T]\!]_\mathcal{B}.$$
If $\mathcal{B} = \{\mathbf{b}_j\}_{j=1}^n$, then column j of $[\![L]\!]_\mathcal{B}[\![T]\!]_\mathcal{B}$ is given by
$$[\![L]\!]_\mathcal{B}[T(\mathbf{b}_j)]_\mathcal{B} = [L(T(\mathbf{b}_j))]_\mathcal{B} = [(L \circ T)(\mathbf{b}_j)]_\mathcal{B}.$$
Since $[(L \circ T)(\mathbf{b}_j)]_\mathcal{B}$ is column j of $[\![L \circ T]\!]_\mathcal{B}$, this is enough to prove the theorem. $\square$

The fact that $\mathcal{L}(V)$ and $M_n(\mathbb{F})$ are isomorphic as $\mathbb{F}$-algebras has implications that we continue to pursue in the next section.

Exercises 3.4.

3.4.1. Let $V \cong \mathbb{F}^n$ and $W \cong \mathbb{F}^m$. Let $\mathcal{B} = \{\mathbf{b}_i\}_{i=1}^n$ be an ordered basis for V and let $\mathcal{C} = \{\mathbf{c}_j\}_{j=1}^m$ be an ordered basis for W. Use $\mathcal{B}$ and $\mathcal{C}$ to find a basis for $\mathcal{L}(V, W)$.

3.4.2. Let V be a vector space over $\mathbb{F}$. For a in $\mathbb{F}$, define $f_a : V \to V$ by $f_a(\mathbf{v}) = a\mathbf{v}$. Let $S = \{f_a \mid a \in \mathbb{F}\}$.
  (a) Show that S is a vector subspace of $\mathcal{L}(V)$.
  (b) Show that S is an $\mathbb{F}$-algebra. This makes S a subalgebra of $\mathcal{L}(V)$.
  (c) Show that S and $\mathbb{F}$ are isomorphic as $\mathbb{F}$-algebras.

3.4.3. Let S be the set of automorphisms in $\mathcal{L}(V)$ for some vector space V. Show that function composition is a binary operation on S. Is S a subalgebra of $\mathcal{L}(V)$?

3.4.4. Finish the proof of Theorem 3.32.

3.4.5. Show that the inverse of an $\mathbb{F}$-algebra isomorphism is also an $\mathbb{F}$-algebra isomorphism.

3.4.6. Let $\mathcal{A}$ be an algebra with identity ι. Show that the zero mapping is an algebra homomorphism. How can we interpret Theorem 3.32(b) in this context?

3.5. Equivalences

What follows is one of the worst uses of the word "equivalent" in all of mathematics. In this case, it is only applied to matrices. As it is in common usage, familiarity is advisable, but we will not need it for long.

Definition 3.34. Equivalent matrices are A and B in $M_{m,n}(\mathbb{F})$ such that $B = PAQ$ for some invertible P in $M_m(\mathbb{F})$ and Q in $M_n(\mathbb{F})$.

When using the word "equivalent" or "equivalence" as it arises out of Definition 3.34, we put it in italics.

Theorem 3.35. *Equivalence is an equivalence relation on* $M_{m,n}(\mathbb{F})$.

Proof. Taking $P = I_m$ and $Q = I_n$, we have $A = PAQ$ for all A in $M_{m,n}(\mathbb{F})$. This shows that *equivalence* is reflexive.

If A and B belong to $M_{m,n}(\mathbb{F})$ and there are P invertible in $M_m(\mathbb{F})$ and Q invertible in $M_n(\mathbb{F})$ with $B = PAQ$, then $A = P^{-1}BQ^{-1}$. This shows that *equivalence* is symmetric.

If $B = P_1 A Q_1$ and $C = P_2 B Q_2$, for A, B, C in $M_{m,n}(\mathbb{F})$, P_i invertible in $M_m(\mathbb{F})$, and Q_i invertible in $M_n(\mathbb{F})$, then

$$C = P_2(P_1 A Q_1)Q_2 = (P_2 P_1) A (Q_1 Q_2).$$

Since the product of invertible matrices is invertible, this shows that *equivalence* is transitive. $\square$

The following is immediate by Theorem 3.23.

Corollary 3.36. *Let V and W be finite-dimensional vector spaces over $\mathbb{F}$. If L belongs to $\mathcal{L}(V, W)$, then different matrix representations of L are equivalent.*

We can now see the following.

Theorem 3.37. *Every $m \times n$ matrix A is equivalent to a matrix of the form*

$$(3.10) \quad \left[\begin{array}{c|c} I_k & O_{k,n-k} \\ \hline O_{m-k,k} & O_{m-k,n-k} \end{array}\right]$$

where $k = \dim \operatorname{Col} A$ and $O_{r,s}$ is the $r \times s$ zero matrix.

Proof. Let A belong to $M_{m,n}(\mathbb{F})$. Take an ordered basis for $\mathbb{F}^n$, $\mathcal{B} = \{\mathbf{v}_i\}_{i=1}^n$, where $\{\mathbf{v}_i\}_{i=k+1}^n$ is a basis for $\operatorname{Nul} A$. We leave it as an exercise to verify that $\{A\mathbf{v}_i\}_{i=1}^k$ is a basis for $\operatorname{Col} A$. Let $\mathcal{C} = \{A\mathbf{v}_1, \ldots, A\mathbf{v}_k, \mathbf{w}_{k+1}, \ldots, \mathbf{w}_m\}$ be an ordered basis for $\mathbb{F}^m$. The matrix representation $[\![L_A]\!]_{\mathcal{B}}^{\mathcal{C}}$ then has the form given in the statement of the theorem. Since A is the matrix representation of L with respect to the natural bases of $\mathbb{F}^n$ and $\mathbb{F}^m$, A and $[\![L_A]\!]_{\mathcal{B}}^{\mathcal{C}}$ are equivalent. This is enough to prove the theorem. $\square$

Definition 3.38. Matrices A and B in $M_{m,n}(\mathbb{F})$ are **rank equivalent** if $\operatorname{rank} A = \operatorname{rank} B$.

Now we can stop saying matrices are *equivalent* and instead say they are rank equivalent.

The case in which matrices represent a linear operator on $V \cong \mathbb{F}^n$ is of special interest.

Definition 3.39. Matrices A and B in $M_n(\mathbb{F})$ are **similar** provided there is an invertible matrix P in $M_n(\mathbb{F})$ such that $B = PAP^{-1}$.

Proof of the following is nearly identical to that of Theorem 3.35. We leave the details as an exercise.

Theorem 3.40. *Similarity is an equivalence relation on $M_n(\mathbb{F})$.*

When A and B are similar matrices, we write $A \underset{s}{\sim} B$.

Similarity is more discerning than rank equivalence, in the following sense. While similar matrices are rank equivalent, rank equivalent matrices are not necessarily similar. This stands to reason if we think about the roles of P and Q when we consider rank equivalent matrices A and B in $M_n(\mathbb{F})$, with $B = PAQ$. The only requirement for P and Q is that they be invertible and have the correct dimensions. When A and B in $M_n(\mathbb{F})$ are similar, we still have $B = PAQ$, for invertible P and Q, but now $Q = P^{-1}$. This makes similarity more restrictive.

The parallel lives of matrices and linear transformations extend to the equivalence relations we have defined on $M_{m,n}(\mathbb{F})$ and $M_n(\mathbb{F})$.

Definition 3.41. Linear transformations L_1 and L_2 in $\mathcal{L}(V, W)$ are **rank equivalent** when $\operatorname{rank} L_1 = \operatorname{rank} L_2$. Linear operators L_1 and L_2 in $\mathcal{L}(V)$ are **similar** when $L_2 = TL_1T^{-1}$ for some invertible T in $\mathcal{L}(V)$.

3.5. Equivalences

Proof of the following is virtually identical to proofs of the analogous results for matrices.

Theorem 3.42. Let V and W be vector spaces over $\mathbb{F}$. Rank equivalence is an equivalence relation on $\mathcal{L}(V,W)$. Similarity is an equivalence relation on $\mathcal{L}(V)$.

We also use $\tilde{s}$ to indicate similarity of linear operators.

We establish the obvious links between $\mathcal{L}(V,W)$ and $M_{m,n}(\mathbb{F})$ when the underlying vector spaces are finite dimesional. We have already seen one link, which we restate here in a different form to emphasize its role in this part of our narrative.

Lemma 3.43. If $V \cong \mathbb{F}^n$ and $W \cong \mathbb{F}^m$, then L_1 and L_2 in $\mathcal{L}(V,W)$ are rank equivalent if and only if their matrix representations in $M_{m,n}(\mathbb{F})$ are rank equivalent.

Since $[\![\ ]\!]_\mathcal{B} : \mathcal{L}(V) \to M_n(\mathbb{F})$ is an $\mathbb{F}$-algebra isomorphism when $V \cong \mathbb{F}^n$, it is invertible.

Theorem 3.44. If $V \cong \mathbb{F}^n$, then L_1 and L_2 in $\mathcal{L}(V)$ are similar if and only if their matrix representations in $M_n(\mathbb{F})$ are in the same similarity class.

Proof. Let V, L_1, and L_2 be as hypothesized.

Suppose $L_2 = TL_1T^{-1}$ for T invertible in $\mathcal{L}(V)$. Let $\mathcal{B}$ be an ordered basis for V. Since $[\![\ ]\!]_\mathcal{B}$ is an algebra isomorphism, we have
$$[\![L_2]\!]_\mathcal{B} = [\![T]\!]_\mathcal{B}[\![L_1]\!]_\mathcal{B}([\![T]\!]_\mathcal{B})^{-1}.$$
This proves that if L_1 and L_2 are similar, then so are their matrix representations with respect to a given basis for V. Since similarity is an equivalence relation on $M_n(\mathbb{F})$, this is enough to prove the theorem in one direction.

Suppose, conversely, that matrix representations for L_1 and L_2 belong to the same similarity class. Since $[\![L_2]\!]_{\mathcal{B}_1}$ and $[\![L_2]\!]_{\mathcal{B}_2}$ are in the same similarity class when $\mathcal{B}_1$ and $\mathcal{B}_2$ are bases for V, we can assume that V has a basis $\mathcal{B}$ such that $[\![L_1]\!]_\mathcal{B} \tilde{s} [\![L_2]\!]_\mathcal{B}$. Say that
$$[\![L_2]\!]_\mathcal{B} = P[\![L_1]\!]_\mathcal{B} P^{-1}$$
for some invertible P in $M_n(\mathbb{F})$. Since $T \mapsto [\![T]\!]_\mathcal{B}$ is an algebra isomorphism, its inverse $\mathcal{L}(V) \to M_n(\mathbb{F})$ is an isomorphism as well. (See Exercise 3.4.5.) This means there must be invertible T in $\mathcal{L}(V)$ so that $[\![T]\!]_\mathcal{B} = P$. By the properties of algebra isomorphisms, we have
$$[\![L_2]\!]_\mathcal{B} = [\![T]\!]_\mathcal{B}[\![L_1]\!]_\mathcal{B}([\![T]\!]_\mathcal{B})^{-1} = [\![TL_1T^{-1}]\!]_\mathcal{B},$$
implying that $L_2 = TL_1T^{-1}$. It follows that $L_2 \tilde{s} L_1$, which proves the theorem. □

The following summarizes what we know about a vector space automorphism on a finite-dimensional vector space and any associated matrix representation.

Theorem 3.45. Let $V \cong \mathbb{F}^n$ and let L belong to $\mathcal{L}(V)$. Suppose $\mathcal{B}$ is an ordered basis for V. If $A = [\![L]\!]_\mathcal{B}$, then the following statements are logically equivalent.

(a) A is invertible.

(b) L is invertible.

(c) The null space of A is trivial.

(d) The kernel of L is trivial.
(e) The column space of A is $\mathbb{F}^n$.
(f) The columns of A form a basis for $\mathbb{F}^n$.
(g) The range of L is V.
(h) L has a right-inverse, i.e., there is T in $\mathcal{L}(V)$ such that $L \circ T$ is the identity mapping on V.
(i) A has a right-inverse, i.e., there is B in $M_n(\mathbb{F})$ such that $AB = I_n$.
(j) L has a left-inverse, i.e., there is T in $\mathcal{L}(V)$ such that $T \circ L$ is the identity mapping on V.
(k) A has a left-inverse, i.e., there is B in $M_n(\mathbb{F})$ such that $BA = I_n$.

Proof. Let A and L be as hypothesized.

Corollary 2.40, the Rank Theorem, Theorem 3.18, and Corollary 3.21 together imply the equivalences of all of (a)–(g).

The equivalence of (h) and (i) follows from Theorem 3.33, as does the equivalence of (j) and (k). By Corollary 3.12, (i) and (k) are equivalent which means that either one implies (a). Since (a) implies (i), any one statement in the theorem is equivalent to any other. □

Exercises 3.5.

3.5.1. Let A belong to $M_{m,n}(\mathbb{F})$. Let $\mathcal{B} = \{\mathbf{v}_i\}_{i=1}^n$ be an ordered basis for $\mathbb{F}^n$, where $\{\mathbf{v}_i\}_{i=k+1}^n$ is a basis for Nul A. Verify that $\{A\mathbf{v}_i\}_{i=1}^k$ is a basis for Col A.

3.5.2. Verify that $\widetilde{s}$ is an equivalence relation on $M_n(\mathbb{F})$.

3.5.3. Verify that $\widetilde{s}$ is an equivalence relation on $\mathcal{L}(V)$.

3.5.4. Let $V \cong \mathbb{F}^n$ and let L in $\mathcal{L}(V)$ be the scalar mapping, $\mathbf{x} \mapsto c\mathbf{x}$ for some fixed c in $\mathbb{F}$. Show that there is only one element in the similarity class determined by L.

3.5.5. Let $A = \begin{bmatrix} 1 & 1 \\ 1 & 1 \end{bmatrix}$ and $B = \begin{bmatrix} 1 & 1 \\ 2 & 2 \end{bmatrix}$ belong to $M_2(\mathbb{R})$.
   (a) Show that A and B are rank equivalent. Find P and Q so that $B = PAQ$.
   (b) Show that A and B are not similar.

3.5.6. A **scalar matrix** is one of the form cI_n for c in the underlying field.
   (a) What elements in $M_n(\mathbb{F})$ are rank equivalent to a given scalar matrix?
   (b) What elements in $M_n(\mathbb{F})$ are similar to a given scalar matrix?

3.5.7. Let A and B belong to $M_n(\mathbb{F})$. Assume $A \widetilde{s} B$. Recall that we define $A^0 := I_n$.
   (a) Show that $A^2 \widetilde{s} B^2$.
   (b) Show that if A is invertible, then B is invertible and $A^{-1} \widetilde{s} B^{-1}$.
   (c) If A is invertible, write A^{-k} for $(A^{-1})^k$. Show that when A and B are both invertible, $A^k \widetilde{s} B^k$ for any k in $\mathbb{Z}$.

3.5.8. Let A and C belong to $M_n(\mathbb{F})$. Show that if A is invertible, then $AC \widetilde{s} CA$.

Chapter 4

Systems of Linear Equations

When solving a system of linear equations by hand, we usually apply row reduction to the associated augmented matrix. While this is familiar, proof that the reduced row echelon form of a matrix is unique may not be. One of our goals in this chapter is to address that gap.

Given that two matrices are row equivalent, what do we know about the associated linear transformations? How is row equivalence related to *equivalence* and similarity? The work we do here goes towards answering these questions.

Introduction

We start by setting some notation and a few conventions.

An $m \times n$ system of linear equations over a field $\mathbb{F}$ has the form

(4.1)
$$\begin{array}{ccccccccc} a_{11}x_1 & + & a_{12}x_2 & + & \cdots & + & a_{1n}x_n & = & b_1 \\ a_{21}x_1 & + & a_{22}x_2 & + & \cdots & + & a_{2n}x_n & = & b_2 \\ & & & & \vdots & & & & \\ a_{m1}x_1 & + & a_{m2}x_2 & + & \cdots & + & a_{mn}x_n & = & b_m \end{array}$$

where the b_is and a_{ij}s are scalars, $(a_{i1}, \ldots, a_{in})$ is a nonzero vector in $\mathbb{F}^n$ for each $i \in \{1, \ldots, m\}$, and $x_1, \ldots, x_n$ are unknowns in $\mathbb{F}$. The a_{ij}s are the coefficients of the system.

The system in (4.1) is **homogeneous** provided $(b_1, \ldots, b_m) = \mathbf{0}_m$.

The coefficient matrix for (4.1) is $A = [a_{ij}]$, which belongs to $M_{m,n}(\mathbb{F})$. Letting $\mathbf{b} = (b_1, \ldots, b_m)$ in $\mathbb{F}^m$, we write

$$[A \,|\, \mathbf{b}] := \begin{bmatrix} \mathbf{a}_1 & \cdots & \mathbf{a}_n & \mathbf{b} \end{bmatrix}$$

77

to indicate the *augmented matrix* for the system of linear equations in (4.1). Taking $\mathbf{x} = (x_1, \ldots, x_n)$ to be an unknown vector in $\mathbb{F}^n$, we have the matrix form of (4.1),

$$A\mathbf{x} = \mathbf{b}.$$

A *solution* to the system in (4.1) is identified with a vector $\mathbf{u} = (u_1, \ldots, u_n)$ in $\mathbb{F}^n$: Each equation in the system is true if we substitute u_j for x_j. The *solution*, or the *solution set*, for the system is the set of all solutions to the system. A *consistent system of linear equations* is one with a nonempty solution set.

Equivalent systems of linear equations have the same number of equations, the same number of unknowns, and the same solution set. For example,

$$\begin{aligned} x + y &= 3 \\ 2x - y &= 0 \end{aligned}$$

and

$$\begin{aligned} x &= 1 \\ y &= 2 \end{aligned}$$

are equivalent systems of linear equations.

We often use row reduction on $[A \mid \mathbf{b}]$ to solve $A\mathbf{x} = \mathbf{b}$, thus, to solve (4.1). Since the solution to (4.1) is identical to the solution to its associated matrix equation, we toggle freely between a given system of linear equations and its matrix form.

Whenever we write $A\mathbf{x} = \mathbf{b}$ without further explanation, assume that A is in $M_{m,n}(\mathbb{F})$, $\mathbf{b}$ is in $\mathbb{F}^m$, and that $\mathbf{x}$ is a variable that takes values in $\mathbb{F}^n$.

4.1. The Solution Set

The ability to solve small systems of linear equations by hand is essential to pulling back the curtain on the mathematics that underlies systems of linear equations and their solutions. It is also important to recognize the solution to a system of linear equations as part of a vector space. The latter is our project in this section.

We will have occasion to use notation that we defined in Chapter 3. Given A in $M_{m,n}(\mathbb{F})$, $L_A : \mathbb{F}^n \to \mathbb{F}^m$ is the associated linear transformation:

$$L_A(\mathbf{v}) = A\mathbf{v}.$$

Lemma 4.1. *A system of linear equations with matrix equation $A\mathbf{x} = \mathbf{b}$ is consistent if and only if $\mathbf{b}$ belongs to Col A.*

Proof. If S is an $m \times n$ system of linear equations with matrix equation $A\mathbf{x} = \mathbf{b}$, then the solution to S is $L_A^{-1}[\{\mathbf{b}\}]$. The preimage of $\{\mathbf{b}\}$ is nonempty if and only if $\mathbf{b}$ is in the range of L_A, that is, the column space of A. □

Now that we see the solution to a system of linear equations as the preimage of a vector in $\mathbb{F}^m$ under a linear mapping $\mathbb{F}^n \to \mathbb{F}^m$, it is easy to see the solution as a coset of the kernel of that mapping.

Lemma 4.2. *If $A\mathbf{x} = \mathbf{b}$ is a consistent matrix equation over $\mathbb{F}$, its full solution set is $\mathbf{u} + \text{Nul } A$ where $\mathbf{u}$ is one solution to the equation.*

4.1. The Solution Set

Proof. Let A belong to $M_{m,n}(\mathbb{F})$ and let $\mathbf{b}$ belong to $\operatorname{Col} A$. Let $S \subseteq \mathbb{F}^n$ be the full solution set to $A\mathbf{x} = \mathbf{b}$. Fix $\mathbf{u}$ in S.

If $\mathbf{y}$ belongs to S, then
$$A(\mathbf{y} - \mathbf{u}) = A\mathbf{y} - A\mathbf{u} = \mathbf{b} - \mathbf{b} = \mathbf{0}_m,$$
which proves that $\mathbf{y} - \mathbf{u}$ belongs to $\operatorname{Nul} A$. It follows that $\mathbf{y}$ is in $\mathbf{u} + \operatorname{Nul} A$ and, from there, that $S \subseteq \mathbf{u} + \operatorname{Nul} A$.

If $\mathbf{w}$ is arbitrary in $\operatorname{Nul} A$, then
$$A(\mathbf{u} + \mathbf{w}) = A\mathbf{u} + A\mathbf{w} = \mathbf{b} + \mathbf{0}_m = \mathbf{b},$$
which proves that $\mathbf{u} + \mathbf{w}$ belongs to S. It follows that $\mathbf{u} + \operatorname{Nul} A \subseteq S$ and, from there, that $\mathbf{u} + \operatorname{Nul} A = S$. $\square$

Assuming $A\mathbf{x} = \mathbf{b}$ is consistent, Lemma 4.2 implies that its solution is unique if and only if L_A is injective.

Systems of linear equations help us see the exact nature of affine sets in $\mathbb{F}^n$.

Theorem 4.3. A subset of $\mathbb{F}^n$ is an affine set if and only if it is the solution set to a system of linear equations over $\mathbb{F}$.

Proof. If $S \subseteq \mathbb{F}^n$ is an affine set, it has the form $S = \mathbf{b} + W$ where W is a subspace of $\mathbb{F}^n$ and $\mathbf{b}$ is a vector in $\mathbb{F}^n$. Given such a set, let W' be a direct complement to W in $\mathbb{F}^n$. Note that we can assume $\mathbf{b}$ is in W'.

Let A be the standard matrix representation of the projection onto W' along W. For $\mathbf{w}'$ in W' and $\mathbf{w}$ in W, we have
$$A(\mathbf{w}' + \mathbf{w}) = \mathbf{w}'.$$
In particular, $\operatorname{Nul} A = W$ and $A\mathbf{b} = \mathbf{b}$. Lemma 4.2 thus ensures that S is the solution set to $A\mathbf{x} = \mathbf{b}$. This completes the proof in one direction.

Lemma 4.2 gives us the proof in the other direction. $\square$

The first course in linear algebra usually presents a theorem that says a system of linear equations over $\mathbb{R}$ has no solutions, exactly one solution, or infinitely many solutions. The generalization to systems over arbitrary fields is a corollary to Theorem 4.3.

Corollary 4.4. A system of linear equations has no solution, exactly one solution, or at least as many solutions as the cardinality of the underlying field.

Proof. Let $A\mathbf{x} = \mathbf{b}$ be the matrix form of a consistent system of linear equations over $\mathbb{F}$. We have established that its solution set, S, is a coset of $\operatorname{Nul} A$. By Exercise 2.2.2, $\operatorname{Nul} A$ and its cosets all have the same cardinality.

If $\operatorname{Nul} A = \{\mathbf{0}_n\}$, then $|\operatorname{Nul} A| = |S| = 1$. If $\dim \operatorname{Nul} A \geq 1$, then $\operatorname{Nul} A$ contains a 1-dimensional subspace
$$\operatorname{Span}\{\mathbf{w}\} = \{c\mathbf{w} \mid c \in \mathbb{F}\},$$
which has cardinality $|\mathbb{F}|$. It follows that if the solution set to the system of equations is nonempty, either $|S| = 1$ or $|S| \geq |\mathbb{F}|$. $\square$

Each equation in (4.1) is a constraint on our choice of a point in $\mathbb{F}^n$. With each constraint, we expect to lose a degree of freedom on the solution set. Where we have n degrees of freedom in choosing a point in $\mathbb{F}^n$, if the coordinates of the point are to satisfy a single linear equation, we expect to have $n-1$ degrees of freedom. Before proving that this is, in fact, the case, we have a definition.

Definition 4.5. A **hyperspace** is a subspace W of a vector space, V, such that $\dim V/W = 1$. A **hyperplane** is a coset of a hyperspace.

While Definition 4.5 applies in spaces of arbitrary dimension, a hyperplane in an n-dimensional space is an $(n-1)$-plane.

Lemma 4.6. *The solution to one linear equation over $\mathbb{F}$ in n unknowns is a hyperplane in $\mathbb{F}^n$.*

Proof. Let A be the coefficient matrix for a single linear equation in n unknowns over $\mathbb{F}$. Notice that A is a nonzero row vector in $\mathbb{F}_n$. Since A is nonzero and L_A maps $\mathbb{F}^n$ into $\mathbb{F}$, the Rank Theorem for matrices ensures that

$$\operatorname{rank} A = \dim \mathbb{F} = 1.$$

It follows that the nullity of A is $n-1$, thus, that any coset of $\operatorname{Nul} A$ is a hyperplane in $\mathbb{F}^n$. $\square$

Consider the solution to $x + y + z = 1$ in $\mathbb{R}^3$. We can take coordinates $x = r$ and $y = s$ arbitrary in $\mathbb{R}$, which forces $z = 1 - r - s$. The solution set is the affine plane in $\mathbb{R}^3$ given by

$$\{(r, s, 1 - r - s) \,|\, r, s \in \mathbb{R}\} = \{(0, 0, 1) + r(1, 0, -1) + s(0, 1, -1), \,|\, r, s \in \mathbb{R}\}$$
$$= (0, 0, 1) + \operatorname{Span}\{(1, 0, -1), (0, 1, -1)\}.$$

Figure 4.1 shows the part of the plane that intersects the first octant.

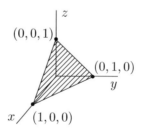

Figure 4.1. The plane in $\mathbb{R}^3$ given by $x + y + z = 1$.

The lemma implies that the solution to an $m \times n$ system of linear equations is the intersection of a collection of m hyperplanes in $\mathbb{F}^n$. Theorem 4.3, then, tells us that the nonempty intersection of a set of hyperplanes in $\mathbb{F}^n$ is itself an affine set. In low dimensions, this means a point, a line, or a plane. Contrast this with the following simple nonlinear system over $\mathbb{R}$:

$$\begin{aligned} x^2 - y &= 0 \\ x - y &= -2. \end{aligned} \tag{4.2}$$

4.1. The Solution Set

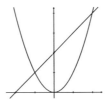

Figure 4.2. The curves described by equations in (4.2).

Each equation represents a curve in $\mathbb{R}^2$. The first equation represents a parabola; the second represents a line. The curves intersect in exactly two points: $(2, 4)$ and $(-1, 1)$. If the curves were hyperplanes, their intersection would contain the entire line determined by those two points.

Exercises 4.1.

4.1.1. The following problem appears on a Babylonian clay tablet possibly dating from as far back as 1700 BCE. We paraphrase from [12].

> One of two fields yields 2/3 *sila* per *sar* and the second yields 1/2 *sila* per *sar*.[1] The first field yields 500 *sila* more than the second. The two fields together total 1800 *sar*. What is the size of each field?

Use simultaneous equations to solve the problem.[2]

4.1.2. Another ancient text, the *Nine Chapters on the Mathematical Art*, compiled during the Han dynasty in China (206 BCE–220 CE), also treated systems of linear equations. The following is again taken from [12].

> There are three classes of grain, of which three bundles of the first, two of the second, and one of the third make 39 measures. Two of the first, three of the second and one of the third make 34 measures. And one of the first, two of the second and three of the third make 26 measures. How many measures of grain are contained in one bundle of each type?

Solve the problem.[3]

4.1.3. Consider the equation $y = x^2$, for (x, y) in $\mathbb{R}^2$. Describe the solution set parametrically and argue that it is 1-dimensional, that is, there is a single degree of freedom in the choice of a point from the solution set.

4.1.4. Consider the equation $z = x^2 + y^2$, for (x, y, z) in $\mathbb{R}^3$. Describe the solution set parametrically and argue that it is 2-dimensional, in the sense of the previous problem.

---

[1] *Sila* is a measure of crop yield, like bushel, and *sar* is a measure of land area, like acre.
[2] The solution on the tablet is found by *false position*, a method whereby a first guess at an answer it used to home in on the correct solution.
[3] The solution in the *Nine Chapters* uses what looks to modern eyes like matrices and back-substitution.

4.1.5. Consider the following system of linear equations over $\mathbb{F}_3$:

$$\begin{aligned} x + y + z &= 1 \\ y &= 2. \end{aligned}$$

(a) Find a parametric description for the full solution set in $\mathbb{F}_3{}^3$.
(b) List the points in the solution set.
(c) Describe the full solution as an affine set in $\mathbb{F}_3{}^3$.

4.1.6. Consider the following system of equations, taking $\mathbf{x} = (x, y, z)$ to be a variable in $\mathbb{R}^3$:

$$\begin{aligned} x + y + z &= 1 \\ 2x + y + z &= 4. \end{aligned}$$

(a) Sketch the region of intersection of the plane given by $2x + y + z = 4$ and the first octant, that is, the region where x, y, z are all positive.
(b) Sketch the intersection of each plane with each of the coordinate planes in $\mathbb{R}^3$.
(c) What equation do you get when you subtract the first equation from the second?
(d) Describe the full solution set parametrically.
(e) Describe the full solution as an affine set in $\mathbb{R}^3$.
(f) Sketch the full solution set, and provide a description of the elements in your sketch.

4.1.7. Describe the full solution to $x_1 + x_3 + x_4 = 1$ as an affine set in $\mathbb{F}^4$. What is its dimension?

4.1.8. Describe the full solution to

$$\begin{aligned} x_1 \phantom{{}+{}} + x_3 + x_4 &= 1 \\ x_2 &= 2 \end{aligned}$$

as an affine set in $\mathbb{F}^4$. What is its dimension?

4.1.9. Let S be the full solution set in $\mathbb{F}_5{}^3$ for the following system:

$$\begin{aligned} x + 2y + 2z &= 1 \\ 2x \phantom{{}+{}} + z &= 2. \end{aligned}$$

(a) Describe S as an affine set.
(b) Find $|S|$.

4.1.10. Show that a system of linear equations is homogeneous if and only if $\mathbf{0}$ is a solution.

4.1.11. Find a 2×2 matrix A over $\mathbb{R}$ and a vector $\mathbf{b}$ in $\mathbb{R}^2$ so that $A\mathbf{x} = \mathbf{0}$ has many solutions and $A\mathbf{x} = \mathbf{b}$ is inconsistent.

4.1.12. Find a matrix A in $M_{m,n}(\mathbb{R})$ and a vector $\mathbf{b}$ in $\mathbb{R}^m$ so that L_A is injective and $A\mathbf{x} = \mathbf{b}$ is inconsistent.

4.1.13. Suppose $A\mathbf{x} = \mathbf{b}$ has a unique solution. What can you say about the solution to $A\mathbf{x} = \mathbf{0}$?

4.1.14. In a 3×2 linear system over $\mathbb{R}$, each equation describes a line in $\mathbb{R}^2$. Sketch the possible configurations of the lines when there is no solution to the system, when there is exactly one solution to the system, and when there are infinitely many solutions to the system.

4.1.15. Sketch the sine curve $y = \sin x$ in $\mathbb{R}^2$ and find a line that intersects the curve in exactly one point, another that intersects the curve in exactly two points, a third line that intersects the curve in exactly three points, and a fourth line that intersects the curve in infinitely many points.

4.1.16. Consider three different systems of linear equations in three unknowns: The first has two equations, the second has three equations, the third has four equations. Describe the objects underlying the equations in each system. Give geometric descriptions for the possible solution sets for each of the three types of system.

4.1.17. Let $W = \mathrm{Span}\{(1,1,1),(-1,1,1)\} \subseteq \mathbb{R}^3$. Find a system of linear equations for which W is the solution set.

4.2. Elementary Matrices

Recall that if P in $M_n(\mathbb{F})$ is invertible, then $\mathbf{v} \mapsto P\mathbf{v}$ is an automorphism on $\mathbb{F}^n$. In particular,
$$P\mathbf{v}_1 = P\mathbf{v}_2 \iff \mathbf{v}_1 = \mathbf{v}_2$$
for any $\mathbf{v}_1$ and $\mathbf{v}_2$ in $\mathbb{F}^n$.

Suppose now that A belongs to $M_{m,n}(\mathbb{F})$. Notice that $A\mathbf{u} = \mathbf{b}$ for some $\mathbf{u}$ in $\mathbb{F}^n$ and $\mathbf{b}$ in $\mathbb{F}^m$, if and only if
$$PA\mathbf{u} = P\mathbf{b}$$
for any invertible P in $M_n(\mathbb{F})$. This is enough to prove the following theorem.

Theorem 4.7. Let S be an $m \times n$ system of linear equations over $\mathbb{F}$ with associated matrix equation $A\mathbf{x} = \mathbf{b}$. Let S' be an $m \times n$ system of linear equations over $\mathbb{F}$ with associated matrix equation $B\mathbf{x} = \mathbf{c}$. The systems S and S' are equivalent if and only if
$$P[A \,|\, \mathbf{b}] = [B \,|\, \mathbf{c}]$$
for some invertible matrix P in $M_n(\mathbb{F})$.

When solving a system of linear equations by hand, we typically apply elementary row operations to the associated augmented matrix in a step-by-step effort to uncover an equivalent system with a more transparent solution set. Our goal now is to put this process on a more secure algebraic footing.

We list the elementary row operations for reference.

Type-1: Switch two rows.

Type-2: Replace one row with a nonzero multiple of the same row.

Type-3: Replace one row with the sum of that row and a nonzero multiple of a second row in the matrix.

Definition 4.8. An **elementary matrix** is the result of applying a single elementary row operation to an identity matrix. An $n \times n$ elementary matrix of **type-k** is one that results from applying a type-k elementary row operation to I_n.

Some examples of elementary matrices over $\mathbb{R}$ follow:

$$E_1 = \begin{bmatrix} 0 & 1 \\ 1 & 0 \end{bmatrix}, \quad E_2 = \begin{bmatrix} 1 & 0 & 0 \\ 0 & -2 & 0 \\ 0 & 0 & 1 \end{bmatrix}, \quad E_3 = \begin{bmatrix} 1 & 0 & 0 & 5 \\ 0 & 1 & 0 & 0 \\ 0 & 0 & 1 & 0 \\ 0 & 0 & 0 & 1 \end{bmatrix}.$$

E_1 is a type-1 matrix, E_2 is a type-2 matrix, and E_3 is a type-3 matrix.

Recall that if P and A are matrices for which PA is defined and $\mathbf{w}_i$ is row i of P, then row i of PA is $\mathbf{w}_i A$. This means that row i of PA is the linear combination of the rows of A that we get using the coordinates of $\mathbf{w}_i$ as weights.

We use $\{\mathbf{e}_1, \ldots, \mathbf{e}_m\}$ to indicate the standard ordered basis for $\mathbb{F}_m$ in the proof of the following.

Lemma 4.9. If A belongs to $M_{m,n}(\mathbb{F})$ and E in $M_m(\mathbb{F})$ is a type-k elementary matrix, then EA is the $m \times n$ matrix that we get by applying the type-k row operation to A that we must apply to I_m to get E.

Proof. Let A and E be as hypothesized. Let $\mathbf{w}_i$ be row i of A and let $\mathbf{e}'_i$ in $\mathbb{F}_m$ be row i of E. When $\mathbf{e}'_i = \mathbf{e}_i$, row i of EA is $\mathbf{w}_i$.

If E is a type-1 elementary matrix, then for some $i \neq j$ in $\{1, \ldots, m\}$, $\mathbf{e}'_i = \mathbf{e}_j$ and $\mathbf{e}'_j = \mathbf{e}_i$. For all other k in $\{1, \ldots, m\}$, $\mathbf{e}'_k = \mathbf{e}_k$. In this case, $\mathbf{e}'_i A = \mathbf{w}_j$, $\mathbf{e}'_j A = \mathbf{w}_i$, and for k different from i or j, $\mathbf{e}'_k A = \mathbf{w}_k$. This shows that the effect of premultiplying A by E is to switch rows i and j.

If E is a type-2 elementary matrix, then $\mathbf{e}'_i = c\mathbf{e}_i$ for some i in $\{1, \ldots, m\}$ and some nonzero scalar, c. For all other k in $\{1, \ldots, m\}$, $\mathbf{e}'_k = \mathbf{e}_k$. In this case, $\mathbf{e}'_i A = c\mathbf{w}_i$, and for $k \neq i$, $\mathbf{e}'_k A = \mathbf{w}_k$. This shows that the effect of premultiplying A by E is to scale row i by c.

If E is a type-3 elementary matrix, then for some $i \neq j$ in $\{1, \ldots, m\}$, $\mathbf{e}'_i = \mathbf{e}_i + c\mathbf{e}_j$, where c is some nonzero scalar. For all other k in $\{1, \ldots, m\}$, $\mathbf{e}'_k = \mathbf{e}_k$. In this case, $\mathbf{e}'_i A = \mathbf{w}_i + c\mathbf{w}_j$, and for $k \neq i$, $\mathbf{e}'_k A = \mathbf{w}_k$. This shows that the effect of premultiplying A by E is the type-3 elementary row operation that changes I_m to E. $\square$

Elementary row operations are reversible. A type-1 operation reverses itself. To reverse a type-2 operation that scales row i by c, scale row i by $1/c$. To reverse a type-3 operation that adds c times row j to row i, apply the type-3 operation that adds $-c$ times row j to row i. These ideas combine with Lemma 4.9 to give us the following.

Theorem 4.10. A type-k elementary matrix is invertible and its inverse is a type-k elementary matrix.

Proof of the next corollary follows Exercise 3.1.7.

4.2. Elementary Matrices

Corollary 4.11. If $E_1, \ldots, E_k$ in $M_n(\mathbb{F})$ are elementary matrices, then $E_1 \cdots E_k$ is invertible with
$$(E_1 \cdots E_k)^{-1} = E_k^{-1} \cdots E_1^{-1}.$$

Theorem 4.7, in tandem with Corollary 4.11, implies that in applying row reduction to the augmented matrix for a system of linear equations, we can be assured that all associated systems of linear equations are equivalent.

When A is in $M_{m,n}(\mathbb{F})$ and P in $M_m(\mathbb{F})$ is the product of elementary matrices, we say that A and PA are **row equivalent**. An easy exercise reveals that row equivalence is an equivalence relation on $M_{m,n}(\mathbb{F})$. We use $A \widetilde{\text{r}} B$ to indicate that A and B are row equivalent. One of our goals going forward is to prove that every invertible matrix is the product of elementary matrices. This will imply that two augmented matrices represent equivalent systems of equations if and only if the matrices are row equivalent.

Exercises 4.2.

4.2.1. Under what conditions is the product of two $m \times m$ elementary matrices commutative?

4.2.2. A type-1 *elementary column operation* on a matrix swaps two columns of the matrix. A type-2 elementary column operation scales a single column of the matrix by a nonzero factor in the underlying field. A type-3 elementary column operation adds a nonzero multiple of one column to a second column.
   (a) Identify an elementary column operation that has the same effect as the elementary row operation that switches rows i and k when applied to I_n.
   (b) Identify an elementary column operation that has the same effect as scaling row i by a factor $c \neq 0$ when applied to I_n.
   (c) Identify an elementary column operation that has the same effect as adding c times row k to row i when applied to I_n.
   (d) What is the effect of postmultiplying an $m \times n$ matrix A by an $n \times n$ elementary matrix E?

4.2.3. Show that row equivalence is an equivalence relation on $M_{m,n}(\mathbb{F})$.

4.2.4. List the matrices in each row equivalence class in $M_2(\mathbb{F}_2)$.

4.2.5. Suppose you apply type-3 elementary row operations to rows 1 and 2 of the following matrix so that you are left with a matrix that has third column $(0, 1, 0)$:
$$A = \begin{bmatrix} 0 & 1 & -1 & 2 \\ 0 & 0 & 1 & 3 \\ 0 & 0 & 2 & 0 \end{bmatrix}.$$
What elementary matrices would you use and in what order would you use them? Does the order matter?

4.2.6. If $A \widetilde{\text{r}} B$, are A and B *equivalent* matrices? If A and B are *equivalent* matrices, are they row equivalent?

4.3. Reduced Row Echelon Form

The key to revealing the full solution set for a linear system is the careful identification of the *row echelon form* of a matrix. The language may be confusing: The row echelon form of a matrix A is not another "version" of A. It is a matrix in the same row equivalence class as A. Our goal in this section is to prove the existence and uniqueness of the row echelon form for any matrix in $M_{m,n}(\mathbb{F})$.

We start with a review of terminology.

The **leading entry** in a nonzero row i of a matrix $A = [a_{ij}]$ is the first nonzero entry in that row, that is, a_{ij} where j is least so that $a_{ij} \neq 0$. When the leading entry in a row is 1, that entry is called a **leading 1**.

A matrix A is in **row echelon form** provided the following conditions hold.

REF1: If row i is a row of zeroes and row k is a nonzero row, then $i > k$.

REF2: If the leading entry in row i is in column j, then for $k > i$, row k is either a row of zeroes or its leading entry is in column $\ell > j$.

If the following two conditions also hold, then A is in **reduced row echelon form**.

REF3: The leading entry in any nonzero row is 1.

REF4: A leading 1 is the only nonzero entry in its column.

Suppose A in $M_{m,n}(\mathbb{F})$ is in reduced row echelon form. When the ij-entry of A is a leading 1, the ij-position of every matrix in the row equivalence class determined by A is a **pivot position**. When column j of A contains a pivot position, column j of every matrix in the row equivalence class determined by A is a **pivot column**. A **nonpivot column** of A is a column without a pivot position.

We will see that the word "pivot" arises both as a noun and as a verb.

It is important to note that a matrix has at most one pivot position per column and at most one pivot position per row. A 3×5 matrix, for example, has at most three pivot positions.

We refer to a *submatrix* in the statement of the next lemma. A matrix A in $M_{m,n}(\mathbb{F})$ has submatrix A' if we obtain A' by excising rows and/or columns from A.

The proof of the following is an exercise.

Lemma 4.12. Let A in $M_{m,n}(\mathbb{F})$ be in reduced row echelon form. Let A' be a submatrix of A.

(a) If we get A' by excising a row from A, then A' is in reduced row echelon form.

(b) If we get A' by excising a nonpivot column from A, then A' is in reduced row echelon form.

(c) If we get A' by excising column n from A, then A' is in reduced row echelon form.

Consider a nonzero matrix $A = [a_{ij}]$. Suppose column j is a pivot column with at least two nonzero entries, one of them in row i. We *pivot* on the ij-entry when we use row i in one or more type-3 elementary row operations to produce a matrix

4.3. Reduced Row Echelon Form

with a zero in the kj-entry for $k \neq i$. We may refer to *pivoting up* when $k < i$, or *pivoting down* when $k > i$. When we apply a sequence of type-3 elementary row operations to produce a matrix with exactly one nonzero entry in a particular column, we may refer to this as *clearing a column*.

Proof of the following existence theorem suggests an algorithm for row reduction.

Theorem 4.13 (Row Reduction Algorithm). *Every matrix is row equivalent to a matrix in reduced row echelon form.*

Proof. Let $A = [a_{ij}]$ belong to $M_{m,n}(\mathbb{F})$. Since the zero matrix is in reduced row echelon form, we may assume A is nonzero.

We prove the theorem by induction on m, the number of rows of the matrix.

Any row vector A in $\mathbb{F}_n$ is already in row echelon form. A type-2 operation puts it into reduced row echelon form. This proves the base case.

Suppose the theorem is true for any matrix with $m \leq k$ rows and say that A has $m = k+1$ rows. Columns of zeroes are unaffected by elementary row operations so we assume the first column of A is nonzero.

Choose $a_{i1} \neq 0$ and apply type-1 and type-2 elementary row operations to A as necessary to produce $B \tilde{\text{r}} A$ with a leading 1 in the 1,1-entry. Pivot down on the 1,1-entry to clear column one. This produces $C \tilde{\text{r}} A$ with the following form:

$$(4.3) \qquad C = \begin{bmatrix} 1 & * & \cdots & * \\ \hline 0 & * & \cdots & * \\ \vdots & \vdots & & \\ 0 & * & \cdots & * \end{bmatrix}$$

where $*$ may be any value in $\mathbb{F}$.

If we excise the first row and the first column of C, then by the induction hypothesis, the resulting $(m-1) \times (n-1)$ submatrix of C is row equivalent to C', a matrix in reduced row echelon form. If we apply to C a sequence of elementary row operations that puts C' into reduced row echelon form, premultiplying by elementary matrices in $M_m(\mathbb{F})$, each with first row $\begin{bmatrix} 1 & 0 & \cdots & 0 \end{bmatrix}$ in $\mathbb{F}_m$, we get $D \tilde{\text{r}} A$, with the following form:

$$D = \begin{bmatrix} 1 & * & \cdots & * \\ \hline 0 & & & \\ \vdots & & C' & \\ 0 & & & \end{bmatrix}.$$

In particular, D is in row echelon form. A finite sequence of type-3 and type-2 row operations puts D into reduced row echelon form.

The theorem follows by the principle of induction. $\square$

The Row Reduction Algorithm allows us to identify a set of well-defined pivot positions, thus pivot columns, in a matrix. It does not address uniqueness. To prove that there is not more than one matrix in a given row equivalence class in $M_{m,n}(\mathbb{F})$ in reduced row echelon form, we associate pivot columns and nonpivot columns of a matrix to variables in an associated system of linear equations.

The following lemma will help us in these efforts. Its proof is an exercise.

Lemma 4.14. Let $A\mathbf{x} = \mathbf{b}$ be the matrix equation associated to an $m \times n$ system of linear equations. $A\mathbf{x} = \mathbf{b}$ is inconsistent if and only if column $n + 1$ of $[A \,|\, \mathbf{b}]$ is a pivot column.

The proof of our next theorem is based on [23].

Theorem 4.15. The reduced row echelon form for a matrix is unique.

Proof. Let A belong to $M_{m,n}(\mathbb{F})$. We prove the theorem by induction on n, the number of columns of A.

If A has one column, then the reduced row echelon form of A is either $(0, \ldots, 0) = \mathbf{0}_m$ or $(1, 0, \ldots, 0)$. Since a zero column is unchanged by row reduction, the two forms cannot be row equivalent. This proves the theorem in the base case.

Assume the theorem is true if $n \leq k$ and suppose A has $n = k + 1$ columns. Let E_i in $M_m(\mathbb{F})$ be elementary matrices so that $E_t \cdots E_1 A = B$ is in reduced row echelon form. Let A' be the $m \times (n-1)$ submatrix of A that we get by excising column n from A. If B' is the submatrix of B that we get by excising column n from B, then $E_t \cdots E_1 A' = B'$. Lemma 4.12 establishes that B' is in reduced row echelon form.

Supppose $C \stackrel{\sim}{r} A$ is different from B and is in reduced row echelon form. By assumption, B and C differ only in column n. Let $\mathbf{c}_n = (c_1, \ldots, c_m)$ be column n of C, and let $\mathbf{b}_n = (b_1, \ldots, b_m)$ be column n of B. We have $C = [B' \,|\, \mathbf{c}_n]$ and $B = [B' \,|\, \mathbf{b}_n]$.

Let $\mathbf{u} = (u_1, \ldots, u_n)$ be in Nul A. As reduced row echelon forms of A, B and C both satisfy
$$B\mathbf{u} = C\mathbf{u} = \mathbf{0}_m$$
so
$$(B - C)\mathbf{u} = [O_{m,n-1} \,|\, \mathbf{b}_n - \mathbf{c}_n](u_1, \ldots, u_n) = ((b_1 - c_1)u_n, \ldots, (b_m - c_m)u_n) = \mathbf{0}_m.$$
Since $\mathbf{b}_n - \mathbf{c}_n$ is nonzero, there is i so that $b_i \neq c_i$. It follows that $u_n = 0$. Since it is not associated to a free variable, column n for all of A, B, and C is a pivot column.

Consider, though, that if the nth columns of B and C are pivot columns, the leading 1s in those columns can only appear in row ℓ, where ℓ is least so that row ℓ of B' is zero. It follows that the leading 1 in row ℓ of both B and C is in the ℓn-entry. This means that
$$\mathbf{b}_n = \mathbf{c}_n = \mathbf{e}_\ell,$$
the ℓth element in the standard ordered basis for $\mathbb{F}^n$, contradicting our assumption that $\mathbf{b}_n \neq \mathbf{c}_n$. We conclude that $B = C$, thus, that the reduced row echelon form of A is unique. □

We have now proved the following theorem.

4.4. Row Equivalence

Theorem 4.16. Two matrices in $M_{m,n}(\mathbb{F})$ are row equivalent if and only if they have the same reduced row echelon form.

Exercises 4.3.

4.3.1. Write down all reduced row echelon forms for a 2×3 matrix over $\mathbb{F}$. Use $*$ to indicate an entry that can have any value in $\mathbb{F}$.

4.3.2. Prove Lemma 4.12.

4.3.3. Recall that the *transpose* of a matrix $A = [a_{ij}]$ is $A^T = [b_{ij}]$ where $b_{ij} = a_{ji}$. In other words, A^T is what we get by switching the rows and columns of A. Find a 3×2 reduced row echelon form matrix A so that A^T is not in reduced row echelon form.

4.3.4. Find all 2×3 matrices over $\mathbb{F}_3$ that are row equivalent to $\begin{bmatrix} 1 & 1 & 1 \\ 0 & 0 & 0 \end{bmatrix}$.

4.3.5. Prove Lemma 4.14.

4.3.6. Suppose A is an $m \times n$ matrix with a pivot in every row. What can you say about possible solutions to a system associated to a matrix equation $A\mathbf{x} = \mathbf{b}$?

4.3.7. Suppose A is an $m \times n$ matrix with a pivot in every column. What can you say about possible solutions to a system associated to a matrix equation $A\mathbf{x} = \mathbf{b}$?

4.3.8. Let A be in $M_{m,n}(\mathbb{F})$, $\mathbf{b}$ be in $\mathbb{F}^m$, and let $\mathbf{x}$ be an unknown vector in $\mathbb{F}^n$.
  (a) If A has the maximum number of pivot positions possible, what can you say about $A\mathbf{x} = \mathbf{b}$?
  (b) If A does not have the maximum number of pivot positions possible, what can you say about $A\mathbf{x} = \mathbf{0}$?
  (c) If A does not have the maximum number of pivot positions possible, what can you say about $A\mathbf{x} = \mathbf{b}$, assuming $\mathbf{b} \neq \mathbf{0}$?

4.3.9. Restate the Row Reduction Algorithm explicitly as an existence theorem.

4.3.10. Let $A = \begin{bmatrix} 1 & 2 & -3 & 1 & 2 \\ 2 & 4 & -4 & 6 & 10 \\ 3 & 6 & -6 & 9 & 12 \end{bmatrix}$ belong to $M_{3,5}(\mathbb{R})$.
  (a) Find the reduced row echelon form for A.
  (b) Solve $A\mathbf{x} = \mathbf{b}$, where $\mathbf{b} = (-1, 2, 3)$.
  (c) Solve $A\mathbf{x} = \mathbf{0}$.
  (d) What can you say about the full solution set to $A\mathbf{x} = \mathbf{b}$ when $\mathbf{b}$ is arbitrary in $\mathbb{R}^3$?

4.4. Row Equivalence

Consider $A = \begin{bmatrix} 1 & 2 \\ 1 & 2 \end{bmatrix}$ and $B = \begin{bmatrix} 1 & 2 \\ 0 & 0 \end{bmatrix}$. Since it only takes one type-3 elementary row operation to get from A to B, we see that $A \widetilde{r} B$ and that rank $A =$ rank $B = 1$. Notice, though, that $\operatorname{Col} A = \operatorname{Span}\{(1,1)\}$ while $\operatorname{Col} B = \operatorname{Span}\{(1,0)\}$. Row equivalent matrices have the same rank, but typically they have different column spaces. They do not, however, have different null spaces.

Lemma 4.17. Row equivalent matrices have identical null spaces.

Proof. Let A belong to $M_{m,n}(\mathbb{F})$ and let P the product of $m \times m$ elementary matrices over $\mathbb{F}$. Since P is invertible, $A\mathbf{u} = \mathbf{0}$ if and only if
$$PA\mathbf{u} = P\mathbf{0} = \mathbf{0}.$$
This is enough to prove the lemma. $\square$

The null space of a matrix is the set of vectors with coordinates that list the coefficients in a dependence relation on the columns of the matrix. In particular, if $A = \begin{bmatrix} \mathbf{a}_1 & \cdots & \mathbf{a}_n \end{bmatrix}$ belongs to $M_{m,n}(\mathbb{F})$, then
$$u_1 \mathbf{a}_1 + \cdots + u_n \mathbf{a}_n = \mathbf{0}_m$$
if and only if $\mathbf{u} = (u_1, \ldots, u_n)$ is in $\operatorname{Nul} A$.

This is what we mean by the following important theorem.

Theorem 4.18. Row equivalence preserves dependence relations on the columns of a matrix.

Example 4.19. Consider $A = \begin{bmatrix} 1 & 2 & -1 & 0 \\ -1 & 1 & 2 & 1 \\ -2 & 0 & 3 & -1 \end{bmatrix}$ over $\mathbb{R}$. The reduced row echelon form for A is
$$B = \begin{bmatrix} 1 & 0 & 0 & 11 \\ 0 & 1 & 0 & -2 \\ 0 & 0 & 1 & 7 \end{bmatrix}.$$
The following dependence relation on the columns of B is immediately evident:
$$(11, -2, 7) = 11(1, 0, 0) - 2(0, 1, 0) + 7(0, 0, 1).$$
A little bit of arithmetic verifies that
$$(0, 1, -1) = 11(1, -1, -2) - 2(2, 1, 0) + 7(0, 1, -1).$$
The first three columns of B form a linearly independent set in $\mathbb{R}^3$, so the first three columns of A form a linearly independent set in $\mathbb{R}^3$. The fourth column in both A and B is a linear combination of the first three columns with respective weights $11, -2, 7$.

When A in $M_{m,n}(\mathbb{F})$ is in reduced row echelon form, its pivot columns comprise a subset of the standard ordered basis for $\mathbb{F}^m$. Since dependence among the columns is preserved in a row equivalence class, the pivot columns in any matrix $B \stackrel{\sim}{r} A$ must also form a linearly independent set in $\mathbb{F}^m$. Evidently, each nonpivot column of A is in the span of the pivot columns of A. Since this is a statement about dependence among columns, it holds for B as well. In particular, any nonpivot column of B is in the span of the pivot columns of B. This is enough to prove the following.

Theorem 4.20. The pivot columns in a matrix form a basis for its column space.

The corollary is immediate.

Corollary 4.21. The rank of a matrix is the number of pivot positions it has.

4.4. Row Equivalence

The following terminology gives us the vocabulary to connect the variables in a system of linear equations and the columns of an associated coefficient matrix.

Definition 4.22. A **basic variable** for a system of linear equations is a variable associated to a pivot column of its coefficient matrix. A **free variable** for a system of linear equations is a variable associated to a nonpivot column of its coefficient matrix.

Theorem 4.23. The dimension of the solution set for a consistent system of linear equations is the number of free variables associated to the system.

Proof. We have already established that the dimension of the solution set for a consistent system of linear equations given by $A\mathbf{x} = \mathbf{b}$ is $\dim \operatorname{Nul} A$. Since Theorem 4.20 tells us that $\dim \operatorname{Col} A$ is the number of pivot columns in A, the Rank Theorem for matrices tells us that $\dim \operatorname{Nul} A$ is the number of nonpivot columns in A, that is, the number of free variables in any associated consistent system of linear equations. $\square$

Example 4.24. Consider the following system of equations over $\mathbb{R}$:

$$\begin{aligned} x_1 + x_2 - 2x_3 + 4x_4 &= 5 \\ 2x_1 + 2x_2 - 3x_3 + x_4 &= 3 \\ 3x_1 + 3x_2 - 4x_3 - 2x_4 &= 1. \end{aligned} \quad (4.4)$$

Via row reduction, we find that the augmented matrix $[A \,|\, \mathbf{b}]$ is row equivalent to

$$\begin{bmatrix} 1 & 1 & 0 & -10 & | & -9 \\ 0 & 0 & 1 & -7 & | & -7 \\ 0 & 0 & 0 & 0 & | & 0 \end{bmatrix} = B. \quad (4.5)$$

Since the last column of B is not a pivot column, the system in (4.4) is consistent, with basic variables x_1 and x_3 and free variables x_2 and x_4. We can describe the solution set for (4.4) by

$$\{(9 - r + 10s, r, -7 + 7s, s) \,|\, r, s \in \mathbb{R}\} \subseteq \mathbb{R}^3.$$

Taking $W = \operatorname{Span}\{(-1, 1, 0, 0), (10, 0, 7, 1)\}$, we can write the solution set as an affine plane in $\mathbb{R}^4$,

$$(9, 0, -7, 0) + W.$$

The next theorem realizes one of our goals.

Theorem 4.25. *Every invertible matrix is a product of elementary matrices.*

Proof. If A in $M_n(\mathbb{F})$ is an invertible matrix, then L_A is an automorphism on $\mathbb{F}^n$. Since $\dim \operatorname{Nul} A = 0$, every column of A is a pivot column so the reduced row echelon form of A is I_n. Since $PA = I_n$ only if $P = A^{-1}$, A^{-1} must be the product of elementary matrices. By Theorem 4.10 and Corollary 4.11, A must be the product of elementary matrices, as well. $\square$

The **row space** of A in $M_{m,n}(\mathbb{F})$ is the subspace of $\mathbb{F}_n$ spanned by its rows. We denote the row space of A by $\operatorname{Row} A$. To see that the space spanned by the rows of a matrix is fixed for a given row similarity class, we need only look at an equation of the form $B = PA$, where A and B belong to $M_{m,n}(\mathbb{F})$ and P in $M_m(\mathbb{F})$

is invertible. Any element of the form $\mathbf{p}A$ for $\mathbf{p}$ belonging to $\mathbb{F}_m$ is in the span of the rows of A. It follows that Row $B \subseteq$ Row A. Since P is invertible, we also have $P^{-1}B = A$, so Row $A \subseteq$ Row B. This is enough to prove the following.

Theorem 4.26. Row equivalent matrices have identical row spaces.

When we want a basis for the column space of a matrix A, we apply row reduction to identify the pivot columns of A. We must then go back to the original matrix, A, because the pivot columns of A — not the pivot columns of its reduced row echelon form — comprise a basis for Col A. Finding a basis for Row A requires fewer steps.

Lemma 4.27. The nonzero rows of the reduced row echelon form for a matrix A comprise a basis for Row A.

Proof. Suppose A is in $M_{m,n}(\mathbb{F})$ and that B is the reduced row echelon form of A. Evidently, the nonzero rows of B span Row $A =$ Row B. We just have to argue that they form a linearly independent set.

Say that $\mathbf{w}_1, \ldots, \mathbf{w}_k$ are the nonzero rows of B. Suppose
$$c_1\mathbf{w}_1 + \cdots + c_k\mathbf{w}_k = \mathbf{0}_n$$
for scalars c_i. If the leading 1 in $\mathbf{w}_i$ is in column j, there can be no nonzero entry in column j among the rest of the $\mathbf{w}_\ell$s. It follows that $c_i = 0$ for $i = 1, \ldots, k$, thus, that the nonzero rows of B are linearly independent. $\square$

Theorem 4.28. The rank of a matrix is the dimension of both its row space and its column space.

Proof. If A in $M_{m,n}(\mathbb{F})$ has rank k, then A has k pivot columns. The reduced row echelon form for A has exactly one nonzero row per pivot, so it has k nonzero rows. By Lemma 4.27, this is enough to prove the theorem. $\square$

Exercises 4.4.

4.4.1. We say that A in $M_{m,n}(\mathbb{F})$ has *full rank* if its rank is $k = \min\{m, n\}$.
  (a) List all the full rank 2×3 reduced row echelon form matrices over $\mathbb{F}_2$.
  (b) Repeat the previous exercise, this time over $\mathbb{R}$ using $*$ to indicate a position that can be filled with any real number.
  (c) Repeat the previous two exercises for full rank 3×2 matrices in reduced row echelon form.

4.4.2. Let A in $M_{m,n}(\mathbb{F})$ have full rank. What can we say about Ker L_A and $L_A[\mathbb{F}^n]$?

4.4.3. How many row equivalence classes in $M_{2,3}(\mathbb{R})$ contain rank 2 matrices?

4.4.4. Define $L : \mathbb{R}[t]_2 \to M_2(\mathbb{R})$ by $L(p(t)) = \begin{bmatrix} p(0) & p(1) \\ p(-1) & p'(0) \end{bmatrix}$.
  (a) Show that L is linear.
  (b) Find a basis for Ker L.
  (c) Find a basis for $L[\mathbb{R}[t]_2]$.
  (d) Find A, the matrix representation of L with respect to the standard ordered basis for $\mathbb{R}[t]_2$ and the standard ordered basis for $M_2(\mathbb{R})$.

(e) Find rank A directly, that is, without using rank $A = \dim L[V]$.
(f) Find bases for Col A and Nul A.
(g) Compare your basis for Ker L and your basis for Nul A. How can you get from one to the other?
(h) Compare your basis for $L[\mathbb{R}[t]_2]$ and Col A. How can you get from one to the other?

4.5. An Early Use of the Determinant

Students learn to hand calculate determinants for 2×2 and 3×3 matrices in precalculus courses but may wonder what a determinant actually measures and how it came to be an object worthy of attention. In this section, we consider the work of Gabriel Cramer (1704–1752), who described and applied determinants to find a formula for the solution of a so-called nonsingular $n \times n$ system of linear equations over $\mathbb{R}$. There are other definitions of the determinant that we look at in later chapters and Cramer's work was not the first known reference to determinants. It was an important influence, though, and it suggests some historical context for what can seem to be a puzzling object.

A determinant, in mathematics, is a number associated to a square matrix. A determinant, in English, is a deciding factor for something: The cost of feeding a dairy cow is a determinant in the price of butter. The determinant of an $n \times n$ system of linear equations is the deciding factor in whether the system has a uniquely determined — that is, well-defined — solution. A system with nonzero determinant has a solution set containing a single vector. Its solution is thus uniquely determined. A system with zero determinant has a solution set that is empty or that contains more than one vector. In either case, its solution is not uniquely determined.

Another curious technical word arises here. An $n \times n$ system of linear equations that has a unique solution may be called **nonsingular**. The coefficient matrix of a nonsingular system of linear equations is also said to be nonsingular. When applied to a matrix, nonsingular is a synonym for invertible. When applied to a system of linear equations, nonsingular means having a single solution. On the face of it, this may seem peculiar, less so if we dig a little deeper.

Singular means rare, or unexpected: The Big Bang was a singular event. Usage of the word nonsingular in the context of $n \times n$ systems of linear equations suggests that running across an $n \times n$ system of linear equations with no solutions, or more than one solution, would be a singular event. In some regard, this is not unreasonable.

Consider a 2×2 system of linear equations over $\mathbb{R}$. Each equation in the system represents a line in the xy-plane. The lines intersect in a unique point if and only if the system has exactly one solution. This is the nonsingular case. When the lines do not intersect, they are parallel, and the solution set is empty. When the two equations represent the same line — $x + y = 1$ and $2x + 2y = 2$, for instance — each point on the line is a solution, so the solution set is infinite. In the latter two settings, the lines have the same slope. This is the singular case.

Choosing a 2×2 system over $\mathbb{R}$ at random is not identical to, but may be associated to, choosing one line in the xy-plane, putting it back into the plane, and then choosing a line again. If we do this, what is the probability of coming up with two lines that have the same slope? The slope of a line in the xy-plane is an element in $\mathbb{R} \cup \{\infty\}$. If we view our experiment as choosing an element in $\mathbb{R} \cup \{\infty\}$ at random, putting that element back, then choosing again at random, the probability of getting the same element twice is, indeed, zero. In this sense, the nonsingular case really does represent the ordinary run of things. Getting a 2×2 system of linear equations over $\mathbb{R}$ with no solution or infinitely many solutions would be a rare event, according to this analysis.

Our experiences with methods of solving systems of linear equations by hand — elimination and back-substitution, or row reduction — suggest that anyone dealing extensively with small nonsingular systems would sooner or later arrive at a formula for the solution. Cramer described a formula that works for any $n \times n$ system over $\mathbb{R}$. In doing so, he defined a measurement that reveals the determinacy of a system, that is, whether the solution to the system is uniquely determined. This is what we call the determinant. We would like to consider Cramer's description of the determinant and its role in his formula.

We start with the lowest rank case, that of a 2×2 system of linear equations over $\mathbb{R}$.

Lemma 4.29. Consider a 2×2 system of linear equations given by

(4.6)
$$\begin{array}{rcl} a_{11}x + a_{12}y & = & b_1 \\ a_{21}x + a_{22}y & = & b_2 \end{array}$$

where a_{ij}, b_i are all in $\mathbb{R}$ and x and y are unknowns. The system is nonsingular if and only if

(4.7)
$$a_{11}a_{22} - a_{12}a_{21} \neq 0.$$

Proof. Assume (4.6) is nonsingular so that the lines represented by the two equations have different slopes.

Suppose one of the lines is vertical so that the coefficient of y in the associated equation is zero. Say $a_{12} = 0$ so that $a_{12}a_{21} = 0$. Since the lines belong to different parallel classes, $a_{22} \neq 0$. Since the equations are linear, $a_{11} \neq 0$ so $a_{11}a_{22} \neq 0$. We conclude that if the system is nonsingular and one of the lines is vertical, (4.7) is true.

If the system is nonsingular and neither line is vertical, then the first line has slope $-a_{11}/a_{12}$ and the second has slope $-a_{21}/a_{22}$. Since the lines belong to different parallel classes,

(4.8)
$$\frac{a_{11}}{a_{12}} \neq \frac{a_{21}}{a_{22}}.$$

Rearranging terms, we conclude that (4.7) holds in this case, as well.

Assume now that (4.7) holds. We will verify that the lines have different slopes, thus, that they intersect in a single point. This will be enough to show that the system in (4.6) is nonsingular.

4.5. An Early Use of the Determinant

If one of the equations in (4.6) represents a vertical line, then say, for the sake of variety, that it is the second line, so that $a_{22} = 0$. By (4.7), $a_{12}a_{21} \neq 0$, so the first line cannot be vertical. If neither line is vertical, then $a_{12}a_{22} \neq 0$, so we can divide (4.7) through by $a_{12}a_{22}$ to establish that (4.8) holds, that is, to show that the lines have different slopes. □

Lemma 4.30. The solution to a nonsingular 2×2 system of linear equations as given in (4.6) is

$$x = \frac{b_1 a_{22} - b_2 a_{12}}{a_{11}a_{22} - a_{12}a_{21}}, \quad y = \frac{b_2 a_{11} - b_1 a_{21}}{a_{11}a_{22} - a_{12}a_{21}}.$$

Proof. Suppose that the system given in (4.6) is nonsingular and that neither of the lines represented is vertical. We use the first equation in (4.6) to solve for y in terms of x:

(4.9) $$y = \frac{b_1}{a_{12}} - \frac{a_{11}}{a_{12}}x.$$

Substituting into the second equation, we get

$$a_{21}x + \frac{b_1 a_{22}}{a_{12}} - \frac{a_{11}a_{22}}{a_{12}}x = b_2.$$

A little bit of high school alegbra applies from here to show that both x and y are as given in the statement of the lemma.

Next suppose that the system is nonsingular but that $a_{12} = 0$. Nonsingularity guarantees that $a_{22} \neq 0$ and since $a_{11}a_{22} - a_{12}a_{21} \neq 0$, a_{11} must also be nonzero.

The first equation in (4.6) gives us $x = b_1/a_{11}$. Substituting into the second equation we get

$$\frac{b_1 a_{21}}{a_{11}} + a_{22}y = b_2.$$

Proof of the result requires just a few more steps that we leave to the reader. □

The statement of Lemma 4.30 is *Cramer's Rule* for a 2×2 system of linear equations over $\mathbb{R}$.

The calculations required to solve arbitrary $n \times n$ systems via eliminating variables and back-substitution would have been exceptionally onerous in the 18th century. Even the sketch of a proof of Lemma 4.30 is enough to suggest that the calculations are best done behind closed doors in any century. When he wrote about the problem, Cramer relegated the details of his calculations to an appendix [19].

We define the **determinant** of the system of linear equations in (4.6) to be

$$a_{11}a_{22} - a_{12}a_{21}.$$

While the determinant for a nonsingular 2×2 system appears in the denominators of the expressions for the variable solutions in Lemma 4.30, the numerators of those expressions are also determinants. Consider

$$\begin{aligned} b_1 x + a_{12} y &= * \\ b_2 x + a_{22} y &= * \end{aligned}$$

where $*$ is any element in $\mathbb{R}$. The determinant for this system is the numerator for x in Cramer's Rule. The determinant for the following system is the numerator for y:

$$\begin{aligned} a_{11}x + b_1 y &= * \\ a_{21}x + b_2 y &= *. \end{aligned}$$

As in the 2×2 case, solutions for the variables in a nonsingular $n \times n$ system of linear equations are ratios of determinants, with the determinant of the given system always appearing in the denominator. Here is Cramer's description[4] of that denominator. This is from [19]. Note that the word *term* is synonymous with *summand*.

> The number of equations and unknowns being n, we will find the value of each unknown by forming n fractions of which the common denominator has as many terms as there are different arrangements of n things. Each term in the denominator is composed of the coefficients $a_{1\_}, a_{2\_}, a_{3\_}, \ldots$, always written in the same order, but to which we distribute, in the second index, the first n positive integers arranged in all possible ways. Thus, when we have three unknowns, the denominator has $1 \times 2 \times 3 = 6$ terms, composed of the three coefficients $a_{1\_}, a_{2\_}, a_{3\_}$, which receive successively the following sequences of second indices: $123, 132, 213, 231, 312, 321$. We give to these terms $+$ and $-$ signs, according to the following rule: When a second index is followed in the same term, mediately or immediately, by a second index smaller than it, I will call that a derangement. We count, for each term, the number of derangements: If it is even or zero, the term will have the $+$ sign; if it is odd, the term will have the $-$ sign. For example in the term $a_{11}a_{22}a_{33}$, there is no derangement: This term will have therefore the $+$ sign. The term $a_{13}a_{21}a_{32}$ also has the $+$ sign, since it has two derangements: 3 before 1 and 3 before 2. But the term $a_{13}a_{22}a_{31}$, which has three derangements — 3 before 2, 3 before 1, 2 before 1 — will have the $-$ sign.

We can use Cramer's ideas to define the determinant of an $n \times n$ system of equations over an arbitrary field. (The only adjustment we make is to view 0 as an even number.) Recall that $n!$ is the number of different arrangements of n items.

Definition 4.31. The **determinant** of an $n \times n$ system of linear equations with the form

$$\begin{aligned} a_{11}x_1 + \cdots + a_{1n}x_n &= b_1 \\ a_{21}x_1 + \cdots + a_{2n}x_n &= b_2 \\ &\vdots \\ a_{n1}x_1 + \cdots + a_{nn}x_n &= b_n \end{aligned}$$

is a sum of $n!$ terms. Each term is a signed product of the form $a_{1\_}a_{2\_}a_{3\_}\ldots a_{n\_}$, to which we distribute, in the second index, the first n positive integers arranged in all possible ways. The sign of a term is given as follows: If the sequence of

---

[4] We paraphrase and use some modern notation.

4.5. An Early Use of the Determinant

second indices has an even number of derangements, the sign is $+$. If the sequence of second indices has an odd number of derangements, the sign is $-$.

The determinant of an $n \times n$ matrix A, det A, is the determinant of any system of linear equations with coefficient matrix equal to A.

We can see the following as a corollary to Cramer's work.

Theorem 4.32. An $n \times n$ system of linear equations is nonsingular if and only if its determinant is nonzero.

While Cramer's Rule is an unlikely approach to solving a system of linear equations, his work gives us a look into the trajectory of the idea of the determinant from the 18th century. In Chapter 6, we will see the determinant from a point of view that developed more recently.

Exercises 4.5.

4.5.1. Find all the vertical lines in $\mathbb{F}_5^2$. Make both a sketch and a list of the points for each line.

4.5.2. How many vertical lines do you expect to find in $\mathbb{F}_p^2$, for p any prime?

4.5.3. Find all the lines with slope 1 in $\mathbb{F}_5^2$. Make both a sketch and a list of the points for each line.

4.5.4. How many lines with a given slope do you expect to find in $\mathbb{F}_p^2$, for p any prime?

4.5.5. How many parallel classes of lines are there in $\mathbb{F}_5^2$?

4.5.6. How many parallel classes of lines are there in $\mathbb{F}_p^2$, for p any prime?

4.5.7. Suppose we choose a line in $\mathbb{F}_p^2$, put it back, then choose a second line at random. What is the probability that the two lines come from the same parallel class? What is the probability that the second line is the same as the first line?

4.5.8. How rare is the singular case for a 2×2 system over $\mathbb{F}_3$?

4.5.9. Finish the calculations to verify the result stated in Lemma 4.30.

4.5.10. Consider the following system over $\mathbb{R}$:

$$(4.10) \qquad \begin{aligned} 2x - 3y &= 1 \\ x + 2y &= -1. \end{aligned}$$

(a) Solve the system using elimination and back-substitution.
(b) Verify that the solution is the same as that given by Cramer's Rule.
(c) Sketch the lines represented by the equations in the original system.

4.5.11. Consider the system in (4.10) over $\mathbb{F}_5$.
(a) Rewrite the system without using minus signs.
(b) Solve the system using elimination and back-substitution.
(c) Verify that Cramer's Rule gives the same solution.
(d) Sketch the lines in $\mathbb{F}_5^2$ given by equations in the original system.

4.5.12. Students often learn mnemonic devices for calculating determinants of small matrices. One trick for calculating the determinant of a 3×3 matrix is captured in Figure 4.3. Apply Definition 4.31 directly to find the

determinant of a 3 × 3 system of linear equations with coefficient matrix
$$A = \begin{bmatrix} a_{11} & a_{12} & a_{13} \\ a_{21} & a_{22} & a_{23} \\ a_{31} & a_{32} & a_{33} \end{bmatrix}.$$
Compare this to what you get by applying the mnemonic in Figure 4.3.

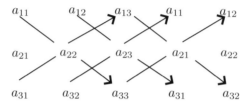

Figure 4.3. Add products along the arrows pointing down; subtract products along the arrows pointing up.

4.5.13. How many terms does the determinant of a 4 × 4 matrix have?

4.5.14. To use Cramer's Rule on a 9 × 9 system of equations, we need the determinant of the system, as well as nine other determinants, as described above. How many multiplications does this require? (Count $a_{11}a_{22}\cdots a_{99}$ as one multiplication.)

4.6. LU-Factorization

LU-factorization is a technique that can be applied to solving systems of linear equations, inverting matrices, and calculating determinants. In this section, we consider the ideas that support LU-factorization and we look at a few low-rank examples.

The optimal setting for solving a system of linear equations is one in which we can get the solution either by back-substitution or by forward-substitution. While this does not make an enormous difference for small systems of equations that we solve by hand, the computational savings this setting affords is magnified considerably in large systems of equations solved using machines. This is the driving idea behind LU-factorization. LU-factorization can be applied anytime we have a coefficient matrix that can be put into row echelon form using only downward pivoting. Some terminology helps facilitate our discussion.

A matrix $A = [a_{ij}]$ in $M_n(\mathbb{F})$ is **lower triangular** provided $a_{ij} = 0$ for $i < j$. When $a_{ij} = 0$ for $i > j$, A is **upper triangular**. The matrix A is a **diagonal matrix** if it is both upper and lower triangular: $a_{ij} = 0$ if $i \neq j$.

The following matrices are, respectively, lower triangular, upper triangular, and diagonal:
$$\begin{bmatrix} 1 & 0 & 0 \\ 2 & 4 & 0 \\ 1 & 2 & 3 \end{bmatrix}, \quad \begin{bmatrix} 1 & 2 & 3 \\ 0 & 1 & 4 \\ 0 & 0 & 3 \end{bmatrix}, \quad \begin{bmatrix} 2 & 0 & 0 \\ 0 & 5 & 0 \\ 0 & 0 & 1 \end{bmatrix}.$$

4.6. LU-Factorization

When the coefficient matrix for a system of linear equations is upper triangular, we can solve the system using back-substitution. When the coefficient matrix for a system of linear equations is lower triangular, we can solve the system using forward-substitution.

Example 4.33. Let $A = \begin{bmatrix} 1 & 2 & 3 \\ 0 & 1 & 4 \\ 0 & 0 & 3 \end{bmatrix}$ and let $\mathbf{b} = (-1, 1, 2)$. We solve $A\mathbf{x} = \mathbf{b}$ for $\mathbf{x} = (x, y, z)$. The last equation in the system of linear equations associated to $A\mathbf{x} = \mathbf{b}$ is $3z = 2$ so $z = 2/3$. Back-substituting into $y + 4z = 1$, we get $y = -5/3$. Back-substituting into the first equation, we get

$$x + 2y + 3z = x + (2)(-5/3) + (3)(2/3) = -1$$

so $x = 1/3$.

A square matrix in row echelon form is upper triangular. An arbitrary matrix $A = [a_{ij}]$ in row echelon form may not be upper triangular, but it still has the property that $a_{ij} = 0$ when $i > j$ and it is still the case that solving an associated system of linear equations only requires back-substitution.

Upper and lower triangular matrices have nice mathematical properties, as well.

Lemma 4.34. The set of lower triangular, respectively upper triangular, matrices in $M_n(\mathbb{F})$ forms an $\mathbb{F}$-subalgebra.

Proof. Let $\mathcal{W}$ be the set of lower triangular matrices in $M_n(\mathbb{F})$. We leave the proof that $\mathcal{W}$ is a vector subspace of $M_n(\mathbb{F})$ as an exercise.

Let A and B belong to $\mathcal{W}$. Write $A = (\mathbf{a}_1, \ldots, \mathbf{a}_n)$ and $B = (\mathbf{b}_1, \ldots, \mathbf{b}_n)$ and consider

$$AB = \begin{bmatrix} A\mathbf{b}_1 & \ldots & A\mathbf{b}_n \end{bmatrix}.$$

To show that matrix multiplication is a binary operation on $\mathcal{W}$, it is sufficient to show that for $j > 1$, coordinates in rows $1 - (j-1)$ of $A\mathbf{b}_j$ are zero.

Since coordinates $1 - (j-1)$ of $\mathbf{b}_j$ are zero, we can view $A\mathbf{b}_j$ as a linear combination of $\mathbf{a}_j, \ldots, \mathbf{a}_n$. Since coordinates $1 - (i-1)$ of $\mathbf{a}_i$ are zero for any $i > 1$, coordinates of any linear combination of $\mathbf{a}_j, \ldots, \mathbf{a}_n$ must be zero in rows $1 - (j-1)$. This is enough to prove the lemma for the lower triangular case.

The proof for the upper triangular case is nearly identical. We leave the details as an exercise. $\square$

A type-2 elementary matrix is a diagonal matrix, while a type-3 elementary matrix is lower or upper triangular, depending on whether it effects a downward or upward pivot. Consider, for example,

$$\begin{bmatrix} 1 & 0 \\ -2 & 1 \end{bmatrix} \begin{bmatrix} 1 \\ 2 \end{bmatrix} = \begin{bmatrix} 1 \\ 0 \end{bmatrix}.$$

Premultiplying by the lower triangular matrix effects a downward pivot on the vector $(1, 2)$.

Since a type-3 elementary matrix E and its inverse effect a change on the same row, the next lemma is immediate.

Lemma 4.35. The inverse of a lower triangular type-3 elementary matrix is a lower triangular type-3 elementary matrix. The inverse of an upper triangular type-3 elementary matrix is an upper triangular type-3 elementary matrix.

Since every invertible matrix is a product of elementary matrices, the next theorem follows the lemma.

Theorem 4.36. The inverse of an invertible upper triangular matrix is an upper triangular matrix. The inverse of an invertible lower triangular matrix is a lower triangular matrix.

Suppose A is an $m \times n$ matrix that can be put into row echelon form using downward pivots exclusively. Let $E_1, \ldots, E_\ell$ be the $m \times m$ type-3 elementary matrices that effect the row reduction, so that
$$U = E_\ell \cdots E_1 A$$
is in row echelon form. Taking $L = E_1^{-1} \cdots E_\ell^{-1}$, we then have $A = LU$. This is what we mean by an **LU-factorization** for A: L stands for "lower" and U for "upper." In this setting, L is a lower triangular matrix, the product of elementary matrices that effect downward pivots, and U is in row echelon form.

Suppose we have an LU-factorization for A in $M_n(\mathbb{F})$. We can then write
$$A\mathbf{x} = LU\mathbf{x}$$
where $L^{-1} = E_\ell \cdots E_1$. If all we want is to find $\mathbf{x}$ so that $A\mathbf{x} = \mathbf{b}$, then we can solve
$$U\mathbf{x} = E_\ell \cdots E_1 \mathbf{b}$$
entirely by back-substitution. This is often what we do when we use row reduction to solve a system of linear equations by hand.

When we want to find the inverse of an invertible matrix, A in $M_n(\mathbb{F})$, we solve $A\mathbf{x}_i = \mathbf{e}_i$ for each $\mathbf{e}_i$ in $\mathcal{E}$, the standard ordered basis for $\mathbb{F}^n$. In each instance, we are solving for column i of A^{-1}. Many industrial applications present the same profile, that is, several different systems of linear equations with the same coefficient matrix.

If we want to solve $A\mathbf{x}_i = \mathbf{b}_i$, for i belonging to some index set, and if A has an LU-factorization, then for each i we can write
$$A\mathbf{x}_i = LU\mathbf{x}_i = \mathbf{b}_i.$$
Letting $U\mathbf{x}_i = \mathbf{y}_i$, we first solve
$$L\mathbf{y}_i = \mathbf{b}_i$$
for $\mathbf{y}_i$ using forward-substitution. We then solve
$$U\mathbf{x}_i = \mathbf{y}_i$$
for $\mathbf{x}_i$ using back-substitution.

There are matrices that require type-1 row operations in row reduction. When the objective is to solve a system of linear equations, though, we can avoid type-1 row operations by rearranging the equations in the system before introducing matrices. From that point, type-3 row operations are sufficient for the row reduction.

The next example follows [15].

4.6. LU-Factorization

Example 4.37. Let $A = \begin{bmatrix} 2 & -4 & -2 & 3 \\ 6 & -9 & -5 & 8 \\ 2 & -7 & -3 & 9 \\ 4 & -2 & -2 & -1 \\ -6 & 3 & 3 & 4 \end{bmatrix}$. We seek an LU-factorization of A.

Pivoting down from the 1,1-position we get $A' = \begin{bmatrix} 2 & -4 & -2 & 3 \\ 0 & 3 & 1 & -1 \\ 0 & -3 & -1 & 6 \\ 0 & 6 & 2 & -7 \\ 0 & -9 & -3 & 13 \end{bmatrix}$. Performing the same pivoting operations on I_4 we get the product of the type-3 elementary matrices that effect the pivoting:

$$E_4 E_3 E_2 E_1 = \begin{bmatrix} 1 & 0 & 0 & 0 & 0 \\ -3 & 1 & 0 & 0 & 0 \\ -1 & 0 & 1 & 0 & 0 \\ -2 & 0 & 0 & 1 & 0 \\ 3 & 0 & 0 & 0 & 1 \end{bmatrix}.$$

Pivoting down from the 2,2-position in A', we get $A'' = \begin{bmatrix} 2 & -4 & -2 & 3 \\ 0 & 3 & 1 & -1 \\ 0 & 0 & 0 & 5 \\ 0 & 0 & 0 & -5 \\ 0 & 0 & 0 & 10 \end{bmatrix}.$

The product of elementary matrices associated to this sequence of pivots is

$$E_7 E_6 E_5 = \begin{bmatrix} 1 & 0 & 0 & 0 & 0 \\ 0 & 1 & 0 & 0 & 0 \\ 0 & 1 & 1 & 0 & 0 \\ 0 & -2 & 0 & 1 & 0 \\ 0 & 3 & 0 & 0 & 1 \end{bmatrix}.$$

Finally, we pivot down from the 3,5-position in A'' to get $U \stackrel{\sim}{r} A$ in row echelon form:

$$U = \begin{bmatrix} 2 & -4 & -2 & 3 \\ 0 & 3 & 1 & -1 \\ 0 & 0 & 0 & 5 \\ 0 & 0 & 0 & 0 \\ 0 & 0 & 0 & 0 \end{bmatrix}.$$

The product of elementary matrices that effect this sequence of pivots is

$$E_9 E_8 = \begin{bmatrix} 1 & 0 & 0 & 0 & 0 \\ 0 & 1 & 0 & 0 & 0 \\ 0 & 0 & 1 & 0 & 0 \\ 0 & 0 & 1 & 1 & 0 \\ 0 & 0 & -2 & 0 & 1 \end{bmatrix}.$$

We have
$$E_9 E_8 \cdots E_1 A = U,$$

so
$$A = E_1^{-1} \cdots E_9^{-1} U.$$

The inverses of the E_ℓ matrices and their products are easy to compute. We get $L = E_1^{-1} \cdots E_9^{-1}$ given by

$$L = \begin{bmatrix} 1 & 0 & 0 & 0 & 0 \\ 3 & 1 & 0 & 0 & 0 \\ 1 & 0 & 0 & 1 & 0 \\ 2 & 0 & 0 & 1 & 0 \\ -3 & 0 & 0 & 0 & 1 \end{bmatrix} \begin{bmatrix} 1 & 0 & 0 & 0 & 0 \\ 0 & 1 & 0 & 0 & 0 \\ 0 & -1 & 1 & 0 & 0 \\ 0 & 2 & 0 & 1 & 0 \\ 0 & -3 & 0 & 0 & 1 \end{bmatrix} \begin{bmatrix} 1 & 0 & 0 & 0 & 0 \\ 0 & 1 & 0 & 0 & 0 \\ 0 & 0 & 1 & 0 & 0 \\ 0 & 0 & -1 & 1 & 0 \\ 0 & 0 & 2 & 0 & 1 \end{bmatrix}$$

$$= \begin{bmatrix} 1 & 0 & 0 & 0 & 0 \\ 3 & 1 & 0 & 0 & 0 \\ 1 & -1 & 1 & 0 & 0 \\ 2 & 2 & -1 & 1 & 0 \\ -3 & -3 & 2 & 0 & 1 \end{bmatrix}.$$

In conclusion,

$$A = LU = \begin{bmatrix} 1 & 0 & 0 & 0 & 0 \\ 3 & 1 & 0 & 0 & 0 \\ 1 & -1 & 1 & 0 & 0 \\ 2 & 2 & -1 & 1 & 0 \\ -3 & -3 & 2 & 0 & 1 \end{bmatrix} \begin{bmatrix} 2 & -4 & -2 & 3 \\ 0 & 3 & 1 & -1 \\ 0 & 0 & 0 & 5 \\ 0 & 0 & 0 & 0 \\ 0 & 0 & 0 & 0 \end{bmatrix}.$$

Now we can solve any system of linear equations given by $A\mathbf{x} = \mathbf{b}$ essentially using only forward- and back-substitution, that is, pivoting down and pivoting up.

Example 4.38. Let $A = \begin{bmatrix} 3 & -7 & -2 & 2 \\ -3 & 5 & 1 & 0 \\ 6 & -4 & 0 & -5 \\ -9 & 5 & -5 & 12 \end{bmatrix}$. Here we find an LU-factorization for A and then use it to solve two different systems with coefficient matrix A.

Performing row reduction on A, we start by pivoting down from the $1,1$-entry, using L_1, a product of three elementary matrices:

$$L_1 A = \begin{bmatrix} 1 & 0 & 0 & 0 \\ 1 & 1 & 0 & 0 \\ -2 & 0 & 1 & 0 \\ 3 & 0 & 0 & 1 \end{bmatrix} A = \begin{bmatrix} 3 & -7 & -2 & 2 \\ 0 & -2 & -1 & 2 \\ 0 & 10 & 4 & -9 \\ 0 & -16 & -11 & 18 \end{bmatrix} = A'.$$

The next step is to pivot down from the $2,2$-entry in A', using L_2, a product of two more elementary matrices:

$$L_2 A' = \begin{bmatrix} 1 & 0 & 0 & 0 \\ 0 & 1 & 0 & 0 \\ 0 & 5 & 1 & 0 \\ 0 & -8 & 0 & 1 \end{bmatrix} A' = \begin{bmatrix} 3 & -7 & -2 & 2 \\ 0 & -2 & -1 & 2 \\ 0 & 0 & -1 & 1 \\ 0 & 0 & -3 & 2 \end{bmatrix} = A''.$$

4.6. LU-Factorization

The last step in row reduction is to pivot down from the 3, 3-entry in A'' using a single elementary matrix, L_3:

$$L_3 A'' = \begin{bmatrix} 1 & 0 & 0 & 0 \\ 0 & 1 & 0 & 0 \\ 0 & 0 & 1 & 0 \\ 0 & 0 & -3 & 1 \end{bmatrix} A'' = \begin{bmatrix} 3 & -7 & -2 & 2 \\ 0 & -2 & -1 & 2 \\ 0 & 0 & -1 & 1 \\ 0 & 0 & 0 & -1 \end{bmatrix} = U.$$

Here, $L = (L_3 L_2 L_1)^{-1} = L_1^{-1} L_2^{-1} L_3^{-1}$. Then $A = LU$, with

$$L = \begin{bmatrix} 1 & 0 & 0 & 0 \\ -1 & 1 & 0 & 0 \\ 2 & 0 & 1 & 0 \\ -3 & 0 & 0 & 1 \end{bmatrix} \begin{bmatrix} 1 & 0 & 0 & 0 \\ 0 & 1 & 0 & 0 \\ 0 & -5 & 1 & 0 \\ 0 & 8 & 0 & 1 \end{bmatrix} \begin{bmatrix} 1 & 0 & 0 & 0 \\ 0 & 1 & 0 & 0 \\ 0 & 0 & 1 & 0 \\ 0 & 0 & 3 & 1 \end{bmatrix}$$

$$= \begin{bmatrix} 1 & 0 & 0 & 0 \\ -1 & 1 & 0 & 0 \\ 2 & -5 & 1 & 0 \\ -3 & 8 & 3 & 1 \end{bmatrix}.$$

Now we solve $A\mathbf{x}_i = \mathbf{b}_i$ in two different cases: $\mathbf{b}_1 = (2, 1, -1, 3)$ and $\mathbf{b}_2 = (2, -1, 2, -3)$.

We can solve the two different systems at once. First we find $\mathbf{y}_i$ so that $L\mathbf{y}_i = \mathbf{b}_i$:

$$[\, L \mid \mathbf{b}_1 \quad \mathbf{b}_2 \,] = \begin{bmatrix} 1 & 0 & 0 & 0 & | & 2 & 2 \\ -1 & 1 & 0 & 0 & | & 1 & -1 \\ 2 & -5 & 1 & 0 & | & -1 & 2 \\ -3 & 8 & 3 & 1 & | & 3 & -3 \end{bmatrix}.$$

We start by pivoting down on the 1, 1-position:

$$[\, L \mid \mathbf{b}_1 \quad \mathbf{b}_2 \,] \tilde{r} \begin{bmatrix} 1 & 0 & 0 & 0 & | & 2 & 2 \\ 0 & 1 & 0 & 0 & | & 3 & 1 \\ 0 & -5 & 1 & 0 & | & -5 & -2 \\ 0 & 8 & 3 & 1 & | & 9 & 3 \end{bmatrix}.$$

We follow through by pivoting down on the 2, 2-position:

$$[\, L \mid \mathbf{b}_1 \quad \mathbf{b}_2 \,] \tilde{r} \begin{bmatrix} 1 & 0 & 0 & 0 & | & 2 & 2 \\ 0 & 1 & 0 & 0 & | & 3 & 1 \\ 0 & 0 & 1 & 0 & | & 10 & 3 \\ 0 & 0 & 3 & 1 & | & -15 & -5 \end{bmatrix}.$$

Now we pivot down on the 3, 3-position:

$$[\, L \mid \mathbf{b}_1 \quad \mathbf{b}_2 \,] \tilde{r} \begin{bmatrix} 1 & 0 & 0 & 0 & | & 2 & 2 \\ 0 & 1 & 0 & 0 & | & 3 & 1 \\ 0 & 0 & 1 & 0 & | & 10 & 3 \\ 0 & 0 & 0 & 1 & | & -45 & -14 \end{bmatrix}.$$

Now we solve $U\mathbf{x}_i = \mathbf{y}_i$ for $\mathbf{y}_1 = (2, 3, 10, -45)$ and $\mathbf{y}_2 = (2, 1, 3, -14)$.

We have
$$[\,U\mid \mathbf{y}_1\ \mathbf{y}_2\,] = \begin{bmatrix} 3 & -7 & -2 & 2 & 2 & 2 \\ 0 & -2 & -1 & 2 & 3 & 1 \\ 0 & 0 & -1 & 1 & 10 & 3 \\ 0 & 0 & 0 & -1 & -45 & -14 \end{bmatrix}.$$

We start by pivoting up on the $4,4$-entry, then the $3,3$-entry, then the $2,2$-entry. These steps give us the following:
$$[\,U\mid \mathbf{y}_1\ \mathbf{y}_2\,]\underset{r}{\sim} \begin{bmatrix} 3 & 0 & 0 & 0 & 164 & 52 \\ 0 & -2 & 0 & 0 & -52 & -16 \\ 0 & 0 & -1 & 0 & -35 & -11 \\ 0 & 0 & 0 & -1 & -45 & -14 \end{bmatrix}.$$

We finish with three type-2 elementary row operations:
$$[\,U\mid \mathbf{y}_1\ \mathbf{y}_2\,]\underset{r}{\sim} \begin{bmatrix} 1 & 0 & 0 & 0 & 164/3 & 52/3 \\ 0 & 1 & 0 & 0 & 26 & 8 \\ 0 & 0 & 1 & 0 & 35 & 11 \\ 0 & 0 & 0 & 1 & 45 & 14 \end{bmatrix}.$$

We then have $\mathbf{x}_1 = (164/3, 26, 35, 45)$ and $\mathbf{x}_2 = (52/3, 8, 11, 14)$.

We consider one more example, this time to find the inverse of a matrix.

Example 4.39. Let $A = \begin{bmatrix} 1 & 2 & 3 \\ 2 & 3 & 4 \\ 1 & 2 & 5 \end{bmatrix}$. If $A = LU$, then $A^{-1} = U^{-1}L^{-1}$. Note, though, that it is easier to find L^{-1} than to find L: L^{-1} is the product of type-3 elementary matrices that we use to reduce A to row echelon form.

Pivoting down from the $1,1$-entry of A we have
$$\begin{bmatrix} 1 & 0 & 0 \\ -2 & 1 & 0 \\ -1 & 0 & 1 \end{bmatrix} A = \begin{bmatrix} 1 & 2 & 3 \\ 0 & -1 & -2 \\ 0 & 0 & 2 \end{bmatrix} = U.$$

This gives us $L^{-1} = \begin{bmatrix} 1 & 0 & 0 \\ -2 & 1 & 0 \\ -1 & 0 & 1 \end{bmatrix}$. We find U^{-1} by row reducing as follows:

$$[U\mid I_3] = \begin{bmatrix} 1 & 2 & 3 & 1 & 0 & 0 \\ 0 & -1 & -2 & 0 & 1 & 0 \\ 0 & 0 & 2 & 0 & 0 & 1 \end{bmatrix}$$

$$\underset{r}{\sim} \begin{bmatrix} 1 & 2 & 0 & 1 & 0 & -3/2 \\ 0 & -1 & 0 & 0 & 1 & 1 \\ 0 & 0 & 2 & 0 & 0 & -1 \end{bmatrix} \underset{r}{\sim} \begin{bmatrix} 1 & 0 & 0 & 1 & 2 & 1/2 \\ 0 & 1 & 0 & 0 & -1 & -1 \\ 0 & 0 & 1 & 0 & 0 & 1/2 \end{bmatrix}.$$

This gives us $U^{-1} = \begin{bmatrix} 1 & 2 & 1/2 \\ 0 & -1 & -1 \\ 0 & 0 & 1/2 \end{bmatrix}$ so that

$$A^{-1} = U^{-1}L^{-1} = \begin{bmatrix} 1 & 2 & 1/2 \\ 0 & -1 & -1 \\ 0 & 0 & 1/2 \end{bmatrix} \begin{bmatrix} 1 & 0 & 0 \\ -2 & 1 & 0 \\ -1 & 0 & 1 \end{bmatrix} = \begin{bmatrix} -7/2 & 2 & 1/2 \\ 1 & -1 & -1 \\ -1/2 & 0 & 1/2 \end{bmatrix}.$$

4.6. LU-Factorization

Exercises 4.6.

4.6.1. Prove that a square matrix in row echelon form is upper triangular.

4.6.2. Finish the proof of Lemma 4.34.

4.6.3. Let $A = \begin{bmatrix} 1 & -2 & 0 & 3 \\ 0 & 0 & 2 & 1 \\ 0 & 0 & 0 & 1 \end{bmatrix}$. Use back-substitution to solve $A\mathbf{x} = (7, 5, -12)$.

4.6.4. In Example 4.38 we solved two different matrix equations with the same coefficient matrix using an LU-factorization of the coefficient matrix. Redo that problem, this time solving $U\mathbf{x} = L^{-1}\mathbf{b}$, where $A = LU$ is an LU-factorization of the coefficient matrix, A. Argue that you can do the problem using upward pivoting on the matrix $\begin{bmatrix} U \mid B \end{bmatrix}$ where $B = L^{-1}\left(\begin{bmatrix} \mathbf{b}_1 & \mathbf{b}_2 \end{bmatrix}\right)$.

4.6.5. Let $A = \begin{bmatrix} 1 & 2 & 1 \\ 2 & 3 & 3 \\ -3 & -10 & 2 \end{bmatrix}$.

(a) Find an LU-factorization for A.

(b) Is A invertible? Argue your answer without attempting to find A^{-1}.

4.6.6. Let A be an invertible 3×3 matrix. Let $\mathbf{b}_1$ be some fixed vector in $\mathbb{F}^3$ and for $i = 1, 2, \ldots$ let $\mathbf{x}_i$ satisfy $A\mathbf{x}_i = \mathbf{b}_i$ for $\mathbf{b}_{i+1} = \mathbf{x}_i + \mathbf{b}_i$. Describe $\mathbf{x}_k$ in terms of A^{-1} and $\mathbf{b}_1$. (Hint: Use induction to find $\mathbf{b}_k$.)

Chapter 5

Introductions

We now know enough about vector spaces, subspaces, linear transformations, and matrices to fuel journeys down many different paths. In this chapter, we study foundational ideas that can inform a few of those journeys. These ideas arise several times in the sequel.

5.1. Dual Spaces

We have already discussed the vector space of linear transformations from one vector space to another. Perhaps counterintuitively, the mappings from a vector space into its scalar field are particularly interesting. They are called **linear functionals**.

Example 5.1. Let $V = C([0,1])$ where $[0,1] \subseteq \mathbb{R}$. We leave it as an exercise to verify that
$$f \mapsto \int_0^1 f(t)\,dt$$
defines a linear functional on V.

The set of all linear functionals on a vector space is its **dual space**. We denote the dual space of V by V^*. Since $V^* = \mathcal{L}(V, \mathbb{F})$, where $\mathbb{F}$ is the scalar field for V, Corollary 3.28 tells us that if $\dim V = n$, then $\dim V^* = n$, as well.

When V is infinite-dimensional, it may not be the case that $V \cong V^*$. In that setting, V^* may be called the *algebraic dual* or the *full algebraic dual* of V and "dual space" may refer to a proper subspace of V^*.

Since a linear mapping $\mathbb{F}^n \to \mathbb{F}$ is effected by premultiplication by an $n \times 1$ matrix, the following is immediate.

Theorem 5.2. *The dual space of $\mathbb{F}^n$ is $\mathbb{F}_n$.*

This is the interesting relationship between $\mathbb{F}^n$ and $\mathbb{F}_n$ that we promised in Section 1.2. It also suggests a deeper relationship between duality and matrix transposition that we explore below. Before going farther into this, we give the transpose of a matrix its official introduction.

Definition 5.3. The **transpose** of a matrix $A = [a_{ij}]$ is $A^T = [c_{ij}]$ where $c_{ij} = a_{ji}$.

If $\mathbf{v} = (a_1, \ldots, a_n)$ is in $\mathbb{F}^n$, then $\mathbf{v}^T = \begin{bmatrix} a_1 & \cdots & a_n \end{bmatrix}$ belongs to $\mathbb{F}_n$.

When we say that the difference between $(a_1, \ldots, a_n)$ in $\mathbb{F}^n$ and $\begin{bmatrix} a_1 & \cdots & a_n \end{bmatrix}$ in $\mathbb{F}_n$ is purely cosmetic, we are saying that $\mathbf{v} \mapsto \mathbf{v}^T$ is a canonical isomorphism $\mathbb{F}^n \to \mathbb{F}_n$: This is an isomorphism defined independently of a basis. In general, we cannot expect to find a canonical isomorphism from a vector space to its dual space or to a subspace of its dual space. We can, though, define an injective linear mapping from V to V^* by means of a *dual basis*.

Let V be a vector space with basis $\mathcal{B} = \{\mathbf{b}_j\}_{j \in J}$. Define each element f_i in the **dual basis**, $\mathcal{B}^* = \{f_i\} \subseteq V^*$, by extending linearly from the rule

(5.1) $$f_i(\mathbf{b}_j) = \delta(i,j).$$

Example 5.4. Let V be the span of $\mathcal{B} = \{\sin t, \cos t, e^t\} \subseteq C(\mathbb{R})$. Let $\mathbf{v} = 2\sin t + 3\cos t - e^t$. Writing $\mathcal{B}^* = \{f_1, f_2, f_3\}$, we have $f_1(\mathbf{v}) = 2$, $f_2(\mathbf{v}) = 3$, and $f_3(\mathbf{v}) = 1$.

We leave it as an exercise to show that if $\mathcal{B}$ is a basis for a vector space, V, then $\mathcal{B}^*$ is linearly independent in V^*.

Example 5.5. Let $\mathcal{E}$ be the standard ordered basis for $\mathbb{F}^n$. For f_i in $\mathcal{E}^*$ defined as in (5.1), we have
$$f_i(\mathbf{e}_j) = \delta(i,j).$$
As an element of $\mathbb{F}_n$, f_i must be given by
$$\begin{bmatrix} \delta(1,i) & \cdots & \delta(n,i) \end{bmatrix}.$$
In other words, $f_i = \mathbf{e}_i^T$.

We designate the *standard ordered basis for* $\mathbb{F}_n$ by
$$\mathcal{E}^* = \{\mathbf{e}_1^T, \ldots, \mathbf{e}_n^T\}.$$
Since $\mathbf{y}$ in $\mathbb{F}_n$ can be written
$$\mathbf{y} = \begin{bmatrix} a_1 & \cdots & a_n \end{bmatrix} = \sum_{i=1}^n a_i \mathbf{e}_i^T,$$
we see that

(5.2) $$[\mathbf{y}]_{\mathcal{E}^*} = \sum_{i=1}^n a_i \mathbf{e}_i = \mathbf{y}^T.$$

Dual bases may also be deployed in infinite-dimensional settings.

Example 5.6. Let $\mathcal{B} = \{1, t, \ldots\} \subseteq \mathbb{R}[t]$ and let $\mathcal{B}^* = \{f_0, f_1, \ldots,\}$ be the dual basis. Consider the mapping $f(p(t)) = \int_0^1 p(t)\,dt$ from Example 5.1, this time applied to $\mathbb{R}[t]$. As
$$\int_0^1 a_0 + a_1 t + \cdots + a_n t^n \, dt = a_0 + \frac{1}{2}a_1 + \cdots + \frac{1}{n+1}a_n,$$

5.1. Dual Spaces

we see that we can write f as a formal sum using elements in $\mathcal{B}^*$:
$$f = f_0 + \frac{1}{2}f_1 + \frac{1}{3}f_2 + \cdots + \frac{1}{n+1}f_n + \cdots.$$

The Rank Theorem ensures that a linear functional on a vector space is either identically zero or onto its scalar field. Theorem 2.35 thus implies that if f is in V^*, $V/\operatorname{Ker} f$ is trivial or 1-dimensional. This is enough for half of the proof of the next result.

Lemma 5.7. A subset of a vector space V is a hyperspace if and only if it is the kernel of a nontrivial element in V^*.

Proof. Suppose V is an arbitrary vector space over $\mathbb{F}$. We have already shown that $V/\operatorname{Ker} f$ is a hyperspace if f in V^* is nontrivial. To prove the result in the other direction, suppose $W \subseteq V$ is a hyperspace. Let $\{\mathbf{b}\}$ be a basis for any direct complement of W. Recall that if $\mathbf{v}$ is in V, there are unique $\mathbf{w}$ in W and c in $\mathbb{F}$ such that
$$\mathbf{v} = \mathbf{w} + c\mathbf{b}.$$
This means we can define f in V^* by extending linearly from $f(\mathbf{b}) = 1$ and $f(\mathbf{w}) = 0$ for all $\mathbf{w}$ in W. Since $W = \operatorname{Ker} f$, this is enough to complete proof. □

We saw in Chapter 4 that every affine set in $\mathbb{F}^n$ is the solution to a system of linear equations. The next theorem applies to all vector spaces. We outline a proof, leaving the details to the exercises.

Theorem 5.8. If V is a vector space with subspace W, such that $\dim V/W = k$, then W is the intersection of k hyperspaces in V.

Proof. Let V and W be as hypothesized. Fix W', a direct complement to W. Theorem 2.35 implies that $\dim W' = k$. Let $\mathcal{B}$ be a basis for W' and let $\mathcal{B}^* = \{f_i\}_{i=1}^k$ be its dual basis. We leave it as an exercise to verify that $W = \bigcap_{i=1}^k \operatorname{Ker} f_i$. □

While the relationship between a vector space V and its dual is intriguing, the dual of the dual space contains a more natural reflection of V.

Definition 5.9. The **bidual space** associated to a vector space, V, is the vector space dual to V^*, $(V^*)^* = V^{**}$.

Let V be an arbitrary vector space. Define
$$\varphi(\mathbf{v})(f) := f(\mathbf{v}), \tag{5.3}$$
for $\mathbf{v}$ in V and f in V^*. In the proof of the next theorem, we show that φ is an injective linear transformation $V \to V^{**}$. We call it the canonical mapping $V \to V^{**}$.

Theorem 5.10. If V is any vector space over a field $\mathbb{F}$, then V^{**} contains a subspace that is canonically isomorphic to V.

Proof. Let V be as hypothesized and let φ be as defined in (5.3). Showing that φ is an injective linear transformation $V \to V^{**}$ is enough to show
$$V \cong \varphi[V] \subseteq V^{**}.$$

First we verify that $\varphi(\mathbf{v})$ is in V^{**} for any $\mathbf{v}$ in V.

If f_1 and f_2 are in V^* and c is in $\mathbb{F}$, then for any $\mathbf{v}$ in V,
$$\varphi(\mathbf{v})(cf_1 + f_2) = (cf_1 + f_2)(\mathbf{v}) = cf_1(\mathbf{v}) + f_2(\mathbf{v}) = c\varphi(\mathbf{v})(f_1) + \varphi(\mathbf{v})(f_2).$$
This verifies that $\varphi(\mathbf{v})$ is linear so $\varphi[V] \subseteq V^{**}$.

Next, we show that $\varphi : V \to V^{**}$ is linear. If $\mathbf{v}_1$ and $\mathbf{v}_2$ are in V and c is in $\mathbb{F}$, then for any f in V^*,
$$\begin{aligned}\varphi(c\mathbf{v}_1 + \mathbf{v}_2)(f) &= f(c\mathbf{v}_1 + \mathbf{v}_2) = cf(\mathbf{v}_1) + f(\mathbf{v}_2) \\ &= c\varphi(\mathbf{v}_1)(f) + \varphi(\mathbf{v}_2)(f) = (c\varphi(\mathbf{v}_1) + \varphi(\mathbf{v}_2))(f).\end{aligned}$$
It follows that
$$\varphi(c\mathbf{v}_1 + \mathbf{v}_2) = c\varphi(\mathbf{v}_1) + \varphi(\mathbf{v}_2),$$
so φ is linear.

Finally, we argue that φ is injective. If $\mathbf{v}$ in V is nonzero, then by Theorem 1.31(a), V has a basis $\mathcal{B}$ containing $\mathbf{v}$. Define f in V^* by extending linearly from the rule $f(\mathbf{v}) = 1$ and $f(\mathbf{x}) = 0$ for all $\mathbf{x}$ in $\mathcal{B} \setminus \{\mathbf{v}\}$. We then have
$$\varphi(\mathbf{v})(f) = f(\mathbf{v}) = 1.$$
It follows that $\operatorname{Ker} \varphi$ contains no nonzero elements, thus, that φ is injective. □

The following is immediate.

Corollary 5.11. A finite-dimensional vector space is canonically isomorphic to its bidual space.

We next consider the dual of a linear transformation.

Definition 5.12. Let V and W be vector spaces over $\mathbb{F}$ with L in $\mathcal{L}(V,W)$. The **dual mapping** of L is L^* defined by
$$L^*(f)(\mathbf{v}) := f(L(\mathbf{v}))$$
for f in W^* and $\mathbf{v}$ in V.

If L is in $\mathcal{L}(V,W)$, then L^* maps W^* to V^*. The next theorem verifies both that L^* is linear and that the mapping $L \mapsto L^*$ is linear.

Theorem 5.13. Let V and W be vector spaces over $\mathbb{F}$. The mapping $L \mapsto L^*$ is a linear transformation $\mathcal{L}(V,W) \to \mathcal{L}(W^*, V^*)$.

Proof. Let V, W, and $\mathbb{F}$ be as hypothesized. Fix L in $\mathcal{L}(V,W)$. First we verify that L^* is linear. Let f and g belong to W^*, c belong to $\mathbb{F}$, and $\mathbf{v}$ belong to V. We have
$$\begin{aligned}L^*(cf + g)(\mathbf{v}) &= (cf + g)(L(\mathbf{v})) = cf(L(\mathbf{v})) + g(L(\mathbf{v})) \\ &= cL^*(f)(\mathbf{v}) + L^*(g)(\mathbf{v}) = (cL^*(f) + L^*(g))(\mathbf{v}).\end{aligned}$$
It follows that
$$L^*(cf + g) = cL^*(f) + L^*(g),$$
proving that L^* is in $\mathcal{L}(W^*, V^*)$.

5.1. Dual Spaces

Next we show that $L \mapsto L^*$ is another linear transformation.

Suppose T is a second element in $\mathcal{L}(V, W)$. Taking f, c, and $\mathbf{v}$ as above, we have
$$(cL + T)^*(f)(\mathbf{v}) = f((cL + T)(\mathbf{v})) = f(cL(\mathbf{v}) + T(\mathbf{v})).$$
Since f is linear,
$$f(cL(\mathbf{v}) + T(\mathbf{v})) = cf(L(\mathbf{v})) + f(T(\mathbf{v}))$$
$$= cL^*(f)(\mathbf{v}) + T^*(f)(\mathbf{v}) = (cL^* + T^*)(f)(\mathbf{v}).$$
It follows that
$$(cL + T)^* = cL^* + T^*,$$
in other words, that $L \mapsto L^*$ is a linear transformation $\mathcal{L}(V, W) \to \mathcal{L}(W^*, V^*)$. □

We look more closely at dualizing linear transformations between finite-dimensional spaces in the next section.

Exercises 5.1.

5.1.1. Show that $f \mapsto \int_0^1 f(t)\,dt$ defines a linear functional on $C([0, 1])$.

5.1.2. Prove that $\mathbf{v} \mapsto \mathbf{v}^T$ is an isomorphism $\mathbb{F}^n \to \mathbb{F}_n$.

5.1.3. Why is a linear functional on a vector space either identically zero or onto its scalar field?

5.1.4. The *trace* of a square matrix $A = [a_{ij}]$ is $\operatorname{Tr} A = \sum_{i=1}^n a_{ii}$.
  (a) Show that Tr is a linear functional on $M_n(\mathbb{F})$.
  (b) Show that $\operatorname{Tr}(A^T) = \operatorname{Tr} A$ for any A in $M_n(\mathbb{F})$.
  (c) Show that $\operatorname{Tr}(AB) = \operatorname{Tr}(BA)$ for A and B in $M_n(\mathbb{F})$.

5.1.5. Let V be a vector space with basis $\mathcal{B}$. Show that $\mathcal{B}^*$ is linearly independent. Conclude that if $\dim V = n$, then $\mathcal{B}^*$ is a basis for V^*.

5.1.6. Let V be a vector space with basis $\mathcal{B} = \{\mathbf{b}_i\}_{i=1}^n$. Let $\mathcal{B}^* = \{f_i\}_{i=1}^n$. Let $\mathbf{v}$ belong to V. What do we usually call the mapping $\mathbf{v} \mapsto (f_1(\mathbf{v}), \ldots, f_n(\mathbf{v}))$?

5.1.7. Let $\mathcal{B} = \{(1, 1), (1, -1)\} \subseteq \mathbb{R}^2$ and let $\mathcal{B}^* = \{f_1, f_2\}$. For arbitrary (x, y) in $\mathbb{R}^2$, find $f_i(x, y)$, $i = 1, 2$.

5.1.8. Let $\mathcal{B} = \{1 + t, 1 + 2t\} \subseteq \mathbb{R}[t]_1$ and let $\mathcal{B}^* = \{f_1, f_2\}$. Find $f_i(a + bt)$ for a, b in $\mathbb{R}$ and $i = 1, 2$.

5.1.9. Let $V = \mathbb{R}[t]_1$. Define f_1 by $f_1(p(t)) = \int_0^1 p(t)\,dt$ and define f_2 by $f_2(p(t)) = \int_0^2 p(t)\,dt$.
  (a) Argue that $\{f_1, f_2\}$ is a basis for V^*. (Hint: Consider Example 1.24.)
  (b) Find a basis for V so that $\{f_1, f_2\}$ is its dual basis.

5.1.10. Suppose $\varphi : V \to V^*$ is an isomorphism and that $\mathcal{B}$ is a basis for V. Is $\varphi[\mathcal{B}]$ the dual basis for $\mathcal{B}$? Explain your answer.

5.1.11. There is a tacit reference to a dual basis in the proof of Lemma 5.7. Find it.

5.1.12. Referring to the proof of Theorem 5.8, explain how Theorem 2.35 applies to show that the dimension of a direct complement to W is k.

5.1.13. Complete the proof of Theorem 5.8 to show that $W = \bigcap_{i=1}^{k} \operatorname{Ker} f_i$.

5.1.14. Let $W = \operatorname{Span}\{(1,1,1)\} \subseteq \mathbb{R}^3$ so that $\mathbb{R}^3/W$ is 2-dimensional. Write W as the intersection of two hyperspaces in $\mathbb{R}^3$.

5.1.15. Let L be defined on $\mathbb{R}^2$ by $L(a,b) = (a, a+2b)$. Find $L^*\left(\begin{bmatrix} a & b \end{bmatrix}\right)$.

5.2. Transposition and Duality

Here we pursue the connection between duality and transposition of matrices. To get started, we check that $A \mapsto A^T$ defines a linear transformation from $M_{m,n}(\mathbb{F})$ to $M_{n,m}(\mathbb{F})$.

Lemma 5.14. If c is in $\mathbb{F}$ and A and B belong to $M_{m,n}(\mathbb{F})$, then
$$(cA + B)^T = cA^T + B^T.$$

Proof. Let A, B, and c be as hypothesized. Write $A = [a_{ij}]$, $B = [b_{ij}]$, and $cA + B = [c_{ij}]$. Since
$$c_{ij} = ca_{ij} + b_{ij},$$
the ij-entry of $(cA+B)^T$ is
$$c_{ji} = ca_{ji} + b_{ji},$$
which is also the ij-entry in $cA^T + B^T$. □

The following is nearly immediate.

Theorem 5.15. Transposition is a vector space isomorphism $M_{m,n}(\mathbb{F}) \to M_{n,m}(\mathbb{F})$ and is equal to its own inverse.

Proof. Since its kernel is the zero matrix, transposition is injective. Since
$$\dim M_{m,n}(\mathbb{F}) = \dim M_{n,m}(\mathbb{F}),$$
transposition is an isomorphism by Corollary 2.40. That $(A^T)^T = A$ for any matrix A is evident. □

Is transposition an algebra automorphism on $M_n(\mathbb{F})$? It is not, according to the next lemma.

Lemma 5.16. If A and B are matrices such that AB is defined, then
$$(AB)^T = B^T A^T.$$

Proof. Let $A = [a_{ij}]$ and $B = [b_{ij}]$ be as hypothesized. The ij-entry of AB is $c_{ij} = \sum_{k=1}^{n} a_{ik} b_{kj}$. The ij-entry of $(AB)^T$ is then
$$c_{ji} = \sum_{k=1}^{n} a_{jk} b_{ki} = \sum_{k=1}^{n} b_{ki} a_{jk}.$$
Row i of B^T is $\begin{bmatrix} b_{1i} & \cdots & b_{ni} \end{bmatrix}$ and column j of A^T is $(a_{j1}, \ldots, a_{jn})$ so the ij-entry of $B^T A^T$ is
$$b_{1i} a_{j1} + \cdots + b_{ni} a_{jn} = \sum_{k=1}^{n} b_{ki} a_{jk}.$$
Since they agree entry by entry, $(AB)^T = B^T A^T$. □

5.2. Transposition and Duality

Evidently, if A is an upper triangular matrix, then A^T is lower triangular. Of course, if A is a diagonal matrix, $A^T = A$.

Theorem 5.17. Let A belong to $M_{m,n}(\mathbb{F})$.

(a) Transposition maps $\operatorname{Col} A$ isomorphically to $\operatorname{Row} A^T$ and maps $\operatorname{Row} A$ isomorphically to $\operatorname{Col} A^T$.

(b) The rank of A is the rank of A^T.

(c) The matrix A is invertible if and only if A^T is invertible in which case $(A^T)^{-1} = (A^{-1})^T$.

Proof. Let A be as hypothesized.

Since transposition is an isomorphism that maps the columns of A to the rows of A^T, and the rows of A to the columns of A^T, part (a) is immediate. Since $\operatorname{rank} A = \dim \operatorname{Col} A = \dim \operatorname{Row} A^T$ for any matrix A, part (b) follows.

Finally, A in $M_n(\mathbb{F})$ is invertible if and only if
$$AA^{-1} = I_n = I_n{}^T = (AA^{-1})^T = (A^{-1})^T A^T.$$
This is enough to prove part (c). $\square$

Example 5.18. Consider $A = \begin{bmatrix} 1 & 1 & -2 & 4 & 5 \\ 2 & 2 & -3 & 1 & 3 \\ 3 & 3 & -4 & -2 & 1 \end{bmatrix}$ in $M_{3,5}(\mathbb{R})$. In this example, we use row reduction on A and A^T to compare the bases for $\operatorname{Col} A$ and $\operatorname{Row} A^T$ and bases for $\operatorname{Row} A$ and $\operatorname{Col} A^T$.

The reduced row echelon form of A is
$$\begin{bmatrix} 1 & 1 & 0 & -10 & -9 \\ 0 & 0 & 1 & -7 & -7 \\ 0 & 0 & 0 & 0 & 0 \end{bmatrix}.$$

This tells us that the set comprised of the first and third columns of A — that is, $\mathcal{B}_c = \{(1,2,3), (-2,-3,-4)\} \subseteq \mathbb{R}^3$ — is a basis for $\operatorname{Col} A$. The set of nonzero rows of the reduced row echelon form of A is a basis for $\operatorname{Row} A$. We designate that set by $\mathcal{B}_r = \{[\,1 \ \ 1 \ \ 0 \ \ -10 \ \ 9\,], [\,0 \ \ 0 \ \ 1 \ \ -7 \ \ -7\,]\} \subseteq \mathbb{R}_5$.

The reduced row echelon form for A^T is
$$\begin{bmatrix} 1 & 0 & -1 \\ 0 & 1 & 2 \\ 0 & 0 & 0 \\ 0 & 0 & 0 \\ 0 & 0 & 0 \end{bmatrix}.$$

This gives us that $\{[\,1 \ \ 0 \ \ -1\,], [\,0 \ \ 1 \ \ 2\,]\} \subseteq \mathbb{R}^3$ is a basis for $\operatorname{Row} A^T$. Taking linear combinations of the vectors in that set, we get $\{[1 \ \ 2 \ \ 3], [-2 \ \ -3 \ \ -4]\}$, the set of transposes of the vectors in $\mathcal{B}_c$.

The pivot columns of A^T are $\{(1,1,-2,4,5), (2,2,-3,1,3)\}$. Taking linear combinations of the vectors in that set, we get $\{(1,1,0,-10,9), (0,0,1,-7,-7)\}$, transposes of the vectors in $\mathcal{B}_r$.

Suppose L is in $\mathcal{L}(\mathbb{F}^n, \mathbb{F}^m)$ and that A_L is its standard matrix representation. We have $L^* : \mathbb{F}_m \to \mathbb{F}_n$ where
$$L^*(\mathbf{y})(\mathbf{v}) = \mathbf{y}(L(\mathbf{v})) = \mathbf{y}(A_L \mathbf{v}) = (\mathbf{y} A_L) \mathbf{v}$$
for any $\mathbf{y}$ in $\mathbb{F}_m$ and $\mathbf{v}$ in $\mathbb{F}^n$. Evidently, $L^*(\mathbf{y}) = \mathbf{y} A_L$ but what is the matrix representation of L^* with respect to the standard ordered bases of $\mathbb{F}_m$ and $\mathbb{F}_n$?

If $\mathbf{w}_i$ is row i of A_L, then
$$L^*(\mathbf{e}_i^T) = \mathbf{e}_i^T A_L = \mathbf{w}_i.$$
This gives us
$$[L^*(\mathbf{e}_i^T)]_{\mathcal{E}^*} = \mathbf{w}_i^T.$$
Since $\mathbf{w}_i^T$ is column i of $A_L{}^T$, we have proved the following.

Theorem 5.19. If L belongs to $\mathcal{L}(\mathbb{F}^n, \mathbb{F}^m)$ and A_L is its standard representation, then the representation of L^* with respect to the standard ordered bases of $\mathbb{F}_m$ and $\mathbb{F}_n$ is $A_L{}^T$.

Let φ be the canonical injection $V \to V^{**}$. Suppose that $\mathcal{B} = \{\mathbf{b}_i\}$ is a basis for V with dual basis $\mathcal{B}^* = \{f_i\}$. Is the basis dual to $\mathcal{B}^*$ in V^{**} then $\{\varphi(\mathbf{b}_i)\}$? For f_j in $\mathcal{B}^*$, we have
$$\varphi(\mathbf{b}_i)(f_j) = f_j(\mathbf{b}_i) = \delta(i,j),$$
which means that the answer to the question is yes. We use this in the proof of our next theorem, which generalizes Theorem 5.19.

Theorem 5.20. Let $V \cong \mathbb{F}^n$ and $W \cong \mathbb{F}^m$. Suppose L belongs to $\mathcal{L}(V, W)$. If $\mathcal{B}$ and $\mathcal{C}$ are, respectively, ordered bases for V and W, and if $A = [\![L]\!]_\mathcal{B}^\mathcal{C}$, then $[\![L^*]\!]_{\mathcal{C}^*}^{\mathcal{B}^*} = A^T$.

Proof. Let everything be as hypothesized with $\mathcal{B} = \{\mathbf{b}_j\}_{j=1}^n$ and $\mathcal{C} = \{\mathbf{c}_i\}_{i=1}^m$. Let $\mathcal{B}^* = \{f_j\}_{j=1}^n$ and $\mathcal{C}^* = \{g_i\}_{i=1}^m$. The ij-entry in A is the ith coordinate of $[L(\mathbf{b}_j)]_\mathcal{C}$, which is
$$g_i(L(\mathbf{b}_j)) = L^*(g_i)(\mathbf{b}_j) = \varphi(\mathbf{b}_j)(L^*(g_i)).$$
By definition of the dual basis for $\mathcal{B}^*$, $\varphi(\mathbf{b}_j)(L^*(g_i))$ is the jth coordinate of $[L^*(g_i)]_{\mathcal{B}^*}$, that is, the ji-entry of $[\![L^*]\!]_{\mathcal{C}^*}^{\mathcal{B}^*}$. We have shown, then, that the ij-entry of $[\![L]\!]_\mathcal{B}^\mathcal{C}$ is the ji-entry of $[\![L^*]\!]_{\mathcal{C}^*}^{\mathcal{B}^*}$. This is enough to prove the result. □

Example 5.21. Consider $L : \mathbb{R}[t]_2 \to M_2(\mathbb{R})$ given by
$$L(a + bt + ct^2) = \begin{bmatrix} a+b & b+c \\ a-b & a-c \end{bmatrix}.$$
Let $\mathcal{B} = \{1, t, t^2\}$ and let $\mathcal{C}$ be the standard ordered basis for $M_2(\mathbb{R})$. To find $[\![L]\!]_\mathcal{B}^\mathcal{C}$, we note that
$$L(1) = \begin{bmatrix} 1 & 0 \\ 1 & 1 \end{bmatrix}, \quad L(t) = \begin{bmatrix} 1 & 1 \\ -1 & 0 \end{bmatrix}, \quad L(t^2) = \begin{bmatrix} 0 & 1 \\ 0 & -1 \end{bmatrix}.$$
From here we get
$$[\![L]\!]_\mathcal{B}^\mathcal{C} = \begin{bmatrix} 1 & 1 & 0 \\ 0 & 1 & 1 \\ 1 & -1 & 0 \\ 1 & 0 & -1 \end{bmatrix}.$$

5.2. Transposition and Duality

If $\mathcal{C}^* = \{g_1, g_2, g_3, g_4\}$, then $g_i\left(\begin{bmatrix} a_1 & a_2 \\ a_3 & a_4 \end{bmatrix}\right) = a_i$. A modest calculation then shows that

$$L^*(g_1)(1) = 1, \quad L^*(g_1)(t) = 1, \quad L^*(g_1)(t^2) = 0,$$

$$L^*(g_2)(1) = 0, \quad L^*(g_2)(t) = 1, \quad L^*(g_2)(t^2) = 1,$$

$$L^*(g_3)(1) = 1, \quad L^*(g_3)(t) = -1, \quad L^*(g_3)(t^2) = 0,$$

and finally, that

$$L^*(g_4)(1) = 1, \quad L^*(g_4)(t) = 0, \quad L^*(g_4)(t^2) = -1.$$

From here we have

$$[\![L^*]\!]_{\mathcal{C}^*}^{\mathcal{B}^*} = \begin{bmatrix} 1 & 0 & 1 & 1 \\ 1 & 1 & -1 & 0 \\ 0 & 1 & 0 & -1 \end{bmatrix},$$

which is, indeed, the matrix transpose of $[\![L]\!]_{\mathcal{B}}^{\mathcal{C}}$.

Exercises 5.2.

5.2.1. Check the calculations that we refer to in Example 5.18.

5.2.2. Let $A = \begin{bmatrix} 1 & 1 & -2 & 3 & 4 \\ 2 & 3 & 3 & -1 & 3 \\ 5 & 7 & 4 & 1 & 5 \end{bmatrix}$. Find bases for $\operatorname{Col} A$ and $\operatorname{Row} A$ using the reduced row echelon form of A. Based on those findings, what are bases for $\operatorname{Col} A^T$ and $\operatorname{Row} A^T$? Verify these using the reduced row echelon form of A^T and comparing the bases it yields for $\operatorname{Col} A^T$ and $\operatorname{Row} A^T$, as in Example 5.18.

5.2.3. Define $L: \mathbb{R}^2 \to \mathbb{R}^3$ by $L(a, b) = (a, -b, a + b)$. Let $\mathcal{B} = \{(1, 1), (0, 1)\}$.
   (a) Find $[\![L]\!]_{\mathcal{B}}^{\mathcal{E}}$.
   (b) Find $L^*(\mathbf{e}_i^T)$ explicitly as an element in $\mathbb{R}_2$, for $i = 1, 2, 3$.
   (c) Find each element in $\mathcal{B}^*$ explicitly as a row vector.
   (d) Find $[\![L^*]\!]_{\mathcal{E}^*}^{\mathcal{B}^*}$ by calculating each column directly.

5.2.4. Check the details in Example 5.21.

5.2.5. Let L in $\mathcal{L}(\mathbb{R}^2, \mathbb{R}[t]_2)$ be defined by

$$L(a, b) = a + (a + b)t + (b - 2a)t^2.$$

   (a) Let $\mathcal{B} = \{1, t, t^2\}$ in $\mathbb{R}[t]_2$. Find $[\![L]\!]_{\mathcal{E}}^{\mathcal{B}}$.
   (b) Let $\mathcal{B}^* = \{f_1, f_2, f_3\}$. Find $f_i(2 - 3t + 5t^2)$ for $i = 1, 2, 3$.
   (c) Find $L^*(-f_1 + 2f_2 - 3f_3)(3, -1)$.
   (d) Find $L^*(c_1 f_1 + c_2 f_2 + c_3 f_3)$ explicitly as an element in $\mathbb{R}_2$, for c_i in $\mathbb{R}$.
   (e) Calculate $[\![L^*]\!]_{\mathcal{B}^*}^{\mathcal{E}^*}$ directly, as we did in Example 5.21, to verify that $[\![L^*]\!]_{\mathcal{B}^*}^{\mathcal{E}^*}$ is equal to the matrix transpose of $[\![L]\!]_{\mathcal{E}}^{\mathcal{B}}$.

5.3. Bilinear Forms, Their Matrices, and Duality

Recall from Exercise 3.1.10 that the mapping $M_n(\mathbb{F}) \times M_n(\mathbb{F}) \to M_n(\mathbb{F})$ given by $(A, B) \mapsto AB$ is **bilinear**. This means that both premultiplication by A, $X \mapsto AX$, and postmultiplication by B, $X \mapsto XB$, are linear transformations $M_n(\mathbb{F}) \to M_n(\mathbb{F})$. We could have defined an $\mathbb{F}$-algebra to be a vector space V over $\mathbb{F}$ with a bilinear mapping $V \times V \to V$ called multiplication. Bilinear mappings come up in many different settings. In this section we introduce *bilinear forms*, which map into the scalar field of a vector space.

Definition 5.22. Let V be a vector space over a field $\mathbb{F}$.

A **bilinear form** on V is a mapping $\sigma : V \times V \to \mathbb{F}$ that satisfies

(a) $\sigma(c\mathbf{v} + \mathbf{w}, \mathbf{x}) = c\sigma(\mathbf{v}, \mathbf{x}) + \sigma(\mathbf{w}, \mathbf{x})$ and
(b) $\sigma(\mathbf{v}, c\mathbf{w} + \mathbf{x}) = c\sigma(\mathbf{v}, \mathbf{w}) + \sigma(\mathbf{v}, \mathbf{x})$

for all $\mathbf{v}, \mathbf{w}$, and $\mathbf{x}$ in V and c in $\mathbb{F}$.

We can define a bilinear form on $\mathbb{F}^n$ by $\sigma(\mathbf{v}, \mathbf{x}) := \mathbf{v}^T \mathbf{x}$. When $\mathbb{F} = \mathbb{R}$, σ is the dot product. (We review the dot product in Section B.5.) For $\mathbf{v}$ and $\mathbf{x}$ in $\mathbb{R}^n$, we write
$$\mathbf{v} \cdot \mathbf{x} := \mathbf{v}^T \mathbf{x}.$$

If $V \cong \mathbb{F}^n$, we can define a bilinear form on V by fixing a basis for V, $\mathcal{B} = \{\mathbf{b}_i\}_{i=1}^n$, a matrix $A = [a_{ij}]$ in $M_n(\mathbb{F})$, and decreeing that
$$\sigma(\mathbf{v}, \mathbf{x}) = [\mathbf{v}]_\mathcal{B}{}^T A [\mathbf{x}]_\mathcal{B},$$
for all $\mathbf{v}$ and $\mathbf{x}$ in V. We could define the same bilinear form by letting
$$\sigma(\mathbf{b}_i, \mathbf{b}_j) = a_{ij}$$
and *extending bilinearly*.

Example 5.23. Let $V = \mathbb{R}[t]_1$, $\mathcal{B} = \{1, t\}$, and $A = \begin{bmatrix} -1 & 2 \\ 1 & 0 \end{bmatrix}$. Define σ on V by
$$\sigma(1, 1) = -1, \quad \sigma(1, t) = 2, \quad \sigma(t, 1) = 1, \quad \text{and} \quad \sigma(t, t) = 0.$$
Bilinear extension gives us, for instance,
$$\begin{aligned}\sigma(2 - 3t, 1 + 4t) &= \sigma(2, 1) + \sigma(2, 4t) + \sigma(-3t, 1) + \sigma(-3t, 4t) \\ &= 2\sigma(1, 1) + 8\sigma(1, t) - 3\sigma(t, 1) - 12\sigma(t, t) \\ &= (2)(-1) + (8)(2) - (3)(1) = 11.\end{aligned}$$

We can go in the other direction, as well.

Let σ be a bilinear form on a vector space $V \cong \mathbb{F}^n$. Fix $\mathcal{B} = \{\mathbf{b}_i\}_{i=1}^n$, an ordered basis for V. The matrix representation of σ with respect to $\mathcal{B}$ is
$$[\sigma]_\mathcal{B} = [a_{ij}]$$
where $a_{ij} = \sigma(\mathbf{b}_i, \mathbf{b}_j)$. It is easy then to verify that if $\mathbf{v}$ and $\mathbf{x}$ belong to V,

(5.4) $$\sigma(\mathbf{v}, \mathbf{x}) = [\mathbf{v}]_\mathcal{B}{}^T [\sigma]_\mathcal{B} [\mathbf{x}]_\mathcal{B}.$$

5.3. Bilinear Forms, Their Matrices, and Duality

Example 5.24. Recall that the trace of a square matrix is the sum of its diagonal entries. Exercise 5.1.4 was to show that the trace mapping, $A \mapsto \operatorname{Tr} A$, is linear and that $\operatorname{Tr}(AB) = \operatorname{Tr}(BA)$ if A and B are in $M_n(\mathbb{F})$. Putting these together, we have that $\sigma : M_n(\mathbb{F}) \times M_n(\mathbb{F}) \to \mathbb{F}$ defined by

$$\sigma(A, B) = \operatorname{Tr}(AB)$$

is a bilinear form with the property $\sigma(A, B) = \sigma(B, A)$.

Here we will find the matrix representation of σ on $M_2(\mathbb{R})$, using the standard ordered basis

$$\mathcal{B} = \{E_{11}, E_{12}, E_{21}, E_{22}\}.$$

A Cayley table helps us track products between matrices in $\mathcal{B}$. We denote the zero matrix in $M_2(\mathbb{R})$ by O.

(5.5)

	E_{11}	E_{12}	E_{21}	E_{22}
E_{11}	E_{11}	E_{12}	O	O
E_{12}	O	O	E_{11}	E_{12}
E_{21}	E_{21}	E_{22}	O	O
E_{22}	O	O	E_{21}	E_{22}

Since $\operatorname{Tr} E_{11} = \operatorname{Tr} E_{22} = 1$ and $\operatorname{Tr} E_{12} = \operatorname{Tr} E_{21} = 0$, we have

$$\sigma(E_{ij}, E_{ji}) = 1, \text{ for } i, j = 1, 2$$

with all other inner products among elements in $\mathcal{B}$ equal to zero. This gives us

$$[\sigma]_\mathcal{B} = \begin{bmatrix} 1 & 0 & 0 & 0 \\ 0 & 0 & 1 & 0 \\ 0 & 1 & 0 & 0 \\ 0 & 0 & 0 & 1 \end{bmatrix}.$$

Let $A = \begin{bmatrix} 1 & 2 \\ 0 & 3 \end{bmatrix}$ and $B = \begin{bmatrix} 0 & 1 \\ 4 & 5 \end{bmatrix}$. We have $[A]_\mathcal{B} = (1, 2, 0, 3)$ and $[B]_\mathcal{B} = (0, 1, 4, 5)$ so

$$[A]_\mathcal{B}^T [\sigma]_\mathcal{B} [B]_\mathcal{B} = \begin{bmatrix} 1 & 2 & 0 & 3 \end{bmatrix} \begin{bmatrix} 1 & 0 & 0 & 0 \\ 0 & 0 & 1 & 0 \\ 0 & 1 & 0 & 0 \\ 0 & 0 & 0 & 1 \end{bmatrix} \begin{bmatrix} 0 \\ 1 \\ 4 \\ 5 \end{bmatrix}$$

$$= \begin{bmatrix} 1 & 0 & 2 & 3 \end{bmatrix} \begin{bmatrix} 0 \\ 1 \\ 4 \\ 5 \end{bmatrix} = 8 + 15 = 22.$$

Calculating $\sigma(A, B)$ directly we have

$$\sigma(A, B) = \operatorname{Tr}\left(\begin{bmatrix} 1 & 2 \\ 0 & 3 \end{bmatrix} \begin{bmatrix} 0 & 1 \\ 4 & 5 \end{bmatrix}\right) = \operatorname{Tr}\begin{bmatrix} 8 & 11 \\ 12 & 15 \end{bmatrix} = 22.$$

How are different matrix representations of the same bilinear form related? Suppose $\mathcal{B} = \{\mathbf{b}_i\}_{i=1}^n$ and $\mathcal{C}$ are ordered bases for $V \cong \mathbb{F}^n$. Let σ be a bilinear form on V and let P be the $\mathcal{B}$ to $\mathcal{C}$ change of basis matrix. For all $\mathbf{v}$ and $\mathbf{x}$ in V, we have

$$[\mathbf{v}]_\mathcal{C}^T [\sigma]_\mathcal{C} [\mathbf{x}]_\mathcal{C} = (P[\mathbf{v}]_\mathcal{B})^T [\sigma]_\mathcal{C} (P[\mathbf{x}]_\mathcal{B}) = [\mathbf{v}]_\mathcal{B}^T P^T [\sigma]_\mathcal{C} P [\mathbf{x}]_\mathcal{B}.$$

Taking $\mathbf{v} = \mathbf{b}_i$ and $\mathbf{x} = \mathbf{b}_j$, we get that the ij-entry of $P^T[\sigma]_\mathcal{C}P$ is $\sigma(\mathbf{b}_i, \mathbf{b}_j)$. This is enough to prove the next lemma.

Lemma 5.25. Let σ be a bilinear form on $V \cong \mathbb{F}^n$. If $\mathcal{B}$ and $\mathcal{C}$ are ordered bases for V, then $[\sigma]_\mathcal{B} = P^T[\sigma]_\mathcal{C}P$ where P is the $\mathcal{B}$ to $\mathcal{C}$ change of basis matrix.

Example 5.26. Here we find the matrix representation of the dot product on $\mathbb{R}^2$ with respect to $\mathcal{B} = \{(-1,1),(1,2)\}$ and then with respect to $\mathcal{C} = \{(1,0),(-1,2)\}$. We have $[\cdot]_\mathcal{B} = \begin{bmatrix} 2 & 1 \\ 1 & 5 \end{bmatrix}$ and $[\cdot]_\mathcal{C} = \begin{bmatrix} 1 & -1 \\ -1 & 5 \end{bmatrix}$.

The $\mathcal{B}$ to $\mathcal{C}$ change of basis matrix is $P = \begin{bmatrix} [(-1,1)]_\mathcal{C} & [(1,2)]_\mathcal{C} \end{bmatrix}$, which turns out to be $P = \begin{bmatrix} -1/2 & 2 \\ 1/2 & 1 \end{bmatrix}$. We can then check that

$$P^T[\cdot]_\mathcal{C}P = \begin{bmatrix} -1/2 & 1/2 \\ 2 & 1 \end{bmatrix} \begin{bmatrix} 1 & -1 \\ -1 & 5 \end{bmatrix} \begin{bmatrix} -1/2 & 2 \\ 1/2 & 1 \end{bmatrix} = \begin{bmatrix} 2 & 1 \\ 1 & 5 \end{bmatrix} = [\cdot]_\mathcal{B},$$

as promised in Lemma 5.25.

Since every invertible matrix can be viewed as a change of basis matrix, we have the following.

Lemma 5.27. Let $V \cong \mathbb{F}^n$ for some field $\mathbb{F}$. Matrices A and B in $M_n(\mathbb{F})$ are matrix representations of the same bilinear form on V if and only if $B = P^TAP$ for some invertible matrix, P in $M_n(\mathbb{F})$.

Proof. We have shown that if A and B are matrix representations of σ, then $B = P^TAP$ for an invertible matrix P. We leave proof of the other direction as an exercise. $\square$

When A is in $M_n(\mathbb{F})$ and P in $M_n(\mathbb{F})$ is invertible, we say that A and P^TAP are **congruent matrices**. We leave it as an exercise to show that congruence is an equivalence relation on $M_n(\mathbb{F})$.

Next we identify some of the connections between bilinear forms and dual spaces.

Let V be a vector space over $\mathbb{F}$ with bilinear form σ. Since σ is linear in its second variable, we can define $f: V \to V^*$ by $f(\mathbf{v})(\mathbf{x}) = \sigma(\mathbf{v}, \mathbf{x})$ for all $\mathbf{x}$ in V. Since σ is linear in its first variable, the mapping f is itself linear. We can play the same trick to get another linear transformation $g: V \to V^*$ defined by $g(\mathbf{v})(\mathbf{x}) = \sigma(\mathbf{x}, \mathbf{v})$.

When f and g are both injective, σ is **nondegenerate**. This means that if $\sigma(\mathbf{v}, \mathbf{x}) = 0$ for all $\mathbf{x}$ in V, then $\mathbf{v} = \mathbf{0}_V$ and if $\sigma(\mathbf{x}, \mathbf{v}) = 0$ for all $\mathbf{x}$ in V, then $\mathbf{v} = \mathbf{0}_V$. When V is finite-dimensional, we only need one of these criteria. This is part of the next theorem.

Theorem 5.28. A bilinear form σ on $V \cong \mathbb{F}^n$ is nondegenerate if and only if $f: V \to V^*$ given by

$$f(\mathbf{v})(\mathbf{x}) = \sigma(\mathbf{v}, \mathbf{x})$$

is injective, which is true if and only if any matrix representation of σ is invertible.

Proof. Suppose σ is a bilinear form defined on $V \cong \mathbb{F}^n$. Let $\mathcal{B}$ be a basis for V and let $A = [\sigma]_\mathcal{B}$. Let $f : V \to V^*$ be as given in the statement of the theorem. For $\mathbf{v}$ and $\mathbf{x}$ in V, we have

$$f(\mathbf{v})(\mathbf{x}) = \sigma(\mathbf{v}, \mathbf{x}) = [\mathbf{v}]_\mathcal{B}{}^T A [\mathbf{x}]_\mathcal{B} = (A^T [\mathbf{v}]_\mathcal{B})^T [\mathbf{x}]_\mathcal{B}.$$

Noting that the $\mathcal{B}$-coordinate mapping and transposition are both isomorphisms, we can complete the proof from here, step by step.

First, f is injective if and only if $(A^T[\mathbf{v}]_\mathcal{B})^T$ is nonzero in $\mathbb{F}_n$ whenever $\mathbf{v}$ is nonzero in V. Second, $(A^T[\mathbf{v}]_\mathcal{B})^T$ is nonzero in $\mathbb{F}_n$ whenever $\mathbf{v}$ is nonzero in V if and only if A^T has a trivial null space. Next, A^T has a trivial null space if and only if A^T is invertible, which is the case if and only if A is invertible. Finally, A is invertible if and only if $A[\mathbf{v}]_\mathcal{B}$ is nonzero for nonzero $\mathbf{v}$ in V, which is the case if and only if there is $\mathbf{x}$ in V such that

$$[\mathbf{x}]_\mathcal{B}{}^T (A[\mathbf{v}]_\mathcal{B}) = [\mathbf{x}]_\mathcal{B}{}^T A [\mathbf{v}]_\mathcal{B} = \sigma(\mathbf{x}, \mathbf{v}) \neq 0.$$

This is enough to prove the theorem. $\square$

Since the matrix representation of the trace bilinear form in Example 5.24 is its own inverse, that bilinear form is nondegenerate.

The next theorem is a variant on the last theorem that applies to spaces of arbitrary dimension. The proof is routine, so we leave it as an exercise.

Theorem 5.29. Let V be a vector space with a nondegenerate bilinear form, σ. The mapping $\mathbf{v} \mapsto f_\mathbf{v}$ where

$$f_\mathbf{v}(\mathbf{x}) := \sigma(\mathbf{v}, \mathbf{x})$$

for $\mathbf{x}$ in V is an injective linear transformation $V \to V^*$.

If V is a vector space with an injective linear transformation $f : V \to V^*$, then σ defined by

$$\sigma(\mathbf{v}, \mathbf{x}) := f(\mathbf{v})(\mathbf{x})$$

for $\mathbf{x}$ in V is a nondegenerate bilinear form on V.

We have already seen that $\mathbb{F}^n$ and $\mathbb{F}_n$ are canonically isomorphic, so the next lemma should come as no surprise. The proof is an exercise.

Lemma 5.30. The mapping $\sigma : \mathbb{F}^n \times \mathbb{F}^n \to \mathbb{F}$ given by

$$\sigma(\mathbf{v}, \mathbf{x}) = \mathbf{v}^T \mathbf{x}$$

is a nondegenerate bilinear form on $\mathbb{F}^n$.

A **symmetric bilinear form** on V is a bilinear form σ such that

$$\sigma(\mathbf{v}, \mathbf{x}) = \sigma(\mathbf{x}, \mathbf{v})$$

for all $\mathbf{v}$ and $\mathbf{x}$ in V. The dot product on $\mathbb{R}^n$ and the trace bilinear form are symmetric.

A **skew-symmetric form** on V is a bilinear form σ such that

$$\sigma(\mathbf{v}, \mathbf{x}) = -\sigma(\mathbf{x}, \mathbf{v})$$

for all $\mathbf{v}$ and $\mathbf{x}$ in V. Defining σ on $\mathbb{R}^2$ by $[\sigma]_{\mathcal{E}} = \begin{bmatrix} 0 & 1 \\ -1 & 0 \end{bmatrix}$ gives us a skew-symmetric form.

When the underlying field has characteristic two, symmetric and skew-symmetric forms are the same.

It should be evident that σ is a symmetric bilinear form on $V \cong \mathbb{F}^n$ if and only if $[\sigma]_{\mathcal{B}}^T = [\sigma]_{\mathcal{B}}$, where $\mathcal{B}$ is any ordered basis of V. Likewise, σ is a skew-symmetric form if and only if $[\sigma]_{\mathcal{B}}^T = -[\sigma]_{\mathcal{B}}$. We leave the justifications of those statements to the exercises.

Definition 5.31. A **symmetric matrix** is an $n \times n$ matrix A such that $A^T = A$. A **skew-symmetric matrix** is an $n \times n$ matrix A such that $A^T = -A$.

The set of symmetric matrices in $M_n(\mathbb{F})$ is a vector subspace of $M_n(\mathbb{F})$. Likewise, the skew-symmetric matrices in $M_n(\mathbb{F})$ comprise a vector space.

We leave it as an exercise to verify that when A is arbitrary in $M_n(\mathbb{F})$ and $\operatorname{char} \mathbb{F} \neq 2$, $A + A^T$ is symmetric and $A - A^T$ is skew-symmetric. This gives us the decomposition

$$A = \frac{1}{2}((A + A^T) + (A - A^T)),$$

which proves the following theorem.

Theorem 5.32. If $\operatorname{char} \mathbb{F} \neq 2$, then every matrix in $M_n(\mathbb{F})$ can be written as the sum of a symmetric matrix and a skew-symmetric matrix.

Exercises 5.3.

5.3.1. Argue directly that σ defined on $\mathbb{F}^n$ by $\sigma(\mathbf{v}, \mathbf{x}) = \mathbf{v}^T \mathbf{x}$ is a nondegenerate symmetric bilinear form.

5.3.2. Show that fixing a basis $\mathcal{B}$ for $V \cong \mathbb{F}^n$, fixing a matrix A in $M_n(\mathbb{F})$, and defining

$$\sigma(\mathbf{v}, \mathbf{x}) := [\mathbf{v}]_{\mathcal{B}}^T A [\mathbf{x}]_{\mathcal{B}}$$

gets us a bilinear form, σ, defined on V.

5.3.3. Find the matrix representations of the dot product on $\mathbb{R}^2$ with respect to $\mathcal{B} = \{(1,1), (0,1)\}$ and with respect to $\mathcal{C} = \{(1,1), (-1,1)\}$. Use a change of basis matrix to verify that the two matrix representations are congruent.

5.3.4. Finish the proof of Lemma 5.27.

5.3.5. Show that congruence is an equivalence relation on $M_n(\mathbb{F})$.

5.3.6. Let $\mathbf{v}$ and $\mathbf{x}$ be nonzero vectors in $\mathbb{R}^2$. Show that $\mathbf{v} \cdot \mathbf{x} = \mathbf{v}^T \mathbf{x} = 0$ if and only if $\operatorname{Span}\{\mathbf{v}\}$ and $\operatorname{Span}\{\mathbf{x}\}$ are perpendicular lines.

5.3.7. Verify that (5.4) holds when σ is a bilinear form on V and $\mathbf{v}$ and $\mathbf{x}$ belong to V.

5.3.8. Prove Theorem 5.29.

5.3.9. Let V be a vector space over $\mathbb{F}$ with a bilinear form, σ. As in the text, define
$$f : V \to V^*$$
by $f(\mathbf{v}) = \sigma(\mathbf{v}, \mathbf{x})$. Let $\mathcal{B} = \{\mathbf{b}_i\}_{i \in I}$ be a basis for V. Is $\{f(\mathbf{b}_i)\}_{i \in I}$ the dual basis, $\mathcal{B}^*$?

5.3.10. Give an example of a 3×3 matrix over $\mathbb{F}_3$ that is skew-symmetric and invertible.

5.3.11. Give an example of a nonzero, 2×2 skew-symmetric matrix over $\mathbb{R}$.

5.3.12. Give an example of a nonzero, 3×3 skew-symmetric matrix over $\mathbb{R}$.

5.3.13. Let A be a matrix representation of a bilinear form σ.
   (a) Show that A is symmetric if and only if σ is symmetric.
   (b) Show that A is skew-symmetric if and only if σ is skew-symmetric.

5.3.14. Let A in $M_n(\mathbb{F})$ be skew-symmetric and assume char $\mathbb{F} \neq 2$.
   (a) What are the diagonal entries of A?
   (b) What is the relationship between $\mathbf{a}_i$, column i of A, and $\mathbf{w}_i$, row i of A?

5.3.15. Let A be in $M_n(\mathbb{F})$. Verify that $A + A^T$ is symmetric and that $A - A^T$ is skew-symmetric.

5.3.16. Let $A = \begin{bmatrix} 1 & -1 & 0 & 2 \\ 4 & 0 & -1 & 3 \\ 0 & 2 & 0 & 6 \\ 1 & 1 & 2 & 0 \end{bmatrix}$. Write A as the sum of a symmetric matrix and a skew-symmetric matrix.

5.3.17. In this problem, $A = [a_{ij}]$ and $B = [b_{ij}]$ are skew-symmetric matrices in $M_4(\mathbb{R})$. Let $\mathbf{a}_i$ be column i of A. Let $\mathbf{b}_i$ be column i of B.
   (a) Find A if $\mathbf{a}_1 = (0, 2, 1, 3)$, $a_{32} = -7$, $a_{24} = 5$, and $a_{43} = -4$.
   (b) Find B if $b_{21} = 4$, $\mathbf{b}_3 = (-1, 2, 0, 6)$, $b_{14} = 3$, and $b_{42} = -2$.
   (c) Calculate $\sigma(A, B) = \frac{1}{2}\operatorname{Tr}(A^T B)$. How many multiplications do you need?

5.4. Linear Operators and Direct Sums

We turn away from duality in this section, to dive a little deeper into relationships between subspaces and linear transformations. All our direct sums here are internal so when we write $V = \bigoplus_{j \in J} V_j$, V_j is a subspace of V for all $j \in J$.

Recall that if L is a linear operator on V, a subspace $U \subseteq V$ is L-invariant provided $L[U] \subseteq U$. The kernel of any linear operator L is L-invariant, as is its image.

Definition 5.33. An *L-decomposition* of a vector space V is a direct sum decomposition, $V = U \oplus W$, where U and W are both L-invariant for L in $\mathcal{L}(V)$. When V has L-decomposition $U \oplus W$, we refer to $U \oplus W$ as an *L-sum* and we say that U and W are *L-complements* in V.

Any projection $L : V \to V$ gives us an L-decomposition because in this case, $V = L[V] \oplus \operatorname{Ker} L$.

When $V = U \oplus W$ is an L-decomposition, $L|_U$ belongs to $\mathcal{L}(U)$ and $L|_W$ belongs to $\mathcal{L}(W)$. In this case, we write $L_U := L|_U$ and $L_W := L|_W$, and

$$L = L_U \oplus L_W,$$

and we say that L is a **direct sum of linear operators** L_U and L_W. We also say that $L_U \oplus L_W$ is a **direct sum decomposition of** L. When both U and W are nontrivial spaces, we say the decomposition $L_U \oplus L_W$ is nontrivial.

If L is the projection onto the y-axis along the line $y = x$ in $\mathbb{R}^2$, then we get a nontrivial direct sum decomposition for L by taking

$$U = \operatorname{Ker} L = \operatorname{Span}\{(1,1)\}$$

and

$$W = L[\mathbb{R}^2] = \operatorname{Span}\{(0,1)\}.$$

We see then that L_U is identically zero and L_W is the identity mapping.

Does every linear operator have a nontrivial direct sum decomposition? The next examples show that the answer is no.

Example 5.34. Consider L in $\mathcal{L}(\mathbb{R}^2)$ given by $L(x,y) = (2x+y, 2y)$. A nontrivial decomposition for L would correspond to a decomposition of $\mathbb{R}^2$ into two L-invariant 1-dimensional subspaces. Each subspace would thus be of the form $\operatorname{Span}\{\mathbf{v}\}$, for some nonzero $\mathbf{v}$ in $\mathbb{R}^2$. Moreover, for $\mathbf{v} = (x, y)$, we would need $L(x, y) = c(x, y)$, for some scalar, c. This would mean $2x = cx$ and $x + 2y = cy$. We leave it as an exercise to verify that

$$\begin{array}{rcl} (2-c)x & = & 0 \\ x + (2-c)y & = & 0 \end{array}$$

has a nontrivial solution only when $c = 2$ and $x = 0$. It follows that the only 1-dimensional L-invariant subspace is $\operatorname{Span}\{(0,1)\}$, thus, L does not have a direct sum decomposition.

Example 5.35. Differentiation, $D(f) = f'$, is a linear operator on $C^\infty(\mathbb{R})$. Let

$$\mathcal{C} = \{e^{at} \mid a \in \mathbb{R}\},$$

and let $U = \operatorname{Span} \mathcal{C}$. It turns out that $\mathcal{C}$ is a basis for U. (We outline the verification that $\mathcal{C}$ is linearly independent in the exercises.) We would like to explore the possibility that U has a D-complement in $C^\infty(\mathbb{R})$.

A typical element in U has the form

$$f(t) = c_1 e^{a_1 t} + \cdots + c_k e^{a_k t}$$

for distinct real numbers $a_1, \ldots, a_k$ and some collection of scalars $c_1, \ldots c_k$. Since

$$f'(t) = a_1 c_1 e^{a_1 t} + \cdots + a_k c_k e^{a_k t},$$

we see that $D[U] \subseteq U$.

Next we attempt to construct a D-complement to U in C^∞.

Let $\mathcal{B}$ be a basis for $C^\infty(\mathbb{R})$ that contains $\mathcal{C}$. Let $W = \operatorname{Span}(\mathcal{B} \setminus \mathcal{C})$. Since $g(t) = te^t$ belongs to $C^\infty(\mathbb{R})$, we must be able to write g as a linear combination of exponentials and some f in W. We then have

$$f(t) = te^t - \sum_{i=1}^{k} c_i e^{a_i t} \in W,$$

so that

$$f'(t) = (t+1)e^t - \sum_{i=1}^{k} a_i c_i e^{a_i t}.$$

If W were D-invariant, then f' would belong to W, thus would $f' - f$. As a linear combination of elements in $\mathcal{C}$, $f' - f$ belongs to U. Since it is in the intersection of W and U, $f - f'$ must be identically zero. From calculus, we know that $f = f'$ only if $f(t) = ce^{at}$ for some c and a in $\mathbb{R}$. Since ce^{at} is in U, we conclude that there is no D-invariant complement to U in $C^\infty(\mathbb{R})$.

Suppose L in $\mathcal{L}(V)$ has a direct sum decomposition $L = L_U \oplus L_W$. Let $\mathcal{B}_U$ and $\mathcal{B}_W$ be respective bases for U and W. The union $\mathcal{B} = \mathcal{B}_U \cup \mathcal{B}_W$ must then be a basis for V. Since U and W are L-invariant, $L[\mathcal{B}_U]$ and $L[\mathcal{B}_W]$ are disjoint sets that span $L[V]$. Not only do we have $V = U \oplus W$ and $L = L_U \oplus L_W$, we also have $L[V] = L[U] \oplus L[W]$.

The finite-dimensional case illuminates why it can be useful to know whether a linear operator has a direct sum decomposition.

Suppose $V \cong \mathbb{F}^n$ and that $L = L_U \oplus L_W$ is in $\mathcal{L}(V)$. Say that $\dim U = k$, so that $\dim W = n - k$. Let

$$\mathcal{B}_U = \{\mathbf{b}_1, \ldots, \mathbf{b}_k\}$$

be an ordered basis for U, and let

$$\mathcal{B}_W = \{\mathbf{b}_{k+1}, \ldots, \mathbf{b}_n\}$$

be an ordered basis for W. We then have an ordered basis for V,

$$\mathcal{B} = \{\mathbf{b}_1, \ldots, \mathbf{b}_k, \mathbf{b}_{k+1}, \ldots, \mathbf{b}_n\}.$$

The $\mathcal{B}$-coordinate vector for each element in $L[\mathcal{B}_U]$ has the form

$$(a_1, \ldots, a_k, 0, \ldots, 0).$$

Likewise, the $\mathcal{B}$-coordinate vector for each element in $L[\mathcal{B}_W]$ has the form $(0, \ldots, 0, a_{k+1}, \ldots, a_n)$. It follows that

$$[\![L]\!]_\mathcal{B} = \left[\begin{array}{c|c} A_U & O_{k,n-k} \\ \hline O_{n-k,k} & A_W \end{array}\right] \tag{5.6}$$

where

$$A_U = [\![L_U]\!]_{\mathcal{B}_U} \in M_k(\mathbb{F})$$

and

$$A_W = [\![L_W]\!]_{\mathcal{B}_W} \in M_{n-k}(\mathbb{F}).$$

The matrix in (5.6) is in **block diagonal form**. The square blocks on the diagonal are A_U and A_W. The off-diagonal zero matrices are not necessarily square. Each of rows $1 - k$ of $[\![L]\!]_\mathcal{B}$ has the form

$$\begin{bmatrix} a_{i1} & \cdots & a_{ik} & 0 & \cdots & 0 \end{bmatrix}.$$

Each of rows $k+1$ through n has the form
$$\begin{bmatrix} 0 & \cdots & 0 & a_{i\,k+1} & \cdots & a_{in} \end{bmatrix}.$$

If we had a further direct sum decomposition of either L_U or L_W, then A_U and/or A_W would itself be in block diagonal matrix form.

When A in $M_n(\mathbb{F})$ can be written in the form

(5.7) $$\begin{bmatrix} A_1 & & \\ & \ddots & \\ & & A_k \end{bmatrix}$$

where each A_i is a square matrix and all entries outside the square blocks are zero, we write
$$A = \mathrm{diag}(A_1, \ldots, A_k).$$
Each A_i is then a **square block submatrix** of A.

We leave proof of the next theorem as an exercise.

Theorem 5.36. Let V be finite-dimensional over $\mathbb{F}$. A linear operator L on V has a block diagonal matrix representation if and only if L has a direct sum decomposition.

The theorem says that a direct sum decomposition of a linear operator on a finite-dimensional space and a block diagonal representation of a linear operator are two ways of looking at the same thing.

Induction on the number of square blocks can be applied to prove the following. We leave the details as an exercise.

Lemma 5.37. Let $A = \mathrm{diag}(A_1, \ldots, A_k)$ and $B = \mathrm{diag}(B_1, \ldots, B_k)$. If A_i and B_i belong to $M_{n_i}(\mathbb{F})$ for each i, then $AB = \mathrm{diag}(A_1 B_1, \ldots, A_k B_k)$.

Lemma 5.37 generalizes. For example, if
$$A = \begin{bmatrix} A_{11} & A_{12} \\ A_{21} & A_{22} \end{bmatrix}$$
and
$$B = \begin{bmatrix} B_{11} & B_{12} \\ B_{21} & B_{22} \end{bmatrix}$$
belong to $M_{m,n}(\mathbb{F})$ and each product $A_{ik} B_{kj}$ is defined, then
$$AB = \begin{bmatrix} A_{11} B_{11} + A_{12} B_{21} & A_{11} B_{12} + A_{12} B_{22} \\ A_{21} B_{11} + A_{22} B_{21} & A_{21} B_{12} + A_{22} B_{22} \end{bmatrix}.$$

Exercises 5.4.

5.4.1. Show that if $L(x,y) = (2x, x+2y)$ is defined on $\mathbb{R}^2$, then the only nontrivial L-invariant subspace in $\mathbb{R}^2$ is the y-axis.

5.4.2. Find the L-decomposition for $\mathbb{R}^2$ and the direct sum decomposition for L in $\mathcal{L}(\mathbb{R}^2)$, when $L(x,y) = (2x, x+3y)$.

5.4.3. For U as given in Example 5.35, is $D[U] = U$?

5.4. Linear Operators and Direct Sums

5.4.4. Let $\mathcal{C} = \{e^{at} \,|\, a \in \mathbb{R}\}$. We use induction here to sketch a proof that $\mathcal{C}$ is linearly independent. Later in the course, we will see a more elegant proof. Suppose that for distinct a_i in $\mathbb{R}$ and scalars c_i in $\mathbb{R}$,
$$f(t) = c_1 e^{a_1 t} + \cdots + c_k e^{a_k t} = 0 \ \forall t \in \mathbb{R}.$$
If $c_1 e^{a_1 t} = 0$, then since $e^{a_1 t} > 0$ for all t, $c_1 = 0$. Assume that if $k < n$, then $f(t) = 0$ implies $c_1 = \cdots = c_k = 0$.

(a) Find $(D - a_k I)(f)$, where $D(f) = f'$ and I is the identity operator.
(b) Why is $(D - a_k I)(f)$ identically zero?
(c) What can you conclude about the coefficients in $(D - a_k I)(f)$?
(d) The coefficients in $(D - a_k I)(f)$ are products. One of the factors in each product is nonzero. Which is it, and why is it nonzero?
(e) Finish the proof that $c_i = 0$ for all i, thus that $\mathcal{C}$ is linearly independent.

5.4.5. This is another approach to showing that $\mathcal{C} = \{e^{at} \,|\, a \in \mathbb{R}\}$ is linearly independent. Let $a_1, \ldots, a_k$ be distinct in $\mathbb{R}$, and suppose
$$f(t) = c_1 e^{a_1 t} + \ldots + c_k e^{a_k t} = 0 \ \forall t \in \mathbb{R}.$$

(a) Argue that for $j = 1, \ldots, k-1$, $D^j(f)$ is identically zero.
(b) Write out the set of equations that you get by setting $f(t)$ and $D^j(f)$ equal to zero for $j = 1, \ldots, k-1$.
(c) Write out the $k \times k$ system of equations that you get by evaluating $f(0)$ and $D^j(f)(0)$, for $j = 1, \ldots, k-1$. We want to find the c_is. What is the coefficient matrix?
(d) If everything is going according to plan, you should be looking at a **Vandermonde matrix**. If we prove that this matrix is nonsingular, then we know the only solution to our system of equations is $c_i = 0$ for $i = 1, \ldots, k$. To reach this conclusion, consider the mapping
$$\chi : \mathbb{R}[t]_{k-1} \to \mathbb{R}^k$$
given by $\chi(p(t)) = (p(a_1), \ldots, p(a_k))$. Let A be the matrix representation for this mapping with respect to the standard bases in $\mathbb{R}[t]_{k-1}$ and $\mathbb{R}^k$. Find A and A^T.
(e) What is the maximum number of distinct zeroes that a degree $k-1$ polynomial over $\mathbb{R}$ can have? Use this observation to argue that χ has trivial kernel and thus is an isomorphism.
(f) Complete the argument that A is nonsingular, thus, that our $k \times k$ homogeneous system only has the trivial solution, thus, that
$$\{e^{a_1 t}, \ldots, e^{a_k t}\}$$
is linearly independent, thus, that $\mathcal{C}$ is linearly independent.

5.4.6. Suppose V is finite-dimensional and that L in $\mathcal{L}(V)$ has a diagonal block matrix form as given in (5.7). Show that L is the direct sum of linear operators on V, $L_1, \ldots, L_k$. Describe the associated subspaces of V.

5.4.7. Prove that if V is finite-dimensional and if L in $\mathcal{L}(V)$ is a direct sum of linear operators in $\mathcal{L}(V)$, then L has a block diagonal matrix representation.

5.4.8. Prove Lemma 5.37.

5.4.9. Let $A = \begin{bmatrix} 2 & 0 & 0 & 0 & 0 \\ 0 & 3 & -4 & 0 & 0 \\ 0 & -1 & 2 & 0 & 0 \\ 0 & 0 & 0 & 5 & 3 \\ 0 & 0 & 0 & -1 & -2 \end{bmatrix}$ in $M_5(\mathbb{R})$.

(a) Write A as a block diagonal matrix.

(b) Find an L-decomposition of $\mathbb{R}^5$ for L in $\mathcal{L}(\mathbb{R}^5)$ given by $L(\mathbf{v}) = A\mathbf{v}$.

5.5. Groups of Matrices

We have seen that upper triangular matrices, lower triangular matrices, symmetric matrices, and skew-symmetric matrices all determine vector subspaces of $M_n(\mathbb{F})$. Here, we identify certain subsets in $M_n(\mathbb{F})$ that are closed under matrix multiplication, while not necessarily closed under linear combination.

The first question we may ask about a matrix, or a linear transformation, is whether it is invertible. While it is easy to see that invertible matrices do not comprise a subspace of $M_n(\mathbb{F})$ and that automorphisms do not comprise a subspace of $\mathcal{L}(V)$, each of these sets is an example of a *group*.

Definition 5.38. A **group** $(G, \circ)$ is a set G with an associative binary operation $\circ$ such that

(a) there is a $\circ$ identity element in G and

(b) every element in G has a $\circ$ inverse in G.

When $(G, \circ)$ is a group, we also say that G is a **group under** $\circ$.

A group $(G, \circ)$ is **abelian** provided $\circ$ is commutative.

A **subgroup** is a subset of a group $(G, \circ)$ that is itself a group under $\circ$.

When discussing abstract groups, we use multiplicative notation, that is, the binary operation is indicated by juxtaposition, and the inverse of g is g^{-1}. We usually indicate the identity element by e. The exception to this rule is when we are working with abelian groups in which case we use additive notation, indicating the binary operation with $+$, using $-g$ for the inverse of g, and indicating the identity element by 0.

Since every element in a group has an inverse, every element in a group is an inverse.

The set of invertible matrices in $M_n(\mathbb{F})$ is designated $GL_n(\mathbb{F})$.[1] We have seen that for all A and B in $GL_n(\mathbb{F})$,

$$(AB)^{-1} = B^{-1}A^{-1}.$$

It follows that matrix multiplication is a binary operation on $GL_n(\mathbb{F})$, with multiplicative identity I_n. By Theorem 3.2, multiplication in $GL_n(\mathbb{F})$ is associative. Since A in $M_n(\mathbb{F})$ is invertible if and only if A^{-1} is invertible, every matrix in $GL_n(\mathbb{F})$ has an inverse in $GL_n(\mathbb{F})$. This is enough to prove that $GL_n(\mathbb{F})$ is a group under matrix multiplication.

---

[1] The notation $GL(n, \mathbb{F})$ is also in common usage.

5.5. Groups of Matrices

Definition 5.39. The **general linear group** associated to $\mathbb{F}^n$ is the set of invertible matrices in $M_n(\mathbb{F})$ under matrix multiplication.

Example 5.40. We can write down all the elements in $GL_2(\mathbb{F}_2)$:
$$\left\{ \begin{bmatrix} 1 & 0 \\ 0 & 1 \end{bmatrix}, \begin{bmatrix} 0 & 1 \\ 1 & 0 \end{bmatrix}, \begin{bmatrix} 1 & 1 \\ 1 & 0 \end{bmatrix}, \begin{bmatrix} 1 & 1 \\ 0 & 1 \end{bmatrix}, \begin{bmatrix} 1 & 0 \\ 1 & 1 \end{bmatrix}, \begin{bmatrix} 0 & 1 \\ 1 & 1 \end{bmatrix} \right\}.$$

It is easy to check that
$$H = \left\{ \begin{bmatrix} 1 & 0 \\ 0 & 1 \end{bmatrix}, \begin{bmatrix} 0 & 1 \\ 1 & 0 \end{bmatrix} \right\}$$
is a subgroup of $GL_2(\mathbb{F}_2)$.

Next, notice that
$$\begin{bmatrix} 1 & 1 \\ 1 & 0 \end{bmatrix}^{-1} = \begin{bmatrix} 0 & 1 \\ 1 & 1 \end{bmatrix}.$$

Note, moreover, that
$$\begin{bmatrix} 1 & 1 \\ 1 & 0 \end{bmatrix} \begin{bmatrix} 1 & 1 \\ 1 & 0 \end{bmatrix} = \begin{bmatrix} 0 & 1 \\ 1 & 1 \end{bmatrix}.$$

From here, it is easy to verify that
$$K = \left\{ \begin{bmatrix} 1 & 0 \\ 0 & 1 \end{bmatrix}, \begin{bmatrix} 1 & 1 \\ 1 & 0 \end{bmatrix}, \begin{bmatrix} 0 & 1 \\ 1 & 1 \end{bmatrix} \right\}$$
is another subgroup of $GL_2(\mathbb{F}_2)$.

We often care less about arbitrary linear operators on a vector space and more about linear operators that respect some kind of additional structure on the space. The dot product, for instance, allows us to define distance and angle in $\mathbb{R}^n$ and in so doing, gives us a link between linear algebra and Euclidean geometry. The length of $\mathbf{v}$ in $\mathbb{R}^n$ is
$$\|\mathbf{v}\| = \sqrt{\mathbf{v} \cdot \mathbf{v}},$$
the distance between points $\mathbf{v}$ and $\mathbf{x}$ in $\mathbb{R}^n$ is $\|\mathbf{v} - \mathbf{x}\|$, and the angle between vectors $\mathbf{v}$ and $\mathbf{x}$ in $\mathbb{R}^n$ is φ where
$$\cos \varphi = \frac{\mathbf{v} \cdot \mathbf{x}}{\|\mathbf{v}\| \|\mathbf{x}\|}.$$

If L in $\mathcal{L}(\mathbb{R}^n)$ satisfies
$$L(\mathbf{v}) \cdot L(\mathbf{x}) = \mathbf{v} \cdot \mathbf{x}$$
for all $\mathbf{v}$ and $\mathbf{x}$ in $\mathbb{R}^n$, then it does not change the length of a vector, the distance between two vectors, or the angle between two vectors in $\mathbb{R}^n$. In respecting the dot product, L respects the geometry of $\mathbb{R}^n$.

Definition 5.41. An **orthogonal operator** on $\mathbb{R}^n$ is L in $\mathcal{L}(\mathbb{R}^n)$ such that
$$L(\mathbf{v}) \cdot L(\mathbf{x}) = \mathbf{v} \cdot \mathbf{x}$$
for all $\mathbf{v}$ and $\mathbf{x}$ in $\mathbb{R}^n$.

If L in $\mathcal{L}(\mathbb{R}^n)$ is orthogonal and A in $M_n(\mathbb{R})$ is the standard matrix representation of L, then

(5.8) $$A\mathbf{v} \cdot A\mathbf{x} = (A\mathbf{v})^T (A\mathbf{x}) = \mathbf{v}^T A^T A \mathbf{x} = \mathbf{v}^T \mathbf{x}$$

for every **v** and **x** in $\mathbb{R}^n$. We leave it as an exercise to verify that (5.8) holds if and only if $A^T A = I_n$. In particular, A is in $GL_n(\mathbb{R})$ and $A^T = A^{-1}$.

This motivates the next definition.

Definition 5.42. An **orthogonal matrix** is a square matrix A such that $A^T = A^{-1}$.

Definition 5.42 applies over any field, though the orthogonal matrices of greatest interest are those over $\mathbb{R}$.

It is easy to show that the product of orthogonal matrices in $M_n(\mathbb{F})$ is also orthogonal and that the inverse of an orthogonal matrix is orthogonal. This is most of the work required to show that the set of orthogonal matrices in $M_n(\mathbb{F})$ is a subgroup of $GL_n(\mathbb{F})$.

Definition 5.43. The **orthogonal group** associated to $\mathbb{F}^n$ is
$$O_n(\mathbb{F}) := \{A \in GL_n(\mathbb{F}) \mid A^T = A^{-1}\}.$$

Example 5.44. How much can we say about A if it belongs to $O_2(\mathbb{R})$? Assume $A = \begin{bmatrix} a & b \\ c & d \end{bmatrix}$ is in $O_2(\mathbb{R})$ and note that we then have
$$AA^T = \begin{bmatrix} a^2 + b^2 & ac + bd \\ ac + bd & c^2 + d^2 \end{bmatrix}.$$

This allows us to say that $a^2 + b^2 = c^2 + d^2 = 1$ and $ac + bd = 0$. Since $a^2 + b^2$ and $c^2 + d^2$ are both 1, our choices for the entries in A amount to $a = \pm\cos\varphi$, $b = \pm\sin\varphi$, $c = \pm\cos\psi$, and $d = \pm\sin\psi$ for some real φ and ψ. This gives us

(5.9)
$$\pm\cos\varphi\cos\psi \pm \sin\varphi\sin\psi = 0.$$

Effectively, there are two possibilities for the sign choices in (5.9). One is
$$\cos\varphi\cos\psi + \sin\varphi\sin\psi = \cos(\varphi - \psi) = 0.$$

The second is
$$\cos\varphi\cos\psi - \sin\varphi\sin\psi = \cos(\varphi + \psi) = 0.$$

In the first case, we can take $\varphi - \psi = \pi/2$ and use trigonometric identities to find that $\cos\psi = \sin\varphi$ and $\sin\psi = -\cos\varphi$. We then have
$$A = \begin{bmatrix} \cos\varphi & \sin\varphi \\ \sin\varphi & -\cos\varphi \end{bmatrix}.$$

Note that $\det A = -1$.

The same approach applies in the second case. We can take $\varphi + \psi = \pi/2$ and use identities to get $\cos\psi = -\sin\varphi$ and $\sin\psi = \cos\varphi$. This gives us
$$A = \begin{bmatrix} \cos\varphi & \sin\varphi \\ -\sin\varphi & \cos\varphi \end{bmatrix}.$$

In this case, $\det A = 1$.

5.5. Groups of Matrices

Some examples of matrices in $O_2(\mathbb{R})$, aside from I_2, are $\begin{bmatrix} 0 & 1 \\ 1 & 0 \end{bmatrix}$, $\begin{bmatrix} 0 & 1 \\ -1 & 0 \end{bmatrix}$, and $\begin{bmatrix} 1 & 0 \\ 0 & -1 \end{bmatrix}$. Later, we will see that each element in $O_2(\mathbb{R})$ is either a rotation or an orthogonal reflection.

We look at the complex analog to the orthogonal group in the next section.

Exercises 5.5.

5.5.1. Show that $GL_n(\mathbb{F})$ is not a vector space.

5.5.2. Give an example to show that the sum of automorphisms on a vector space is not necessarily an automorphism.

5.5.3. If V is any vector space, the collection of automorphisms on V is often designated $GL(V)$. Show that $GL(V)$ is a group.

5.5.4. Can the empty set be considered a group?

5.5.5. Explain the statement, "Since every element in a group has an inverse, every element in a group is an inverse."

5.5.6. Using the following methods, show that in $GL_2(\mathbb{F}_2)$, $\begin{bmatrix} 1 & 1 \\ 1 & 0 \end{bmatrix}^{-1} = \begin{bmatrix} 0 & 1 \\ 1 & 1 \end{bmatrix}$:

  (a) Apply the formula in (3.4).
  (b) Row reduce $\left[\begin{array}{cc|cc} 1 & 1 & 1 & 0 \\ 1 & 0 & 0 & 1 \end{array}\right]$.
  (c) Show that $\begin{bmatrix} 1 & 1 \\ 1 & 0 \end{bmatrix}\begin{bmatrix} a \\ b \end{bmatrix} = \begin{bmatrix} a+b \\ b \end{bmatrix}$ and that $\begin{bmatrix} 0 & 1 \\ 1 & 1 \end{bmatrix}\begin{bmatrix} a+b \\ b \end{bmatrix} = \begin{bmatrix} a \\ b \end{bmatrix}$, for all a, b in $\mathbb{F}_2$.

5.5.7. Show that if g and h belong to a group and $gh = g$, then $h = e$.

5.5.8. Show that if g and h belong to a group and $gh = e$, then $h = g^{-1}$.

5.5.9. Show that if g and h belong to a group, then $(gh)^{-1} = h^{-1}g^{-1}$.

5.5.10. Show that $(g^{-1})^{-1} = g$ if g is an element in any group.

5.5.11. In reference to Example 5.40, show that H and K are subgroups of $GL_2(\mathbb{F}_2)$.

5.5.12. Show that the diagonal matrices in $GL_n(\mathbb{F})$ form a subgroup of $GL_n(\mathbb{F})$.

5.5.13. Let A belong to $M_n(\mathbb{R})$ and let L_A be the associated linear operator on $\mathbb{R}^n$. Show that L_A is an orthogonal operator if and only if $A^T A = I_n$.

5.5.14. Verify that $O_n(\mathbb{F})$ is a subgroup of $GL_n(\mathbb{F})$.

5.5.15. Give a geometric description of the mappings in $O_2(\mathbb{R})$ associated to each of $\begin{bmatrix} 0 & 1 \\ 1 & 0 \end{bmatrix}$, $\begin{bmatrix} 0 & 1 \\ -1 & 0 \end{bmatrix}$, and $\begin{bmatrix} 1 & 0 \\ 0 & -1 \end{bmatrix}$. (Hint: Where are e_1 and e_2 mapped in each case?)

5.5.16. Use polar coordinates on elements in $\mathbb{R}^2$ to argue that the linear transformation determined by $\begin{bmatrix} \cos\varphi & -\sin\varphi \\ \cos\varphi & \sin\varphi \end{bmatrix}$ is a rotation through φ radians.

5.5.17. Let $A = \begin{bmatrix} \cos\varphi & \sin\varphi \\ \sin\varphi & -\cos\varphi \end{bmatrix}$, $\mathbf{v} = \begin{bmatrix} 1+\cos\varphi \\ \sin\varphi \end{bmatrix}$, and $\mathbf{x} = \begin{bmatrix} \sin\varphi \\ -1-\cos\varphi \end{bmatrix}$.

(a) Show that $A\mathbf{v}$ is a scalar multiple of $\mathbf{v}$ and that $A\mathbf{x}$ is a scalar multiple of $\mathbf{x}$.
(b) Prove that if $\sin\varphi \neq 0$, then $\{\mathbf{v}, \mathbf{x}\}$ is a basis for $\mathbb{R}^2$.
(c) Suppose $\sin\varphi = 0$. Find a basis for $\mathbb{R}^2$, $\mathcal{B} = \{\mathbf{v}_1, \mathbf{v}_2\}$, so that $A\mathbf{v}_1$ is a scalar multiple of $\mathbf{v}_1$ and $A\mathbf{v}_2$ is a scalar multiple of $\mathbf{v}_2$.
(d) Conclude that in either case, A determines an orthogonal reflection on $\mathbb{R}^2$.

5.6. Self-Adjoint and Unitary Matrices

Complex vector spaces, in some sense, occupy their own world. Here we embark on a gentle introduction to two classes of complex matrices that are analogous, respectively, to real symmetric matrices and real orthogonal matrices.

Definition 5.45. The **conjugate transpose**, or **adjoint**, of a complex matrix, A, is $A^\dagger = \overline{A^T}$.

If $A = \begin{bmatrix} 2+i & -3i & 0 \\ 1 & 4-2i & -1 \\ i & 1+i & 0 \end{bmatrix}$, then $A^\dagger = \begin{bmatrix} 2-i & 1 & -i \\ 3i & 4+2i & 1-i \\ 0 & -1 & 0 \end{bmatrix}$.

Properties of the conjugate transpose of a complex matrix mirror properties of the transpose of a real matrix. To see this, we start by detailing how conjugation of matrices gets along with matrix operations.

Lemma 5.46. *Let A and B be complex matrices and let c belong to $\mathbb{C}$.*

(a) *If $A + B$ is defined, then $\overline{cA + B} = \bar{c}\bar{A} + \bar{B}$.*
(b) *If AB is defined, then $\overline{AB} = \bar{A}\bar{B}$.*
(c) $\overline{A^T} = \bar{A}^T$.

Proof. Let $A = [a_{ij}]$ and $B = [b_{ij}]$ be complex matrices. If $A + B$ is defined, the ij-entry of $cA + B$ is $ca_{ij} + b_{ij}$. By Lemma B.2, the ij-entry of $\overline{cA + B}$ is

$$\overline{ca_{ij} + b_{ij}} = \overline{ca_{ij}} + \overline{b_{ij}} = \bar{c}\,\overline{a_{ij}} + \overline{b_{ij}},$$

which is the ij-entry of $\bar{c}\bar{A} + \bar{B}$. This proves part (a) of the lemma.

Proof of the rest of the lemma is an exercise. □

The next lemma is specific to invertible matrices.

Lemma 5.47. *If A is an invertible complex matrix, then so is $\bar{A}$ and $\bar{A}^{-1} = \overline{A^{-1}}$.*

Proof. If A is an $n \times n$ invertible complex matrix, then by Lemma 5.46,

$$\overline{AA^{-1}} = \bar{A}\,\overline{A^{-1}} = \overline{I_n}.$$

Since I_n is real, $\overline{I_n} = I_n$. This is enough to establish that $\bar{A}$ is invertible with inverse $\overline{A^{-1}}$. □

The first statement of Lemma 5.46 says that complex conjugation is nearly, but not quite, linear on $M_{m,n}(\mathbb{C})$. There is a name for mappings like this.

5.6. Self-Adjoint and Unitary Matrices

Definition 5.48. When V and W are complex vector spaces, $L : V \to W$ is **conjugate-linear**, or **antilinear**, provided
$$L(c\mathbf{v} + \mathbf{x}) = \bar{c}L(\mathbf{v}) + L(\mathbf{x})$$
for $\mathbf{v}, \mathbf{x}$ in V and c in $\mathbb{C}$.

Applying Lemma 5.16, Theorem 5.17, Lemma 5.46, and Lemma 5.47, we get the following properties of the conjugate transpose operation, $A \mapsto A^\dagger = \overline{A^T}$.

Lemma 5.49. Let A and B be complex matrices and let c be a complex constant.

(a) $(A^\dagger)^T = \bar{A}$.
(b) $\overline{A^\dagger} = A^T$.
(c) $(A^\dagger)^\dagger = A$.
(d) If $A + B$ is defined, then $(cA + B)^\dagger = \bar{c}A^\dagger + B^\dagger$.
(e) If AB is defined, then $(AB)^\dagger = B^\dagger A^\dagger$.
(f) If A is invertible, then $A^\dagger$ is invertible with $(A^\dagger)^{-1} = (A^{-1})^\dagger$.

Lemma 5.49(d) says that $A \mapsto A^\dagger$ is conjugate-linear on $M_{m,n}(\mathbb{C})$.

Definition 5.50. A **self-adjoint** or **Hermitian matrix** is A in $M_n(\mathbb{C})$ such that $A^\dagger = A$.

Self-adjoint matrices are the complex analog to symmetric matrices. Note that if A in $M_n(\mathbb{R})$ is symmetric, then since $\bar{A} = A$, we can say that
$$A^\dagger = A^T = A.$$
In particular, a real symmetric matrix is self-adjoint. A complex symmetric matrix, though, is not self-adjoint, in general.

The complex analog to an orthogonal matrix is a *unitary matrix*.

Definition 5.51. A **unitary matrix** is A in $M_n(\mathbb{C})$ such that
$$A^\dagger = A^{-1}.$$

Notice once again that if A in $M_n(\mathbb{R})$ is orthogonal, then
$$A^\dagger = A^T = A^{-1}$$
so A is unitary. A complex orthogonal matrix, though, is not unitary, in general.

The definition makes it clear that A is unitary if and only if $A^{-1} = A^\dagger$ is unitary. We leave it as an exercise to verify that the product of unitary matrices is unitary. This will justify the next definition.

Definition 5.52. The **unitary group** associated to $\mathbb{C}^n$ is
$$U_n(\mathbb{C}) := \{A \in GL_n(\mathbb{C}) \,|\, A^\dagger = A^{-1}\}.$$

Unitary matrices come up frequently in applications to geometry and physics. A quick check reveals that the matrix $\begin{bmatrix} 0 & i \\ 1 & 0 \end{bmatrix}$ in $M_2(\mathbb{C})$ is unitary.

While we can define orthogonal matrices over any field, unitary matrices are associated to $\mathbb{C}$. We meet up again with orthogonal and unitary matrices, and the associated linear transformations, in Chapters 7 and 10.

Exercises 5.6.

5.6.1. Finish the proof of Lemma 5.46.

5.6.2. Let $A = \begin{bmatrix} i & 1 \\ -1 & i \end{bmatrix}$ belong to $M_2(\mathbb{C})$.
  (a) Find $\bar{A}$, A^T, and $A^\dagger$.
  (b) Find A^{-1}, if it exists.

5.6.3. Apply the instructions from the last problem to $A = \begin{bmatrix} i & -3 \\ 1 & i \end{bmatrix}$.

5.6.4. Define σ on $\mathbb{C}^n \times \mathbb{C}^n$ by
$$\sigma(\mathbf{v}, \mathbf{x}) = \mathbf{v}^\dagger \mathbf{x}.$$
  (a) Show that σ is conjugate-linear in its first variable and linear in its second variable.
  (b) Show that $\sigma(\mathbf{x}, \mathbf{v}) = \overline{\sigma(\mathbf{v}, \mathbf{x})}$.
  (c) Let $\mathbf{v}$ belong to $\mathbb{C}$. Show that $\sigma(\mathbf{v}, \mathbf{v})$ is the squared length of $\mathbf{v}$ if we identify $\mathbb{C}$ with $\mathbb{R}^2$.

5.6.5. Let A in $M_n(\mathbb{C})$ be self-adjoint. Define $\sigma : \mathbb{C}^n \times \mathbb{C}^n \to \mathbb{C}$ by
$$\sigma(\mathbf{v}, \mathbf{x}) = \mathbf{v}^\dagger A \mathbf{x}.$$
  (a) Show that σ is conjugate-linear in its first variable and linear in its second variable.
  (b) Show that $\sigma(\mathbf{x}, \mathbf{v}) = \overline{\sigma(\mathbf{v}, \mathbf{x})}$.

5.6.6. Show that the set of $n \times n$ self-adjoint matrices is a real subspace of the real vector space $M_n(\mathbb{C})$.

5.6.7. Is the product of self-adjoint matrices in $M_n(\mathbb{C})$ self-adjoint?

5.6.8. Do unitary matrices form a vector subspace of $M_n(\mathbb{C})$?

5.6.9. Show that $U_n(\mathbb{C})$ is actually a subgroup of $GL_n(\mathbb{C})$.

5.6.10. Find A in $M_2(\mathbb{C})$ that is self-adjoint but not symmetric.

5.6.11. Find A in $M_2(\mathbb{C})$ that is symmetric but not self-adjoint.

5.6.12. Find A in $M_2(\mathbb{C})$ that is orthogonal but not unitary.

5.6.13. Find A in $M_2(\mathbb{C})$ that is unitary but not orthogonal.

Chapter 6

The Determinant Is a Multilinear Mapping

Multilinear algebra is a broad subject with myriad applications in mathematics and physics. A proper development of the subject requires a dedicated course and involves a level of abstraction well beyond anything we have seen to this point. Here we get a taste of the subject on the way to our goal, which is to define the determinant as an alternating n-linear form on $\mathbb{F}^n$.

6.1. Multilinear Mappings

When V is a vector space, we use V^k to designate the Cartesian product of k copies of V:
$$V^k := \overbrace{V \times \cdots \times V}^{k \text{ factors}}.$$
Elements in V^k are ordered k-tuples of vectors in V. As sets, V^k and
$$V^{\oplus k} = \overbrace{V \oplus \cdots \oplus V}^{k \text{ summands}}$$
are identical, but $V^{\oplus k}$ is a vector space and V^k is not. Multilinear mappings on V are defined in terms of V^k.

Let V and W be vector spaces over $\mathbb{F}$. A k-**linear mapping on** V into W is a mapping
$$\mu : V^k \to W$$
that is linear in each of its k variables, taken one at a time. This means that for each $i \in \{1, \ldots, k\}$,
$$\mu(\mathbf{v}_1, \ldots, c\mathbf{v}_i + \mathbf{x}_i, \ldots, \mathbf{v}_k) = c\mu(\mathbf{v}_1, \ldots, \mathbf{v}_i, \ldots, \mathbf{v}_k) + \mu(\mathbf{v}_1, \ldots, \mathbf{x}_i, \ldots, \mathbf{v}_k)$$
when the vectors $\mathbf{v}_1, \ldots, \mathbf{v}_k, \mathbf{x}_i$ are in V and c is in $\mathbb{F}$. A k-**linear form on** V is a k-linear mapping from V into $\mathbb{F}$.

The mappings at hand have domain V^k, but we say they are k-linear on V.

A 1-linear mapping is a linear mapping, $V \to W$. A 2-linear mapping is a bilinear mapping, $V \times V \to W$. A 3-linear mapping is *trilinear*. The locution *multilinear mappings* refers to k-linear mappings, either when k is unspecified or when there is more than one positive integer k under consideration.

Our goal in this and the following three sections is to lay the groundwork for understanding the determinant as a multilinear form.

When V is a vector space over $\mathbb{F}$, the **kth tensor power of V, $V^{\otimes k}$**, is a vector space we define using V^k. Elements of $V^{\otimes k}$ are linear combinations of **pure tensors**, which we write as
$$\mathbf{v}_1 \otimes \cdots \otimes \mathbf{v}_k$$
for $\mathbf{v}_i$ in V. Arithmetic with pure tensors in $V^{\otimes k}$ arises from defining

(6.1) $$(\mathbf{v}_1, \ldots, \mathbf{v}_k) \mapsto \mathbf{v}_1 \otimes \cdots \otimes \mathbf{v}_k$$

to be a k-linear mapping on V. This means that addition and scaling of pure tensors follow the rule

(6.2)
$$\mathbf{v}_1 \otimes \cdots \mathbf{v}_{i-1} \otimes (c\mathbf{v}_i + \mathbf{x}_i) \otimes \mathbf{v}_{i+1} \otimes \cdots \otimes \mathbf{v}_k$$
$$= c(\mathbf{v}_1 \otimes \cdots \otimes \mathbf{v}_i \otimes \cdots \otimes \mathbf{v}_k) + \mathbf{v}_1 \otimes \cdots \otimes \mathbf{v}_{i-1} \otimes \mathbf{x}_i \otimes \mathbf{v}_{i+1} \otimes \cdots \otimes \mathbf{v}_k,$$

for $\mathbf{v}_i$, $\mathbf{x}_i$ in V and c in $\mathbb{F}$.

While $V^{\otimes k}$ is not an algebra, it may help to think of adding and scaling pure tensors as being like adding and scaling noncommutative products in an algebra. If a, b, c, and d belong to some algebra, $a(2b+c)d = 2abd + acd$ and, generally, we cannot simplify expressions such as $ab + cd$ or even $ab + ba$. If a, b, c, and d belong to a vector space V, then according to (6.2),
$$a \otimes (2b + c) \otimes d = 2(a \otimes b \otimes d) + a \otimes c \otimes d$$
in $V^{\otimes 3}$.

Keep in mind that an arbitrary element in $V^{\otimes k}$ is not a pure tensor, but a linear combination of pure tensors.

The deeper abstract material that underlies multilinear algebra guarantees the existence and uniqueness of $V^{\otimes k}$ as a vector space over $\mathbb{F}$. It also obviates the difficulties of justifying our next theorem, the proof of which we omit.

Theorem 6.1. *If $\mathcal{B} = \{\mathbf{b}_i\}_{i \in I}$ is an indexed basis for V, then the set of all pure tensors of the form*
$$\mathbf{b}_{i_1} \otimes \cdots \otimes \mathbf{b}_{i_k},$$
where $\mathbf{b}_{i_j}$ are (not necessarily distinct) elements in $\mathcal{B}$, is a basis for $V^{\otimes k}$. In particular, if $\dim V = n$, then $\dim V^{\otimes k} = n^k$.

Example 6.2. A basis for $\mathbb{R}^2 \otimes \mathbb{R}^2$ is
$$\{\mathbf{e}_1 \otimes \mathbf{e}_1, \mathbf{e}_1 \otimes \mathbf{e}_2, \mathbf{e}_2 \otimes \mathbf{e}_1, \mathbf{e}_2 \otimes \mathbf{e}_2\}$$
where $\{\mathbf{e}_1, \mathbf{e}_2\}$ is the standard ordered basis for $\mathbb{R}^2$. We can then write
$$(1, 2) \otimes (-1, 1) = (\mathbf{e}_1 + 2\mathbf{e}_2) \otimes (-\mathbf{e}_1 + \mathbf{e}_2)$$
$$= -\mathbf{e}_1 \otimes \mathbf{e}_1 + \mathbf{e}_1 \otimes \mathbf{e}_2 - 2\mathbf{e}_2 \otimes \mathbf{e}_1 + 2\mathbf{e}_2 \otimes \mathbf{e}_2.$$

6.1. Multilinear Mappings

It is important to appreciate differences between $V^{\oplus k}$ and $V^{\otimes k}$. If we write elements in $V \oplus V$ as ordered pairs of vectors from V, then the zero vector in $V \oplus V$ is $(\mathbf{0}_V, \mathbf{0}_V)$. In $V \otimes V$, both $\mathbf{0}_V \otimes \mathbf{v}$ and $\mathbf{x} \otimes \mathbf{0}_V$ are equal to the zero vector, for any $\mathbf{v}$ and $\mathbf{x}$ in V. (We leave the verification of this as an exercise.) If $\mathbf{v}$ and $\mathbf{x}$ are in V and c is in the underlying field, then in $V \oplus V$,

$$c(\mathbf{v}, \mathbf{x}) = (c\mathbf{v}, c\mathbf{x})$$

while in $V \otimes V$,

$$c(\mathbf{v} \otimes \mathbf{x}) = (c\mathbf{v}) \otimes \mathbf{x} = \mathbf{v} \otimes (c\mathbf{x}).$$

The next theorem generalizes Theorem 3.26.

Theorem 6.3. *Let V and W be vector spaces over a field $\mathbb{F}$. For k fixed in $\mathbb{Z}_+$, the collection of k-linear mappings from V into W is a vector space over $\mathbb{F}$.*

Proof. Let V and W be as hypothesized. By Theorem 1.14, the set of all mappings from V^k to W is a vector space. We just have to show that the k-linear mappings are a subspace.

Mapping all $(\mathbf{v}_1, \ldots, \mathbf{v}_k)$ in V^k to $\mathbf{0}_W$, we get a k-linear mapping on V. It follows that the set of k-linear mappings from V into W is nonempty.

We leave it as an exercise to show that if μ_1 and μ_2 are k-linear mappings from V into W, then so is $c\mu_1 + \mu_2$ when c is in $\mathbb{F}$. This will be enough to complete proof of the theorem. $\square$

We use $\mathrm{Klin}(V)$ to denote the vector space of k-linear forms on V.

There is a natural association between $V^{\otimes k}$ and $\mathrm{Klin}(V)^*$, the dual space of $\mathrm{Klin}(V)$. Specifically, we define $f : V^{\otimes k} \to \mathrm{Klin}(V)^*$ by

$$f(\mathbf{v}_1 \otimes \cdots \otimes \mathbf{v}_k)(\mu) := \mu(\mathbf{v}_1, \ldots, \mathbf{v}_k) \in \mathbb{F},$$

for μ in $\mathrm{Klin}(V)$. We leave it as an exercise to show that f is an injective linear transformation. This is enough to prove that $V^{\otimes k}$ is canonically isomorphic to a subspace of $\mathrm{Klin}(V)$. When V is finite-dimensional, $V^{\otimes k} \cong \mathrm{Klin}(V)^*$.

We can go in the other direction, as well. Any k-linear form on V determines an element in $(V^{\otimes k})^*$. In particular, define $g : \mathrm{Klin}(V) \to (V^{\otimes k})^*$ by

$$g(\mu)(\mathbf{v}_1 \otimes \cdots \otimes \mathbf{v}_k) := \mu(\mathbf{v}_1, \ldots, \mathbf{v}_k) \in \mathbb{F},$$

for $\mathbf{v}_1 \otimes \cdots \otimes \mathbf{v}_k$ in $V^{\otimes k}$. We leave it as an exercise to show that g, too, is a canonical injective linear transformation. This will show that $\mathrm{Klin}(V)$ is canonically isomorphic to a subspace of $(V^{\otimes k})^*$. When V is finite-dimensional, $\mathrm{Klin}(V) \cong (V^{\otimes k})^*$.

We identify an important subspace of $\mathrm{Klin}(V)$ in the next section.

Exercises 6.1.

6.1.1. Show that any $\mathbf{v}_1 \otimes \cdots \otimes \mathbf{v}_k$ for which $\mathbf{v}_i = \mathbf{0}_V$ is the zero vector in $V^{\otimes k}$.

6.1.2. How many different ways can we express the additive inverse of $\mathbf{v}_1 \otimes \cdots \otimes \mathbf{v}_k$ in $V^{\otimes k}$ as a pure tensor?

6.1.3. Write $(2, -3) \otimes (-5, 1)$ in $\mathbb{R}^2 \otimes \mathbb{R}^2$ as a linear combination of elements in $\mathcal{B} = \{\mathbf{e}_1 \otimes \mathbf{e}_1, \mathbf{e}_1 \otimes \mathbf{e}_2, \mathbf{e}_2 \otimes \mathbf{e}_1, \mathbf{e}_2 \otimes \mathbf{e}_2\}$.

6.1.4. Write $(1,-1) \otimes (-1,3) + (2,-2) \otimes (3,-2)$ in $\mathbb{R}^2 \otimes \mathbb{R}^2$ as a pure tensor.

6.1.5. A dimension argument ensures that $\mathbb{R}^2 \oplus \mathbb{R}^2 \cong \mathbb{R}^2 \otimes \mathbb{R}^2$. Use the natural ordered basis for $\mathbb{R}^2$ to identify an isomorphism $L: \mathbb{R}^2 \oplus \mathbb{R}^2 \to \mathbb{R}^2 \otimes \mathbb{R}^2$. Find $L((-1,3),(2,7))$.

6.1.6. Finish the proof of Theorem 6.3 by showing that if μ_1 and μ_2 are k-linear mappings from V into W, then for any scalar c, $c\mu_1 + \mu_2$ is a k-linear mapping from V into W.

6.1.7. Fix a vector space, V, and define $L: V \otimes V \to \mathcal{L}(V^*, V)$ by
$$L(\mathbf{v} \otimes \mathbf{x})(f) = f(\mathbf{v})\mathbf{x}$$
for any f in V^*.
  (a) Show that L is a linear injection.
  (b) Show that L is an isomorphism when V is finite-dimensional.
  (c) For $V = \mathbb{R}^2$, find $L((2,-3) \otimes (-5,1))(f)$ where $f = \begin{bmatrix} a & b \end{bmatrix}$.

6.1.8. Let V be a vector space over $\mathbb{F}$. Show that, for μ in Klin(V) and $\mathbf{v}_i$ in V, $f: V^{\otimes k} \to \text{Klin}(V)^*$ defined by
$$f(\mathbf{v}_1 \otimes \cdots \otimes \mathbf{v}_k)(\mu) := \mu(\mathbf{v}_1, \ldots, \mathbf{v}_k)$$
determines an injective linear mapping. Conclude that $V^{\otimes k}$ is canonically isomorphic to a subspace of Klin(V)*.

6.1.9. Let V be a vector space over $\mathbb{F}$. Show that, for μ in Klin(V) and $\mathbf{v}_i$ in V, $g: \text{Klin}(V) \to (V^{\otimes k})^*$ defined by
$$g(\mu)(\mathbf{v}_1 \otimes \cdots \otimes \mathbf{v}_k) := \mu(\mathbf{v}_1, \ldots, \mathbf{v}_k)$$
determines an injective linear transformation. Conclude that Klin(V) is canonically isomorphic to a subspace of $(V^{\otimes k})^*$.

6.2. Alternating Multilinear Mappings

Introductory linear algebra courses often cite the theorem that says the determinant of a matrix changes by a factor of -1 when we apply a type-1 elementary row operation to the matrix or when we switch two columns of the matrix. This tells us that the determinant depends on the order of the columns or rows of a matrix, in a very specific way. The notion of *alternating mappings* captures this idea.

Definition 6.4. A k-linear mapping
$$\mu: V \times \cdots \times V \to W$$
is **alternating** provided
$$\mu(\mathbf{v}_1, \ldots, \mathbf{v}_k) = \mathbf{0}_W$$
when there are $i \neq j$ such that $\mathbf{v}_i = \mathbf{v}_j$.

Suppose $\mu: V \to W$ is an alternating k-linear map. Let $\mathbf{x}, \mathbf{y}, \mathbf{v}_3, \ldots, \mathbf{v}_k$ be arbitrary in V. Consider that
$$\mu(\mathbf{x} + \mathbf{y}, \mathbf{x} + \mathbf{y}, \mathbf{v}_3, \ldots, \mathbf{v}_k) = \mathbf{0}_W.$$

6.2. Alternating Multilinear Mappings

Using k-linearity, we have

$$\begin{aligned}\mathbf{0}_W &= \mu(\mathbf{x}+\mathbf{y},\mathbf{x}+\mathbf{y},\mathbf{v}_3,\ldots,\mathbf{v}_k) \\ &= \mu(\mathbf{x},\mathbf{x},\mathbf{v}_3,\ldots,\mathbf{v}_k) + \mu(\mathbf{x},\mathbf{y},\mathbf{v}_3,\ldots,\mathbf{v}_k) \\ &\quad + \mu(\mathbf{y},\mathbf{x},\mathbf{v}_3,\ldots,\mathbf{v}_k) + \mu(\mathbf{y},\mathbf{y},\mathbf{v}_3,\ldots,\mathbf{v}_k).\end{aligned}$$

Since both $\mu(\mathbf{x},\mathbf{x},\mathbf{v}_3,\ldots,\mathbf{v}_k)$ and $\mu(\mathbf{y},\mathbf{y},\mathbf{v}_3,\ldots,\mathbf{v}_k)$ are zero, we see that

$$\mu(\mathbf{x},\mathbf{y},\mathbf{v}_3,\ldots,\mathbf{v}_k) = -\mu(\mathbf{y},\mathbf{x},\mathbf{v}_3,\ldots,\mathbf{v}_k).$$

There is nothing special about the first two components of μ. The argument we just advanced applies to show that when we swap any two components of an element in V^k, the effect is to multiply μ by a factor of -1.

Definition 6.5. A k-linear mapping μ is **skew-symmetric** if its value changes by a factor of -1 when two of its input vectors are switched.

We have shown that an alternating k-linear mapping is skew-symmetric.

We saw in Section 5.3 that when the underlying field has characteristic 2, skew-symmetric and symmetric bilinear forms are the same. The same principle applies here. Over a field of characteristic 2, any skew-symmetric mapping is *symmetric*, that is, unchanged by a reordering of input vectors. When the underlying field has characteristic different from 2, though, skew-symmetric k-linear mappings and alternating k-linear mappings are the same.

Lemma 6.6. *Suppose V is a vector space over $\mathbb{F}$, where $\operatorname{char}\mathbb{F} \neq 2$. A k-linear mapping on V is alternating if and only if it is skew-symmetric.*

Proof. We have already argued that under the hypotheses of the lemma, an alternating k-linear mapping on V is skew-symmetric. Suppose, then, that μ is a skew-symmetric k-linear mapping from V into W.

Let $\mathbf{u}$ in V^k be given by $\mathbf{u} = (\mathbf{v}_1,\ldots,\mathbf{v}_k)$ where $\mathbf{v}_i$ is in V, and for some $i \neq j$, $\mathbf{v}_i = \mathbf{v}_j$. If we switch the positions of $\mathbf{v}_i$ and $\mathbf{v}_j$ in $\mathbf{u}$, $\mathbf{u}$ does not change but since μ is skew-symmetric,

$$\mu(\mathbf{u}) = -\mu(\mathbf{u}).$$

It follows that $2\mu(\mathbf{u}) = \mathbf{0}_W$. If the underlying field does not have characteristic 2, then $\mu(\mathbf{u}) = \mathbf{0}_W$. This proves that μ is alternating. $\square$

Let $\operatorname{Alt}k(V)$ denote the set of alternating k-linear forms on a vector space V. We leave it as an exercise to verify that $\operatorname{Alt}k(V)$ is a subspace of $\operatorname{Klin}(V)$. Since $V^{\otimes k}$ is canonically isomorphic to a subspace of $\operatorname{Klin}(V)^*$, we might expect some vector space associated to $V^{\otimes k}$ to be canonically isomorphic to $\operatorname{Alt}k(V)^*$. The item we are looking for is actually a quotient space of $V^{\otimes k}$.

Let $\mathbf{U}$ be the subspace of $V^{\otimes k}$ spanned by all expressions of the form

$$\mathbf{v}_1 \otimes \cdots \otimes \mathbf{v}_k, \mathbf{v}_i \in V, \text{ where } \mathbf{v}_i = \mathbf{v}_j, \text{ for some } i \neq j.$$

The k**th exterior power** of V is

$$\Lambda^k V := V^{\otimes k}/\mathbf{U}.$$

We call $\Lambda^2 V$ the **exterior square** of V.

We use $\mathbf{v}_1 \wedge \cdots \wedge \mathbf{v}_k$ to designate the coset in $\Lambda^k V$ that contains
$$\mathbf{v}_1 \otimes \cdots \otimes \mathbf{v}_k.$$
We can think of an arbitrary element in $\Lambda^k V$ as a linear combination of expressions
$$\mathbf{v}_1 \wedge \cdots \wedge \mathbf{v}_k$$
where
$$(\mathbf{v}_1, \ldots, \mathbf{v}_k) \mapsto \mathbf{v}_1 \wedge \cdots \wedge \mathbf{v}_k$$
is an alternating k-linear mapping on V. Where we think of elements in $V^{\otimes k}$ as linear combinations of noncommutative products of k vectors from V, we think of the elements in $\Lambda^k V$ as linear combinations of alternating products of k vectors from V.

Example 6.7. What is $\Lambda^2 V$ when $V = \mathbb{R}^3$? We have to be careful in looking for a basis of $\mathbf{U}$, described as above. Taking elements from the standard ordered basis for $\mathbb{R}^3$, $\mathbf{U}$ certainly contains
$$\text{Span}\{\mathbf{e}_i \otimes \mathbf{e}_i \,|\, i = 1, 2, 3\}$$
but it also contains all linear combinations of expressions of the form
$$(\mathbf{e}_i \otimes \mathbf{e}_j) + (\mathbf{e}_j \otimes \mathbf{e}_i) \quad \text{for } i < j \in \{1, 2, 3\}.$$
In fact,
$$\{\mathbf{e}_i \otimes \mathbf{e}_i, \mathbf{e}_i \otimes \mathbf{e}_j + \mathbf{e}_j \otimes \mathbf{e}_i \,|\, i < j \in \{1, 2, 3\}\}$$
is a basis for $\mathbf{U}$. Since $\dim(V \otimes V) = 9$ and $\dim \mathbf{U} = 6$, $\dim \Lambda^2 V = 3$. A basis for $\Lambda^2 V$ can be written
$$\{\mathbf{e}_1 \wedge \mathbf{e}_2, \mathbf{e}_1 \wedge \mathbf{e}_3, \mathbf{e}_2 \wedge \mathbf{e}_3\}.$$

Example 6.7 suggests how to approach calculating the dimension of an exterior power of a vector space. Next we see how $\Lambda^k V$ and its dual are related respectively to $\text{Altk}(V)^*$ and $\text{Altk}(V)$.

If we define f on $\Lambda^k V$ by
$$f(\mathbf{v}_1 \wedge \cdots \wedge \mathbf{v}_k)(\mu) := \mu(\mathbf{v}_1, \ldots, \mathbf{v}_k)$$
for μ in $\text{Altk}(V)$, then $f : \Lambda^k V \to \text{Altk}(V)^*$ is a well-defined, injective linear transformation. Likewise, $g : \text{Altk}(V) \to (\Lambda^k V)^*$ defined by
$$g(\mu)(\mathbf{v}_1 \wedge \cdots \mathbf{v}_k) := \mu(\mathbf{v}_1, \ldots, \mathbf{v}_k),$$
for $\mathbf{v}_i$ in V, is a linear injection. It follows that when V is finite-dimensional, $\Lambda^k V$ is canonically isomorphic to $\text{Altk}(V)^*$ and that $\text{Altk}(V)$ is canonically isomorphic to $(\Lambda^k V)^*$. We leave verifications of these statements to the exercises.

Permutations start to play a role in the next part of our story. We take the next two sections to study the permutations on a finite set.

Exercises 6.2.

6.2.1. Show that when V is any vector space, the set of alternating k-linear forms, $\text{Altk}(V)$, is a subspace of the space of all k-linear forms, $\text{Klin}(V)$.

6.2.2. Show that defining
$$f(\mathbf{v}_1 \wedge \cdots \wedge \mathbf{v}_k)(\mu) := \mu(\mathbf{v}_1, \ldots, \mathbf{v}_k),$$
for $\mathbf{v}_1 \wedge \cdots \wedge \mathbf{v}_k$ in $\Lambda^k V$ and μ in $\text{Altk}(V)$, gives us a well-defined, injective linear mapping $f : \Lambda^k V \to \text{Altk}(V)^*$.

6.2.3. Show that defining
$$g(\mu)(\mathbf{v}_1 \wedge \cdots \wedge \mathbf{v}_k) := \mu(\mathbf{v}_1, \ldots, \mathbf{v}_k),$$
for μ in $\text{Altk}(V)$ and $\mathbf{v}_1 \wedge \cdots \wedge \mathbf{v}_k$ in $\Lambda^k V$, gives us an injective linear mapping $g : \text{Altk}(V) \to (\Lambda^k V)^*$.

6.2.4. Let μ belong to $\text{Altk}(V)$ for some vector space, V. Show that
$$\mu(\mathbf{a}_1, \ldots, \mathbf{a}_k) = 0$$
if there is a dependence relation on $\mathbf{a}_1, \ldots, \mathbf{a}_k$.

6.2.5. In Example 6.7, we note that $\mathbf{U}$ contains all linear combinations of expressions of the form
$$(\mathbf{e}_i \otimes \mathbf{e}_j) + (\mathbf{e}_j \otimes \mathbf{e}_i) \quad \text{for } i < j \in \{1, 2, 3\}.$$
Why is this true?

6.2.6. Suppose $\dim V = 4$. What are $\dim \Lambda^2 V$ and $\dim \Lambda^3 V$?

6.2.7. Suppose $\dim V = n$. What is $\dim \Lambda^n V$?

6.2.8. Let V be an n-dimensional vector space. Show that $\Lambda^k V$ is the trivial vector space when $k > n$.

6.2.9. Let $\dim V = n$. What is the formula for $\dim \Lambda^k V$? (Section A.4 contains a discussion of counting.)

6.3. Permutations, Part I

A **permutation** is a bijection from a set to itself. The collection of permutations on $S = \{1, \ldots, k\}$ is designated $\mathcal{S}_k$.[1] We start by establishing that $\mathcal{S}_k$ is a group.

The **order of a finite group** is the number of elements it contains.

Theorem 6.8. $\mathcal{S}_k$ is a group of order $k!$.

Proof. We denote the identity mapping on $\{1, \ldots, k\}$ by ι. Since ι is a bijection, $\mathcal{S}_k$ is nonempty. If ρ and σ belong to $\mathcal{S}_k$, then by Lemma A.30, the composition $\rho \circ \sigma = \rho\sigma$ is also a bijection on S. Exercise A.6.6 is to show that if ρ is a bijection, it is invertible and ρ^{-1} is a bijection. Since $\mathcal{S}_k$ is nonempty and closed under composition and inversion, it is a group.

We can denote ρ in $\mathcal{S}_k$ by
$$\begin{pmatrix} 1 & 2 & \cdots & k \\ \rho(1) & \rho(2) & \cdots & \rho(k) \end{pmatrix}.$$
This makes it evident that we can identify ρ with an ordering on the elements in S. Since Lemma A.16 tells us that there are $k!$ ways to order the elements in S, $|\mathcal{S}_k| = k!$. □

---

[1] When the concern is the permutations on a set, what matters is how many elements the set contains, not the names of the elements in the set.

The **symmetric group** on k letters is $\mathcal{S}_k$.

Like the groups of matrices we saw in Section 5.5, $\mathcal{S}_k$ is a *multiplicative group*, meaning that we treat its binary operation, function composition, as a type of (noncommutative) multiplication.

We saw in Chapter 1 that we could use exponential notation meaningfully on any nonzero element in a field, or invertible element in an associative algebra, as long as the exponents are integers. The same principle applies to the elements in a group. If G is a multiplicative group with identity element e, then, as in a field, we define $g^0 := e$ for all g in G, and we denote the inverse of g by g^{-1}. For a positive integer n, we define

$$g^n := \overbrace{g \cdots g}^{n \text{ factors}}$$

and

$$g^{-n} = \overbrace{g^{-1} \cdots g^{-1}}^{n \text{ factors}}.$$

Example 6.9. Consider ρ in $\mathcal{S}_4$ given by

$$\rho = \begin{pmatrix} 1 & 2 & 3 & 4 \\ 3 & 1 & 4 & 2 \end{pmatrix}.$$

Applying ρ repeatedly to the elements in $S = \{1, 2, 3, 4\}$, we have

$$\rho^2 = \begin{pmatrix} 1 & 2 & 3 & 4 \\ 4 & 3 & 2 & 1 \end{pmatrix}, \quad \rho^3 = \begin{pmatrix} 1 & 2 & 3 & 4 \\ 2 & 4 & 1 & 3 \end{pmatrix}, \quad \rho^4 = \begin{pmatrix} 1 & 2 & 3 & 4 \\ 1 & 2 & 3 & 4 \end{pmatrix}.$$

This is enough to show that $\rho^{-1} = \rho^3$.

The example suggests the next lemma. We leave the proof as an exercise.

Lemma 6.10. Let G be a multiplicative group with identity element e. If g belongs to G and m and n are integers, then $g^m g^n = g^{m+n}$ and $(g^m)^n = g^{mn}$.

A **cycle** is a mapping in $\mathcal{S}_k$ with the form $a_1 \mapsto a_2 \mapsto \cdots \mapsto a_t \mapsto a_1$ for a_i in $\{1, \ldots, k\}$, where any element in the complement of $\{a_1, \ldots, a_t\}$ in $\{1, \ldots, k\}$ is fixed. We designate a cycle using the notation

$$\begin{pmatrix} a_1 & \cdots & a_t \end{pmatrix}.$$

A cycle of the form $\begin{pmatrix} a & b \end{pmatrix}$ is a **transposition**.

Example 6.11. Since $|\mathcal{S}_3| = 6$ and since the following are distinct elements in $\mathcal{S}_3$,

$$\begin{pmatrix} 1 & 2 \end{pmatrix}, \begin{pmatrix} 1 & 3 \end{pmatrix}, \begin{pmatrix} 2 & 3 \end{pmatrix}, \begin{pmatrix} 1 & 2 & 3 \end{pmatrix}, \begin{pmatrix} 1 & 3 & 2 \end{pmatrix},$$

we see that every nonidentity element in $\mathcal{S}_3$ is a cycle.

Observation 6.12. The description of a cycle $\rho = \begin{pmatrix} a_1 & \cdots & a_t \end{pmatrix}$ as the mapping on $\{1, \ldots, k\}$ given by

$$a_1 \mapsto a_2 \mapsto \cdots \mapsto a_t \mapsto a_1$$

makes it evident that

$$\rho = \begin{pmatrix} a_2 & \cdots & a_t & a_1 \end{pmatrix} = \begin{pmatrix} a_3 & \cdots & a_t & a_1 & a_2 \end{pmatrix} = \cdots = \begin{pmatrix} a_t & a_1 & \cdots & a_{t-1} \end{pmatrix}.$$

Proof of the following is an exercise.

6.3. Permutations, Part I

Lemma 6.13. If $\rho = \begin{pmatrix} a_1 & \cdots & a_t \end{pmatrix}$ is a cycle in $\mathcal{S}_k$, then
$$\rho = \begin{pmatrix} a & \rho(a) & \rho^2(a) & \cdots & \rho^{t-1}(a) \end{pmatrix}$$
and $\rho^t(a) = a$ for any a in $\{a_1, \ldots, a_t\}$.

Generally, a nonidentity element in $\mathcal{S}_k$ is not a cycle. We will show, however, that every element in $\mathcal{S}_k$ is a product of cycles.

Cramer's description of the determinant involved counting the number of derangements in a permutation. In the modern theory, transpositions replace derangements. In particular, we will see that every permutation in $\mathcal{S}_k$ can be written as a product of either an even number of transpositions or an odd number of transpositions, but not as both. This allows us to identify a given permutation as either even or odd, thus, to assign a $+$ or $-$ to a term in a determinant associated to a given permutation.

It is easy to write a given permutation as a product of transpositions — establish that every permutation is a product of cycles and then sort out how to write a cycle as a product of transpositions — but proof of the result about classifying a permutation as even or odd takes a bit of doing. Since the study of permutations is fundamental to group theory, which is ubiquitous in mathematics and its applications, we take the time to work through what we need here, though not all of it is essential to what comes later in the course.

Definition 6.14. If ρ is in $\mathcal{S}_k$ and a is in $\{1, \ldots, k\}$, the set
$$\mathcal{O}_{a,\rho} = \{\rho^n(a) \mid n \in \mathbb{Z}\}$$
is the **orbit of a under ρ**. A **singleton orbit** is an orbit containing exactly one element.

Observation 6.12 and Lemma 6.13 together imply that if $\rho = \begin{pmatrix} a_1 & \cdots & a_t \end{pmatrix}$ is in $\mathcal{S}_k$, then
$$\mathcal{O}_{a,\rho} = \{a_1, \ldots, a_t\}$$
for any a in $\{a_1, \ldots, a_t\}$.

Lemma 6.15. A cycle is a permutation with exactly one nonsingleton orbit.

Proof. We have shown that if ρ is a cycle, then it has one nonsingleton orbit. Suppose $\mathcal{O}_{a,\rho} = \{\rho^n(a) \mid n \in \mathbb{Z}\}$ is the only nonsingleton orbit for ρ in $\mathcal{S}_k$. As a subset of $\{1, \ldots, k\}$, $\{\rho^n(a) \mid n \in \mathbb{Z}_+\}$ must be finite. This means that we can identify n least in $\mathbb{Z}_+$ such that $\rho^n(a) = \rho^m(a)$, for some $m < n$ in $\mathbb{Z}_+$. We then have $\rho^{n-m}(a) = a$. Letting $t = n - m > 0$, we can write
$$\rho = \begin{pmatrix} a & \rho(a) & \ldots \rho^{t-1}(a) \end{pmatrix}. \qquad \square$$

The **length of a cycle** $\rho = \begin{pmatrix} a_1 & \cdots & a_t \end{pmatrix}$ is t. **Disjoint cycles** are cycles with disjoint orbits.

Proof of the following is a routine verification that we leave to the reader.

Lemma 6.16. Disjoint cycles commute.

Since the binary operation in $\mathcal{S}_k$ is function composition, we work from right to left to calculate the effect of composing nondisjoint cycles. This takes practice.

Example 6.17. We would like to understand ρ in $\mathcal{S}_5$ given by
$$\rho = \begin{pmatrix} 3 & 4 & 5 & 2 \end{pmatrix} \begin{pmatrix} 2 & 1 & 5 \end{pmatrix}.$$
We can start with any element in the cycle on the right, making sure to find $\rho(a)$ for each a in $\{1, 2, 3, 4, 5\}$. In the first cycle, $1 \mapsto 5$, and in the second cycle, $5 \mapsto 2$, so $\rho(1) = 2$. In the first cycle, $2 \mapsto 1$, and in the second, $1 \mapsto 1$, so $\rho(2) = 1$. This gives us $\begin{pmatrix} 1 & 2 \end{pmatrix}$ as one factor of ρ. Since we do not yet know $\rho(3)$, we next note that in the first cycle, $3 \mapsto 3$ and in the second, $3 \mapsto 4$, so $\rho(3) = 4$. In the first cycle, $4 \mapsto 4$, and in the second, $4 \mapsto 5$, so $\rho(4) = 5$. In the first cycle, $5 \mapsto 2$, and in the second, $2 \mapsto 3$, so $\rho(5) = 3$. Now we know that ρ is the product of disjoint cycles
$$\rho = \begin{pmatrix} 1 & 2 \end{pmatrix} \begin{pmatrix} 3 & 4 & 5 \end{pmatrix}.$$

As the product of disjoint cycles, ρ has orbits
$$\mathcal{O}_{1,\rho} = \{1, 2\}, \quad \mathcal{O}_{3,\rho} = \{3, 4, 5\}.$$

We leave it as an exercise to show that the orbits of a permutation in $\mathcal{S}_k$ are the cells in a partition on $\{1, \ldots, k\}$.

Theorem 6.18. *Every permutation on a finite set is a product of disjoint cycles.*

Proof. Let $S = \{1, \ldots, k\}$, and let ρ belong to $\mathcal{S}_k$. Let U be the union of nonsingleton orbits of ρ. We prove the lemma by induction on $|U|$.

If $|U| = 0$, ρ is the identity mapping which can be written
$$(1)(2) \cdots (k).$$
This proves the theorem in the base case.

If ρ is not the identity mapping, let $\mathcal{O}_{a,\rho}$ be a nonsingleton orbit of ρ. By Lemma 6.13, the associated cycle is
$$\eta = \begin{pmatrix} a & \rho(a) & \cdots & \rho^n(a) \end{pmatrix}$$
for some positive integer n.

Let $S' = S \setminus \mathcal{O}_{a,\rho}$ and consider $\rho|_{S'}$. Let U' be the union of nonsingleton orbits of $\rho|_{S'}$. Notice that $U = U' \cup \mathcal{O}_{a,\rho}$. Since $|U'| < |U|$, the induction hypothesis is that $\rho|_{S'}$ is a product of disjoint cycles, $\eta_1, \ldots, \eta_r$. Since the orbits associated to $\eta, \eta_1, \ldots, \eta_r$ are pairwise disjoint and $\rho = \eta \eta_1 \cdots \eta_r$, this is enough to prove the theorem. $\square$

Theorem 6.19. *If $k \geq 2$, then every permutation in $\mathcal{S}_k$ is a product of transpositions.*

Proof. We can write $\iota = \begin{pmatrix} 1 & 2 \end{pmatrix} \begin{pmatrix} 1 & 2 \end{pmatrix}$ in $\mathcal{S}_k$ as long as $k \geq 2$. Given any cycle in $\mathcal{S}_k$, we can write
$$\begin{pmatrix} a_1 & \cdots & a_t \end{pmatrix} = \begin{pmatrix} a_1 & a_t \end{pmatrix} \begin{pmatrix} a_1 & a_{t-1} \end{pmatrix} \cdots \begin{pmatrix} a_1 & a_2 \end{pmatrix}.$$
By Theorem 6.18, this is enough to prove the theorem. $\square$

6.3. Permutations, Part I

Consider now that
$$\begin{pmatrix}1 & 2 & 3\end{pmatrix} = \begin{pmatrix}1 & 3\end{pmatrix}\begin{pmatrix}1 & 2\end{pmatrix} = \begin{pmatrix}2 & 3\end{pmatrix}\begin{pmatrix}1 & 3\end{pmatrix} = \begin{pmatrix}2 & 1\end{pmatrix}\begin{pmatrix}2 & 3\end{pmatrix}.$$
This is to emphasize that a given permutation may be written in different ways as a product of transpositions.

The approach we use to prove our main theorem is the one Paul Halmos described in [8].

Theorem 6.20. If ρ in $\mathcal{S}_k$ is given by
$$\rho = \tau_1 \cdots \tau_r = \tau'_1 \cdots \tau'_s$$
where $\tau_1, \ldots, \tau_r$ and $\tau'_1, \ldots, \tau'_s$ are transpositions, then r and s are both even or both odd.

Proof. Let $t_1, \ldots, t_k$ be distinct commuting indeterminates. Define $q(t_1, \ldots, t_k) = q(\mathbf{t})$, a polynomial in $t_1, \ldots, t_k$, by

(6.3) $\quad q(\mathbf{t}) = (t_1 - t_2)(t_1 - t_3) \cdots (t_1 - t_n)(t_2 - t_3) \cdots (t_2 - t_n) \cdots (t_{k-1} - t_k).$

Note that $q(\mathbf{t})$ is a product of differences, $t_i - t_j$, where $1 \leq i < j \leq k$. We can write
$$q(\mathbf{t}) = \prod_{1 \leq i < j \leq k} t_i - t_j$$
and we can view $q(\mathbf{t})$ as defining a mapping $q : \mathbb{R}^k \to \mathbb{R}$ given by
$$q(a_1, \ldots, a_k) = \prod_{1 \leq i < j \leq k} a_i - a_j.$$

Fix ρ in $\mathcal{S}_k$ and define $\rho(q(\mathbf{t}))$ to be the polynomial we get by applying ρ to each subscript in (6.3). We have
$$\rho(q(\mathbf{t})) = \prod_{1 \leq i < j \leq k} t_{\rho(i)} - t_{\rho(j)}.$$
Notice that $\rho(q(\mathbf{t})) = \pm q(\mathbf{t})$.

We claim that if τ is a transposition, then $\tau(q(\mathbf{t})) = -q(\mathbf{t})$.

We can write any transposition as $\tau = \begin{pmatrix}i & j\end{pmatrix}$, for $i < j$ in $S = \{1, \ldots, k\}$. Vis-à-vis i and j, there are three types of factors of $q(\mathbf{t})$: $t_i - t_j$, those with neither t_i nor t_j, and those with exactly one of t_i or t_j. Under τ, $t_i - t_j \mapsto -(t_i - t_j)$, while factors with neither t_i nor t_j are unaffected by τ. Factors of $q(\mathbf{t})$ that have exactly one of t_i or t_j occur in what we call *twin-pairs*: One twin has t_i and the other twin has t_j. For instance, if $i < \ell < j$, then $t_i - t_\ell$ and $t_\ell - t_j$ are twin-pair factors of $q(\mathbf{t})$. In this case,
$$\tau(t_i - t_\ell) = t_j - t_\ell = -(t_\ell - t_j)$$
and
$$\tau(t_\ell - t_j) = t_\ell - t_i = -(t_i - t_\ell).$$
The product of twin-pairs of the form $(t_i - t_\ell)(t_\ell - t_j)$ is unchanged by τ.

There are two remaining types of twin-pair factors of $q(\mathbf{t})$. The first type has the form $t_\ell - t_i$ and $t_\ell - t_j$. This type is associated to indices $\ell < i < j$. The second type has the form $t_i - t_\ell$ and $t_j - t_\ell$. This type is associated to indices $i < j < \ell$. In each case, τ swaps the twins, leaving the product of the twin-pair unchanged.

This proves that $\tau(q(\mathbf{t})) = -q(\mathbf{t})$, as claimed.

It follows from here that if ρ is arbitrary in $\mathcal{S}_k$, then $\rho(q(\mathbf{t})) = (-1)^r q(\mathbf{t})$ when ρ can be written as the product of r transpositions. When r is even, $(-1)^r = 1$, and when r is odd, $(-1)^r = -1$. Since the effect of a permutation on $q(\mathbf{t})$ is independent of how it is factored into transpositions, we conclude that ρ can be written as the product of either an even number of transpositions or an odd number of transpositions, but not both. □

Definition 6.21. A permutation in $\mathcal{S}_k$ is **even** provided it can be written as the product of an even number of transpositions. A permutation in $\mathcal{S}_k$ is **odd** provided it can be written as the product of an odd number of transpositions. The **sign of a permutation** is 1 if the permutation is even, and it is -1 if the permutation is odd.

We use $\mathrm{sgn}(\rho)$ to indicate the sign of ρ in $\mathcal{S}_k$.

Proof of the following is an exercise.

Lemma 6.22. If ρ_1 and ρ_2 belong to $\mathcal{S}_k$, then $\mathrm{sgn}(\rho_1 \rho_2) = \mathrm{sgn}(\rho_1) \mathrm{sgn}(\rho_2)$.

Corollary 6.23. For any ρ in $\mathcal{S}_k$, $\mathrm{sgn}(\rho) = \mathrm{sgn}(\rho^{-1})$.

Proof. The proof follows immediately from the lemma once we realize that since ι is the product of zero transpositions and zero is an even number, $\mathrm{sgn}\,\iota = 1$. □

Exercises 6.3.

6.3.1. Write each of the following permutations as a product of disjoint cycles:
(a) $\rho = \begin{pmatrix} 1 & 2 & 3 & 4 & 5 & 6 \\ 2 & 3 & 5 & 1 & 6 & 4 \end{pmatrix}$.
(b) $\rho = \begin{pmatrix} 1 & 2 & 3 & 4 & 5 & 6 & 7 & 8 \\ 3 & 5 & 1 & 2 & 4 & 8 & 6 & 7 \end{pmatrix}$.
(c) $\rho = \begin{pmatrix} 1 & 2 & 3 & 4 & 5 & 6 & 7 & 8 \\ 8 & 3 & 2 & 4 & 7 & 1 & 5 & 6 \end{pmatrix}$.

6.3.2. Write $\begin{pmatrix} 1 & 3 & 4 & 5 \end{pmatrix}\begin{pmatrix} 1 & 3 & 2 \end{pmatrix}\begin{pmatrix} 1 & 2 & 4 & 3 \end{pmatrix}$ as a cycle.

6.3.3. Let $\nu = \begin{pmatrix} 1 & 3 & 5 \end{pmatrix}$, $\varphi = \begin{pmatrix} 2 & 5 & 4 \end{pmatrix}$, and $\psi = \begin{pmatrix} 2 & 3 \end{pmatrix}$ in $\mathcal{S}_5$. Write each of $\nu\varphi\psi$ and $\psi\varphi\nu$ as a product of disjoint cycles.

6.3.4. Prove that disjoint cycles commute.

6.3.5. Write $\begin{pmatrix} 2 & 3 & 4 & 1 \end{pmatrix}$ as a product of transpositions.

6.3.6. Write $\rho = \begin{pmatrix} 1 & 2 \end{pmatrix}\begin{pmatrix} 1 & 3 \end{pmatrix}\begin{pmatrix} 1 & 4 \end{pmatrix}\begin{pmatrix} 1 & 5 \end{pmatrix}$ as a cycle. Find ρ^{-1}.

6.3.7. Prove Lemma 6.10.

6.3.8. Show that the orbits of an element in $\mathcal{S}_k$ form a partition of $S = \{1, \ldots, k\}$.

6.3.9. Let $S = \{1, 2, \ldots, 9\}$.
(a) Write down two different partitions on S, each containing three cells with three elements.

(b) Using disjoint cycles, write down two different permutations associated to each of the two partitions that you gave in part (a).

6.3.10. Prove Lemma 6.13.

6.3.11. Suppose $\rho = \tau_1 \cdots \tau_\ell$, where the τ_is are transpositions. Show that $\rho^{-1} = \tau_\ell \cdots \tau_1$.

6.3.12. Prove that if A is a matrix, then any matrix we get by permuting the rows of A is row equivalent to A.

6.3.13. Consider a cycle $\rho = \begin{pmatrix} a_1 & \cdots & a_t \end{pmatrix}$ in $\mathcal{S}_k$. For what values of t is ρ even?

6.3.14. Find the sign of each permutation in $\mathcal{S}_3$.

6.3.15. Prove that the sum of two even integers is even, that the sum of two odd integers is even, and that the sum of an even and an odd integer is odd.

6.3.16. Prove Lemma 6.22.

6.4. Permutations, Part II

We need more about certain subgroups of $\mathcal{S}_k$ before taking on the determinant. Our first result is inessential for our goals, but irresistible.

Theorem 6.24. *The even permutations form a subgroup of $\mathcal{S}_k$.*

Proof. Let A_k be the collection of even permutations in $\mathcal{S}_k$. Since ι belongs to A_k, A_k is nonempty. By Lemma 6.22, the product of even permutations is even. By the corollary to the lemma, when ρ is even, ρ^{-1} is even. Since A_k is nonempty and closed under inverses and composition, it is a subgroup of $\mathcal{S}_k$. □

The group of even permutations in $\mathcal{S}_k$ is called the **alternating group on k letters**.

We have seen vector space and algebra analogs of the following definition.

Definition 6.25. A mapping $\varphi : G \to G'$ is a **group homomorphism** provided $(G, \circ)$ and $(G', \circ')$ are groups and φ satisfies

(6.4) $$\varphi(g \circ h) = \varphi(g) \circ' \varphi(h),$$

for all g, h in G.

A **group isomorphism** is a bijective group homomorphism. A **group automorphism** is an isomorphism from a group to itself.

Groups G and G' are **isomorphic** provided there is a group isomorphism $\varphi : G \to G'$. In this case, we write $G \cong G'$.

The language to express the notion of isomorphic groups is enough to suggest, if not prove, the following.

Lemma 6.26. *Fix k in $\mathbb{Z}_+$. If k' in $\mathbb{Z}_+$ satisfies $k' \leq k$, then $\mathcal{S}_k$ contains a subgroup isomorphic to $\mathcal{S}_{k'}$.*

We do not have to look further than $\mathcal{S}_3$ to see that $\mathcal{S}_k$ will generally have several subgroups isomorphic to $\mathcal{S}_{k'}$ when $k' < k$. The next definition identifies the most obvious ones.

Definition 6.27. The **stabilizer** of $i \in \{1, \ldots, k\}$ is the set of permutations ρ in $\mathcal{S}_k$ such that $\rho(i) = i$.

We use G_i to indicate the stabilizer in $\mathcal{S}_k$ of $i \in \{1, \ldots, k\}$ and leave it as an exercise to verify that G_i is a subgroup of $\mathcal{S}_k$.

Example 6.28. Since we can identify $G_3 \subseteq \mathcal{S}_4$ with the set of permutations on $\{1, 2, 4\}$, it is evident that $G_3 \cong \mathcal{S}_3$. Knowing that $|\mathcal{S}_3| = 6$ makes it easy for us to list the elements in G_3:

$$G_3 = \{\iota, \begin{pmatrix} 1 & 2 \end{pmatrix}, \begin{pmatrix} 1 & 4 \end{pmatrix}, \begin{pmatrix} 2 & 4 \end{pmatrix}, \begin{pmatrix} 1 & 2 & 4 \end{pmatrix}, \begin{pmatrix} 1 & 4 & 2 \end{pmatrix}\}.$$

A vector space is an abelian group under addition and its subspaces are subgroups. The definition of the coset of a subspace in a vector space applies in any group, as long as we replace addition with the group binary operation, and subspace with subgroup. Since the binary operation in an arbitrary group is not commutative, we must distinguish between left and right cosets.

Definition 6.29. A **left** H-**coset** in a group G is a set of the form

$$aH = \{ah \mid h \in H\}$$

where H is a subgroup of G and a is a fixed element in G. A **right** H-**coset**[2] in a group G is a set of the form

$$Ha = \{ha \mid h \in H\}$$

where H is a subgroup of G and a is a fixed element in G.

Theorem 6.30. If G is a group with subgroup H, then the left H-cosets partition G.

Proof. Under the hypotheses, we must show that every element in G is in a left H-coset and that the intersection of distinct left H-cosets is empty.

Since e is in H, g in G belongs to gH.

Suppose $g_1 H \cap g_2 H$ is nonempty with

$$g_1 h_1 = g_2 h_2$$

for some h_1, h_2 in H. We then have

$$g_1 = g_2 h_2 h_1^{-1}$$

so that for all h in H,

$$g_1 h = g_2(h_2 h_1^{-1} h) \in g_2 H.$$

It follows that $g_1 H \subseteq g_2 H$. Since we also have

$$g_1(h_1 h_2^{-1} h) = g_2 h \in g_1 H,$$

$g_2 H \subseteq g_1 H$. This proves that if $g_1 H \cap g_2 H$ is not empty, then $g_1 H = g_2 H$. In other words, distinct left H-cosets are nonintersecting. □

---

[2]Some authors prefer to call gH a right H-coset and Hg a left-H coset.

6.4. Permutations, Part II

Of course the result holds when we replace left H-cosets with right H-cosets. We leave the proof as an exercise.

It is important to understand that in a nonabelian group, the left H-cosets and the right H-cosets determine different partitions on the ambient group.

Example 6.31. Consider $G = \mathcal{S}_3$. Notice that $G = \{\iota, \tau_1, \tau_2, \tau_3, \rho, \rho^2\}$ where τ_i is the transposition that fixes $i \in \{1, 2, 3\}$ and $\rho = \begin{pmatrix} 1 & 2 & 3 \end{pmatrix}$. The following Cayley table gives us the products in $\mathcal{S}_3$.

$\circ$	ι	τ_1	τ_2	τ_3	ρ	ρ^2
ι	ι	τ_1	τ_2	τ_3	ρ	ρ^2
τ_1	τ_1	ι	ρ	ρ^2	τ_2	τ_3
τ_2	τ_2	ρ^2	ι	ρ	τ_3	τ_1
τ_3	τ_3	ρ	ρ^2	ι	τ_1	τ_2
ρ	ρ	τ_3	τ_1	τ_2	ρ^2	ι
ρ^2	ρ^2	τ_2	τ_3	τ_1	ι	ρ

Consider $H = \{\iota, \tau_1\}$. The left H-cosets are

$$\tau_2 H = \{\tau_2, \rho^2\}, \quad \tau_3 H = \{\tau_3, \rho\}.$$

The right H-cosets are $H\rho = \{\rho, \tau_2\}$ and $H\rho^2 = \{\rho^2, \tau_3\}$. The only left H-coset equal to a right H-coset is H itself.

The following notation is in common usage.

When we write an indexed list as

$$\mathbf{v}_1, \ldots, \widehat{\mathbf{v}_j}, \ldots, \mathbf{v}_k$$

or an indexed product as

$$a_1 \cdots \widehat{a_j} \cdots a_k,$$

it means we are omitting the term under the circumflex.

The next theorem is a tool in the proof of our main theorem in the next section. Note that our usage of the notation τ_i changes here.

Theorem 6.32. Let $G_i \subseteq \mathcal{S}_k$ be the stabilizer of $i \in \{1, \ldots, k\}$. Let $\tau_j = \begin{pmatrix} i & j \end{pmatrix}$ for $j \in \{1, \ldots, n\}$, where we take $\tau_i = \iota$. The coset $\tau_j G_i$ is then the set of bijections from $\{1, \ldots, \widehat{i}, \ldots, k\}$ to $\{1, \ldots, \widehat{j}, \ldots, k\}$. Moreover, the cosets $\{\tau_j G_i\}_{j=1}^k$ partition $\mathcal{S}_k$.

Proof. Let G_i and τ_j be as hypothesized. Designate the set of bijections

$$\{1, \ldots, \widehat{i}, \ldots, k\} \to \{1, \ldots, \widehat{j}, \ldots, k\}$$

by $H_{ij} \subseteq \mathcal{S}_k$.

If ρ is in G_i, then $\tau_j \rho(i) = \tau_j(i) = j$. Since $\tau_j \rho$ is a bijection on $\{1, \ldots, k\}$ that maps i to j, it must map $\{1, \ldots, \widehat{i}, \ldots, k\}$ to $\{1, \ldots, \widehat{j}, \ldots, k\}$. This shows that $\tau_j G_i \subseteq H_{ij}$.

Next suppose that ρ is in H_{ij}. As a bijection on $\{1, \ldots, k\}$ that maps $\{1, \ldots, \widehat{i}, \ldots, k\}$ to $\{1, \ldots, \widehat{j}, \ldots, k\}$, ρ must map i to j. It follows that

$$\tau_j \rho(i) = \tau_j(j) = i,$$

so $\tau_j \rho$ belongs to G_i. Since $\tau_j^2 = \iota$,
$$\tau_j \rho \in G_i \implies \tau_j^2 \rho = \rho \in \tau_j G_i.$$
This shows that $H_{ij} \subseteq \tau_j G_i$. We conclude that $H_{ij} = \tau_j G_i$.

Let ρ be arbitrary in $\mathcal{S}_k$. If $\rho(i) = j$, then ρ is in $\tau_j G_i$. This is enough to show that

(6.5) $$\bigcup_{j=1}^n \tau_j G_i = \mathcal{S}_k.$$

If ρ is in both $\tau_j G_i$ and $\tau_n G_i$, then $\rho(i) = j = n$ so $\tau_j = \tau_n$. It follows that (6.5) is a disjoint union and, from there, that $\{\tau_j G_i\}_{j=1}^n$ is a partition on $\mathcal{S}_k$. □

Example 6.33. Here we find $H_{23} = \tau_3 G_2$ in $\mathcal{S}_4$, $\tau_3 = \begin{pmatrix} 2 & 3 \end{pmatrix}$. We have
$$G_2 = \{\iota, \begin{pmatrix} 1 & 3 \end{pmatrix}, \begin{pmatrix} 1 & 4 \end{pmatrix}, \begin{pmatrix} 3 & 4 \end{pmatrix}, \begin{pmatrix} 1 & 3 & 4 \end{pmatrix}, \begin{pmatrix} 1 & 4 & 3 \end{pmatrix}\}$$
so that
$$\tau_3 G_2 = \{\begin{pmatrix} 2 & 3 \end{pmatrix}, \begin{pmatrix} 1 & 2 & 3 \end{pmatrix}, \begin{pmatrix} 2 & 3 \end{pmatrix}\begin{pmatrix} 1 & 4 \end{pmatrix}, \begin{pmatrix} 3 & 4 & 2 \end{pmatrix},$$
$$\begin{pmatrix} 1 & 2 & 3 & 4 \end{pmatrix}, \begin{pmatrix} 1 & 4 & 2 & 3 \end{pmatrix}\}.$$
For comparison, we have
$$G_2 \tau_3 = \{\begin{pmatrix} 2 & 3 \end{pmatrix}, \begin{pmatrix} 2 & 1 & 3 \end{pmatrix}, \begin{pmatrix} 1 & 4 \end{pmatrix}\begin{pmatrix} 2 & 3 \end{pmatrix}, \begin{pmatrix} 2 & 4 & 3 \end{pmatrix},$$
$$\begin{pmatrix} 2 & 4 & 1 & 3 \end{pmatrix}, \begin{pmatrix} 2 & 1 & 4 & 3 \end{pmatrix}\}.$$
This shows that the left and right G_2-cosets define different partitions on $\mathcal{S}_4$.

We now know enough about $\mathcal{S}_k$ to prove our main theorem about determinants in the next section.

Exercises 6.4.

6.4.1. Show that G_i, the stabilizer of $i \in \{1, \ldots, k\}$, is a subgroup of $\mathcal{S}_k$.

6.4.2. Let ρ be any element in $\mathcal{S}_k$. Show that $H = \{\rho^k \mid k \in \mathbb{Z}\}$ is a subgroup of $\mathcal{S}_k$.

6.4.3. Let A_k be the alternating group on k letters.
  (a) Show that $\rho A_k \rho^{-1} = A_k$ for any ρ in $\mathcal{S}_k$.
  (b) Let τ be any transposition in $\mathcal{S}_k$. Describe the coset τA_k.
  (c) Prove that $|A_k| = k!/2$.

6.4.4. Verify that the set $\{-1, 1\}$ is an abelian group under multiplication. Show that
$$\text{sgn} : \mathcal{S}_k \to \{-1, 1\}$$
is a group homomorphism. What is its kernel, that is, $\text{sgn}^{-1}[\{1\}]$?

6.4.5. Apply Theorem 6.32 to find the set of bijections in $\mathcal{S}_4$ that map $\{2, 3, 4\}$ to $\{1, 2, 3\}$.

6.4.6. Identify every subgroup of $\mathcal{S}_3$ that is isomorphic to $\mathcal{S}_2$.

6.4.7. Let G be a group and let g be fixed in G. Show that $x \mapsto gx$ is a bijection on G. When is it a group automorphism?

6.4.8. Let G be a group and let $\mathcal{S}(G)$ be the collection of permutations on the set underlying G. Use the last problem to define an injective group homomorphism $f : G \to \mathcal{S}(G)$.

6.4.9. Show that the mapping $g \mapsto g^{-1}$ is a bijection on any group. When is it an automorphism?

6.4.10. Let H be a subgroup of any group, G. Show that, for any g in G, $|gH| = |Hg| = |H|$.

6.4.11. Verify the Cayley table in Example 6.31. Find both the left and the right H-cosets for $H = \{\iota, \tau_3\}$.

6.4.12. Repeat the last exercise for $H = \{\iota, \rho, \rho^2\}$.

6.4.13. Suppose G is a group with $|G| = 2n$ for some positive integer n. Let H be a subgroup of G with $|H| = n$. What can you say about the left and right H-cosets in G?

6.4.14. We identified $H_{ij} \subseteq \mathcal{S}_k$ as the set of bijections from $\{1, \ldots, \widehat{i}, \ldots, k\}$ to $\{1, \ldots, \widehat{j}, \ldots, k\}$.
 (a) Is H_{ij} a subgroup of $\mathcal{S}_k$?
 (b) Let $G_i \subseteq \mathcal{S}_k$ be the stabilizer of $i \in \{1, \ldots, k\}$, and let $\tau_j = \begin{pmatrix} i & j \end{pmatrix}$. Show that $H_{ij} = G_j \tau_i$.

6.5. The Determinant

Using what we now know about $\mathcal{S}_n$, we can define the determinant of $A = [a_{ij}]$ in $M_n(\mathbb{F})$ as Cramer understood it:

$$(6.6) \qquad \det A := \sum_{\rho \in \mathcal{S}_n} \operatorname{sgn}(\rho)\, a_{1\rho(1)} \cdots a_{k\rho(k)}.$$

We are also ready to look at the determinant of a matrix in $M_n(\mathbb{F})$ as an element in $\operatorname{Altn}(\mathbb{F}^n)$. Note that if $\dim V = n$, then $\dim \Lambda^n V = 1$. Since $\operatorname{Altn}(V) \cong (\Lambda^n V)^*$, it follows that when $\dim V = n$, $\dim \operatorname{Altn}(V) = 1$.

Definition 6.34. The **determinant** of the matrix $A = \begin{bmatrix} \mathbf{a}_1 & \cdots & \mathbf{a}_n \end{bmatrix}$ in $M_n(\mathbb{F})$ is
$$\varepsilon(\mathbf{a}_1, \ldots, \mathbf{a}_n)$$
where ε in $\operatorname{Altn}(\mathbb{F}^n)$ is defined by $\varepsilon(\mathbf{e}_1, \ldots, \mathbf{e}_n) = 1$, for $\{\mathbf{e}_1, \ldots, \mathbf{e}_n\}$, the standard ordered basis for $\mathbb{F}^n$.

Our current goal is to verify that this definition of determinant is the same as the one given in (6.6).

Example 6.35. We apply Definition 6.34 to calculate $\varepsilon(\mathbf{a}_1, \mathbf{a}_2)$ where
$$\mathbf{a}_i = a_{i1}\mathbf{e}_1 + a_{i2}\mathbf{e}_2 \in \mathbb{F}^2.$$
Our objective is to see that
$$\varepsilon(\mathbf{a}_1, \mathbf{a}_2) = a_{11}a_{22} - a_{12}a_{21}.$$
We start by using bilinearity to expand $\varepsilon(\mathbf{a}_1, \mathbf{a}_2)$.

Linearity in the first argument gives us

(6.7) $\quad \varepsilon(\mathbf{a}_1, \mathbf{a}_2) = \varepsilon(a_{11}\mathbf{e}_1 + a_{12}\mathbf{e}_2, \mathbf{a}_2) = a_{11}\varepsilon(\mathbf{e}_1, \mathbf{a}_2) + a_{12}\varepsilon(\mathbf{e}_2, \mathbf{a}_2).$

Expanding terms in (6.7) and remembering that $\varepsilon(\mathbf{e}_i, \mathbf{e}_i) = 0$, we get
$$a_{11}\varepsilon(\mathbf{e}_1, \mathbf{a}_2) = a_{11}a_{22}\varepsilon(\mathbf{e}_1, \mathbf{e}_2)$$
and
$$a_{12}\varepsilon(\mathbf{e}_2, \mathbf{a}_2) = a_{12}a_{21}\varepsilon(\mathbf{e}_2, \mathbf{e}_1).$$
Since $\varepsilon(\mathbf{e}_2, \mathbf{e}_1) = -1$, we have
$$\varepsilon(\mathbf{a}_1, \mathbf{a}_2) = a_{11}a_{22} - a_{12}a_{22}.$$

The next lemma states the effect of a permutation on the value of any alternating multilinear mapping.

Lemma 6.36. Let V be a vector space and let μ belong to Altk(V). If ρ is in $\mathcal{S}_k$, then
$$\mu(\mathbf{v}_{\rho(1)}, \ldots, \mathbf{v}_{\rho(k)}) = \text{sgn}(\rho)\,\mu(\mathbf{v}_1, \ldots, \mathbf{v}_k).$$

Proof. Let V and μ be as hypothesized. As an alternating map, μ is skew-symmetric so if τ in $\mathcal{S}_k$ is a transposition, then
$$\mu(\mathbf{v}_{\tau(1)}, \ldots, \mathbf{v}_{\tau(k)}) = -\mu(\mathbf{v}_1, \ldots, \mathbf{v}_k).$$
The proof now follows Theorem 6.19. $\square$

Next is our main theorem here. It leads directly to our goal.

Theorem 6.37. Let V be a vector space with basis $\mathcal{B} = \{\mathbf{b}_1, \ldots, \mathbf{b}_n\}$. If μ is an alternating n-linear mapping on V and $\mathbf{a}_i = \sum_{j=1}^{n} a_{ij}\mathbf{b}_j$ is in V, then
$$\mu(\mathbf{a}_1, \ldots, \mathbf{a}_n) = \sum_{\rho \in \mathcal{S}_n} a_{1\rho(1)}a_{2\rho(2)} \cdots a_{n\rho(n)}\mu(\mathbf{b}_{\rho(1)}, \ldots, \mathbf{b}_{\rho(n)}).$$

Proof. Let V, $\mathcal{B}$, and $\mu: V^n \to W$ be as hypothesized.

We proceed by induction on n, noting that our work in Example 6.35 effectively applies to prove the base case, $n = 2$.

Assume the result holds when $\dim V = k$, for any $2 < k < n$. Let $\mathbf{a}_1, \ldots, \mathbf{a}_n$ in V be as hypothesized. By linearity in the first argument, we have
$$\mu(\mathbf{a}_1, \ldots, \mathbf{a}_n) = \sum_{j=1}^{n} a_{1j}\mu(\mathbf{b}_j, \mathbf{a}_2, \ldots, \mathbf{a}_n).$$
Since μ is alternating, we leave $\mu(\mathbf{b}_j, \mathbf{a}_2, \ldots, \mathbf{a}_n)$ unchanged if we substitute
$$\mathbf{a}_k - a_{kj}\mathbf{b}_j = \sum_{\substack{i=1 \\ i \neq j}}^{n} a_{ki}\mathbf{b}_i$$
for $\mathbf{a}_k$, $k \in \{2, \ldots, n\}$. This gives us
$$\mu(\mathbf{a}_1, \ldots, \mathbf{a}_n) = \sum_{j=1}^{n} a_{1j}\mu\left(\mathbf{b}_j, \sum_{\substack{i=1 \\ i \neq j}}^{n} a_{2i}\mathbf{b}_i, \ldots, \sum_{\substack{i=1 \\ i \neq j}}^{n} a_{ni}\mathbf{b}_i\right).$$

6.5. The Determinant

Fixing $j \in \{1, \ldots, n\}$, we can define an alternating $n-1$-linear mapping on $\mathrm{Span}(\mathcal{B} \setminus \{\mathbf{b}_j\}) \subseteq V$,

$$\mu_j(\mathbf{a}_2, \ldots, \mathbf{a}_n) := \mu(\mathbf{b}_j, \mathbf{a}_2, \ldots, \mathbf{a}_n).$$

Applying the induction hypothesis to μ_j, we get

$$\mu_j(\mathbf{a}_2, \ldots, \mathbf{a}_n) = \mu_j\left(\sum_{\substack{i=1 \\ i \neq j}}^n a_{2i}\mathbf{b}_i, \ldots, \sum_{\substack{i=1 \\ i \neq j}}^n a_{ni}\mathbf{b}_i\right)$$

$$= \sum_{\substack{i=2 \\ \rho \in H_{1j}}}^n a_{2\rho(2)} \cdots a_{n\rho(n)} \mu_j(\mathbf{b}_{\rho(2)}, \ldots, \mathbf{b}_{\rho(n)}),$$

where $H_{1j} = \begin{pmatrix} 1 & j \end{pmatrix} G_1$ is the set of elements in $\mathcal{S}_n$ that map $\{2, \ldots, n\}$ to $\{1, \ldots, \widehat{j}, \ldots, n\}$.

Putting all this together, we have

$$\mu(\mathbf{a}_1, \ldots, \mathbf{a}_n) = \sum_{\substack{j=1 \\ \rho \in H_{1j}}}^n a_{1j} a_{2\rho(2)} \cdots, a_{n\rho(n)} \mu(\mathbf{b}_j, \mathbf{b}_{\rho(2)}, \ldots, \mathbf{b}_{\rho(n)}).$$

We are summing over $j = 1, \ldots, n$ and Theorem 6.32 established that $\bigcup_{j=1}^n H_{1j} = \mathcal{S}_n$. This means we can write

$$\mu(\mathbf{a}_1, \ldots, \mathbf{a}_n) = \sum_{\rho \in \mathcal{S}_n} a_{1\rho(1)} \cdots a_{n\rho(n)} \mu(\mathbf{b}_{\rho(1)}, \ldots, \mathbf{b}_{\rho(n)}). \qquad \square$$

The determinant mapping ε has the property that

$$\varepsilon(\mathbf{e}_1, \ldots, \mathbf{e}_n) = 1,$$

so if we designate column j of $A = [a_{ij}]$ in $M_n(\mathbb{F})$ by $\mathbf{a}_j$, then according to Definition 6.34 and Theorem 6.37,

$$\det A = \sum_{\rho \in \mathcal{S}_n} \mathrm{sgn}(\rho) a_{1\rho(1)} \cdots a_{n\rho(n)}.$$

This is the modern rendition of Cramer's definition.

The next lemma assists in the proof of an important property of the determinant.

Lemma 6.38. *The sets of ordered pairs $R = \{(i, \rho(i)) \mid \rho \in \mathcal{S}_n\}_{i=1}^n$ and $L = \{(\rho(i), i) \mid \rho \in \mathcal{S}_n\}_{i=1}^n$ are identical.*

Proof. Fix $i \in \{1, \ldots, n\}$ and ρ in $\mathcal{S}_n$. If $\rho(i) = j$, then we can write $(i, \rho(i)) = (\rho^{-1}(j), j)$. It follows that $(i, \rho(i))$ in R is in L. Since this holds for all $i \in \{1, \ldots, n\}$ and all ρ in $\mathcal{S}_n$, $R \subseteq L$. Likewise, $(\rho(i), i)$ in L can be written $(j, \rho^{-1}(j))$ when $\rho(i) = j$. It follows that $L \subseteq R$, thus, that $L = R$. $\qquad \square$

Theorem 6.39. *If A is an $n \times n$ matrix, then $\det A = \det A^T$.*

Proof. If $A = [a_{ij}]$, then
$$\det A = \sum_{\rho \in \mathcal{S}_n} \operatorname{sgn}(\rho) a_{1\rho(1)} \cdots a_{n\rho(n)}.$$
Since the ij-entry in A^T is a_{ji}, we also have
$$\det A^T = \sum_{\rho \in \mathcal{S}_n} \operatorname{sgn}(\rho) a_{\rho(1)1} \cdots a_{\rho(n)n}.$$
By Lemma 6.38, the sets of subscripts in the two sums are identical. The result follows. □

Going forward, we view the determinant mapping as applicable to either $\mathbb{F}^n$ or $\mathbb{F}_n$.

Exercises 6.5.

6.5.1. Explain the step in the proof of Theorem 6.37 that has us evaluate μ_j by summing over ρ in H_{1j}.

6.5.2. Let A belong to $M_n(\mathbb{F})$.
   (a) Show that $\det B = -\det A$ if we obtain B by applying a type-1 elementary row, or column, operation to A.
   (b) Show that $\det B = c \det A$ if we obtain B by multiplying one row, or one column, of A by a constant c.
   (c) Show that if we obtain B by applying a type-3 elementary row, or column, operation to A, then $\det B = \det A$.

6.5.3. Show that $\det A = 0$ if A in $M_n(\mathbb{F})$ has two identical rows, or two identical columns.

6.5.4. Suppose A in $M_n(\mathbb{R})$ has determinant equal to 3. Let $B = 2A$. What is $\det B$?

6.5.5. If E is a type-1 elementary matrix, what is $\det E$?

6.5.6. If E is a type-2 elementary matrix, what is $\det E$?

6.5.7. If E is a type-3 elementary matrix, what is $\det E$?

6.6. Properties of the Determinant

Many of the properties of the determinant flow practically unimpeded from what we have established to this point.

Theorem 6.40. If $A = [a_{ij}]$ in $M_n(\mathbb{F})$ is upper or lower triangular, then $\det A = a_{11} \ldots a_{nn}$.

Proof. Let $A = [a_{ij}]$ in $M_n(\mathbb{F})$ be upper triangular so that $a_{ij} = 0$ if $i > j$. Suppose ρ is different from ι in $\mathcal{S}_n$. Let i in $\{1, \ldots, n\}$ be maximal so that $\rho(i) \neq i$. If $\rho(i) = j$, then since ρ is a bijection, $\rho(j) \neq j$, so $i > j$. It follows that $a_{i\rho(i)} = 0$. From here we see that when ρ is in $\mathcal{S}_n$, the only nonzero expression of the form
$$a_{1\rho(1)} \cdots a_{n\rho(n)}$$
is the one for which $\rho = \iota$. This proves the lemma when A is upper triangular. We leave proof of the lower triangular case as an exercise. □

6.6. Properties of the Determinant

Notice that $\det I_n = 1$.

Example 6.41. Consider $A = \begin{bmatrix} 2 & -4 & -2 & 3 \\ 6 & -9 & -5 & 8 \\ 2 & 2 & 1 & 9 \\ 4 & -2 & -2 & -1 \end{bmatrix}$. We find $\det A$ using Theorem 6.40 and results cited in Exercise 6.5.2.

Applying type-3 elementary row operations to A, we get
$$A \underset{r}{\sim} \begin{bmatrix} 2 & -4 & -2 & 3 \\ 0 & 3 & 1 & -1 \\ 0 & 6 & 3 & 6 \\ 0 & 6 & 2 & -7 \end{bmatrix} \underset{r}{\sim} \begin{bmatrix} 2 & -4 & -2 & 3 \\ 0 & 3 & 1 & -1 \\ 0 & 0 & 1 & 8 \\ 0 & 0 & 0 & -5 \end{bmatrix}.$$

By Theorem 6.40, $\det A = -45$.

The next theorem states one of the most important properties of the determinant.

Theorem 6.42. If A and B belong to $M_n(\mathbb{F})$, then $\det(AB) = (\det A)(\det B)$.

Proof. Let $A = [a_{ij}]$ and $B = [b_{ij}]$ belong to $M_n(\mathbb{F})$. Let column j of A be $\mathbf{a}_j$ and column j of B be $\mathbf{b}_j$. We have
$$AB = \begin{bmatrix} A\mathbf{b}_1 & \cdots & A\mathbf{b}_n \end{bmatrix}$$
so that for the determinant mapping ε,
$$\det(AB) = \varepsilon(A\mathbf{b}_1, \ldots, A\mathbf{b}_n).$$
Since $A\mathbf{b}_j = b_{1j}\mathbf{a}_1 + b_{2j}\mathbf{a}_2 + \cdots + b_{nj}\mathbf{a}_n$, Theorem 6.37 implies
$$\det(AB) = \sum_{\rho \in \mathcal{S}_n} \operatorname{sgn}(\rho)\, b_{1\rho(1)}b_{2\rho(2)} \cdots b_{n\rho(n)} \varepsilon(\mathbf{a}_1, \ldots, \mathbf{a}_n) = (\det B)(\det A).$$
The result follows the observation that $(\det B)(\det A) = (\det A)(\det B)$. $\square$

Theorem 6.42 has many corollaries. A few follow. Look for more among the exercises.

Corollary 6.43. A square matrix is invertible if and only if its determinant is nonzero. When A is an invertible matrix, $\det A^{-1} = 1/\det A$.

Proof. If A in $M_n(\mathbb{F})$ is invertible, then since
$$\det(AA^{-1}) = \det I_n = 1$$
and $\det(AA^{-1}) = \det A \det A^{-1}$, it follows that neither $\det A$ nor $\det A^{-1}$ is zero and that $\det A^{-1} = 1/\det A$.

Now suppose that A in $M_n(\mathbb{F})$ has nonzero determinant. Let B be the reduced row echelon form of A. Since $B = PA$, for invertible P in $M_n(\mathbb{F})$,
$$\det B = \det(PA) = \det P \det A \neq 0.$$
Since $B = [b_{ij}]$ is in upper triangular form, Theorem 6.40 guarantees that $b_{ii} \neq 0$ for all i. It follows that b_{ii} is the leading 1 in row i, thus, that rank $A = n$, so A is invertible. $\square$

Corollary 6.44. Similar matrices have the same determinant.

Corollary 6.44 says that the determinant is defined, not just for square matrices, but for linear operators on finite-dimensional vector spaces. This is significant because in studying linear algebra, we want to distinguish cases in which an apparent reliance on a particular coordinate system is merely apparent and not central to the story. Consider for example, the class of mappings called reflections. The idea behind a linear reflection on $V \cong \mathbb{F}^n$ is straightforward: All vectors in W, a hyperspace in V, are fixed, while the vectors in some complement to W are scaled by -1. If we construct an ordered basis $\mathcal{B}$ for V that is extended from an ordered basis for W, the matrix representation of the reflection with respect to $\mathcal{B}$ is

$$A = [\mathbf{e}_1 \ \ldots \ \mathbf{e}_{n-1} \ -\mathbf{e}_n].$$

Since $\det A = -1$, Corollary 6.44 guarantees that any reflection on $V \cong \mathbb{F}^n$, for any n, has determinant -1. This is entirely independent of a choice of basis.

Observation 6.45. The determinant of a linear transformation L on a finite-dimensional vector space V is the determinant of any matrix representation of L.

The following is an application to diagonal block matrices, which we introduced in Section 5.4. We can see it as a generalization of Theorem 6.40.

Theorem 6.46. If $A = \text{diag}(A_1, \ldots, A_k)$, then $\det A = \det A_1 \cdots \det A_k$.

Proof. We start by considering $A = \text{diag}(A_1, I_{n-k})$, where A_1 is in $M_k(\mathbb{F})$. If we write $A = [a_{ij}]$, then

$$\det A = \sum_{\rho \in \mathcal{S}_n} \text{sgn}(\rho)\, a_{1\rho(1)} a_{2\rho(2)} \cdots a_{n\rho(n)}.$$

Since $a_{ij} = 0$ when $i \in \{k+1, \ldots, n\}$ and $j \neq i$,

$$a_{1\rho(1)} a_{2\rho(2)} \cdots a_{n\rho(n)} = 0,$$

unless $\rho|_{\{k+1,\ldots,n\}}$ is the identity permutation. It follows that we need not sum over all permutations in $\mathcal{S}_n$, only over the permutations in $\mathcal{S}_k$. We have then that

$$\det A = \sum_{\rho \in \mathcal{S}_k} \text{sgn}(\rho)\, a_{1\rho(1)} \cdots a_{k\rho(k)} = \det A_1.$$

A similiar argument shows that if $A = \text{diag}(I_k, A_2)$, then $\det A = \det A_2$.

Lemma 5.37 gives us that $A = \text{diag}(A_1, A_2)$ is the product of $\text{diag}(A_1, I_{n-k})$ and $\text{diag}(I_k, A_2)$. By Theorem 6.42, then,

$$\det \text{diag}(A_1, A_2) = \det A_1 \det A_2.$$

If $A = \text{diag}(A_1, A_2, \ldots, A_k)$, we may view A as having the form $A = \text{diag}(A_1, B)$, where $B = \text{diag}(A_2, \ldots, A_k)$. We have shown that

$$\det A = \det A_1 \det B.$$

By the induction hypothesis, $\det B = \det A_2 \cdots \det A_k$ so that

$$\det A = \det A_1 \cdots \det A_k. \qquad \square$$

Exercises 6.6.

6.6.1. Let A belong to $M_n(\mathbb{F})$ where $\mathbb{F}$ has characterisitic zero. Show that if A is skew-symmetric $(A^T = -A)$ and nonsingular, then n is even.

6.6.2. Show that if A is an orthogonal matrix $(A^T = A^{-1})$, then $\det A = \pm 1$.

6.6.3. Show that $\overline{\det A} = \det \bar{A}$ for any A in $M_n(\mathbb{C})$.

6.6.4. Show that if A is unitary $(A^\dagger = A^{-1})$, then $|\det A| = 1$. Does this mean that $\det A = \pm 1$?

6.6.5. Let $A = \mathrm{diag}(A_1, \ldots, A_k)$ and $B = \mathrm{diag}(B_1, \ldots, B_k)$, where A_i and B_i belong to $M_{n_i}(\mathbb{F})$ for each i. Show that the determinant of AB is the product $\prod_{i=1}^k \det A_i \det B_i$.

6.6.6. In Example 4.38 we found an LU-factorization of
$$A = \begin{bmatrix} 3 & -7 & -2 & 2 \\ -3 & 5 & 1 & 0 \\ 6 & -4 & 0 & -5 \\ -9 & 5 & -5 & 12 \end{bmatrix}.$$

(a) Use the LU-factorization to find $\det A$ and $\det A^{-1}$.

(b) Use row reduction directly to find $\det A$.

6.6.7. Use type-2 and type-3 elementary row operations, and work by hand to find
$$\det \begin{bmatrix} 1 & a & a^2 & a^3 \\ 1 & b & b^2 & b^3 \\ 1 & c & c^2 & c^3 \\ 1 & d & d^2 & d^3 \end{bmatrix}.$$

6.6.8. Let $A = \begin{bmatrix} 2 & 0 & 0 & 0 & 0 \\ 0 & 3 & -4 & 0 & 0 \\ 0 & -1 & 2 & 0 & 0 \\ 0 & 0 & 0 & 5 & 3 \\ 0 & 0 & 0 & -1 & -2 \end{bmatrix}$ in $M_5(\mathbb{R})$. Part of Exercise 5.4.9 was to describe the blocks if we write A in block diagonal form. Use that form to find $\det A$.

6.6.9. Recall that the set of nonzero elements in a field $\mathbb{F}$ is denoted $\mathbb{F}^*$.

(a) Show that $\mathbb{F}^*$ is an abelian group under multiplication.

(b) Show that $\det : GL_n(\mathbb{F}) \to \mathbb{F}^*$ is a group homomorphism.

Chapter 7

Inner Product Spaces

A vector space is a nonempty set closed under linear combination. Most of what we have studied to this point has been about the consequences of that definition, and that definition alone. In this chapter, we consider properties peculiar to real and complex vector spaces that arise from the additional structure that comes of generalizing the dot product.

It is often productive to treat real and complex vector spaces in one discussion. To establish ground rules, we start by addressing questions that arise out of the differences between the two types of spaces.

Recall that when z belongs to $\mathbb{C}$ and we write $z = x + iy$, or $z = a + ib$, you are to assume x and y, respectively a and b, are real numbers.

Recall that $\mathbb{R} = \{z \in \mathbb{C} \,|\, \bar{z} = z\}$.

7.1. The Dot Product: Under the Hood

The Pythagorean Theorem — a linchpin of Euclidean geometry — gives us that $a^2 + b^2$ is the square of the length of $\mathbf{v} = (a, b)$ in $\mathbb{R}^2$ when we view $\mathbf{v}$ as an arrow in the xy-plane. The distance from the point (a, b) in $\mathbb{R}^2$ to the origin is then $\|\mathbf{v}\| = \sqrt{\mathbf{v} \cdot \mathbf{v}}$. In this sense, the dot product articulates the connection between linear algebra and Euclidean geometry in $\mathbb{R}^n$.

We can see that we do not get the same connection in $\mathbb{C}^n$, just by considering $\mathbf{v} = (i, 1)$ in $\mathbb{C}^2$. With the same dot product that we use on $\mathbb{R}^n$, we have

$$\mathbf{v}^T \mathbf{v} = i^2 + 1^2 = 0.$$

If we want to use this to define distance in $\mathbb{C}^2$, we are forced to say that the distance between the origin and the point $(i, 1)$ is zero. Since different objects in a Euclidean plane cannot occupy the same point, either $(i, 1) = (0, 0)$, which is absurd, or $\mathbb{C}^2$ is not Euclidean under the dot product.

Since $\mathbb{C}^n$ and $\mathbb{R}^{2n}$ are canonically isomorphic as vector spaces, we should be able to interpret $\mathbb{C}^n$ as having Euclidean structure without going too far astray of

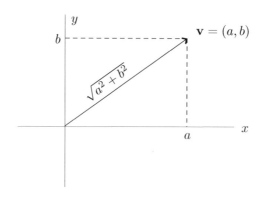

Figure 7.1. The dot product links coordinates of points in $\mathbb{R}^2$ to the Pythagorean Theorem.

the dot product. The trick is to replace the transpose with the conjugate transpose in the definition of the dot product.

Definition 7.1. The **complex dot product**, also called the **standard Hermitian inner product on** $\mathbb{C}^n$, is

(7.1) $$\mathbf{v} \cdot_{\mathbb{C}} \mathbf{w} := \mathbf{v}^\dagger \mathbf{w}$$

for $\mathbf{v}$ and $\mathbf{w}$ in $\mathbb{C}^n$.[1]

Using the complex dot product on $\mathbb{C}^2$, we have $(i,1) \cdot_{\mathbb{C}} (i,1) = (-i)(i) + 1 = 2$. More generally, if $\mathbf{v} = (z_1, \ldots, z_n)$ is in $\mathbb{C}^n$, then

$$\mathbf{v} \cdot_{\mathbb{C}} \mathbf{v} = |z_1|^2 + \cdots + |z_n|^2$$

is a nonnegative real number. This means we can use the complex dot product to define length and distance in $\mathbb{C}^n$ in a way that comports with Euclidean geometry.

Can we make an adjustment to the dot product to get Euclidean geometry on $\mathbb{F}^n$ for an arbitrary field $\mathbb{F}$? We cannot. The problem is that most fields are not ordered fields, meaning that we cannot order the field in a way that respects the field axioms. If we say $1 > 0$ in $\mathbb{F}_3$, for instance, then is $1 + 1 = 2 > 1$? If so, then $2 + 1 = 0 > 1$. These sorts of inconsistencies are unavoidable in a field with positive characteristic. While $\mathbb{C}$ is not itself an ordered field, it contains an ordered field, $\mathbb{R}$. Moreover, $\mathbf{v} \cdot_{\mathbb{C}} \mathbf{v}$ is in $\mathbb{R}$ for $\mathbf{v}$ in $\mathbb{C}^n$. This allows us to compare the length of $\mathbf{v} + \mathbf{w}$ to the lengths of $\mathbf{v}$ and $\mathbf{w}$ and thus to arrive at the triangle inequality, a key property of what we usually think of as a measure of distance.

The rationals are also an ordered field. Would they be enough to get us Euclidean geometry? Not if we want lengths to be in the underlying field. For instance, if we view $\mathbf{v} = (1,1)$ as an element in $\mathbb{Q}^2$, its length is $\sqrt{2}$ which is not defined in $\mathbb{Q}$.

The most obvious way to ensure that Euclidean geometry applies on a given vector space is to require a copy of the continuum — the real line — inside the

---

[1] The complex dot product is often defined as $\mathbf{v}^T \overline{\mathbf{w}}$.

7.1. The Dot Product: Under the Hood

scalar field. This is why, for the rest of this chapter, we restrict our attention to real and complex vector spaces.

Since $\mathbf{v}^\dagger = \mathbf{v}^T$ for $\mathbf{v}$ in $\mathbb{R}^n$, we write

$$\mathbf{v} \cdot \mathbf{w} := \mathbf{v}^\dagger \mathbf{w}$$

for $\mathbf{v}$ and $\mathbf{w}$ in $\mathbb{R}^n$, or $\mathbf{v}$ and $\mathbf{w}$ in $\mathbb{C}^n$. Going forward, we use $\cdot$ in this manner, as convenient.[2] When we want to distinguish the real from the complex dot product, we use $\cdot_\mathbb{R}$ and $\cdot_\mathbb{C}$ or we specify the field of scalars explicitly.

Definition 7.2. The **length** of $\mathbf{v}$ in real or complex n-space is

$$\|\mathbf{v}\| = \sqrt{\mathbf{v} \cdot \mathbf{v}}.$$

We leave it as an exercise to show that in $\mathbb{C}^n$, Definition 7.2 gives us the same notion of length that we get if we identify $\mathbb{C}^n$ with $\mathbb{R}^{2n}$ under a canonical isomorphism $\mathbb{C}^n \to \mathbb{R}^{2n}$.

What follows generalizes Lemma B.6, applying to the complex dot product as well. Its proof is an exercise.

Lemma 7.3. The following properties hold for all $\mathbf{v}, \mathbf{w}$, and $\mathbf{x}$ in real or complex n-space, when c is a scalar:

(a) $\mathbf{v} \cdot (c\mathbf{w} + \mathbf{x}) = c(\mathbf{v} \cdot \mathbf{w}) + \mathbf{v} \cdot \mathbf{x}$.

(b) $\mathbf{v} \cdot \mathbf{w} = \overline{\mathbf{w} \cdot \mathbf{v}}$.

Though we can define the Euclidean inner product and the complex dot product the same way, the two dot products have different algebraic properties. In particular, the complex dot product is neither symmetric nor bilinear.

Definition 7.4. When V is a complex vector space, a mapping

$$\sigma : V \times V \to \mathbb{C}$$

is **Hermitian symmetric** provided

$$\sigma(\mathbf{v}, \mathbf{w}) = \overline{\sigma(\mathbf{w}, \mathbf{v})},$$

for all $\mathbf{v}$ and $\mathbf{w}$ in V.

Where the Euclidean inner product on $\mathbb{R}^n$ is symmetric, the complex dot product is Hermitian symmetric.

Both the Euclidean inner product and the complex dot product are linear in the second variable. When V is a complex vector space and

$$\sigma : V \times V \to \mathbb{C}$$

is Hermitian symmetric and linear in its second variable, it is easy to check that σ is conjugate-linear in its first variable: $(c\mathbf{v}) \cdot \mathbf{w} = \bar{c}(\mathbf{v} \cdot \mathbf{w})$. A mapping $\sigma : V \times V \to \mathbb{C}$ that is linear in one variable and conjugate-linear in the other variable is sometimes called a sesquilinear form: *Sesqui* means one and a half.

Definition 7.5. A **Hermitian form** on a complex vector space V is a mapping $\sigma : V \times V \to \mathbb{C}$ that is Hermitian symmetric and linear in one variable.

---

[2]In the literature, the $\cdot$ notation is usually reserved for $\mathbb{R}^n$.

Example 7.6. Consider $(M_n(\mathbb{F}), \sigma)$ where $\mathbb{F}$ is either $\mathbb{R}$ or $\mathbb{C}$ and $\sigma(A, B) = \operatorname{Tr} A^\dagger B$, the trace of $A^\dagger B$. Exercise 5.1.4 was to show that $(A, B) \mapsto \operatorname{Tr} AB$ is a symmetric bilinear form. We leave it as an exercise to check from there that σ is a Hermitian form on complex matrices (thus, a symmetric bilinear form on real matrices).

We saw in Section 5.3 that if $V \cong \mathbb{F}^n$ has ordered basis $\mathcal{B}$ and $\sigma : V \times V \to \mathbb{F}$ is bilinear, then
$$\sigma(\mathbf{v}, \mathbf{w}) = [\mathbf{v}]_\mathcal{B}^T [\sigma]_\mathcal{B} [\mathbf{w}]_\mathcal{B}$$
for all $\mathbf{v}$ and $\mathbf{w}$ in V. Evidently, σ is symmetric if and only if $[\sigma]_\mathcal{B}$ is a symmetric matrix for any basis $\mathcal{B}$ in V.

We coordinatize a Hermitian form σ defined on $V \cong \mathbb{C}^n$ exactly the same way. Fixing an ordered basis for V, $\mathcal{B} = \{\mathbf{b}_i\}_{i=1}^n$, we let $a_{ij} = \sigma(\mathbf{b}_i, \mathbf{b}_j)$. Define
$$[\sigma]_\mathcal{B} = [a_{ij}] \in M_n(\mathbb{F}).$$

Lemma 7.7. Let $V \cong \mathbb{C}^n$. Let $\sigma : V \times V \to \mathbb{C}$ be a Hermitian form. If $\mathcal{B} = \{\mathbf{b}_i\}_{i=1}^n$ is an ordered basis for V and $A = [\sigma]_\mathcal{B}$, then A is self-adjoint. If σ is linear in its second variable, then

(7.2) $$\sigma(\mathbf{v}, \mathbf{w}) = [\mathbf{v}]_\mathcal{B}^\dagger A [\mathbf{w}]_\mathcal{B}$$

for all $\mathbf{v}$ and $\mathbf{w}$ in V. If σ is linear in its first variable, then
$$\sigma(\mathbf{v}, \mathbf{w}) = [\mathbf{v}]_\mathcal{B}^T A \overline{[\mathbf{w}]_\mathcal{B}}.$$

Proof. Let V, $\mathcal{B}$, σ, and $A = [a_{ij}]$ be as hypothesized. Since
$$a_{ij} = \sigma(\mathbf{b}_i, \mathbf{b}_j) = \overline{\sigma(\mathbf{b}_j, \mathbf{b}_i)} = \overline{a_{ji}},$$
$A = A^\dagger$.

Let $\mathbf{v}$ and $\mathbf{w}$ belong to V, with $[\mathbf{v}]_\mathcal{B} = (c_1, \ldots, c_n)$ and $[\mathbf{w}]_\mathcal{B} = (d_1, \ldots, d_n)$. If σ is linear in its second variable, then it is conjugate-linear in its first variable. This means
$$\sigma(\mathbf{v}, \mathbf{w}) = \sum_{j=1}^n d_j \sigma(\mathbf{v}, \mathbf{b}_j) = \sum_{i,j=1}^n \overline{c_i} d_j \sigma(\mathbf{b}_i, \mathbf{b}_j).$$
If we designate column j of A by $\mathbf{a}_j$, we have
$$A[\mathbf{w}]_\mathcal{B} = d_1 \mathbf{a}_1 + \cdots + d_n \mathbf{a}_n$$
so that
$$[\mathbf{v}]_\mathcal{B}^\dagger A [\mathbf{w}]_\mathcal{B} = \sum_{j=1}^n [\overline{c_1} \cdots \overline{c_n}] d_j \mathbf{a}_j = \sum_{j=1}^n d_j \sum_{i=1}^n \overline{c_i} a_{ij} = \sum_{i,j=1}^n \overline{c_i} d_j \sigma(\mathbf{b}_i, \mathbf{b}_j).$$
This shows that when σ is linear in its second variable, (7.2) holds.

A similar calculation verifies the result for the case when σ is linear in its first variable. $\square$

The next lemma says we can go in the other direction as well. Its proof is an exercise.

7.1. The Dot Product: Under the Hood

Lemma 7.8. If $V \cong \mathbb{C}^n$ and A in $M_n(\mathbb{C})$ is self-adjoint, then $\sigma : V \times V \to \mathbb{C}$ defined by
$$\sigma(\mathbf{v}, \mathbf{w}) := [\mathbf{v}]_\mathcal{B}{}^\dagger A [\mathbf{w}]_\mathcal{B},$$
for $\mathbf{v}$ and $\mathbf{w}$ in V, is Hermitian and linear in its second variable.

Example 7.9. Let $A = \begin{bmatrix} 1 & -i \\ i & -1 \end{bmatrix}$. Since A is self-adjoint,
$$\sigma(\mathbf{v}, \mathbf{w}) := \mathbf{v}^\dagger A \mathbf{w}$$
is a Hermitian form on $\mathbb{C}^2$. If $\mathbf{v} = (z_1, z_2)$ is in $\mathbb{C}^2$, then it is easy to check that
$$\sigma(\mathbf{v}, \mathbf{v}) = \overline{\begin{bmatrix} z_1 & z_2 \end{bmatrix}} A \begin{bmatrix} z_1 \\ z_2 \end{bmatrix}$$
$$= \begin{bmatrix} \overline{z_1} & \overline{z_2} \end{bmatrix} \begin{bmatrix} 1 & -i \\ i & -1 \end{bmatrix} \begin{bmatrix} z_1 \\ z_2 \end{bmatrix} = |z_1|^2 - |z_2|^2.$$

We streamline our discussion going forward by taking all of our Hermitian forms to be linear in the second variable.

Exercises 7.1.

7.1.1. Let V be real or complex and let σ be, respectively, a bilinear or Hermitian form defined on V. Show that $\sigma(\mathbf{0}, \mathbf{v}) = 0$ for all $\mathbf{v}$ in V.

7.1.2. We have defined the complex dot product as $\mathbf{v} \cdot_\mathbb{C} \mathbf{w} := \mathbf{v}^\dagger \mathbf{w}$, while noting that it is sometimes defined as $\mathbf{v} \cdot_\mathbb{C} \mathbf{w} := \mathbf{v}^T \overline{\mathbf{w}}$. Show that for any $\mathbf{v}$ in $\mathbb{C}^n$,
$$\mathbf{v}^\dagger \mathbf{v} = \mathbf{v}^T \overline{\mathbf{v}}.$$

7.1.3. In an ordered field, $\mathbb{F}$, any two elements are comparable, meaning if a and b belong to $\mathbb{F}$, then either $a < b$, $a = b$, or $a > b$. The ordering must also be transitive: If $a < b$ and $b < c$, then $a < c$.
   (a) How can we define positive and negative numbers in an ordered field?
   (b) If $\mathbb{F}$ is an ordered field and $a < b$, then for any c in $\mathbb{F}$, $a + c < b + c$. If c is positive, then $ac < bc$. If c is negative, then $bc < ac$. Suppose we attempt to turn $\mathbb{F}_5$ into an ordered field starting with the decree that $1 > 0$. Where do the ordered field axioms start to break down? Can we do any better if we say $1 < 0$?

7.1.4. The mapping $L : \mathbb{C}^n \to \mathbb{R}^{2n}$ given by
$$L(z_1, \ldots, z_n) = (\operatorname{Re} z_1, \operatorname{Im} z_1, \ldots, \operatorname{Re} z_n, \operatorname{Im} z_n)$$
is a canonical isomorphism allowing us to identify $\mathbb{C}^n$, as a real vector space, with $\mathbb{R}^{2n}$.
   (a) Verify that if $\mathbf{v}$ belongs to $\mathbb{C}^n$, then $\mathbf{v} \cdot_\mathbb{C} \mathbf{v} = L(\mathbf{v}) \cdot_\mathbb{R} L(\mathbf{v})$. This shows that the complex dot product of a vector in $\mathbb{C}^n$ with itself is the square length of the line segment from $\mathbf{0}$ to the head of $\mathbf{v}$, in the Euclidean sense.
   (b) Is $\mathbf{v} \cdot_\mathbb{C} \mathbf{w} = L(\mathbf{v}) \cdot_\mathbb{R} L(\mathbf{w})$ for all $\mathbf{v}, \mathbf{w}$ in $\mathbb{C}^n$?

7.1.5. Prove Lemma 7.3.

7.1.6. Suppose V is a complex vector space with $\sigma : V \times V \to \mathbb{C}$ Hermitian symmetric and linear in one variable. Show that σ is conjugate-linear in the other variable.

7.1.7. Show that if σ is a Hermitian form on V, then $\sigma(\mathbf{v}, \mathbf{v})$ is in $\mathbb{R}$ for all $\mathbf{v}$ in V.

7.1.8. This problem refers to the inner product space in Example 7.6.
(a) Prove that σ is Hermitian symmetric.
(b) Let $A = \begin{bmatrix} 1+i & -2 \\ 3-i & 2i \end{bmatrix}$ and $B = \begin{bmatrix} 0 & i \\ -1+i & 2 \end{bmatrix}$. Find $\sigma(A, A)$, $\sigma(B, B)$, and $\sigma(A, B)$.
(c) Find the matrix representation of σ with respect to the standard basis for $M_2(\mathbb{C})$.

7.1.9. Prove Lemma 7.7 for the case of a Hermitian form that is linear in its first variable.

7.1.10. Prove Lemma 7.8.

7.2. Inner Products

If we tried to use the Hermitian form in Example 7.9 to measure length or distance on $\mathbb{C}^2$, we would have to say that every point in $\mathrm{Span}\{(1,1)\} \subseteq \mathbb{C}^2$ was zero units from the origin. This is not a feature we usually want to see in something we call distance. We need more to bring us closer to the properties of the dot product.

Recall from Section 5.3 that a bilinear form $\sigma : V \times V \to \mathbb{F}$ is nondegenerate provided $\sigma(\mathbf{v}, \mathbf{w}) = 0$ for all $\mathbf{w}$ in V only when $\mathbf{v} = \mathbf{0}_V$, and $\sigma(\mathbf{v}, \mathbf{w}) = 0$ for all $\mathbf{v}$ in V only when $\mathbf{w} = \mathbf{0}_V$. That definition applies more generally to any mapping $\sigma : V \times V \to \mathbb{F}$, when $\mathbb{F}$ is the field underlying the vector space V. In case σ is Hermitian, the proof of Theorem 5.28 can be adjusted to show that when the underlying vector space is finite-dimensional, σ is nondegenerate if and only if any matrix representation for σ is nonsingular.

When $\mathbb{F}$ is an ordered field, we say that σ is **positive definite** provided $\sigma(\mathbf{v}, \mathbf{v}) > 0$ for all nonzero $\mathbf{v}$ in V. Evidently, a positive definite mapping is nondegenerate. The next example shows that a nondegenerate mapping need not be positive definite.

Example 7.10. Define a symmetric bilinear form, σ, on $\mathbb{C}^2$ by

$$\sigma(\mathbf{v}, \mathbf{w}) = \mathbf{v}^T A \mathbf{w},$$

where $A = \begin{bmatrix} 1 & i \\ i & 1 \end{bmatrix}$. Since $\det A = 2$, A is invertible, which implies that σ is nondegenerate. Letting $\mathbf{v} = (1,1)$, we have

$$\sigma(\mathbf{v}, \mathbf{v}) = \begin{bmatrix} 1 & 1 \end{bmatrix} A \begin{bmatrix} 1 \\ 1 \end{bmatrix} = 2 + 2i.$$

Since $2 + 2i$ is not a positive real number, σ is not positive definite.

Definition 7.11. A **Hermitian inner product** is a positive definite Hermitian form on a complex vector space V.

7.2. Inner Products

A **real inner product** is a positive definite symmetric bilinear form on a real vector space.

An **inner product** is a Hermitian or real inner product. An **inner product space**, (V, σ), is a complex vector space V with a Hermitian inner product σ or a real vector space V with a real inner product σ.

Both the Euclidean inner product on $\mathbb{R}^n$ and the complex dot product on $\mathbb{C}^n$ are inner products.

When considering an inner product space, (V, σ), we may say simply that V is an inner product space, or that V is an inner product space under σ.

Whenever we can ignore the differences between real and complex inner product spaces, we do.

Going forward, $\mathbb{F}$ can be either $\mathbb{R}$ or $\mathbb{C}$.

Since an inner product is nondegenerate, Theorem 5.28 implies that in the finite-dimensional setting, any matrix representation for an inner product is invertible. The next lemma puts a finer point on it.

Lemma 7.12. If $A = [a_{ij}]$ in $M_n(\mathbb{F})$ determines an inner product on $V \cong \mathbb{F}^n$, then $a_{ii} > 0$ for all $i \in \{1, \ldots, n\}$.

Proof. Let A and V be as hypothesized. Let $\mathcal{B} = \{\mathbf{b}_i\}_{i=1}^n$ be a basis for V and suppose
$$\sigma(\mathbf{v}, \mathbf{w}) = [\mathbf{v}]_\mathcal{B}{}^\dagger A [\mathbf{w}]_\mathcal{B}$$
is an inner product on V. We then have
$$a_{ii} = \mathbf{e}_i^T A \mathbf{e}_i = [\mathbf{b}_i]_\mathcal{B}{}^\dagger A [\mathbf{b}_i]_\mathcal{B} = \sigma(\mathbf{b}_i, \mathbf{b}_i).$$
Since $\sigma(\mathbf{v}, \mathbf{v})$ is positive for any nonzero $\mathbf{v}$ in V, the result follows. $\square$

A given real or complex vector space may have many different inner products.

Lemma 7.13. Let V be a real or complex vector space. A positive scalar multiple of an inner product on V is an inner product on V. The sum of inner products on V is an inner product on V.

Proof. Let (V, σ) be an inner product space with underlying field $\mathbb{F}$, and let c be a positive real number. If $\zeta : V \times V \to \mathbb{F}$ is given by
$$\zeta(\mathbf{v}, \mathbf{w}) = c\sigma(\mathbf{v}, \mathbf{w}),$$
then $\zeta = c\sigma$ is a positive scalar multiple of σ. We leave it as an exercise to verify that ζ is an inner product on V.

If σ_1 and σ_2 are inner products on V and $\zeta : V \times V \to \mathbb{F}$ is given by
$$\zeta(\mathbf{v}, \mathbf{w}) := \sigma_1(\mathbf{v}, \mathbf{w}) + \sigma_2(\mathbf{v}, \mathbf{w}),$$
then $\zeta = \sigma_1 + \sigma_2$ is a sum of inner products on V. We leave it as an exercise to verify that ζ is an inner product on V. $\square$

The next lemma says that a subspace of an inner product space is an inner product space. This is an important result but its proof is an easy exercise.

Lemma 7.14. If (V, σ) is an inner product space and $U \subseteq V$ is a nontrivial subspace, then $(U, \sigma|_{U \times U})$ is an inner product space.

When we use a matrix to define an inner product on $V \cong \mathbb{F}^n$, we are exploiting the fact that mapping V to $\mathbb{F}^n$ via coordinate vectors is an isomorphism. The next lemma generalizes this idea. The proof is an exercise.

Lemma 7.15. If (V, σ) is an inner product space and $L : V \to W$ is an isomorphism, then
$$\zeta(L(\mathbf{v}), L(\mathbf{x})) := \sigma(\mathbf{v}, \mathbf{x})$$
defines an inner product on W.

While Lemma 7.15 allows us to define an inner product on any real or complex vector space, more typically we look for inner products to arise naturally out of a given context.

Example 7.16. We have seen that $f \mapsto \int_a^b f(t)\,dt$ is a linear functional on the real vector space, $C([a, b])$. Here we look at σ defined on $C([a, b])$ by
$$\sigma(f, g) := \int_a^b f(t)g(t)\,dt,$$
which is an inner product. We leave the verification as an exercise.

Since a definite integral is a generalized sum, the inner product defined in Example 7.16 is a natural extension of the definition of the dot product on $\mathbb{R}^n$.

Exercises 7.2.

7.2.1. Adjust the proof of Theorem 5.28 to argue that a Hermitian form on a finite-dimensional vector space is nondegenerate if and only if any matrix representation of the form is nonsingular.

7.2.2. Prove that the trace bilinear form defined in Example 7.6 is an inner product.

7.2.3. Finish the proof of Lemma 7.13.

7.2.4. Prove Lemma 7.14.

7.2.5. Prove Lemma 7.15.

7.2.6. Use Lemma 7.15 to define an inner product on $\mathbb{R}[t]_2$ via the dot product on $\mathbb{R}^3$.

7.2.7. Prove that $\sigma(f, g) := \int_a^b f(t)g(t)\,dt$ is an inner product on the space of real functions $C([a, b])$.

7.3. Length and Angle

Here we generalize the definitions of the length of a vector, the distance between two vectors, and the angle between two vectors, so that they apply in any inner product space. One of the tools we need is the Cauchy-Schwarz inequality. Before we get to that, we briefly discuss convergence in inner product spaces. We start with the obvious definition.

7.3. Length and Angle

Definition 7.17. The **length** or **norm** of a vector $\mathbf{v}$ in an inner product space (V, σ) is
$$\|\mathbf{v}\| := \sqrt{\sigma(\mathbf{v}, \mathbf{v})}.$$
The **distance** between $\mathbf{v}$ and $\mathbf{w}$ in V is $\|\mathbf{v} - \mathbf{w}\|$.

When we use the word *norm* in this section, we always mean the norm associated to an inner product. This is also called the *induced norm*. It is not unusual to run across vector spaces with norms that are not associated to inner products. In the exercises, we define norm in this more general sense.

The *triangle inequality* holds in any inner product space:
$$\|\mathbf{v} + \mathbf{w}\| \leq \|\mathbf{v}\| + \|\mathbf{w}\|. \tag{7.3}$$
The verification is a routine exercise.

Remain aware of context. We use $|\ |$ to indicate the absolute value or modulus of a scalar in $\mathbb{R}$ or $\mathbb{C}$, and we use $\|\ \|$ to indicate the norm associated to an inner product. Since the dot product makes both $\mathbb{R}$ and $\mathbb{C}$ inner product spaces, the induced norm is
$$\|a\| = \sqrt{a^\dagger a} = \sqrt{\bar{a} a} = |a|.$$
In this sense, a norm is a generalization of the absolute value on $\mathbb{R}$.

If V is an inner product space, the norm of any vector in V is in $\mathbb{R}$. This allows us to define convergence in V using convergence in $\mathbb{R}$.

Definition 7.18. Let V be an inner product space with elements f_n, $n \in \mathbb{Z}_+$. The sequence $\{f_n\}$ **converges** to g in V provided $\|f_n - g\|$ converges to 0 in $\mathbb{R}$.

We may say that a sequence that converges in the sense of Definition 7.18 converges *normwise* or *converges in the norm*. This distinguishes normwise convergence from pointwise convergence. In our next example, the two types of convergence coincide but in Section 7.6, we will see examples of sequences of functions that converge in the norm to $f(t)$ but that do not converge to $f(t)$ pointwise at every point under consideration.

Example 7.19. Here we consider the real inner product space $(C([0,1]), \sigma)$ where
$$\sigma(f, g) = \int_0^1 f(t) g(t)\, dt.$$
Let $f_n(t) = t/n$, for $t \in [0, 1]$. We then have, for each $n \in \mathbb{Z}_+$,
$$\|f_n\|^2 = \int_0^1 \frac{t^2}{n^2}\, dt = \frac{1}{3n^2}.$$
Since $\{\|f_n(t)\|\} = \left\{\frac{1}{\sqrt{3n}}\right\}$ converges to 0 in $\mathbb{R}$, the sequence $\{f_n\}$ converges to the zero function in $C([0,1])$.

The Cauchy-Schwarz inequality shows up in nearly any conversation involving inner product spaces. We use it to define angle in an inner product space.

Theorem 7.20 (Cauchy-Schwarz Inequality). *Let (V, σ) be an inner product space with induced norm $\|\ \|$. If $\mathbf{v}$ and $\mathbf{w}$ belong to V, then*
$$|\sigma(\mathbf{w}, \mathbf{v})|^2 \leq \|\mathbf{v}\|^2 \|\mathbf{w}\|^2.$$

Proof. Let (V, σ), $\mathbf{v}$, and $\mathbf{w}$ be as hypothesized. If either $\mathbf{v}$ or $\mathbf{w}$ is the zero vector, then the theorem is obviously true so assume that both $\mathbf{v}$ and $\mathbf{w}$ are nonzero.

A quick calculation reveals that if
$$\mathbf{x} = \mathbf{v} - \frac{\sigma(\mathbf{w}, \mathbf{v})}{\|\mathbf{w}\|^2}\mathbf{w},$$
then $\sigma(\mathbf{w}, \mathbf{x}) = 0$. This gives us
$$\|\mathbf{x}\|^2 = \sigma(\mathbf{x}, \mathbf{x}) = \sigma\left(\mathbf{v} - \frac{\sigma(\mathbf{w}, \mathbf{v})}{\|\mathbf{w}\|^2}\mathbf{w}, \mathbf{x}\right) = \sigma(\mathbf{v}, \mathbf{x}) \geq 0.$$

We also have
$$\sigma(\mathbf{v}, \mathbf{x}) = \sigma\left(\mathbf{v}, \mathbf{v} - \frac{\sigma(\mathbf{w}, \mathbf{v})}{\|\mathbf{w}\|^2}\mathbf{w}\right) = \|\mathbf{v}\|^2 - \frac{\overline{\sigma(\mathbf{w}, \mathbf{v})}}{\|\mathbf{w}\|^2}\sigma(\mathbf{v}, \mathbf{w}) = \|\mathbf{v}\|^2 - \frac{|\sigma(\mathbf{w}, \mathbf{v})|^2}{\|\mathbf{w}\|^2},$$
allowing us to say
$$\|\mathbf{v}\|^2 - \frac{|\sigma(\mathbf{w}, \mathbf{v})|^2}{\|\mathbf{w}\|^2} \geq 0,$$
thus that
$$\frac{|\sigma(\mathbf{w}, \mathbf{v})|^2}{\|\mathbf{w}\|^2} \leq \|\mathbf{v}\|^2.$$

The theorem follows when we multiply through by $\|\mathbf{w}\|^2 > 0$. $\square$

When $\mathbf{v}$ and $\mathbf{w}$ belong to $\mathbb{R}^2$, $\mathbf{v} \cdot \mathbf{w} = \|\mathbf{v}\|\|\mathbf{w}\|\cos\varphi$. (See Exercise B.5.2.) We get the analogous formula in any inner product space (V, σ), if we replace $\mathbf{v} \cdot \mathbf{w}$ not with $\sigma(\mathbf{v}, \mathbf{w})$, but with the real part of $\sigma(\mathbf{w}, \mathbf{v})$.

Theorem 7.21. If (V, σ) is an inner product space with induced norm $\|\ \|$ and $\mathbf{v}$ and $\mathbf{w}$ belong to V, then $\frac{\operatorname{Re}(\sigma(\mathbf{w},\mathbf{v}))}{\|\mathbf{v}\|\|\mathbf{w}\|}$ lies in the interval $[-1, 1]$.

Proof. Assume everything is as hypothesized. The Cauchy-Schwarz inequality says
$$|\sigma(\mathbf{w}, \mathbf{v})|^2 \leq \|\mathbf{v}\|^2 \|\mathbf{w}\|^2.$$
Writing $\sigma(\mathbf{w}, \mathbf{v}) = a + ib$, we have
$$|\operatorname{Re}(\sigma(\mathbf{w}, \mathbf{v}))|^2 = a^2 \leq a^2 + b^2 = |\sigma(\mathbf{w}, \mathbf{v})|^2.$$
It follows that
$$\frac{\operatorname{Re}(\sigma(\mathbf{w}, \mathbf{v}))}{\|\mathbf{v}\|\|\mathbf{w}\|} \in [-1, 1]. \quad \square$$

Theorem 7.21 legitimizes our next definition.

Definition 7.22. Let (V, σ) be an inner product space with induced norm $\|\ \|$. The **angle** between nonzero vectors $\mathbf{v}$ and $\mathbf{w}$ in V is $\varphi = \cos^{-1}\left(\frac{\operatorname{Re}(\sigma(\mathbf{w},\mathbf{v}))}{\|\mathbf{v}\|\|\mathbf{w}\|}\right)$.

As an inverse cosine, the angle between two vectors as given in Definition 7.22 is in the interval $[0, \pi]$.

7.3. Length and Angle

Example 7.23. We find the angle between $\mathbf{v} = (1, -i)$ and $\mathbf{w} = (i, 1)$ in $\mathbb{C}^2$ under the complex dot product. Since

$$(1, -i) \cdot (i, 1) = 2i,$$

the real part of $\mathbf{w} \cdot \mathbf{v}$ is zero. This means that the angle between $(1, -i)$ and $(i, 1)$ is $\pi/2$: The vectors are at right angles.

Example 7.24. Consider the real inner product space $(\mathbb{R}[t], \sigma)$, where

$$\sigma(p(t), q(t)) = \int_0^1 p(t) q(t)\, dt.$$

For $p(t) = 1 - t$ and $q(t) = 1 + t$,

$$\sigma(p(t), q(t)) = \int_0^1 1 - t^2\, dt = t - \frac{1}{3}t^3 \Big|_0^1 = \frac{2}{3}.$$

We also have

$$\sigma(p(t), p(t)) = \int_0^1 1 - 2t + t^2\, dt = \frac{1}{3}$$

and

$$\sigma(q(t), q(t)) = \int_0^1 1 + 2t + t^2\, dt = \frac{7}{3}.$$

Letting φ be the angle between $p(t)$ and $q(t)$, we have

$$\varphi = \cos^{-1}\left(\frac{2}{\sqrt{7}}\right) = 0.71372437\ldots,$$

which is approximately 41 degrees.

We have seen that if there is one inner product defined on a vector space V, then there are many inner products defined on V. This means that even on $\mathbb{R}^n$, for instance, the notion of distance is not a feature of the vector space. It varies from one inner product to the next. For instance, we can define an inner product σ on $\mathbb{R}^2$ by

$$\sigma(\mathbf{v}, \mathbf{w}) = \mathbf{v}^T \begin{bmatrix} 1 & 0 \\ 0 & 2 \end{bmatrix} \mathbf{w}.$$

Under σ, $(1, 0)$ is one unit long while $(0, 1)$ is two units long. As a statement about the lengths of legs in a right triangle, the Pythagorean Theorem may seem vulnerable to a change in the norm on $\mathbb{R}^2$. In fact, its generalization, the Law of Cosines, holds in any inner product space. The proof is an easy exercise.

Theorem 7.25 (Law of Cosines). *Let (V, σ) be an inner product space. If $\mathbf{v}$ and $\mathbf{w}$ belong to V, then*

$$\|\mathbf{v} - \mathbf{w}\|^2 = \|\mathbf{v}\|^2 + \|\mathbf{w}\|^2 - 2\|\mathbf{v}\|\|\mathbf{w}\| \cos \varphi$$

where φ is the angle between $\mathbf{v}$ and $\mathbf{w}$.

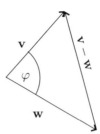

Figure 7.2. The Law of Cosines holds in any inner product space.

Exercises 7.3.

7.3.1. Let (V, σ) be an inner product space. Write $\|\mathbf{v} + \mathbf{w}\|^2$ using $\|\mathbf{v}\|$, $\|\mathbf{w}\|$, and $\operatorname{Re} \sigma(\mathbf{v}, \mathbf{w})$, when $\mathbf{v}$ and $\mathbf{w}$ belong to V.

7.3.2. Let $\mathbb{F}$ be $\mathbb{R}$ or $\mathbb{C}$, and let V be any vector space over $\mathbb{F}$. A **norm** on V is a mapping $\| \ \| : V \to \mathbb{F}$ that satisfies three conditions: the triangle inequality, (7.3); $\|\mathbf{v}\| = 0$ only if $\mathbf{v} = \mathbf{0}$; and $\|c\mathbf{v}\| = |c|\|\mathbf{v}\|$. When V is a vector space with $\| \ \|$, we say that $(V, \| \ \|)$ is a **normed linear space**.
  (a) Show that $| \ |$ is a norm, both on $\mathbb{R}$ and on $\mathbb{C}$.
  (b) Show that if (V, σ) is an inner product space, then $\|\mathbf{v}\| = \sqrt{\sigma(\mathbf{v}, \mathbf{v})}$ is a norm.
  (c) Prove that the **reverse triangle inequality** holds in any normed linear space, $(V, \| \ \|)$, that is, for all $\mathbf{v}$ and $\mathbf{w}$ in V
  $$\|\mathbf{v} + \mathbf{w}\| \geq \|\mathbf{v}\| - \|\mathbf{w}\|.$$

7.3.3. Let $\mathbf{v}$ and $\mathbf{w}$ belong to an inner product space (V, σ) with induced norm $\| \ \|$. Defining
$$\mathbf{u} = \mathbf{v} - \frac{\sigma(\mathbf{w}, \mathbf{v})}{\|\mathbf{w}\|^2} \mathbf{w},$$
show that $\sigma(\mathbf{w}, \mathbf{u}) = 0$.

7.3.4. Prove that the Law of Cosines holds in any inner product space.

7.3.5. The **parallelogram law** says that if (V, σ) is an inner product space with norm $\|\mathbf{v}\| = \sqrt{\sigma(\mathbf{v}, \mathbf{v})}$, then for all $\mathbf{v}$ and $\mathbf{w}$ in V
$$\|\mathbf{v} - \mathbf{w}\|^2 + \|\mathbf{v} + \mathbf{w}\|^2 = 2(\|\mathbf{v}\|^2 + \|\mathbf{w}\|^2).$$
  (a) Prove the parallelogram law.
  (b) Sketch a picture of a parallelogram and its diagonals to explain where the name comes from.

7.4. Orthonormal Sets

The words *orthogonal* and *perpendicular* are synonyms to describe objects at right angles. *Ortho* means "right," as in "correct"; *gon* comes from the Greek for "angle." (An orthodontist corrects the alignment of teeth; a polygon has many angles.) The root of the word perpendicular is *pend*, which means "hang." (A pendant is a

7.4. Orthonormal Sets

bauble that hangs from a chain.) Items that hang near the surface of the earth do so at right angles to level ground.

Definition 7.26. Let (V, σ) be an inner product space. A vector $\mathbf{v}$ in V is **orthogonal** to $\mathbf{w}$ in V provided $\sigma(\mathbf{w}, \mathbf{v}) = 0$.

Orthogonality is a symmetric relation on the elements in an inner product space. If $\mathbf{v}$ is orthogonal to $\mathbf{w}$ in an inner product space (V, σ), then

$$\sigma(\mathbf{v}, \mathbf{w}) = 0 \implies \overline{\sigma(\mathbf{v}, \mathbf{w})} = 0 \implies \sigma(\mathbf{w}, \mathbf{v}) = 0,$$

so $\mathbf{w}$ is orthogonal to $\mathbf{v}$. For this reason, we usually just say that $\mathbf{v}$ and $\mathbf{w}$ are orthogonal.

The orthogonal group is so named because it preserves orthogonality: If $\mathbf{v} \cdot \mathbf{w} = 0$, for $\mathbf{v}$ and $\mathbf{w}$ in $\mathbb{R}^n$, then $A\mathbf{v} \cdot A\mathbf{w} = 0$ for any A in $O_n(\mathbb{R})$.

The zero vector is orthogonal to every vector in an inner product space. This is a convenience. It means, for instance, that the set of all vectors orthogonal to a given vector in an inner product space is a subspace. (The proof is an easy exercise.)

When $\mathbf{v}$ and $\mathbf{w}$ are vectors in any vector space V, we can write $\mathbf{v}$ as a sum of $\mathbf{w}$ and some other vector, namely, $\mathbf{v} - \mathbf{w}$. In an inner product space, we can write $\mathbf{v}$ as a sum of a vector in the direction of $\mathbf{w}$ and a vector orthogonal to $\mathbf{w}$. To understand why this is true, we start with the notion of the component of one vector in the direction of another vector. This should be familiar from $\mathbb{R}^n$. Indeed, Definition 7.27 generalizes Definition B.9 to apply to any inner product space.

Definition 7.27. Let (V, σ) be an inner product space. Let $\mathbf{v}$ belong to V and let $\mathbf{w}$ be nonzero in V. The **component of v in the direction of w**, or the **projection of v onto w**, is

$$\text{proj}_{\mathbf{w}}(\mathbf{v}) := \frac{\sigma(\mathbf{v}, \mathbf{w})}{\|\mathbf{w}\|^2} \mathbf{w}.$$

When $\mathbf{w}$ is nonzero, we identify

$$\mathbf{u}_{\mathbf{w}} := \frac{1}{\|\mathbf{w}\|} \mathbf{w},$$

the **unit vector in the direction of w**. For $\mathbf{v}$ and $\mathbf{w}$ in V, we can then say

$$\text{proj}_{\mathbf{w}}(\mathbf{v}) = \sigma(\mathbf{v}, \mathbf{u}_{\mathbf{w}}) \mathbf{u}_{\mathbf{w}}.$$

Because the length of a unit vector is 1, unit vectors are sometimes treated simply as directions.

Exercise 7.3.3 was to prove the next lemma in a slightly different guise. It came up in the proof of the Cauchy-Schwarz inequality.

Lemma 7.28. If $\mathbf{v}$ and $\mathbf{w}$ belong to an inner product space (V, σ), then

$$\mathbf{v} - \text{proj}_{\mathbf{w}}(\mathbf{v})$$

is orthogonal to $\mathbf{w}$.

Example 7.29. Let $\mathbf{v} = (1, 2, -1)$ and $\mathbf{w} = (1, 1, 1)$ in $(\mathbb{R}^3, \cdot)$. We find $\text{proj}_{\mathbf{w}}(\mathbf{v})$ first by calculating

$$\mathbf{u}_{\mathbf{w}} = \frac{1}{\|\mathbf{w}\|} \mathbf{w} = \left(\frac{1}{\sqrt{3}}, \frac{1}{\sqrt{3}}, \frac{1}{\sqrt{3}} \right)$$

and
$$\mathbf{v} \cdot \mathbf{u_w} = \frac{1}{\sqrt{3}} + \frac{2}{\sqrt{3}} - \frac{1}{\sqrt{3}} = \frac{2}{\sqrt{3}}.$$
This gives us
$$\operatorname{proj}_{\mathbf{w}}(\mathbf{v}) = (\mathbf{v} \cdot \mathbf{u_w})\mathbf{u_w} = \left(\frac{2}{3}, \frac{2}{3}, \frac{2}{3}\right).$$
Notice next that
$$\mathbf{v} - \operatorname{proj}_{\mathbf{w}}(\mathbf{v}) = \left(\frac{1}{3}, \frac{4}{3}, -\frac{5}{3}\right).$$
From there we see that
$$\left(\frac{2}{3}, \frac{2}{3}, \frac{2}{3}\right) \cdot \left(\frac{1}{3}, \frac{4}{3}, -\frac{5}{3}\right) = 0,$$
as guaranteed by Lemma 7.28.

The notion of orthogonality is especially powerful when applied to sets of vectors.

Definition 7.30. Let (V, σ) be an inner product space. An **orthogonal set** $S \subseteq V$ is a set of pairwise orthogonal vectors in V: $\sigma(\mathbf{v}, \mathbf{w}) = 0$ for any $\mathbf{v}$ and $\mathbf{w}$ in S, when $\mathbf{v} \neq \mathbf{w}$. An **orthonormal set** in V is an orthogonal set of unit vectors in V.

If S is an orthonormal set of vectors in an inner product space (V, σ), then $\sigma(\mathbf{u}_i, \mathbf{u}_j) = \delta(i, j)$ for $\mathbf{u}_i, \mathbf{u}_j$ in S.

The next theorem captures an important property of an orthogonal set of vectors.

Theorem 7.31. An orthogonal set of nonzero vectors in any inner product space is linearly independent.

Proof. Let (V, σ) be an inner product space and suppose $S \subseteq V$ is an orthogonal set of nonzero vectors. Let $\{\mathbf{v}_1, \ldots, \mathbf{v}_t\} \subseteq S$. If there are scalars c_i such that $\sum_{i=1}^t c_i \mathbf{v}_i = \mathbf{0}$, then

(7.4) $$0 = \sigma(\mathbf{0}, \mathbf{v}_j) = \sum_{i=1}^t \overline{c_i}\sigma(\mathbf{v}_i, \mathbf{v}_j) = \overline{c_j}\|\mathbf{v}_j\|^2.$$

Since $\mathbf{v}_j$ is nonzero, $\|\mathbf{v}_j\|^2 \neq 0$, implying that $\overline{c_j} = 0$. We conclude that $c_i = 0$ for all coefficients in (7.4). Since this is true for any finite set of vectors in S, S must be linearly independent. □

The theorem justifies our next definition.

Definition 7.32. An **algebraic orthonormal basis** for an inner product space V is an orthonormal spanning set for V.

When the context makes it clear — for instance, when we are dealing with a finite-dimensional inner product space — we drop the adjective *algebraic* and refer simply to orthonormal bases.

The standard bases for $\mathbb{R}^n$ and $\mathbb{C}^n$ are orthonormal bases with respect to the dot product.

7.4. Orthonormal Sets

Every finite-dimensional inner product space has an orthonormal basis. The algorithm for constructing an orthonormal set from a finite spanning set is called the **Gram-Schmidt process**.

7.4.1. The Gram-Schmidt Process. Suppose (V, σ) is an inner product space and that we are given a set of nonzero vectors in V, $S = \{\mathbf{v}_1, \ldots, \mathbf{v}_n\}$. We are concerned with Span S so we may assume that S is linearly independent.

Let $\mathbf{u}_1$ be the unit vector in the direction of $\mathbf{v}_1$. Let

$$\mathbf{w}_2 = \mathbf{v}_2 - \sigma(\mathbf{u}_1, \mathbf{v}_2)\mathbf{u}_1.$$

We have

$$\sigma(\mathbf{u}_1, \mathbf{w}_2) = \sigma(\mathbf{u}_1, \mathbf{v}_2) - \sigma(\mathbf{u}_1, \mathbf{v}_2) = 0.$$

Take $\mathbf{u}_2$ to be the unit vector in the direction of $\mathbf{w}_2$. This gives us $\{\mathbf{u}_1, \mathbf{u}_2\}$, an orthonormal set that spans Span$\{\mathbf{v}_1, \mathbf{v}_2\}$.

Repeating this process, we reach a point where we have constructed an orthonormal set $\{\mathbf{u}_i\}_{i=1}^{t-1}$ that spans Span$\{\mathbf{v}_i\}_{i=1}^{t-1}$. We next define

$$\mathbf{w}_t = \mathbf{v}_t - \sum_{i=1}^{t-1} \sigma(\mathbf{u}_i, \mathbf{v}_t)\mathbf{u}_i.$$

Given $j \in \{1, \ldots, t-1\}$, we have

$$\sigma(\mathbf{u}_j, \mathbf{w}_t) = \sigma(\mathbf{u}_j, \mathbf{v}_t) - \sum_{i=1}^{t-1} \sigma(\mathbf{u}_i, \mathbf{v}_t)\sigma(\mathbf{u}_j, \mathbf{u}_i) = \sigma(\mathbf{u}_j, \mathbf{v}_t) - \sigma(\mathbf{u}_j, \mathbf{v}_t) = 0.$$

Ultimately, we wind up with an orthonormal basis for Span S, $\{\mathbf{u}_1, \ldots, \mathbf{u}_n\}$.

Since we can do this with any finite set of vectors in an inner product space, we have proved the following.

Theorem 7.33. Let (V, σ) be an inner product space with finite-dimensional subspace U. If U is nontrivial and $\mathbf{v}$ is a nonzero vector in U, then U has an orthonormal basis containing $\mathbf{u_v}$, the unit vector in the direction of $\mathbf{v}$.

When we meet an inner product space, the first thing that should come to mind is that every finite-dimensional subspace has an orthonormal basis. The next lemma is probably the second thing that should come to mind. Its proof is an exercise.

Lemma 7.34. Let (V, σ) be an inner product space with subspace U. If $\mathcal{B} = \{\mathbf{u}_i\}_{i=1}^{k}$ is an orthonormal basis for U, then for each $\mathbf{v}$ in U,

$$\mathbf{v} = \sum_{i=1}^{k} \sigma(\mathbf{u}_i, \mathbf{v})\mathbf{u}_i.$$

Example 7.35. Let U be the subspace of $(\mathbb{R}^4, \cdot)$ spanned by

$$\{(1, 1, -1, 1), (-1, 1, -1, 1), (1, 1, 1, 1)\}.$$

We use the Gram-Schmidt process to construct an orthonormal basis for U, starting with a vector in the direction of $(1, 1, -1, 1)$. Since the length of each vector in our spanning set is 2, we have
$$\mathbf{u}_1 = \left(\frac{1}{2}, \frac{1}{2}, -\frac{1}{2}, \frac{1}{2}\right).$$
Next, let
$$\mathbf{v}_2 = (-1, 1, -1, 1) - \left((-1, 1, -1, 1) \cdot \left(\frac{1}{2}, \frac{1}{2}, -\frac{1}{2}, \frac{1}{2}\right)\right) \left(\frac{1}{2}, \frac{1}{2}, -\frac{1}{2}, \frac{1}{2}\right)$$
$$= (-1, 1, -1, 1) - \left(\frac{1}{2}, \frac{1}{2}, -\frac{1}{2}, \frac{1}{2}\right) = \left(-\frac{3}{2}, \frac{1}{2}, -\frac{1}{2}, \frac{1}{2}\right).$$
Since $\|\mathbf{v}_2\| = \sqrt{3}$, the next element in our orthonormal basis is
$$\mathbf{u}_2 = \left(-\frac{\sqrt{3}}{2}, \frac{\sqrt{3}}{6}, -\frac{\sqrt{3}}{6}, \frac{\sqrt{3}}{6}\right).$$
Next we have
$$\mathbf{v}_3 = (1, 1, 1, 1) - \sum_{i=1}^{2} ((1, 1, 1, 1) \cdot \mathbf{u}_i) \mathbf{u}_i = (1, 1, 1, 1) - \mathbf{u}_1 + \frac{1}{\sqrt{3}} \mathbf{u}_2 = \left(0, \frac{2}{3}, \frac{4}{3}, \frac{2}{3}\right).$$
Since $\|(0, 2/3, 4/3, 2/3)\| = \sqrt{8/3}$, we take
$$\mathbf{u}_3 = \left(0, \frac{\sqrt{6}}{6}, \frac{\sqrt{6}}{3}, \frac{\sqrt{6}}{6}\right).$$
Our orthonormal basis for U is then
$$\mathcal{B} = \left\{ \left(\frac{1}{2}, \frac{1}{2}, -\frac{1}{2}, \frac{1}{2}\right), \left(-\frac{\sqrt{3}}{2}, \frac{\sqrt{3}}{6}, -\frac{\sqrt{3}}{6}, \frac{\sqrt{3}}{6}\right), \left(0, \frac{\sqrt{6}}{6}, \frac{\sqrt{6}}{3}, \frac{\sqrt{6}}{6}\right) \right\}.$$

It happens that $\mathbf{v} = (0, 2, 1, 2)$ belongs to U. Lemma 7.34 gives us the coefficients when we write $\mathbf{v}$ as a linear combination of the elements in $\mathcal{B}$:
$$\mathbf{v} \cdot \left(\frac{1}{2}, \frac{1}{2}, -\frac{1}{2}, \frac{1}{2}\right) = \frac{3}{2}, \quad \mathbf{v} \cdot \left(-\frac{\sqrt{3}}{2}, \frac{\sqrt{3}}{6}, -\frac{\sqrt{3}}{6}, \frac{\sqrt{3}}{6}\right) = \frac{\sqrt{3}}{2},$$
$$\mathbf{v} \cdot \left(0, \frac{\sqrt{6}}{6}, \frac{\sqrt{6}}{3}, \frac{\sqrt{6}}{6}\right) = \sqrt{6}.$$
Finally we can check that
$$(0, 2, 1, 2) = \frac{3}{2} \left(\frac{1}{2}, \frac{1}{2}, -\frac{1}{2}, \frac{1}{2}\right)$$
$$+ \frac{\sqrt{3}}{2} \left(-\frac{\sqrt{3}}{2}, \frac{\sqrt{3}}{6}, -\frac{\sqrt{3}}{6}, \frac{\sqrt{3}}{6}\right) + \sqrt{6} \left(0, \frac{\sqrt{6}}{6}, \frac{\sqrt{6}}{3}, \frac{\sqrt{6}}{6}\right).$$

The orthonormal basis we end up with when we apply the Gram-Schmidt process depends on the ordering of the original basis. We see an example of this in the exercises.

7.4. Orthonormal Sets

Exercises 7.4.

7.4.1. Give a direct proof that if $\mathbf{v}$ belongs to an inner product space, then the set of vectors orthogonal to $\mathbf{v}$ is a subspace.

7.4.2. Let $n > 1$ and say that $V = \mathbb{F}^n$ is real or complex n-space under the dot product. Find a nonzero vector orthogonal to nonzero $\mathbf{v} = (a_1, \ldots, a_n)$ in V.

7.4.3. Find the components of $(2, -1, 3)$ in $\mathbb{R}^3$ in the direction of, and orthogonal to, $(1, 2, -1)$ with respect to the dot product.

7.4.4. Prove Lemma 7.34.

7.4.5. Let U be the subspace of $(\mathbb{R}^4, \cdot)$ spanned by the set
$$\mathcal{B} = \{(1, 1, -1, 1), (-1, 1, 1, 1), (1, 1, 1, 1)\}.$$

  (a) Apply the Gram-Schmidt process to find an orthonormal basis for U, treating $\mathcal{B}$ as an ordered basis.
  (b) Redo the last problem, this time working with the ordered basis $\mathcal{B}' = \{(1, 1, 1, 1), (-1, 1, 1, 1), (1, 1, -1, 1)\}$.

7.4.6. Find an orthonormal basis for the subspace of $(\mathbb{R}^3, \cdot)$ given by
$$U = \{(a, a+b, b) \mid a, b \in \mathbb{R}\}.$$

7.4.7. Let $\mathcal{B} = \{(1/\sqrt{5}, 0, 2/\sqrt{5}), (-2/\sqrt{5}, 0, 1/\sqrt{5}), (0, 1, 0)\}$.
  (a) Verify that $\mathcal{B}$ is an orthonormal basis for $(\mathbb{R}^3, \cdot)$.
  (b) Find $[\mathbf{e}_i]_\mathcal{B}$ for $i = 1, 2, 3$.
  (c) Find $[(-1, 3, 2)]_\mathcal{B}$.

7.4.8. Verify the calculations in Example 7.35.

7.4.9. What happens when you apply the Gram-Schmidt algorithm to a linearly dependent set of vectors?

7.4.10. Let V be finite-dimensional with subspace U. Let $\{\mathbf{u}_1, \ldots, \mathbf{u}_t\}$ be an orthonormal basis for U and let
$$\mathcal{B} = \{\mathbf{u}_1, \ldots, \mathbf{u}_t, \mathbf{v}_{t+1}, \ldots, \mathbf{v}_n\}$$
be a basis for V. Suppose we apply the Gram-Schmidt process to $\mathcal{B}$ to get an orthonormal basis for all of V, $\{\mathbf{u}_1, \ldots, \mathbf{u}_n\}$. Let $W = \mathrm{Span}\{\mathbf{u}_{t+1}, \ldots \mathbf{u}_n\}$. Is $W = \mathrm{Span}\{\mathbf{v}_{t+1}, \ldots, \mathbf{v}_n\}$?

7.4.11. Let (V, σ) be a (real or complex) inner product space with orthonormal basis $\mathcal{B} = \{\mathbf{u}_i\}_{i=1}^n$. Let $\mathbf{v}$ and $\mathbf{w}$ belong to V. Show that
$$\sigma(\mathbf{v}, \mathbf{w}) = \sum_{i=1}^n \sigma(\mathbf{v}, \mathbf{u}_i)\sigma(\mathbf{u}_i, \mathbf{w}).$$

7.4.12. The **generalized Pythagorean theorem** says that if $\{\mathbf{u}_1, \ldots, \mathbf{u}_n\}$ is an orthonormal set in an inner product space (V, σ), then under the norm induced by σ,
$$\|\mathbf{u}_1 + \cdots + \mathbf{u}_n\|^2 = \|\mathbf{u}\|^2 + \cdots + \|\mathbf{u}_n\|^2.$$
Prove it.

7.5. Orthogonal Complements

A subspace in a vector space may have many direct complements. We will see in short order that in an inner product space, a subspace has exactly one *orthogonal complement*.

Definition 7.36. Let (V, σ) be an inner product space with subset S. The **orthogonal complement** of S is
$$S^\perp = \{\mathbf{v} \in V \mid \sigma(\mathbf{u}, \mathbf{v}) = 0 \, \forall \mathbf{u} \in S\}.$$

Read $S^\perp$ as S *perp*.

Lemma 7.37. If (V, σ) is an inner product space with subset S, then $S^\perp$ is a subspace of V.

Proof. Let (V, σ) and S be as hypothesized. Since it contains $\mathbf{0}$, $S^\perp$ is nonempty. If $\mathbf{v}_1$ and $\mathbf{v}_2$ are in $S^\perp$ and c is a scalar, then for each $\mathbf{u}$ in S,
$$\sigma(\mathbf{u}, c\mathbf{v}_1 + \mathbf{v}_2) = c\sigma(\mathbf{u}, \mathbf{v}_1) + \sigma(\mathbf{u}, \mathbf{v}_2) = 0$$
so $S^\perp$ is a subspace of V. $\square$

An easy exercise reveals that $S^\perp = W^\perp$ if $W = \operatorname{Span} S$.

We leave proof of the following lemma as an exercise.

Lemma 7.38. If (V, σ) is an inner product space with subspace W, then
$$W \cap W^\perp = \{\mathbf{0}\}.$$

When $S = \{\mathbf{v}\}$, we usually write $\mathbf{v}^\perp$ instead of $\{\mathbf{v}\}^\perp$ or $S^\perp$.

Sometimes when $\mathbf{v}$ is a nonzero vector orthogonal to all elements in a subspace W, we say that $\mathbf{v}$ is *normal* to W or is a *normal vector* to W.

Example 7.39. Let $\mathbf{v} = (1, 1, 1)$ in $\mathbb{R}^3$. Notice that
$$\mathbf{v}^\perp = \{(x, y, z) \in \mathbb{R}^3 \mid x + y + z = 0\}.$$
This is the plane through the origin in $\mathbb{R}^3$ with normal vector $(1, 1, 1)$.

The orthogonal complement of a subspace is uniquely determined in any inner product space. Unless the ambient space is finite-dimensional, though, the orthogonal complement of a subspace is not necessarily a direct complement of the subspace.

Example 7.40. Consider the real inner product space $(C([0, 1]), \sigma)$ with
$$\sigma(f, g) = \int_0^1 f(t)g(t) \, dt.$$
Let $W = \{f \in C([0, 1]) \mid f(0) = 0\}$. It is clear that W is nonempty. Moreover, given f and g in W, $cf + g$ is in W for any scalar c. This shows that W is a subspace of $C([0, 1])$.

7.5. Orthogonal Complements

Let f be arbitrary in $W^\perp$. Since $g(0) = 0$ when $g(t) = tf(t)$, g is in W. Since g and f are orthogonal,

$$\int_0^1 g(t)f(t)\,dt = \int_0^1 tf^2(t)\,dt = 0.$$

Next note that both t and $f^2(t)$ are nonnegative for all t in $[0,1]$. It follows that $tf^2(t) \geq 0$ on $[0,1]$. Since $tf^2(t)$ is continuous, nonnegative, and has an integral of 0 over $[0,1]$, it must be the case that $tf^2(t)$ is identically zero. It follows that $f(t)$ itself is identically zero, thus, that $W^\perp = \{\mathbf{0}\}$. In this case, $W^\perp$ is not a direct complement to W.

The finite-dimensional setting affords less room for mischief.

Theorem 7.41. If (V, σ) is a finite-dimensional inner product space with subspace W, then
$$V = W \oplus W^\perp.$$

Proof. Let (V, σ) and W be as hypothesized. Starting with a basis for V that contains a basis for W, we can use the Gram-Schmidt process to construct an orthonormal basis for V,

$$\mathcal{B} = \{\mathbf{u}_1, \ldots, \mathbf{u}_k, \mathbf{u}_{k+1}, \ldots, \mathbf{u}_n\},$$

where $W = \mathrm{Span}\{\mathbf{u}_1, \ldots, \mathbf{u}_k\}$. Let $U = \mathrm{Span}\{\mathbf{u}_{k+1}, \ldots, \mathbf{u}_n\}$. We have

$$V = W \oplus U$$

and $U \subseteq W^\perp$: Every vector in U is orthogonal to every vector in W.

Suppose there is $\mathbf{v}$ in $W^\perp$ such that

$$\mathbf{v} = \mathbf{w} + \mathbf{u}$$

for some $\mathbf{w}$ in W and $\mathbf{u}$ in $U \subseteq W^\perp$. We then have

$$0 = \sigma(\mathbf{w}, \mathbf{v}) = \|\mathbf{w}\|^2 + \sigma(\mathbf{w}, \mathbf{u}) = \|\mathbf{w}\|^2,$$

implying that $\mathbf{w} = \mathbf{0}$. We conclude that $U = W^\perp$. $\square$

Referring to Example 7.39, we see that

$$\mathbb{R}^3 = \mathrm{Span}\{(1,1,1)\} \oplus \mathrm{Span}\{(1,0,-1), (1,-1,0)\}$$

is one way to write $\mathbb{R}^3$ as the orthogonal direct sum of subspaces.

Example 7.42. Let $U = \mathrm{Span}\{1+t, 1-t\}$ in $\mathbb{R}_2[t]$. We find the orthogonal complement to U under the inner product

$$\sigma(p, q) = \int_{-1}^1 p(t)q(t)\,dt.$$

Because $\dim U = 2$, its orthogonal complement in $\mathbb{R}[t]_2$ is 1-dimensional. Any nonzero $q(t)$ orthogonal to both $1+t$ and $1-t$ under σ thus comprises a basis for $U^\perp$.

If $q(t) = a_0 + a_1 t + a_2 t^2$, then

$$\sigma(1+t, q(t)) = 2a_0 + \frac{2}{3}(a_1 + a_2)$$

and
$$\sigma(1-t, q(t)) = 2a_0 + \frac{2}{3}(-a_1 + a_2).$$
We get coefficients for $q(t)$ by identifying one solution to the following system of equations:
$$a_0 + \tfrac{1}{3}a_1 + \tfrac{1}{3}a_2 = 0$$
$$a_0 - \tfrac{1}{3}a_1 + \tfrac{1}{3}a_2 = 0.$$
Taking $a_0 = 1$, $a_1 = 0$, and $a_2 = -3$ we get
$$U^\perp = \text{Span}\{1 - 3t^2\}.$$

Exercises 7.5.

For all of these exercises, (V, σ) is an inner product space.

7.5.1. Let $S \subseteq V$. Show that $S^\perp = W^\perp$ when $W = \text{Span}\, S$.

7.5.2. Let V be an arbitrary inner product space. Show that $\{\mathbf{0}\}^\perp = V$ and that $V^\perp = \{\mathbf{0}\}$.

7.5.3. Prove Lemma 7.38.

7.5.4. Let V be finite-dimensional with subspace U. Show that $(U^\perp)^\perp = U$.

7.5.5. Let $U = \text{Span}\{(1,1,1), (1,-1,1)\} \subseteq \mathbb{R}^3$.
  (a) Show that $W = \text{Span}\{(1,1,0)\}$ is a direct complement of U.
  (b) Find a basis for $U^\perp$.
  (c) Find a basis for $W^\perp$.

7.5.6. Let V have subspaces U_1 and U_2. Let $W = \text{Span}(U_1 \cup U_2)$. Show that $W^\perp = U_1^\perp \cap U_2^\perp$.

7.5.7. Apply the previous exercise to the inner product space in Example 7.42 by letting $U_1 = \text{Span}\{1+t\}$ and $U_2 = \text{Span}\{1-t\}$ in $\mathbb{R}[t]_2$ and verifying that $U_1^\perp \cap U_2^\perp = U^\perp$, where $U = \text{Span}\{1+t, 1-t\}$.

7.5.8. We saw in Exercise 5.3.1 that $\sigma(\mathbf{v}, \mathbf{w}) := \mathbf{v}^T \mathbf{w}$ is nondegenerate on $V = \mathbb{F}^n$, for any field $\mathbb{F}$. If U is a subspace of $\mathbb{F}^n$, there is nothing to stop us from defining $U^\perp$ exactly as we did in an inner product space:
$$U^\perp = \{\mathbf{v} \in \mathbb{F}^n \mid \mathbf{v}^T \mathbf{u} = 0 \,\forall\, \mathbf{u} \in U\}.$$
  (a) Is $U^\perp$ a subspace of $\mathbb{F}^n$?
  (b) Find a subspace $U \subseteq \mathbb{F}_2^{\,2}$ such that $U \cap U^\perp$ is nontrivial.
  (c) More generally, find a subspace $U \subseteq \mathbb{F}_p^{\,p}$ such that $U \cap U^\perp$ is nontrivial when p is any prime number.

7.6. Inner Product Spaces of Functions

Orthogonal bases play an important role in the study of real and complex inner product spaces of functions. Our goal in this section is to provide enough background for an appreciation of how calculations and terminology apply in those infinite-dimensional settings. We state several theorems from complex analysis without proof.

7.6. Inner Product Spaces of Functions

We start with a brief discussion of complex functions. We already know that a willingness to embrace the complex numbers can open up a view of real polynomials that is otherwise limited. More generally, some background on complex functions can shed light on the nature of real functions.

A *function of a complex variable* — also called a *complex function* or *complex mapping* — maps a subset of $\mathbb{C}$ to $\mathbb{C}$. Just as any complex number has a real and an imaginary part, so does any complex mapping.

Let f be a complex function defined on some domain $D \subseteq \mathbb{C}$. Think of z in D as having the form $z = x + iy$. When it suits us, we can view D as a subset of $\mathbb{R}^2 = \{(x, y) \mid x, y \in \mathbb{R}\}$. The real and imaginary parts of f are denoted $\operatorname{Re} f(x, y)$ and $\operatorname{Im} f(x, y)$: Each is a function of two real variables.

Example 7.43. Consider the complex polynomial $p(z) = 2 + i - 5iz + 3z^2$, where z is an indeterminate in $\mathbb{C}$. Writing $z = x + iy$, we have

$$p(x + iy) = 2 + i - 5i(x + iy) + 3(x + iy)^2 = 2 + i - 5ix + 5y + 3x^2 - 3y^2 + 6ixy$$
$$= 2 + 5y + 3x^2 - 3y^2 + i(1 - 5x + 6xy).$$

From here we have $\operatorname{Re} p(x, y) = 2 + 5y + 3x^2 - 3y^2$ and $\operatorname{Im} p(x, y) = 1 - 5x + 6xy$.

We concern ourselves here with complex functions defined on an interval on the real line. If we think of a complex function $f(z)$ in terms of its real and imaginary parts, then when (x, y) belongs to some real interval $[a, b]$, x ranges over $[a, b]$ and $y = 0$. In this case, we usually designate the single real variable by t, the real part of f by $u(t)$, and the imaginary part of f by $v(t)$. All of our complex functions here, then, have the form

(7.5) $$f(t) = u(t) + iv(t), \quad t \in [a, b] \subseteq \mathbb{R},$$

where $u(t)$ and $v(t)$ are functions of a real variable.

A complex function f given by (7.5) is continuous on $[a, b]$ if and only if the real functions $u(t)$ and $v(t)$ are continuous on $[a, b]$. When $u'(t)$ and $v'(t)$ both exist, the derivative $f'(t)$ is defined by

$$f'(t) := u'(t) + iv'(t).$$

The Fundamental Theorem of Calculus carries over to this setting, allowing us to connect the definite integral to antiderivatives. When $f(t) = u(t) + iv(t)$ is defined on $[a, b] \subseteq \mathbb{R}$, $\int_a^b f(t)\, dt$ is defined as long as both $\int_a^b u(t)\, dt$ and $\int_a^b v(t)\, dt$ are defined. In that case,

$$\int_a^b f(t)\, dt := \int_a^b u(t)\, dt + i \int_a^b v(t)\, dt.$$

Example 7.44. Consider $p(z)$ as given in Example 7.43, this time with z restricted to the real interval $[-1, 1]$. We let $x = t$ so that, here, our concern is with $p(t) = 2 + 3t^2 + i(1 - 5t)$. Applying the techniques of calculus, we find that $p'(t) = 6t - 5i$ and

$$\int_{-1}^{1} p(t)\, dt = \int_{-1}^{1} 2 + 3t^2\, dt + i \int_{-1}^{1} 1 - 5t\, dt = 2t + t^3 \Big|_{-1}^{1} + i \left(t - \frac{5}{2} t^2 \right) \Big|_{-1}^{1} = 6 + i.$$

The *complex conjugate of a function* $f : D \subseteq \mathbb{C} \to \mathbb{C}$ is defined by taking the complex conjugate of $f(z)$ at each point $z \in D$:
$$\bar{f}(z) := \overline{f(z)}.$$
When $f(t) = u(t) + iv(t)$, we have
$$\bar{f}(t) = u(t) - iv(t).$$

Example 7.45. The complex exponential function, $f(z) = e^z$, is ubiquitous in applications. For any t in $\mathbb{R}$, Euler's formula says that
$$e^{it} = \cos t + i \sin t.$$
Applying this to $e^{zt} = e^{(x+iy)t}$, where z is a complex constant, x and y are real constants, and t is a real variable, we have
$$e^{zt} = e^{(x+iy)t} = e^{xt} e^{iyt} = e^{xt}(\cos yt + i \sin yt).$$
The real part of e^{zt} is then $u(t) = e^{xt} \cos yt$ and the imaginary part is $v(t) = e^{xt} \sin yt$. The complex conjugate of e^{zt} is
$$\overline{e^{zt}} = \overline{e^{(x+iy)t}} = e^{xt} \cos yt - i e^{xt} \sin yt.$$
Since $\cos yt$ is an even function and $\sin yt$ is an odd function, we have
$$\overline{e^{(x+iy)t}} = e^{xt}(\cos(-yt) + i \sin(-yt)) = e^{(x-iy)t} = e^{\bar{z}t}.$$
This shows that complex conjugation commutes with exponentiation, just as it commutes with addition and multiplication.

The product rule and the chain rule give us
$$\frac{d}{dt} e^{zt} = \frac{d}{dt}\left(e^{xt}(\cos yt + i \sin yt)\right) = e^{xt}(-y \sin yt + iy \cos yt) + x e^{xt}(\cos yt + i \sin yt)$$
$$= e^{xt}\left((x+iy)\cos yt + (ix - y)\sin yt\right).$$
Noting that $ix - y = i(x + iy)$, we get
$$\frac{d}{dt} e^{zt} = (x+iy) e^{xt}(\cos yt + i \sin yt) = z e^{zt}.$$
From there we can say
$$\int_a^b e^{zt}\, dt = \frac{1}{z}(e^{zb} - e^{za}).$$

We denote the collection of continuous complex functions on $[a,b] \subseteq \mathbb{R}$ by $\mathscr{C}([a,b])$. The properties of real continuous functions imply that sums and scalar multiples of functions in $\mathscr{C}([a,b])$ are in $\mathscr{C}([a,b])$: $\mathscr{C}([a,b])$ is a vector space.

The complex inner product spaces of functions that we look at are all subspaces of $\mathscr{C}([a,b])$, with σ given by

(7.6) $$\sigma(f, g) = \int_a^b \bar{f}(t) g(t)\, dt.$$

We have seen this inner product already, as defined on spaces of continuous real functions.

7.6. Inner Product Spaces of Functions

Properties of real definite integrals apply here to show that σ is Hermitian symmetric and linear in its second variable. An easy exercise verifies that $\bar{f}(t)f(t)$ is a nonnegative real function on $[a,b]$. Its integral over $[a,b]$ must then be positive, unless f is identically zero on $[a,b]$. This is enough to verify that σ is an inner product.

We can identify geometric relationships in $(\mathscr{C}([a,b]), \sigma)$ just as we do in $(\mathbb{R}^n, \cdot)$.

Example 7.46. Let $f(z) = z^2 + iz$ and let $g(z) = z+1$. As polynomials, f and g belong to $\mathscr{C}([0,1])$. We want to find φ, the angle between f and g in $(\mathscr{C}([0,1]), \sigma)$. We have

$$\sigma(f,g) = \int_0^1 (t^2 - it)(t+1)\, dt = \int_0^1 t^3 + (1-i)t^2 - it\, dt$$

$$= \frac{1}{4}t^4 + \frac{1-i}{3}t^3 - \frac{i}{2}t^2 \Big|_0^1 = \frac{7}{12} - i\frac{5}{6}.$$

We also find that

$$\|f\|^2 = \int_0^1 t^4 + t^2\, dt = \frac{8}{15}$$

and

$$\|g\|^2 = \int_0^1 t^2 + 2t + 1\, dt = \frac{7}{3}.$$

This gets us

$$\varphi = \cos^{-1}\left(\frac{7/12}{\sqrt{8/15}\sqrt{7/3}}\right)$$

which is approximately 1.02 radians, or about 58.5 degrees.

The induced norm makes convergence meaningful in any inner product space. Convergence, in turn, allows us to extend consideration from (finite) linear combinations to infinite sums. This brings us to a notion of orthonormal basis that is more general than that of Definition 7.32.

Definition 7.47. Let (V, σ) be an infinite-dimensional inner product space. An **analytic orthonormal basis** for V is a sequence $\{f_n\}_{n=1}^\infty \subseteq V$, such that when we define, for f in V,

(7.7) $$F_N := \sum_{n=1}^N \sigma(f_n, f)f_n,$$

$\|F_N - f\|$ converges to zero.

When the setting is one in which an algebraic basis is uncountable, the term *orthonormal basis* — without the modifier "analytic" — is used in the sense of Definition 7.47.

There is one case in which there appears to be a possibility for confusion between the algebraic and the analytic versions of orthonormal bases. On closer inspection, though, the fog disappears.

Example 7.48. Consider $(\mathbb{C}[t], \sigma)$ where σ is given by (7.6) on $[-1, 1]$:
$$\sigma(p(t), q(t)) = \int_{-1}^{1} \bar{p}(t) q(t) \, dt.$$

When we apply the Gram-Schmidt process to the standard ordered basis for $\mathbb{C}[t]$, $\{1, t, t^2, \ldots\}$, we get (normalized) **Legendre polynomials**.[3] Writing out the first few Legendre polynomials, we have
$$\mathcal{B} = \left\{ \frac{1}{\sqrt{2}}, \sqrt{\frac{3}{2}} t, \sqrt{\frac{5}{8}} (3t^2 - 1), \sqrt{\frac{7}{8}} (5t^3 - 3t), \ldots \right\}.$$

Since every polynomial in $\mathbb{C}[t]$ is a linear combination of elements in the standard ordered basis, $\mathcal{B}$ has to be an algebraic orthonormal basis for $(\mathbb{C}[t], \sigma)$.

Legendre polynomials are important in analytic applications. In those contexts, we may see $\mathcal{B}$ described as an orthonormal basis for $(\mathscr{C}([-1, 1]), \sigma)$, σ still defined as in (7.6). Since any algebraic basis for $\mathscr{C}([-1, 1])$ is uncountable, $\mathcal{B}$ can only be an analytic orthonormal basis for $\mathscr{C}([-1, 1])$.

The idea of a power series expansion associated to a real continuous function should be familiar from calculus. In that setting, for instance, we might calculate the Maclaurin series for $\cos t$,
$$1 - \frac{t^2}{2} + \frac{t^4}{4!} - \frac{t^6}{6!} + \cdots.$$
When we write
$$\cos t = 1 - \frac{t^2}{2} + \frac{t^4}{4!} - \frac{t^6}{6!} + \cdots,$$
it means that for any value of t in $\mathbb{R}$, the series on the right converges to the value of $\cos t$. This is pointwise convergence. With an analytic orthogonal basis, though, convergence is not pointwise. It is via the norm determined by the inner product.

Example 7.49. The set $\mathcal{B} = \{\sqrt{2/\pi} \sin nt\}_{n=1}^{\infty}$ is an orthonormal basis for $(\mathscr{C}([0, \pi]), \sigma)$. Proof that $\mathcal{B}$ is a spanning set for $(\mathscr{C}([0, \pi]), \sigma)$ is beyond the scope of our work here, but we can verify that $\mathcal{B}$ is an orthonormal set. Note first that $\int_0^\pi \cos nt \, dt = 0$ for any positive integer n. We then have
$$\left\| \sqrt{\frac{2}{\pi}} \sin nt \right\|^2 = \int_0^\pi \left(\sqrt{\frac{2}{\pi}} \sin nt \right)^2 dt = \frac{2}{\pi} \int_0^\pi \frac{1 - \cos 2nt}{2} \, dt = \frac{2}{\pi} \frac{\pi}{2} = 1.$$

This establishes that each element in $\mathcal{B}$ is a unit vector. We check for orthogonality by considering
$$\sigma\left(\sqrt{\frac{2}{\pi}} \sin nt, \sqrt{\frac{2}{\pi}} \sin mt \right) = \frac{2}{\pi} \int_0^\pi \sin nt \sin mt \, dt$$
when $m \neq n$. We leave it as an exercises below to establish that the integrand here can be rewritten as a difference of cosine functions, the integral of each of which is zero. This establishes that $\mathcal{B}$ is an orthogonal set.

---

[3] The term *Legendre polynomials* can mean a few slightly different things. Often, it means polynomials P_n normalized to satisfy $P_n(1) = 1$.

7.6. Inner Product Spaces of Functions

When f belongs to $\mathscr{C}([0,\pi])$, the expression

$$\sum_{n=1}^{\infty} \sigma\left(\sqrt{\frac{2}{\pi}}\sin nt, f(t)\right)\sqrt{\frac{2}{\pi}}\sin nt$$

may be called the *Fourier expansion* for f on $[0,\pi]$. The coefficients,

$$a_n = \sigma\left(\sqrt{\frac{2}{\pi}}\sin nt, f(t)\right),$$

are then *Fourier coefficients* for f. Consider, for instance, that when $f(t) = t$,

$$a_n = \sigma\left(\sqrt{\frac{2}{\pi}}\sin nt, t\right) = \int_0^\pi t\sqrt{\frac{2}{\pi}}\sin nt\, dt.$$

Using integration by parts, we have

$$a_n = \sqrt{\frac{2}{\pi}}\left(-\frac{t}{n}\cos nt\bigg|_0^\pi\right) = -\sqrt{\frac{2}{\pi}}\frac{\cos n\pi}{n}.$$

Since $\cos n\pi = (-1)^n$, we see that

$$a_n = \sqrt{\frac{2}{\pi}}(-1)^{n+1}\frac{1}{n}.$$

This means that

$$\left\|t - 2\left(\sin t - \frac{\sin 2t}{2} + \frac{\sin 3t}{3} - \frac{\sin 4t}{4} + \cdots\right)\right\| \to 0.$$

Notice that at both endpoints of the interval $[0,\pi]$, the value of the Fourier expansion for $f(t) = t$ is zero. This may be alarming, but not once we recognize that we measure the difference between the Fourier expansion and f using integration: Two functions that differ only at isolated points on $[a,b]$ have the same integral over $[a,b]$.

Once we get away from the endpoints of the interval, the Fourier expansion here does converge to $f(t) = t$ pointwise. When $t = \pi/2$, for example, the Fourier expansion is

$$2\left(\sin(\pi/2) - \frac{\sin \pi}{2} + \frac{\sin(3\pi/2)}{3} - \frac{\sin(2\pi)}{4} + \frac{\sin(5\pi/2)}{5} - \cdots\right)$$
$$= 2\left(1 - \frac{1}{3} + \frac{1}{5} - \frac{1}{7} + \cdots\right).$$

By the well-known identity

$$\sum_{n=1}^{\infty}\frac{(-1)^{n+1}}{2n+1} = \frac{\pi}{4},$$

we see that

$$2\left(1 - \frac{1}{3} + \frac{1}{5} - \frac{1}{7} + \cdots\right) = \frac{\pi}{2},$$

the value of $f(t)$ at $\pi/2$.

Exercises 7.6.

7.6.1. Let $f(z) = (2z^2 + iz)e^{3z-1}$. Find $\bar{f}(z)$.

7.6.2. Euler understood the connections shared by exponentials, sines, and cosines via his work on differential equations. In particular, he knew that $y_1(t) = 2\cos t$ and $y_2(t) = e^{it} + e^{-it}$ satisfy the same initial value problem:
$$y'' + y = 0, \quad y(0) = 2, \quad y'(0) = 0.$$
From this observation, he concluded that $2\cos t = e^{it} + e^{-it}$ [12].
   (a) Verify Euler's observation about the two functions satisfying the same initial value problem.
   (b) Show that Euler's formula follows from $2\cos t = e^{it} + e^{-it}$.
   (c) Show that $\sin t = \frac{e^{it} - e^{-it}}{2i}$.
   (d) Show that $e^{i\pi} + 1 = 0$. This is sometimes called *Euler's identity*.

7.6.3. When a and b are distinct real numbers, integrals of the form $\int \sin at \sin bt \, dt$, $\int \sin at \cos bt \, dt$, and $\int \cos at \cos bt \, dt$ are particularly troublesome to calculus students. With complex exponentials, we can turn each of these integrands from a product into a sum, simplifying the calculations considerably. Use complex exponentials to show the following:
   (a) $\sin at \sin bt = \frac{1}{2}(\cos(a-b)t - \cos(a+b)t)$,
   (b) $\sin at \cos bt = \frac{1}{2}(\sin(a+b)t + \sin(a-b)t)$, and
   (c) $\cos at \cos bt = \frac{1}{2}(\cos(a+b)t + \cos(a-b)t)$.

7.6.4. Prove that for any $k \in \mathbb{Z}$, $\int_0^1 e^{2ik\pi t} \, dt = 0$.

7.6.5. Verify that if $f(t) = u(t) + iv(t)$, where t in $[a,b]$ is a real variable, then $\overline{\int_a^b f(t) \, dt} = \int_a^b \bar{f}(t) \, dt$.

7.6.6. Let $f(t) = u(t) + iv(t)$, where $u(t)$ and $v(t)$ are real functions defined on some interval $[a,b] \subseteq \mathbb{R}$. Show that $\bar{f}(t)f(t)$ is a nonnegative real function on $[a,b]$.

7.6.7. Let $\mathcal{B} = \{e^{2in\pi t}\}_{n \in \mathbb{Z}}$. A fact we will not prove is that $\mathcal{B}$ is an orthonormal basis for $(\mathscr{C}([0,1]), \sigma)$, σ as given in (7.6).
   (a) Show that $\mathcal{B}$ is an orthonormal set.
   (b) Let $f(t) = t$. Find $a_n = \sigma(e^{2in\pi t}, f(t))$.
   (c) The expansion for any function in $\mathscr{C}([0,1])$ with respect to $\mathcal{B}$ is a doubly infinite series. Assume you can rearrange the terms when $f(t) = t$ and use that to write the series in terms of sines and/or cosines.

7.7. Unitary Transformations

We have seen that the complex analog to an orthogonal matrix is a unitary matrix. In this section, we look more closely at that idea. Throughout this discussion, $\mathbb{F}$ is $\mathbb{C}$ or $\mathbb{R}$. This means that the results in this section apply to orthogonal transformations and to orthogonal matrices.

Definition 7.50. Let (V, σ) and (W, ζ) be inner product spaces over the same field. A **unitary transformation** is a vector space isomorphism L in $\mathcal{L}(V, W)$

7.7. Unitary Transformations

that satisfies
$$\sigma(\mathbf{v}, \mathbf{x}) = \zeta(L(\mathbf{v}), L(\mathbf{x}))$$
for all $\mathbf{v}$ and $\mathbf{x}$ in V.

A **unitary operator** is a unitary transformation $V \to V$.

An **orthogonal transformation** is a unitary transformation on a real inner product space. An **orthogonal operator** is a unitary operator on a real inner product space.

Proof of the following is an exercise.

Lemma 7.51. (a) The inverse of a unitary transformation is a unitary transformation.

(b) If (V, σ) is an inner product space, with orthonormal bases $\mathcal{B} = \{\mathbf{u}_i\}_{i=1}^n$ and $\mathcal{C} = \{\mathbf{x}_i\}_{i=1}^n$, then the linear extension of $\mathbf{u}_i \mapsto \mathbf{x}_i$ is a unitary operator on V.

(c) If (V, σ), (W, ζ), and (U, χ) are inner product spaces with unitary transformations L_1 in $\mathcal{L}(V, W)$ and L_2 in $\mathcal{L}(W, U)$, then $L_2 \circ L_1$ is a unitary transformation in $\mathcal{L}(V, U)$.

Theorem 7.52. The unitary operators on an inner product space comprise a group.

Proof. It should be evident that the identity mapping is a unitary operator on any inner product space. By Lemma 7.51, the unitary operators on an inner product space form a set that is closed under composition and inversion. Since function composition is associative, this is enough to prove the theorem. $\square$

Vector space isomorphisms are linear mappings that send any basis to a basis. What distinguishes unitary transformations among arbitrary vector space isomorphisms?

Theorem 7.53. Let (V, σ) and (W, ζ) be inner product spaces and let $\mathcal{B}$ be an orthonormal basis for V. A mapping L in $\mathcal{L}(V, W)$ is unitary if and only if $L[\mathcal{B}]$ is an orthonormal basis for W.

Proof. Let (V, σ) and (W, ζ) be as hypothesized and let L belong to $\mathcal{L}(V, W)$. Let $\mathcal{B} = \{\mathbf{u}_i\}$ be an orthonormal basis for V.

Suppose L is a unitary transformation. Since L is an isomorphism, $L[\mathcal{B}]$ is a basis for W. Moreover, since
$$\zeta(L(\mathbf{u}_i), L(\mathbf{u}_j)) = \sigma(\mathbf{u}_i, \mathbf{u}_j) = \delta(i, j),$$
$L[\mathcal{B}]$ is an orthonormal set in W. This proves the theorem in one direction.

Suppose next that $L[\mathcal{B}]$ is an orthonormal basis for W. Since L maps a basis of V to a basis of W, it is an isomorphism. We must show that L respects the inner products on V and W. For any $\mathbf{w}$ in W,

$$(7.8) \qquad \mathbf{w} = \sum_{i=1}^n \zeta(L(\mathbf{u}_i), \mathbf{w}) L(\mathbf{u}_i).$$

For any **v** in V,
$$\mathbf{v} = \sum_{i=1}^{n} \sigma(\mathbf{u}_i, \mathbf{v}) \mathbf{u}_i.$$

Fix **w** in W and let $\mathbf{v} = L^{-1}(\mathbf{w})$. Now we can write
$$\mathbf{w} = L(\mathbf{v}) = L\left(\sum_{i=1}^{n} \sigma(\mathbf{u}_i, \mathbf{v}) \mathbf{u}_i\right) = \sum_{i=1}^{n} \sigma(\mathbf{u}_i, \mathbf{v}) L(\mathbf{u}_i).$$

Putting this together with (7.8) and using the fact that $L[\mathcal{B}]$ is a basis, we get

(7.9) $\qquad \sigma(\mathbf{u}_i, \mathbf{v}) = \zeta(L(\mathbf{u}_i), \mathbf{w}) = \zeta(L(\mathbf{u}_i), L(\mathbf{v})),$

for each $i \in \{1, \ldots, n\}$. This holds for all **v** in V.

Given **v** and **x** in V, Exercise 7.4.11 guarantees that
$$\sigma(\mathbf{v}, \mathbf{x}) = \sum_{i=1}^{n} \sigma(\mathbf{v}, \mathbf{u}_i) \sigma(\mathbf{u}_i, \mathbf{x}).$$

We can also write
$$\sigma(\mathbf{v}, \mathbf{x}) = \sum_{i=1}^{n} \overline{\sigma(\mathbf{u}_i, \mathbf{v})} \sigma(\mathbf{u}_i, \mathbf{x}).$$

Putting this together with (7.9), we get
$$\sigma(\mathbf{v}, \mathbf{x}) = \sum_{i=1}^{n} \overline{\zeta(L(\mathbf{u}_i), L(\mathbf{v}))} \zeta(L(\mathbf{u}_i), L(\mathbf{x}))$$
$$= \sum_{i=1}^{n} \zeta(L(\mathbf{v}), L(\mathbf{u}_i)) \zeta(L(\mathbf{u}_i), L(\mathbf{x})) = \zeta(L(\mathbf{v}), L(\mathbf{x})).$$

This proves the theorem in the other direction. $\qquad \square$

While the invertible linear operators on a vector space are the linear operators that map ordered bases to ordered bases, the unitary operators on an inner product space are the linear operators that map ordered orthonormal bases to ordered orthonormal bases.

The unitary transformations that map (V, σ) to $(\mathbb{F}^n, \cdot)$ are of particular interest. The following corollary is nearly immediate by Theorem 7.53. We leave the details as an exercise.

Corollary 7.54. *Let $\mathcal{B}$ be an ordered basis for a finite-dimensional inner product space (V, σ). The mapping $\mathbf{v} \mapsto [\mathbf{v}]_\mathcal{B}$ is a unitary transformation if and only if $\mathcal{B}$ is an orthonormal basis for V.*

Just as the matrix representation of an invertible linear operator on a finite-dimensional vector space with respect to a given ordered basis is an invertible matrix, the matrix representation of a unitary operator on a finite-dimensional inner product space with respect to a given orthonormal basis is a unitary matrix. To better appreciate this part of the story, we take a moment to wallow in the properties of unitary matrices.

7.7. Unitary Transformations

Observation 7.55. A unitary matrix is defined to be A in $M_n(\mathbb{F})$ that satisfies $A^\dagger A = I_n$, but there are several equivalent ways to say this, among them are the following:

(a) $A^\dagger = A^{-1}$;

(b) $A^T = \bar{A}^{-1} = \overline{A^{-1}}$;

(c) $\bar{A} = (A^{-1})^T = (A^T)^{-1}$;

(d) $A = \overline{(A^{-1})^T} = \overline{(A^T)^{-1}} = (\bar{A}^T)^{-1}$.

Remember the theorem that said a matrix in $M_n(\mathbb{F})$ was invertible if and only if its columns formed a basis for $\mathbb{F}^n$? Here is the analog for unitary matrices.

Theorem 7.56. A matrix in $M_n(\mathbb{F})$ is unitary if and only if its columns form an orthonormal basis for $\mathbb{F}^n$.

Proof. Let $A = \begin{bmatrix} \mathbf{a}_1 & \cdots & \mathbf{a}_n \end{bmatrix}$ belong to $M_n(\mathbb{F})$. Row i of $A^\dagger$ is $\mathbf{a}_i^\dagger$ so the ij-entry of $A^\dagger A$ is $\mathbf{a}_i^\dagger \mathbf{a}_j$. It follows that A is unitary if and only if

(7.10) $$\mathbf{a}_i^\dagger \mathbf{a}_j = \mathbf{a}_i \cdot \mathbf{a}_j = \delta(i,j),$$

for i and j in $\{1, \ldots, n\}$. The theorem follows immediately. $\square$

Let $\mathbf{v}$ and $\mathbf{w}$ belong to $\mathbb{F}^n$ and suppose A in $M_n(\mathbb{F})$ is unitary. By Lemma 5.49,

$$A\mathbf{v} \cdot A\mathbf{w} = (A\mathbf{v})^\dagger A\mathbf{w} = \mathbf{v}^\dagger A^\dagger A\mathbf{w} = \mathbf{v} \cdot \mathbf{w}.$$

This is enough to prove the following theorem in one direction.

Theorem 7.57. A matrix A in $M_n(\mathbb{F})$ is unitary if and only if $A\mathbf{v} \cdot A\mathbf{w} = \mathbf{v} \cdot \mathbf{w}$ for all $\mathbf{v}$ and $\mathbf{w}$ in $\mathbb{F}^n$.

Proof. We have left to show that if A in $M_n(\mathbb{F})$ satisfies

$$A\mathbf{v} \cdot A\mathbf{w} = \mathbf{v} \cdot \mathbf{w}$$

for all $\mathbf{v}$ and $\mathbf{w}$ in $\mathbb{F}^n$, then A is unitary. We do this by exploiting the fact that if $\mathbf{e}_i$ is in the standard ordered basis for $\mathbb{F}^n$, then $A\mathbf{e}_i = \mathbf{a}_i$, column i of A. This gives us

$$\delta(i,j) = \mathbf{e}_i \cdot \mathbf{e}_j = A\mathbf{e}_i \cdot A\mathbf{e}_j = \mathbf{a}_i \cdot \mathbf{a}_j,$$

establishing that the columns of A form an orthonormal basis for $\mathbb{F}^n$. By Theorem 7.56, A is unitary. $\square$

The next corollary is immediate.

Corollary 7.58. A matrix A in $M_n(\mathbb{F})$ is unitary if and only if $\mathbf{v} \mapsto A\mathbf{v}$ is a unitary operator on $\mathbb{F}^n$.

Proof of the following theorem is now straightforward.

Theorem 7.59. Let (V, σ) be a finite-dimensional inner product space. An operator L in $\mathcal{L}(V)$ is unitary if and only if $[\![L]\!]_\mathcal{B}$ is a unitary matrix whenever $\mathcal{B}$ is an orthonormal basis for V.

Proof. Let (V, σ) be as hypothesized. Let $\mathcal{B} = \{\mathbf{u}_i\}_{i=1}^n$ be an orthonormal basis for V.

If L in $\mathcal{L}(V)$ is unitary, then by Theorem 7.53, $L[\mathcal{B}]$ is an orthonormal basis for V. By Corollary 7.54, $\{[L(\mathbf{u}_i)]_\mathcal{B}\}_{i=1}^n$ is an orthonormal basis for $\mathbb{F}^n$ under the dot product. It follows by Theorem 7.56 that $[\![L]\!]_\mathcal{B}$ is a unitary matrix.

Suppose now that L in $\mathcal{L}(V)$ is arbitary and that $[\![L]\!]_\mathcal{B}$ is a unitary matrix. Theorem 7.56 guarantees that

$$\{[L(\mathbf{u}_i)]_\mathcal{B}\}_{i=1}^n$$

is an orthonormal basis for $\mathbb{F}^n$ under the dot product. Since it maps an orthonormal basis to an orthonormal basis, L is a unitary operator by Theorem 7.53. □

The following theorem sums up what we know now about unitary matrices and operators. We have proved most of it. The rest is left as an exercise.

Theorem 7.60. Let $\mathcal{B}$ be an orthonormal basis for a finite-dimensional inner product space (V, σ). Let L belong to $\mathcal{L}(V)$ and let $A = [\![L]\!]_\mathcal{B}$. The following statements are logically equivalent.

(a) L is a unitary operator.

(b) A is a unitary matrix.

(c) A^{-1} is a unitary matrix.

(d) L^{-1} is a unitary operator.

(e) $A^\dagger = A^{-1}$.

(f) The columns of A form an orthonormal basis for $\mathbb{F}^n$.

(g) The rows of A form an orthonormal basis for $\mathbb{F}_n$.

Exercises 7.7.

7.7.1. Prove Lemma 7.51.

7.7.2. Let (V, σ) be a finite-dimensional inner product space with orthonormal basis $\mathcal{B}$. Prove that $L \mapsto [\![L]\!]_\mathcal{B}$ is a group isomorphism from the unitary operators on V to the group of unitary matrices in $GL_n(\mathbb{F})$, $U_n(\mathbb{F})$. (Hint: Use Theorem 3.33.)

7.7.3. Write out a proof for Corollary 7.54.

7.7.4. Let (V, σ) be a finite-dimensional inner product space with ordered bases $\mathcal{B}$ and $\mathcal{C}$. Show that the $\mathcal{B}$ to $\mathcal{C}$ change of basis matrix is unitary if and only if $\mathcal{B}$ and $\mathcal{C}$ are orthonormal bases.

7.7.5. Verify the equivalent formulations of the definition of a unitary matrix as given in Observation 7.55.

7.7.6. Let A and P be $n \times n$ unitary matrices. Show that $PAP^\dagger$ is also unitary.

7.7.7. Let L be a unitary operator on (V, σ). Show that if W, a subspace of V, is L-invariant, then so is $W^\perp$.

7.7.8. Verify that the statements in Theorem 7.60 are logically equivalent.

7.7.9. Can a projection be a unitary operator?

7.7.10. Can a linear reflection be a unitary operator?

7.7.11. The standard matrix representation of the rotation on $\mathbb{R}^2$ given by
$$R_\varphi(\|\mathbf{v}\|(\cos\theta, \sin\theta)) = \|\mathbf{v}\|(\cos(\theta+\varphi), \sin(\theta+\varphi))$$
is $\begin{bmatrix} \cos\varphi & -\sin\varphi \\ \sin\varphi & \cos\varphi \end{bmatrix}$.

(a) Find $A = [\![R_\varphi]\!]_\mathcal{B}$ for $\mathcal{B} = \{(1,1), (0,1)\}$.
(b) Is A an orthogonal matrix? How could you tell the answer to that without finding A?

7.8. The Adjoint of an Operator

We saw in the last section that the generalization of unitary matrices is the class of operators on an inner product space that respect the inner product. Here we look at the generalization of the conjugate transpose of a matrix.

Consider that when A is any matrix in $M_n(\mathbb{F})$,
$$A\mathbf{v} \cdot \mathbf{w} = (A\mathbf{v})^\dagger \mathbf{w} = \mathbf{v}^\dagger A^\dagger \mathbf{w} = \mathbf{v}^\dagger A \mathbf{w} = \mathbf{v} \cdot A\mathbf{w}$$
for all $\mathbf{v}$ and $\mathbf{w}$ in $\mathbb{F}^n$. This prompts the following question: When (V, σ) is an arbitrary inner product space with L in $\mathcal{L}(V)$, is there a way to define $L^\dagger$ in $\mathcal{L}(V)$ so that
$$\sigma(L^\dagger(\mathbf{v}), \mathbf{w}) = \sigma(\mathbf{v}, L(\mathbf{w}))$$
for all $\mathbf{v}$ and $\mathbf{w}$ in V?

We answer the question by exploiting the connection between the dual space of an inner product space and the inner product. Our work in Section 5.3 applies to real inner product spaces because a real inner product is bilinear. Here we broaden the discussion to include Hermitian inner products.

Proof of the next lemma requires some minor adjustments to the proof of Theorem 5.29 so we leave it as an exercise.

Lemma 7.61. If (V, σ) is a finite-dimensional inner product space, then there is a conjugate-linear bijection $V \to V^*$, $\mathbf{v} \mapsto f_\mathbf{v}$, determined by $f_\mathbf{v}(\mathbf{w}) = \sigma(\mathbf{v}, \mathbf{w})$ for $\mathbf{w}$ in V.

Theorem 7.62. Let (V, σ) be a finite-dimensional inner product space. If L belongs to $\mathcal{L}(V)$, there is a unique $L^\dagger$ in $\mathcal{L}(V)$ such that
$$\sigma(\mathbf{v}, L(\mathbf{w})) = \sigma(L^\dagger(\mathbf{v}), \mathbf{w})$$
for all $\mathbf{v}$ and $\mathbf{w}$ in V.

Proof. Let (V, σ) be as hypothesized. Fix $\mathbf{v}$ in V and let f be the mapping
$$f(\mathbf{w}) = \sigma(\mathbf{v}, L(\mathbf{w})).$$
Since f belongs to V^*, Lemma 7.61 ensures that there is a unique $\mathbf{x}$ in V so that $f = f_\mathbf{x}$. We then have
$$f(\mathbf{w}) = \sigma(\mathbf{v}, L(\mathbf{w})) = f_\mathbf{x}(\mathbf{w}) = \sigma(\mathbf{x}, \mathbf{w}).$$

This allows us to define $L^\dagger(\mathbf{v}) := \mathbf{x}$ so that
$$\sigma(\mathbf{v}, L(\mathbf{w})) = \sigma(L^\dagger(\mathbf{v}), \mathbf{w})$$
for all $\mathbf{v}$ and $\mathbf{w}$ in V.

The only thing we know about $L^\dagger$ is that it is a mapping $V \to V$. We must show that it is linear.

Given c in the underlying field and $\mathbf{v}_1$ and $\mathbf{v}_2$ in V, we have
$$\sigma(L^\dagger(c\mathbf{v}_1 + \mathbf{v}_2), \mathbf{w}) = \sigma(c\mathbf{v}_1 + \mathbf{v}_2, L(\mathbf{w})) = \bar{c}\sigma(\mathbf{v}_1, L(\mathbf{w})) + \sigma(\mathbf{v}_2, L(\mathbf{w}))$$
$$= \bar{c}\sigma(L^\dagger(\mathbf{v}_1), \mathbf{w}) + \sigma(L^\dagger(\mathbf{v}_2), \mathbf{w}) = \sigma(cL^\dagger(\mathbf{v}_1) + L^\dagger(\mathbf{v}_2), \mathbf{w}).$$
Since this is true for all $\mathbf{w}$ in V, $L^\dagger$ is linear on V.

Next we address uniqueness.

Suppose T in $\mathcal{L}(V)$ also satisfies
$$\sigma(\mathbf{v}, L(\mathbf{w})) = \sigma(T(\mathbf{v}), \mathbf{w})$$
for all $\mathbf{v}$ and $\mathbf{w}$ in V. We then have
$$\sigma(L^\dagger(\mathbf{v}) - T(\mathbf{v}), \mathbf{w}) = 0,$$
for all $\mathbf{w}$ in V. Since σ is nondegenerate, $L^\dagger(\mathbf{v}) = T(\mathbf{v})$ for all $\mathbf{v}$ in V. This is enough to verify that $L^\dagger$ is unique. $\square$

The following definition applies to all inner product spaces.

Definition 7.63. Let (V, σ) be an inner product space and let L belong to $\mathcal{L}(V)$. A mapping $L^\dagger$ in $\mathcal{L}(V)$ that satisfies
$$\sigma(\mathbf{v}, L(\mathbf{w})) = \sigma(L^\dagger(\mathbf{v}), \mathbf{w})$$
for all $\mathbf{v}$ and $\mathbf{w}$ in V is the **adjoint of** L.

The argument for uniqueness of the adjoint in the proof of Theorem 7.62 applies to show that whenever the adjoint of a linear transformation does exist, it must be unique.

Lemma 7.64. Let (V, σ) be an inner product space and let L belong to $\mathcal{L}(V)$. If the adjoint of L is defined, then
$$\sigma(L(\mathbf{v}), \mathbf{w}) = \sigma(\mathbf{v}, L^\dagger(\mathbf{w}))$$
for all $\mathbf{v}$ and $\mathbf{w}$ in V.

Proof. With everything as hypothesized, we have
$$\sigma(L(\mathbf{v}), \mathbf{w}) = \overline{\sigma(\mathbf{w}, L(\mathbf{v}))} = \overline{\sigma(L^\dagger(\mathbf{w}), \mathbf{v})} = \sigma(\mathbf{v}, L^\dagger(\mathbf{w})). \quad \square$$

Properties of the adjoint of a linear operator shadow those of the conjugate transpose of a matrix.

7.8. The Adjoint of an Operator

Lemma 7.65. Let L and T be linear operators on an inner product space. If the adjoints exist, then

(a) $(L + cT)^\dagger = L^\dagger + \bar{c}T^\dagger$, for any scalar c;
(b) $(L^\dagger)^\dagger = L$;
(c) $(L \circ T)^\dagger = T^\dagger \circ L^\dagger$.

Proof. Let (V, σ) be an inner product space. Assume that for L and T in $\mathcal{L}(V)$, both $L^\dagger$ and $T^\dagger$ exist. We then have, for any scalar c and $\mathbf{v}$ and $\mathbf{w}$ in V,

$$\sigma(\mathbf{v}, (L+cT)(\mathbf{w})) = \sigma(\mathbf{v}, L(\mathbf{w})) + c\sigma(\mathbf{v}, T(\mathbf{w}))$$
$$= \sigma(L^\dagger(\mathbf{v}), \mathbf{w}) + c\sigma(T^\dagger(\mathbf{v}), \mathbf{w})$$
$$= \sigma(L^\dagger(\mathbf{v}), \mathbf{w}) + \sigma(\bar{c}T^\dagger(\mathbf{v}), \mathbf{w})$$
$$= \sigma((L^\dagger + \bar{c}T^\dagger)(\mathbf{v}), \mathbf{w}).$$

The first statement of the lemma now follows by the uniqueness of the adjoint, when it exists.

The rest of the proof is left as an exercise. $\square$

The relationship between the adjoint of a linear operator on a finite-dimensional space and the conjugate transpose of its matrix representation is what we would expect, as long as we use an orthonormal basis.

Theorem 7.66. Let (V, σ) be a finite-dimensional inner product space with orthonormal basis $\mathcal{B}$. If L belongs to $\mathcal{L}(V)$ and $A = [\![L]\!]_\mathcal{B}$, then $[\![L^\dagger]\!]_\mathcal{B} = A^\dagger$.

Proof. Let everything be as hypothesized. If $\mathcal{B} = \{\mathbf{u}_i\}_{i=1}^n$, then column j of $A = [\![L]\!]_\mathcal{B}$ is $[L(\mathbf{u}_j)]_\mathcal{B}$. Letting $A = [a_{ij}]$, we know that a_{ij} is the coefficient of $\mathbf{u}_i$ when we write $L(\mathbf{u}_j)$ as a linear combination of the elements in $\mathcal{B}$:

$$a_{ij} = \sigma(\mathbf{u}_i, L(\mathbf{u}_j)) = \sigma(L^\dagger(\mathbf{u}_i), \mathbf{u}_j) = \overline{\sigma(\mathbf{u}_j, L^\dagger(\mathbf{u}_i))}.$$

We use this to write the ji-entry of $(A^\dagger)^T = \bar{A}$ as

$$\overline{\sigma(\mathbf{u}_j, L(\mathbf{u}_i))} = \sigma(\mathbf{u}_i, L^\dagger(\mathbf{u}_j)),$$

which is the ij-entry of $[\![L^\dagger]\!]_\mathcal{B}$. This proves that $A^\dagger = [\![L^\dagger]\!]_\mathcal{B}$. $\square$

Thinking about the matrix theory, we are led to the following definition.

Definition 7.67. A **self-adjoint operator** is an operator L on an inner product space (V, σ) such that $L^\dagger$ exists and $L = L^\dagger$.

Example 7.68. The spin states of an electron can be represented by 1-dimensional subspaces in $\mathbb{C}^2$. Self-adjoint operators on $\mathbb{C}^2$ represent transitions between spin states. The *Pauli spin matrices* are given by

$$\begin{bmatrix} 0 & 1 \\ 1 & 0 \end{bmatrix}, \begin{bmatrix} 0 & -i \\ i & 0 \end{bmatrix}, \begin{bmatrix} 1 & 0 \\ 0 & -1 \end{bmatrix}.$$

Notice that these are both unitary and self-adjoint.

When the underlying field is $\mathbb{R}$, a self-adjoint operator may also be called a **symmetric operator**.

Our last result here is immediate by Theorem 7.66.

Corollary 7.69. Let (V, σ) be a finite-dimensional inner product space. An operator L in $\mathcal{L}(V)$ is self-adjoint if and only if $[\![L]\!]_\mathcal{B}$ is a self-adjoint matrix when $\mathcal{B}$ is an orthonormal basis for V.

Example 7.70. Let $V = \mathbb{C}^3$ under the dot product. Consider the projection $L : V \to V$ given by
$$L(z_1, z_2, z_3) = (z_1, 0, z_3).$$
For (z_1, z_2, z_3) and (w_1, w_2, w_3) in V, we have
$$L(z_1, z_2, z_3) \cdot (w_1, w_2, w_3) = \overline{z_1} w_1 + \overline{z_3} w_3 = (z_1, z_2, z_3) \cdot L(w_1, w_2, w_3)$$
so L is a self-adjoint operator on V.

Exercises 7.8.

7.8.1. Prove Lemma 7.61.

7.8.2. Explain this step in the proof of Theorem 7.62: "Since this is true for all $\mathbf{w}$ in V, $L^\dagger$ is linear on V."

7.8.3. Let (V, σ) be an inner product space over $\mathbb{F}$. Fix c in $\mathbb{F}$ and let $L(\mathbf{v}) = c\mathbf{v}$. Find $L^\dagger$, if it exists.

7.8.4. Finish the proof of Lemma 7.65.

7.8.5. Prove that if A in $M_n(\mathbb{F})$ satisfies $A\mathbf{v} \cdot \mathbf{w} = \mathbf{v} \cdot A\mathbf{w}$ for all $\mathbf{v}$ and $\mathbf{w}$ in $\mathbb{F}^n$, then A is self-adjoint.

7.8.6. Define L in $\mathcal{L}(\mathbb{C}^3)$ by $L(z_1, z_2, z_3) = (z_1, 0, z_3)$. Show that the standard matrix representation of L is self-adjoint.

7.8.7. (a) In Exercise 5.6.6, you should have verified that the collection of $n \times n$ Hermitian matrices forms a vector space over $\mathbb{R}$. Why is this not a vector space over $\mathbb{C}$?
(b) Show that if the entries of A in $M_n(\mathbb{C})$ are real numbers, then A is Hermitian if and only if it is symmetric.
(c) Show that the Pauli spin matrices, together with the identity matrix, comprise a basis for the real vector space of 2×2 Hermitian matrices.

7.8.8. Let $B = [\![L]\!]_\mathcal{B}$ for L as given in Example 7.70 where
$$\mathcal{B} = \{(1,1,0), (1,-1,0), (1,1,1)\}.$$

(a) Let P be the $\mathcal{B}$ to $\mathcal{E}$ change of basis matrix. Use P to explain why B is not self-adjoint.
(b) What property must P have if $[\![L]\!]_\mathcal{B}$ is to be self-adjoint?

7.8.9. Argue that, for any positive integer n, $f(t) \mapsto t^n f(t)$ is a self-adjoint linear operator on $\mathscr{C}([a, b])$ under the inner product $\sigma(f, g) = \int_a^b \bar{f}(t) g(t)\, dt$.

7.9. A Fundamental Theorem

Here we apply what we have learned to discuss results described in "The Fundamental Theorem of Linear Algebra," a paper written by Gilbert Strang [21]. Dating from 1993 and accessible to students, this is among the most frequently cited papers to have appeared in The American Mathematical Monthly. The Fundamental Theorem is not about inner product spaces, per se, but orthogonality is an important element in the story.

The titular theorem of the paper concerns the four fundamental subspaces naturally associated to a real $m \times n$ matrix A: Row A, Col A, Nul A, and Nul A^T. We have studied Col $A \subseteq \mathbb{F}^m$ and Nul $A \subseteq \mathbb{F}^n$, vis-à-vis the mapping $\mathbf{v} \mapsto A\mathbf{v}$, for an arbitrary field, $\mathbb{F}$. Though it is not immediately evident where Nul A^T fits into the narrative, we have established that Row A is canonically isomorphic to Col $A^T \subseteq \mathbb{F}^n$.

Figure 7.3, which is based on [21], shows the relationships among these subspaces in $\mathbb{R}^n$ and $\mathbb{R}^m$. In this rendering, we do not distinguish $\mathbb{R}^n$ and $\mathbb{R}_n$, so for instance, both Row A and Nul A are subspaces of $\mathbb{R}^n$. Here, $\mathbf{x} = \mathbf{x}_p + \mathbf{x}_h$ is an arbitrary nonzero element in $\mathbb{R}^n$, with $\mathbf{x}_p$ nonzero in Row A and $\mathbf{x}_h$ in Nul A. We have $A\mathbf{x} = A\mathbf{x}_p$, a nonzero vector in $\mathbb{R}^m$. Since the product of any row in A with any element in Nul A is 0, the null space of A is the orthogonal complement of Row A in $\mathbb{R}^n$. Likewise, Nul A^T is the orthogonal complement of Col A.

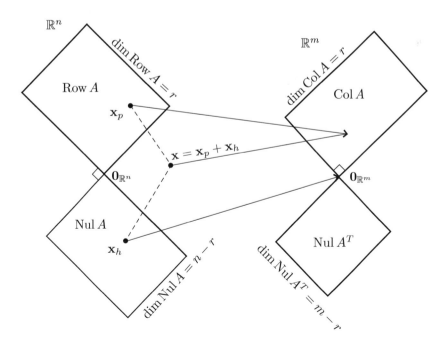

Figure 7.3. Orthogonal complements in the domain and codomain of a matrix transformation $\mathbb{R}^n \to \mathbb{R}^m$.

We would like to understand complex matrices in these terms. In particular, we want to account for the row, column, and null spaces associated to $\bar{A}$ and $A^\dagger$. Our goal is to see these spaces in terms of $\operatorname{Col} A$ and $\operatorname{Nul} A$, the ones we allegedly understand.

One technical result breaches the dam. We call it a theorem, since all the other results can be interpreted as its corollaries.

Theorem 7.71. If A is in $M_{m,n}(\mathbb{C})$, then $\operatorname{Nul} A = (\operatorname{Col} A^\dagger)^\perp$.

Proof. Let $\mathbf{x} = (x_1, \ldots, x_n)$ belong to $\mathbb{C}^n$ and let $A = (\mathbf{w}_1, \ldots, \mathbf{w}_m)$. Entry i of $A\mathbf{x}$ is $\mathbf{w}_i \mathbf{x}$ so $\mathbf{x}$ belongs to $\operatorname{Nul} A$ if and only if $\mathbf{w}_i \mathbf{x} = 0$ for $i = 1, \ldots, m$. We see then that $\mathbf{x}$ is in $\operatorname{Nul} A$ if and only if

$$0 = \mathbf{w}_i \mathbf{x} = (\mathbf{w}_i \mathbf{x})^\dagger = \mathbf{x}^\dagger \mathbf{w}_i^\dagger = \mathbf{x} \cdot \mathbf{w}_i^\dagger.$$

Taking the complex conjugate of $\mathbf{x} \cdot \mathbf{w}_i^\dagger$, we can say that $\mathbf{x}$ is in $\operatorname{Nul} A$ if and only if

$$\mathbf{w}_i^\dagger \cdot \mathbf{x} = 0.$$

Since $\mathbf{w}_i^\dagger$ is column i of $A^\dagger$, the result follows. $\square$

Reversing the roles of $A^\dagger$ and A in Theorem 7.71, we get the following corollary.

Corollary 7.72. If A is in $M_{m,n}(\mathbb{C})$, then $\operatorname{Nul} A^\dagger = (\operatorname{Col} A)^\perp$.

If we substitute A^T for A in Corollary 7.72, we get $\operatorname{Nul} \bar{A} = (\operatorname{Col} A^T)^\perp$. Noting that $\operatorname{Row} A = \operatorname{Col} A^T$, we get the next corollary.

Corollary 7.73. If A is in $M_{m,n}(\mathbb{C})$, then $\operatorname{Nul} \bar{A} = (\operatorname{Row} A)^\perp$.

The picture that tells the story in the more general context is Figure 7.4.

Another topic in [21] is *least-square solutions*. "Least-squares" is the name we use for an algorithm that identifies the next best thing to a solution when $A\mathbf{x} = \mathbf{b}$ is inconsistent. This scenario arises frequently in applications, especially when there are more equations than unknowns in the associated system of equations. The least-squares approach uses geometry to find $\mathbf{x}$ in $\mathbb{F}^n$ that minimizes the distance between $A\mathbf{x}$ and $\mathbf{b}$. Distance is measured using the norm induced by the dot product. The underlying field may be $\mathbb{R}$ or $\mathbb{C}$.

Our next definition captures the main concept underlying least-squares approximation.

Definition 7.74. Let (V, σ) be an inner product space with finite-dimensional subspace U. Let $\mathbf{v}$ belong to V. The **projection of v onto** U, also called the **component of v in** U, is

$$\operatorname{proj}_U(\mathbf{v}) = \sum_{i=1}^k \sigma(\mathbf{u}_i, \mathbf{v}) \mathbf{u}_i$$

where $\{\mathbf{u}_i\}_{i=1}^k$ is an orthonormal basis for U.

When the vector space is real and dimensions are low, we know that $\operatorname{proj}_U \mathbf{v}$ is the closest point in U to $\mathbf{v}$. For instance, to travel the shortest route from a point P to a plane α in $\mathbb{R}^3$, we drop a perpendicular to the plane from P. The point

7.9. A Fundamental Theorem

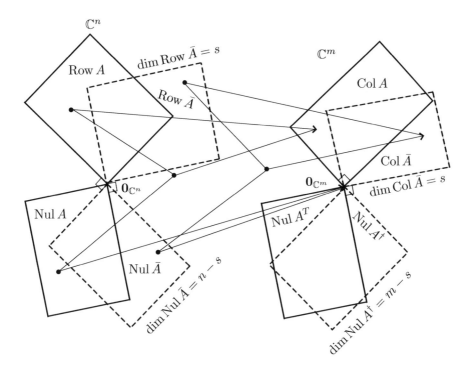

Figure 7.4. Orthogonal complements in the domain and codomain of a matrix transformation $\mathbb{C}^n \to \mathbb{C}^m$.

in the plane where the perpendicular lands is the projection of P onto α, so it is the point in the plane closest to P. The next theorem is the general result. Before stating it, we note the following consequence of Lemma 7.34.

If $\mathcal{B} = \{\mathbf{u}_i\}_{i=1}^n$ is an orthonormal basis for an inner product space (V, σ) and $\mathbf{v}$ is arbitrary in V, then

$$\|\mathbf{v}\|^2 = \sigma(\mathbf{v}, \mathbf{v}) = \sigma\left(\sum_{i=1}^n \sigma(\mathbf{u}_i, \mathbf{v})\mathbf{u}_i, \sum_{j=1}^n \sigma(\mathbf{u}_j, \mathbf{v})\mathbf{u}_j\right)$$

$$= \sum_{i=1}^n \sigma(\sigma(\mathbf{u}_i, \mathbf{v})\mathbf{u}_i, \sigma(\mathbf{u}_i, \mathbf{v})\mathbf{u}_i) = \sum_{i=1}^n \overline{\sigma(\mathbf{u}_i, \mathbf{v})}\sigma(\mathbf{u}_i, \mathbf{v}) = \sum_{i=1}^n |\sigma(\mathbf{u}_i, \mathbf{v})|^2.$$

Theorem 7.75. *If $\mathbf{v}$ belongs to a finite-dimensional inner product space (V, σ) and U is a subspace in V, then the minimal distance between $\mathbf{v}$ and any vector in U is $\|\mathbf{v} - \mathrm{proj}_U \mathbf{v}\|$.*

Proof. Suppose everything is as hypothesized. Let $\mathcal{B} = \{\mathbf{u}_i\}_{i=1}^n$ be an orthonormal basis for V where $\{\mathbf{u}_i\}_{i=1}^k$ is an orthonormal basis for U. We have

$$\mathbf{v} = \sum_{i=1}^n \sigma(\mathbf{u}_i, \mathbf{v})\mathbf{u}_i.$$

If $\mathbf{x}$ is arbitrary in U, then $\sigma(\mathbf{x}, \mathbf{u}_i) = 0$ for $i \in \{k+1, \ldots, n\}$. This means
$$\|\mathbf{v} - \mathbf{x}\|^2 = \sum_{i=1}^{k} |\sigma(\mathbf{v} - \mathbf{x}, \mathbf{u}_i)|^2 + \sum_{i=k+1}^{n} |\sigma(\mathbf{v}, \mathbf{u}_i)|^2.$$
Comparing to
$$\|\mathbf{v} - \mathrm{proj}_U \mathbf{v}\|^2 = \sum_{i=k+1}^{n} |\sigma(\mathbf{v}, \mathbf{u}_i)|^2,$$
we see that for arbitrary $\mathbf{x}$ in U, $\|\mathbf{v} - \mathrm{proj}_U \mathbf{v}\|^2 \leq \|\mathbf{v} - \mathbf{x}\|^2$, which proves the result. $\square$

The following corollary comes out of the proof of Theorem 7.75.

Corollary 7.76. If (V, σ) is a finite-dimensional inner product space with subspace U, then for any $\mathbf{v}$ in V,
$$\mathbf{v} - \mathrm{proj}_U \mathbf{v} \in U^\perp.$$

Let A belong to $M_{m,n}(\mathbb{F})$, where $\mathbb{F}$ is either $\mathbb{R}$ or $\mathbb{C}$. If $A\mathbf{x} = \mathbf{b}$ does not have a solution, it is because $\mathbf{b}$ is not in $\mathrm{Col}\, A$. We are guaranteed a solution to $A\mathbf{x} = \mathrm{proj}_U \mathbf{b}$, though, when $U = \mathrm{Col}\, A$. Moreover, if $A\mathbf{y} = \mathrm{proj}_U \mathbf{b}$, then $\|A\mathbf{y} - \mathbf{b}\|$, hence, $\|A\mathbf{y} - \mathbf{b}\|^2$, is minimized.

Definition 7.77. A **least-squares solution** to $A\mathbf{x} = \mathbf{b}$ is $\mathbf{y}$ that satisfies $A\mathbf{y} = \mathrm{proj}_U \mathbf{b}$, where $U = \mathrm{Col}\, A$. A least-squares solution to $A\mathbf{x} = \mathbf{b}$ is also called a **least-squares approximation** to $\mathbf{x}$ so that $A\mathbf{x} = \mathbf{b}$.

Suppose $\mathbf{y}$ satisfies $A\mathbf{y} = \mathrm{proj}_U \mathbf{b}$, where $U = \mathrm{Col}\, A$. Since $\mathrm{Nul}\, A^\dagger = (\mathrm{Col}\, A)^\perp$, Corollary 7.76 implies that
$$A^\dagger(\mathbf{b} - A\mathbf{y}) = \mathbf{0},$$
thus, that
(7.11) $$A^\dagger \mathbf{b} = A^\dagger A \mathbf{y}.$$
We do not expect A to be invertible, but we will see in a moment that when the columns of A are linearly independent, $A^\dagger A$ is invertible. In a system of linear equations in which there are more equations than unknowns, this is often the case.

Theorem 7.78. If A is an $m \times n$ matrix with rank n, then $A^\dagger A$ is nonsingular. In this case, the least-squares solution to $A\mathbf{x} = \mathbf{b}$ is $\mathbf{y} = (A^\dagger A)^{-1} A^\dagger \mathbf{b}$.

Proof. Let A be as hypothesized, with rank n. Suppose $\mathbf{x}$ in $\mathbb{F}^n$ is arbitrary with
$$A^\dagger A \mathbf{x} = \mathbf{0}.$$
Multiplying both sides of the equation on the left by $\mathbf{x}^\dagger$, we have
$$\mathbf{x}^\dagger A^\dagger A \mathbf{x} = \mathbf{x}^\dagger \mathbf{0} = \mathbf{0}.$$
From here we have
$$\mathbf{x}^\dagger A^\dagger A \mathbf{x} = (A\mathbf{x})^\dagger (A\mathbf{x}) = \|A\mathbf{x}\|^2 = 0.$$
It follows that $A\mathbf{x} = \mathbf{0}$. Since the columns of A are linearly independent, this means $\mathbf{x} = \mathbf{0}$. We conclude that the only solution to $A^\dagger A \mathbf{x} = \mathbf{0}$ is $\mathbf{x} = \mathbf{0}$, thus, that $A^\dagger A$ is nonsingular. This proves the first statement of the theorem.

7.9. A Fundamental Theorem

Let $U = \operatorname{Col} A$. Appealing to (7.11) we have $A^\dagger \mathbf{b} = A^\dagger A \mathbf{y}$, when $\mathbf{y}$ is a least-squares solution to $A\mathbf{x} = \mathbf{b}$. Multiplying both sides of the equation by $(A^\dagger A)^{-1}$, we get
$$(A^\dagger A)^{-1} A^\dagger \mathbf{b} = \mathbf{y}.$$
This proves the second statement of the theorem. □

A least-squares solution is always available but is not always unique. We are guaranteed exactly one least-squares solution to $A\mathbf{x} = \mathbf{b}$ precisely when $A^\dagger A$ is invertible, that is, when the columns of A are linearly independent.

Example 7.79. Here we find the least-squares solution to $A\mathbf{x} = \mathbf{b}$ if
$$A = \begin{bmatrix} 1 & 0 \\ 0 & 1 \\ 1 & 1 \end{bmatrix} \in M_{3,2}(\mathbb{R})$$
and $\mathbf{b} = (-1, 1, 3)$ in $\mathbb{R}^3$. Since the field is $\mathbb{R}$, we have $A^\dagger = A^T$. Since A has rank 2, we can use $\mathbf{y} = (A^T A)^{-1} A^T \mathbf{b}$:
$$A^T A = \begin{bmatrix} 1 & 0 & 1 \\ 0 & 1 & 1 \end{bmatrix} \begin{bmatrix} 1 & 0 \\ 0 & 1 \\ 1 & 1 \end{bmatrix} = \begin{bmatrix} 2 & 1 \\ 1 & 2 \end{bmatrix}.$$
It follows that $(A^T A)^{-1} = \dfrac{1}{3} \begin{bmatrix} 2 & -1 \\ -1 & 2 \end{bmatrix} = \begin{bmatrix} 2/3 & -1/3 \\ -1/3 & 2/3 \end{bmatrix}.$

We also have $A^T \mathbf{b} = (2, 4)$, so $\mathbf{y} = \begin{bmatrix} 2/3 & -1/3 \\ -1/3 & 2/3 \end{bmatrix} \begin{bmatrix} 2 \\ 4 \end{bmatrix} = \begin{bmatrix} 0 \\ 2 \end{bmatrix}.$ This means, for instance, that if $U = \operatorname{Col} A$, $\operatorname{proj}_U \mathbf{b} = (0, 2, 2)$.

One algorithm for finding the least-squares solution to $A\mathbf{x} = \mathbf{b}$ is to calculate an orthonormal basis for $\operatorname{Col} A = U$, then find $\operatorname{proj}_U \mathbf{b}$, and finally, to solve $A\mathbf{y} = \operatorname{proj}_U \mathbf{b}$. This is easy to remember and to describe but too calculation intensive to be of practical use. In applications, other algorithms are used.

The system of equations associated to the matrix equation $A^\dagger A \mathbf{y} = A^\dagger \mathbf{b}$ is the system of **normal equations** associated to $A\mathbf{x} = \mathbf{b}$.

Example 7.80. Let $A = \begin{bmatrix} 1 & 2 \\ 0 & -2 \\ 3 & 2 \end{bmatrix}$ belong to $M_{3,2}(\mathbb{R})$. Let $\mathbf{b} = (3, 3, 1)$ be in $\mathbb{R}^3$. We can see that rank $A = 2$ and that $\mathbf{b}$ does not belong to $\operatorname{Col} A$. Here we identify and solve the system of normal equations associated to $A\mathbf{x} = \mathbf{b}$.

We have $A^T A = \begin{bmatrix} 10 & 8 \\ 8 & 12 \end{bmatrix}$. We also find $A^T \mathbf{b} = (6, 2)$. The augmented matrix associated to $A^T A \mathbf{y} = A^T \mathbf{b}$ is then
$$\left[\begin{array}{cc|c} 10 & 8 & 6 \\ 8 & 12 & 2 \end{array}\right].$$
The normal equations are then
$$\begin{array}{rcl} 10y_1 + 8y_2 & = & 6 \\ 8y_1 + 12y_2 & = & 2. \end{array}$$

The solution is $\mathbf{y} = (1, -1/2)$.

We could also use
$$\mathbf{y} = (A^T A)^{-1} A^T \mathbf{b}.$$

We have $(A^T A)^{-1} = \dfrac{1}{28} \begin{bmatrix} 6 & -4 \\ -4 & 5 \end{bmatrix}$ so that

$$\mathbf{y} = \frac{1}{28} \begin{bmatrix} 6 & -4 \\ -4 & 5 \end{bmatrix} \begin{bmatrix} 6 \\ 2 \end{bmatrix} = \begin{bmatrix} 1 \\ -1/2 \end{bmatrix},$$

as it should be!

Exercises 7.9.

7.9.1. Suppose we are to solve a system of linear equations with matrix equation $A\mathbf{x} = \mathbf{b}$, where $\mathbf{b}$ is orthogonal to $\operatorname{Col} A$. What system of equations is associated to a least-squares solution?

7.9.2. Let $A = \begin{bmatrix} 1 & 2 & -1 & 3 \\ 2 & 4 & 1 & -2 \\ 3 & 6 & 3 & -7 \end{bmatrix}$.
  (a) Find an orthogonal basis for $\operatorname{Col} A$.
  (b) Find an orthogonal basis for $\operatorname{Nul} A$.
  (c) Find an orthogonal basis for $\operatorname{Row} A$.
  (d) Find an orthogonal basis for $\operatorname{Nul} A^T$.

7.9.3. Find a basis for each of the four fundamental subspaces associated to
$$A = \begin{bmatrix} 1 & 2 & 1 & 3 & 1 & 2 \\ 2 & 5 & 5 & 6 & 4 & 5 \\ 3 & 7 & 6 & 11 & 6 & 9 \\ 1 & 5 & 10 & 8 & 9 & 9 \\ 2 & 6 & 8 & 11 & 9 & 12 \end{bmatrix}.$$

7.9.4. Redo Example 7.79, this time calculating $\operatorname{proj}_U \mathbf{b}$ and solving $A\mathbf{y} = \operatorname{proj}_U \mathbf{b}$ using row reduction.

7.9.5. Find the normal equations associated to the matrix equation in Example 7.79.

7.9.6. In Example 7.79, we saw that $A^T A$ was symmetric. Show that $A^T A$ is symmetric for every $m \times n$ matrix A.

7.9.7. Let A belong to $M_{m,n}(\mathbb{C})$. Show that $A^\dagger A$ is invertible if and only if the columns of A are linearly independent.

7.9.8. Verify that $\mathbf{v} - \operatorname{proj}_U \mathbf{v}$ is in $U^\perp$, when U is a subspace of a finite-dimensional inner product space (V, σ) and $\mathbf{v}$ is in V.

7.9. A Fundamental Theorem

7.9.9. Let $A = \begin{bmatrix} 1 & 1 & 0 \\ 1 & 1 & 0 \\ 1 & 0 & 1 \\ 1 & 0 & 1 \\ 1 & 0 & 1 \end{bmatrix}$, and let $\mathbf{b} = (-3, -1, 2, 0, 5)$.

(a) Find an orthonormal basis for $U = \operatorname{Col} A$.
(b) Find $\operatorname{proj}_U \mathbf{b}$.
(c) Find $A^T A$.
(d) Find all least-squares solutions to $A\mathbf{x} = \mathbf{b}$.

Chapter 8

The Life of a Linear Operator

Paul Halmos noted in [8] that linear transformations are "the objects that really make linear algebra interesting." Here we study linear transformations as individual mappings, introducing the notion of an eigenspace, a tool that can help reveal the nature of a linear operator. We also look at some classes of linear transformations that do not yield readily to application of this device.

Coordinates allow us to work with linear transformations on finite-dimensional vector spaces arithmetically. Arithmetic is a powerful tool, powerful enough to enable us to see that when $V \cong \mathbb{F}^n$, $\mathcal{L}(V)$ and $M_n(\mathbb{F})$ are isomorphic $\mathbb{F}$-algebras. A given linear transformation, though, may be better understood geometrically or analytically. For instance $f \to f'$, the differentiation operator, is a linear transformation, as is any rotation of $\mathbb{R}^3$ about the origin. While coordinates facilitate calculations, they may obscure the essential nature of a given linear transformation. How do the numbers in the matrix indicate whether the linear transformation it represents is a rotation or reflection or projection? One of our jobs here is to get closer to understanding how a matrix representation distills the nature of a linear transformation.

The subalgebra of $\mathcal{L}(V)$ generated by single linear transformation, L, is an important device in our analysis. We can think of this subalgebra as the set of polynomials in L, over the underlying field. Since polynomials play such a large role here, we start with a brief review of the terminology and major theorems associated to factoring polynomials.

8.1. Factoring Polynomials

We have discussed the fact that $\mathbb{F}[t]$ is an $\mathbb{F}$-algebra. Here we review some details about multiplication, division, and factoring in $\mathbb{F}[t]$. This will enable us to use polynomials more effectively in our study of linear operators. Our goal is to compile statements of the major theorems without proofs. We start with familiar terminology.

Definition 8.1. A **root** or **zero** of a polynomial $p(t)$ in $\mathbb{F}[t]$ is c in $\mathbb{F}$ such that $p(c) = 0$.

Division in an $\mathbb{F}$-algebra requires a way to compare elements in the algebra. In $\mathbb{F}[t]$, we can use degree. The degree of a nonzero polynomial is a nonnegative integer, and degree is additive across multiplication. By this we mean that if $\deg p(t) = n$ and $\deg q(t) = m$, where $p(t)$ and $q(t)$ belong to $\mathbb{F}[t]$, then $\deg p(t)q(t) = m + n$.

The *division algorithm* guarantees that we can perform division with remainder in $\mathbb{F}[t]$, just as we do in $\mathbb{R}[t]$. When $q(t)$ is a nonzero polynomial, with $\deg q(t) \leq \deg p(t)$, dividing $p(t)$ by $q(t)$ yields a remainder with degree strictly less than $\deg q(t)$. We learn to divide polynomials in precalculus. Nothing about this changes when we work over arbitrary fields.

Example 8.2. Here we divide $2t - 1$ into $3t^2 + t - 2$ over $\mathbb{F}_5$. Since $(2)(4) = 3$, we get $4t$ when we divide $2t$ into $3t^2$. The remainder is then
$$3t^2 + t - 2 - (3t^2 - 4t) = t - 2 + 4t = -2 = 3.$$
This means that
$$\frac{3t^2 + t - 2}{2t - 1} = 4t + \frac{3}{2t - 1}.$$
We may also say that the quotient is $4t$ with remainder 3.

A **nonconstant factor** of a polynomial $p(t)$ in $\mathbb{F}[t]$ is $q(t)$ in $\mathbb{F}[t]$ with positive degree such that there is a remainder of zero when we divide $p(t)$ by $q(t)$. In this case, we may say that $q(t)$ **divides** $p(t)$.

One of the most important theorems that we see in high school algebra is that c is a root of a polynomial $p(t)$ if and only if $t - c$ is a factor of $p(t)$. This tells us, for instance, that since $1 + t^2$ has no root in $\mathbb{R}$, it cannot be factored over $\mathbb{R}$. As a consequence of the division algorithm, this theorem applies to polynomials in $\mathbb{F}[t]$ for an arbitrary field $\mathbb{F}$: A root of any polynomial in $\mathbb{F}[t]$ corresponds to a degree one factor of the polynomial.

If $q(t)$ is a nonconstant factor of $p(t)$, then we say that $q(t)$ is a factor of $p(t)$ with **multiplicity** k, when k is the maximal positive integer such that $q(t)^k$ divides $p(t)$. If $t - c$ is a factor of $p(t)$ with multiplicity k, we say that c has multiplicity k as a root of $p(t)$. When we count the irreducible factors, or the zeroes of a polynomial, we may count multiplicity. This means that if a factor or root has multiplicity k, we count it k times. For example, over $\mathbb{R}$, the roots of $p(t) = t^3 - t^2 - t + 1$ are 1 with multiplicity 2, and -1.

Definition 8.3. A field $\mathbb{F}$ is **algebraically closed** provided every polynomial of degree one or more in $\mathbb{F}[t]$ has zeroes in $\mathbb{F}$.

We can also say that a field $\mathbb{F}$ is algebraically closed provided every polynomial in $\mathbb{F}[t]$ of degree $n > 0$ has n degree one factors in $\mathbb{F}[t]$, counting multiplicity, or n zeroes in $\mathbb{F}$, counting multiplicity.

Definition 8.4. A nonconstant polynomial $p(t)$ in $\mathbb{F}[t]$ is **irreducible over** $\mathbb{F}$ provided $p(t)$ has no nonconstant factor $q(t)$ in $\mathbb{F}[t]$ for which $\deg q(t) < \deg p(t)$. A nonconstant polynomial $p(t)$ in $\mathbb{F}[t]$ is **reducible** if it is not irreducible.

8.2. The Minimal Polynomial

Degree one polynomials are, by definition, irreducible over any field. When $\mathbb{F}$ is algebraically closed, the only irreducible polynomials over $\mathbb{F}$ are those of degree one.

Since $p(t) = t^2 + 1$ has no degree one factors over $\mathbb{R}$, it is irreducible in $\mathbb{R}[t]$. Evidently, $\mathbb{R}$ is not algebraically closed. The **Fundamental Theorem of Algebra** says that $\mathbb{C}$ is algebraically closed. The complex numbers are the **algebraic closure** of $\mathbb{R}$, meaning that $\mathbb{R} \subseteq \mathbb{C}$, $\mathbb{C}$ is algebraically closed, and there is no proper subfield of $\mathbb{C}$ containing $\mathbb{R}$, which is algebraically closed.

An important theorem in algebra says that every field $\mathbb{F}$ has a well-defined algebraic closure. We designate the algebraic closure of $\mathbb{F}$ by $\overline{\mathbb{F}}$. For instance,
$$\mathbb{C} = \overline{\mathbb{R}}.$$

A second important theorem in algebra says that the algebraic closure of any field is infinite. In particular, when p is prime, $\mathbb{F}_p$ is not algebraically closed.

Factoring polynomials is notoriously difficult but by extending the underlying field, we can guarantee that a degree n polynomial has n degree one factors, counting multiplicity. In our pursuit here, we stick with polynomials of low degree, or ones that have easily accessible zeroes. Over $\mathbb{C}$, for example, $t^2 + 1 = (t-i)(t+i)$.

Exercises 8.1.

8.1.1. Factor $3t^2 + t - 2$ over $\mathbb{F}_5$.

8.1.2. Show that $t^2 + 2$ and $t^2 + 3$ are irreducible over $\mathbb{F}_5$.

8.1.3. Find the quotient and remainder when we divide $p(t) = 2 + t + 2t^2$ by $t + 1$, first in $\mathbb{F}_3[t]$, then in $\mathbb{F}_5[t]$, then in $\mathbb{R}[t]$.

8.1.4. Find any zeroes of $p(t) = 2 + t + 2t^2$, first in $\mathbb{F}_3$, then in $\mathbb{F}_5$, then in $\mathbb{R}$, and finally, in $\mathbb{C}$.

8.2. The Minimal Polynomial

A polynomial combines the operations that define an algebra: addition, scaling, and multiplication. Given $p(t)$ in $\mathbb{F}[t]$ and $\mathbf{a}$ in an associative $\mathbb{F}$-algebra with identity, $\mathcal{A}$, $p(\mathbf{a})$ is a well-defined element in $\mathcal{A}$. We may say that $p(\mathbf{a})$ is a *polynomial in* $\mathbf{a}$. In this section, we look at polynomials in linear operators and in square matrices. We start with formalities and some general observations about associative $\mathbb{F}$-algebras with identity.

Let $\mathcal{A}$ be an associative $\mathbb{F}$-algebra with identity element, ι. Given $\mathbf{a}$ in $\mathcal{A}$, we can define a linear transformation $\mathbb{F}[t] \to \mathcal{A}$ by extending linearly from the rule
$$1 \mapsto \iota = \mathbf{a}^0, \quad t^k \mapsto \mathbf{a}^k, \forall k \in \mathbb{Z}_+.$$

We denote the image of $p(t)$ under the mapping by $p(\mathbf{a})$. Evidently, $p(t) \mapsto p(\mathbf{a})$ is an $\mathbb{F}$-algebra homomorphism. The range of the mapping is $\mathbb{F}[\mathbf{a}] \subseteq \mathcal{A}$, the space of polynomials in $\mathbf{a}$. Since powers of $\mathbf{a}$ commute, Theorem 3.32 implies that $\mathbb{F}[\mathbf{a}]$ is a commutative subalgebra of $\mathcal{A}$ with identity. The kernel of the mapping is the **annihilator** of $\mathbf{a}$ in $\mathbb{F}[t]$. Any $p(t)$ in the annihilator of $\mathbf{a}$ has the property $p(\mathbf{a}) = \mathbf{0}_\mathcal{A}$; we say that $p(t)$ annihilates $\mathbf{a}$.

When we defined a projection as a nontrivial linear operator L on a vector space V such that $L^2 = L$, we were defining a projection as a mapping annihilated by $p(t) = t^2 - t$.

Example 8.5. Consider $A = \begin{bmatrix} 0 & 1 \\ 0 & 0 \end{bmatrix}$ in $M_2(\mathbb{R}^2)$. If c is in $\mathbb{R}$ and $p(t) = t - c$, then

$$p(A) = A - cI = \begin{bmatrix} -c & 1 \\ 0 & -c \end{bmatrix}$$

is never equal to the zero matrix. On the other hand, A^2 is the zero matrix. In fact, the annihilator of A is

$$\{t^2 q(t) \mid q(t) \in \mathbb{R}[t]\}.$$

Lemma 8.6. Let $\mathcal{A}$ be an associative $\mathbb{F}$-algebra with identity, and let f be an algebra homomorphism defined on $\mathcal{A}$. If $p(t)$ belongs to $\mathbb{F}[t]$, then

$$f(p(\mathbf{a})) = p(f(\mathbf{a}))$$

for all $\mathbf{a}$ in $\mathcal{A}$.

Proof. The proof is immediate since an algebra homomorphism commutes with scaling, addition, and multiplication. □

Example 8.7. Consider L in $\mathcal{L}(\mathbb{R}^2)$ given by

$$L(x, y) = (y, 0).$$

The standard matrix representation of L is A, as given in Example 8.5. Since $L \mapsto [\![L]\!]_\mathcal{B}$ is an $\mathbb{F}$-algebra isomorphism, the lemma guarantees that

$$[\![p(L)]\!]_\mathcal{B} = p([\![L]\!]_\mathcal{B})$$

for all $p(t)$ in $\mathbb{R}[t]$. It follows that $p(A)$ is the zero matrix in $M_2(\mathbb{R})$ if and only if $p(L)$ is the zero mapping on $\mathbb{R}^2$, thus, that A and L have the same annihilator in $\mathbb{R}[t]$. Notice in fact that

$$L^2(x, y) = L(y, 0) = (0, 0).$$

We need not restrict ourselves to finite-dimensional spaces.

Example 8.8. Consider the differentiation operator, D, on $V = C^\infty(\mathbb{R})$. Note that D^n maps a function to its nth derivative. If $p(t) = t^2 - 5t + 6$ in $\mathbb{R}[t]$, then $p(D) = D^2 - 5D + 6I$ in $\mathbb{R}[D]$. For example, $p(D)(e^{-t}) = 12e^{-t}$ and $p(D)(e^{2t}) = 0$.

Proof of the following is an exercise.

Lemma 8.9. If $\mathcal{A}$ is an associative algebra with identity and $\mathbf{a}$ in $\mathcal{A}$ is invertible, then the mapping $\mathbf{x} \mapsto \mathbf{axa}^{-1}$ is an algebra automorphism on $\mathcal{A}$.

The mapping described in Lemma 8.9, $\mathbf{x} \mapsto \mathbf{axa}^{-1}$, is called **conjugation by a**. Putting this together with Lemma 8.6, we see that if $A \tilde{s} B$ for A and B in $M_n(\mathbb{F})$, then A and B have the same annihilators in $\mathbb{F}[t]$. The next theorem says a bit more but follows likewise on the heels of Lemma 8.6 and Lemma 8.9.

8.2. The Minimal Polynomial

Theorem 8.10. Let V be a vector space over $\mathbb{F}$. If L belongs to $\mathcal{L}(V)$, its annihilator in $\mathbb{F}[t]$ is the annihilator for every linear operator in the similarity class determined by L in $\mathcal{L}(V)$. If $V \cong \mathbb{F}^n$, then L and any matrix representation of L have the same annihilator in $\mathbb{F}[t]$.

We designate the zero mapping in both $\mathcal{L}(V)$ and $M_n(\mathbb{F})$ by O. Context will clarify whether the reference is to a linear operator or a square zero matrix.

Example 8.11. Let L in $\mathcal{L}(\mathbb{R}^3)$ be given by
$$L(x,y,z) = (-x, x-y, -z).$$
We have
$$L^2(x,y,z) = L(-x, x-y, -z) = (x, -2x+y, z)$$
so that
$$(L^2 + 2L + I)(x,y,z) = (x, -2x+y, z) + (-2x, 2x-2y, -2z) + (x,y,z) = (0,0,0).$$
We then have
$$L^2 + 2L + I = O$$
so $p(t) = t^2 + 2t + 1$ annihilates L, every operator similar to L, and every matrix representation of L. We can check this against the standard matrix representation of L,
$$A = \begin{bmatrix} -1 & 0 & 0 \\ 1 & -1 & 0 \\ 0 & 0 & -1 \end{bmatrix}.$$
We have
$$A^2 + 2A + I_3 = \begin{bmatrix} 1 & 0 & 0 \\ -2 & 1 & 0 \\ 0 & 0 & 1 \end{bmatrix} + \begin{bmatrix} -2 & 0 & 0 \\ 2 & -2 & 0 \\ 0 & 0 & -2 \end{bmatrix} + \begin{bmatrix} 1 & 0 & 0 \\ 0 & 1 & 0 \\ 0 & 0 & 1 \end{bmatrix} = O,$$
as promised by Theorem 8.10.

Suppose $V \cong \mathbb{F}^n$ so that $\dim \mathcal{L}(V) = n^2$. As a subspace of $\mathcal{L}(V)$, $\mathbb{F}[L]$ has dimension less than or equal to n^2. This means that for some positive integer k, $\{I, L, L^2, \ldots, L^k\}$ is linearly dependent. Let k in $\mathbb{Z}_+$ be minimal so that, for some a_i in $\mathbb{F}$, a_i not all zero,
$$a_0 I + a_1 L + \cdots + a_{k-1} L^{k-1} + L^k = O.$$
Our next lemma justifies calling $q(t) = a_0 + a_1 t + \cdots + a_{k-1} t^{k-1} + t^k$ the **minimal polynomial** for L.

Note that we define the minimal polynomial to be monic.

Lemma 8.12. If V is finite-dimensional and L belongs to $\mathcal{L}(V)$, then the minimal polynomial for L is unique.

Proof. Suppose V and L are as hypothesized and that both $q_1(t)$ and $q_2(t)$ are minimal polynomials for L. If $\deg q_1(t) \neq \deg q_2(t)$, then one has smaller degree than the other which is impossible by the definition of minimal polynomial. It must then be that $\deg q_1(t) = \deg q_2(t)$. If $q_1(t)$ and $q_2(t)$ are different, then since both are monic, $q_1(t) - q_2(t) = q(t)$ is nonzero with $q(L) = O$ and $\deg q(t) < \deg q_i(t)$,

$i = 1, 2$. This again contradicts the definition of $q_1(t)$ and $q_2(t)$ as minimal polynomials. We conclude that $q_1(t) = q_2(t)$, proving the result. $\square$

Consider L as given in Example 8.11. Since $p(t) = t^2 + 2t + 1 = (t+1)^2$ annihilates L, the minimal polynomial for L has degree two or less. A degree one monic polynomial has the form $t - c$ for some scalar, c. An operator L annihilated by $t - c$ satisfies $L - cI = O$, so $L = cI$. This is a **scalar mapping**. Since L is not a scalar mapping, uniqueness dictates that $t^2 + 2t + 1$ must be its minimal polynomial.

The next result is immediate by Theorem 8.10.

Corollary 8.13. The minimal polynomial for L in $\mathcal{L}(V)$ is the minimal polynomial for every linear operator on V that is similar to L and, when V is finite-dimensional, for every matrix representation of L.

While similar matrices must have the same minimal polynomial, the converse is not true.

Example 8.14. Let $A = \begin{bmatrix} 0 & 1 & 0 & 0 \\ 0 & 0 & 0 & 0 \\ 0 & 0 & 0 & 0 \\ 0 & 0 & 0 & 0 \end{bmatrix}$. It is easy to see that t^2 is the minimal polynomial for A.

Let $B = \begin{bmatrix} 0 & 1 & 0 & 0 \\ 0 & 0 & 0 & 0 \\ 0 & 0 & 0 & 1 \\ 0 & 0 & 0 & 0 \end{bmatrix}$. Like A, B has minimal polynomial t^2.

We can see that A and B are not similar by noting that $\operatorname{rank} A = 1$ and $\operatorname{rank} B = 2$.

We see in the next section that there is a subspace of $V \cong \mathbb{F}^n$ associated to each root of the minimal polynomial for L in $\mathcal{L}(V)$. Since we cannot expect $\mathbb{F}$ to contain the roots of any polynomial, unless $\mathbb{F}$ is algebraically closed, we want to understand how it affects V and L should we feel called upon to extend the underlying scalar field.

Suppose V is defined over $\mathbb{F}$ and that $\widetilde{\mathbb{F}}$ is a field that properly contains $\mathbb{F}$. We define $\widetilde{V}$ to be the span of V over $\widetilde{\mathbb{F}}$. For example, if $V = \mathbb{Q}^2$ and $\widetilde{\mathbb{Q}} = \mathbb{R}$, then $\widetilde{V} = \mathbb{R}^2$. While $(\sqrt{2}, 0)$ is in $\widetilde{V}$ and not in V, $\widetilde{V}$ and V have the same dimensions over their respective fields. One is not a subspace of the other, but we can use a basis for V over $\mathbb{F}$ to define a basis for $\widetilde{V}$ over $\widetilde{\mathbb{F}}$. We can take the same approach to defining $\widetilde{L}$ in $\mathcal{L}(\widetilde{V})$. Given L in $\mathcal{L}(V)$, we define $\widetilde{L}$ in $\mathcal{L}(\widetilde{V})$ by extending linearly from $L(\mathbf{v})$ over $\widetilde{\mathbb{F}}$ for $\mathbf{v}$ in V.

We call this process of passing from V to $\widetilde{V}$, and from L to $\widetilde{L}$, extending the scalar field.

This sounds more complicated than it is. For instance, we can view $A = \begin{bmatrix} 1 & -3 \\ 1 & 1 \end{bmatrix}$ as defining a linear transformation on $\mathbb{Q}^2$, $\mathbb{R}^2$, or $\mathbb{C}^2$. Whatever the underlying field, the minimal polynomial for A is $t^2 - 2t + 4$.

8.2. The Minimal Polynomial

The next theorem shows that the annihilator of L in $\mathbb{F}[t]$ is always the same as the annihilator of $\widetilde{L}$ in $\widetilde{\mathbb{F}}[t]$.

Theorem 8.15. *Let V be a vector space over $\mathbb{F}$ and let L belong to $\mathcal{L}(V)$. Let $\widetilde{\mathbb{F}}$ be a field that contains $\mathbb{F}$ as a subfield. Extend the scalar field to define $\widetilde{V}$ and $\widetilde{L}$ over $\widetilde{\mathbb{F}}$. If $p(t)$ is in $\mathbb{F}[t]$, then $p(L) = O$ if and only if $p(\widetilde{L}) = O$.*

Proof. Let everything be as hypothesized. Given $p(t) = a_0 + a_1 t + \cdots + a_r t^r$ in $\mathbb{F}[t]$, $p(L) = O$ if and only if
$$(8.1) \qquad p(L)(\mathbf{v}) = a_0 L(\mathbf{v}) + \cdots + a_r L^r(\mathbf{v}) = \mathbf{0}_V,$$
for all $\mathbf{v}$ in V. Since V is a spanning set for $\widetilde{V}$ over $\widetilde{\mathbb{F}}$ and because
$$\widetilde{L}^j(\mathbf{v}) = a_j L^j(\mathbf{v})$$
for all $\mathbf{v}$ in V and j in $\mathbb{Z}_{\geq 0}$, we see that (8.1) holds for all $\mathbf{v}$ in V if and only if
$$p(\widetilde{L})(\mathbf{x}) = \mathbf{0}_{\widetilde{V}}$$
for all $\mathbf{x}$ in $\widetilde{V}$. It follows that
$$p(L) = O \Leftrightarrow p(\widetilde{L}) = O. \qquad \square$$

The theorem shows that in extending the base field from $\mathbb{F}$, we neither lose nor gain polynomials over $\mathbb{F}$ that annihilate a given operator.

Exercises 8.2.

8.2.1. Why do we assume associativity and the existence of an identity element in $\mathcal{A}$ when we want to evaluate a polynomial at an element in an algebra $\mathcal{A}$?

8.2.2. Let $\mathcal{A}$ be an associative $\mathbb{F}$-algebra with identity. Show that if $\mathcal{W}$ is a subalgebra containing $\mathbf{a}$ in $\mathcal{A}$, then $\mathcal{W}$ contains $\mathbb{F}[\mathbf{a}]$. This is what we mean when we say that $\mathbb{F}[\mathbf{a}]$ is the subalgebra of $\mathcal{A}$ generated by $\mathbf{a}$.

8.2.3. Prove that the annihilator of A in Example 8.5 is $\{t^2 q(t) \mid q(t) \in \mathbb{R}[t]\}$.

8.2.4. Referring to Example 8.8, calculate $p(D)(2t - 3t^2)$ and $p(D)(\sin(2t))$.

8.2.5. Let $\varphi_1 : \mathcal{A}_1 \to \mathcal{A}_2$ and $\varphi_2 : \mathcal{A}_2 \to \mathcal{A}_3$ be $\mathbb{F}$-algebra homomorphisms. Show that $\varphi_2 \circ \varphi_1$ is an $\mathbb{F}$-algebra homomorphism, $\mathcal{A}_1 \to \mathcal{A}_3$. Conclude that under the appropriate hypotheses, the composition of $\mathbb{F}$-algebra isomorphisms is an $\mathbb{F}$-algebra isomorphism.

8.2.6. Prove Lemma 8.9.

8.2.7. Exercise 1.5.11 was to show that $R = \{tp(t) \mid p(t) \in \mathbb{F}[t]\}$ is a subalgebra of $\mathbb{F}[t]$. Prove that for any $q(t)$ in $\mathbb{F}[t]$, $q(t)R \subseteq R$. A subalgebra of $\mathbb{F}[t]$ with this property is an **ideal** in $\mathbb{F}[t]$.

8.2.8. Let $\mathcal{A}$ be an associative $\mathbb{F}$-algebra with identity. Show that if $\mathbf{a}$ is in $\mathcal{A}$, then the annihilator of $\mathbf{a}$ in $\mathbb{F}[t]$ is an ideal in $\mathbb{F}[t]$.

8.2.9. Show that $q(t) = t^2 - \sqrt{2}t + 1$ is the minimal polynomial for the rotation in $\mathbb{R}^2$ through $\pi/4$. (See Exercise 5.5.16.)

8.2.10. Find the minimal polynomial for L in $\mathcal{L}(\mathbb{R}^2)$ given by
$$L(x, y) = (2x + y, 2y).$$

8.2.11. Let $A = \begin{bmatrix} 1 & 2 \\ -1 & 3 \end{bmatrix}$ belong to $\mathcal{L}(\mathbb{R}^2)$. Prove that $q(t) = t^2 - 4t + 5$ is the minimal polynomial for A.

8.2.12. Let $\mathbb{F}$ be any field. Let L_1 in $\mathcal{L}(\mathbb{F}^3)$ be given by

$$L_1 = \begin{bmatrix} 0 & 1 & 0 \\ 0 & 0 & 0 \\ 0 & 0 & 0 \end{bmatrix}.$$

Let L_2 in $\mathcal{L}(\mathbb{F}^3)$ be given by

$$L_2 = \begin{bmatrix} 1 & 0 & 0 \\ 0 & 0 & 0 \\ 0 & 0 & 0 \end{bmatrix}.$$

(a) Show that L_1 and L_2 are not similar.

(b) Show that L_1 is similar to $L_3 = \begin{bmatrix} 0 & 0 & 0 \\ 1 & 0 & 0 \\ 0 & 0 & 0 \end{bmatrix}$.

(c) Show that L_2 is similar to $L_4 = \begin{bmatrix} 0 & 0 & 0 \\ 0 & 0 & 0 \\ 0 & 0 & 1 \end{bmatrix}$.

8.2.13. Let V be a vector space over $\mathbb{F}$ where char $\mathbb{F} \neq 2$. Let W be a hyperspace in V and let W' be any complement to W. The reflection through W along W' is L in $\mathcal{L}(V)$ such that $L(\mathbf{w}) = \mathbf{w}$ for all $\mathbf{w}$ in W and $L(\mathbf{w}') = -\mathbf{w}'$ for $\mathbf{w}'$ in W'. What is the minimal polynomial for L?

8.2.14. We noted that when V is a vector space over $\mathbb{F}$ and $\widetilde{\mathbb{F}}$ is a field containing $\mathbb{F}$, then $\dim_{\mathbb{F}} V = \dim_{\widetilde{\mathbb{F}}} \widetilde{V}$. How is this not a contradiction of Theorem 1.49, which says that if V is a complex vector space, then $\dim_{\mathbb{R}} V = 2 \dim_{\mathbb{C}} V$?

8.3. Eigenvalues

The L-invariant subspaces of V reveal important information about L in $\mathcal{L}(V)$. The simplest L-invariant subspace is spanned by an L-*eigenvector*.

Definition 8.16. Let L belong to $\mathcal{L}(V)$, where V is any vector space over $\mathbb{F}$. An L-**eigenvalue** is a scalar λ in $\mathbb{F}$, such that, for some nonzero $\mathbf{v}$ in V,

(8.2) $$L(\mathbf{v}) = \lambda \mathbf{v}.$$

An L-**eigenvector** is any nonzero $\mathbf{v}$ in V satisfying (8.2) for some L-eigenvalue λ.

Suppose L in $\mathcal{L}(V)$ is a projection onto a proper nontrivial subspace, W. The L-eigenvectors are the nonzero vectors in W — these have eigenvalue 1 — and the nonzero vectors in Ker L — these have eigenvalue 0.

If L in $\mathcal{L}(V)$ is the reflection through W along W', then every nonzero vector in W is an L-eigenvector with eigenvalue 1, and every nonzero vector in W' is an L-eigenvector with eigenvalue -1.

8.3. Eigenvalues

Sometimes we refer to eigenvalues and eigenvectors without emphasizing the association to a particular linear operator. If $\mathbf{v}$ is an L-eigenvector with eigenvalue λ, we may also say that $\mathbf{v}$ is a λ-eigenvector.

Example 8.17. Consider L in $\mathcal{L}(\mathbb{R}^2)$ given by
$$L(x,y) = \left(\frac{-x-4y}{3}, \frac{x+4y}{3}\right).$$
We have
$$L^2(x,y) = L\left(\frac{-x-4y}{3}, \frac{x+4y}{3}\right)$$
$$= \left(\frac{-(-x-4y) - 4(x+4y)}{9}, \frac{(-x-4y) + 4(x+4y)}{9}\right).$$
After sorting through the calculations, we find that
$$L^2 = L$$
so L is a projection. Like all projections, L has eigenvalues 0 and 1. The kernel of L is the set of points on the line $x + 4y = 0$. The nonzero elements in $\operatorname{Ker} L$ are the 0-eigenvectors for L. We can find a 1-eigenvector for L by solving the system of equations
$$x = \frac{-x-4y}{3}, \quad y = \frac{x+4y}{3}$$
which, on rearranging, looks like
$$\begin{aligned} -4x - 4y &= 0 \\ x + y &= 0. \end{aligned}$$
The nonzero points on the line $x + y = 0$ are thus 1-eigenvectors for L.

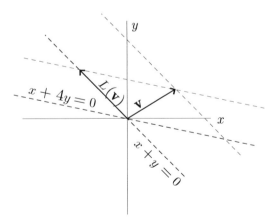

Figure 8.1. L is the projection onto the line $x + y = 0$ along the line $x + 4y = 0$.

The next lemma assures us that an eigenvector is (almost) never alone. The proof is an exercise.

Lemma 8.18. Let L belong to $\mathcal{L}(V)$, where V is defined over $\mathbb{F}$. A nonzero vector $\mathbf{v}$ is a λ-eigenvector for L if and only if every nonzero multiple of $\mathbf{v}$ is a λ-eigenvector for L.

Example 8.19. Let $A = \begin{bmatrix} 1 & 2 \\ 0 & 1 \end{bmatrix}$. The mapping $\mathbf{v} \mapsto A\mathbf{v}$ on $\mathbb{R}^2$ is called a **shear** in the x-direction. (See Figure 8.2.) Since $A\mathbf{e}_1 = \mathbf{e}_1$, $c\mathbf{e}_1$ is an eigenvector for A with eigenvalue 1, for any nonzero c in $\mathbb{R}$.

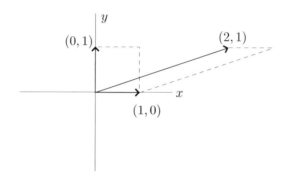

Figure 8.2. The shear on $\mathbb{R}^2$ that fixes $\mathbf{e}_1$ and maps $\mathbf{e}_2$ to $(2,1)$.

Example 8.20. Let D be the differentiation operator on $C^\infty(\mathbb{R})$. Given λ in $\mathbb{R}$,
$$D\left(e^{\lambda t}\right) = \lambda e^{\lambda t}.$$
It follows that $f(t) = e^{\lambda t}$ is a λ-eigenvector for D. The implication is that the differentiation operator on $C^\infty(\mathbb{R})$ has a λ-eigenvector for each real number λ.

While zero may be an eigenvalue for a linear operator, we are careful to insist that the zero vector is not an eigenvector. This is a convention observed throughout the literature, but there is a good reason for it that will reveal itself in due course.

The next result is both evident and central to our approach to finding eigenvalues. The proof is immediate by the definition of eigenvector.

Lemma 8.21. Let V be a vector space over $\mathbb{F}$. Let L belong to $\mathcal{L}(V)$. A scalar λ is an L-eigenvalue if and only if $L - \lambda I$ has nontrivial kernel.

While a single eigenvector spans a 1-dimensional subspace of the ambient vector space, we need only consider a scalar operator, $L(\mathbf{v}) = c\mathbf{v}$ for some fixed c in the underlying scalar field, to see that the set of eigenvectors associated to a single eigenvalue may span a larger subspace.

Definition 8.22. Let L belong to $\mathcal{L}(V)$ and let λ be an L-eigenvalue. The λ-**eigenspace** associated to L is
$$V_\lambda = \operatorname{Ker}(L - \lambda I).$$
The **geometric multiplicity** of λ is $\dim V_\lambda$.

8.3. Eigenvalues

Notice that $V_\lambda = \text{Ker}(L - \lambda I)$ is the union of $\mathbf{0}_V$ and all the λ-eigenvectors for L in $\mathcal{L}(V)$.

The word *eigenspace* may seem exotic but the idea is not. We have defined linear operators in terms of eigenspaces, without using that terminology. This is what we did in Exercise 8.2.13, when we defined a reflection on V as the identity mapping on some hyperspace $W \subseteq V$ and multiplication by -1 on a direct complement of W.

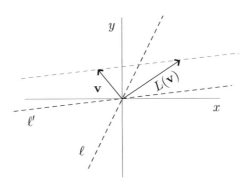

Figure 8.3. A nonorthogonal reflection on $\mathbb{R}^2$ determined by ℓ.

An L-eigenspace is a particular type of L-invariant subspace. As such, an L-eigenspace has properties enjoyed by any L-invariant subspace. For instance, if W is an L-invariant subspace, then for any $\mathbf{w}$ in W, $L(L(\mathbf{w}))$ is also in W. In fact, for any nonnegative integer k, $L^k(\mathbf{w})$ is in W. Since any subspace is closed under linear combination, any L-invariant subspace is $p(L)$-invariant for $p(L)$ in $\mathbb{F}[L]$. When studying L-eigenspaces, or any other L-invariant subspaces, we can toggle freely inside $\mathbb{F}[L]$ without disturbing the underlying invariant subspaces. We frame this as a lemma, leaving the full proof as an exercise.

Lemma 8.23. Let V be a vector space over $\mathbb{F}$ and let L belong to $\mathcal{L}(V)$. If λ is an L-eigenvalue, then both the λ-eigenspace, $\text{Ker}(L - \lambda I)$, and $(L - \lambda I)[V]$ are $p(L)$-invariant for any $p(t)$ in $\mathbb{F}[t]$.

Example 8.24. Let L in $\mathcal{L}(\mathbb{R}^2)$ be given by $L(x, y) = (3x - y, 2x)$. We then have, for instance, that $L(x, x) = (2x, 2x)$ so that $(1, 1)$ is an L-eigenvector with eigenvalue 2. If $L(\mathbf{v}) = 2\mathbf{v}$, then for $\mathbf{v} = (x, y)$, we have

$$3x - y = 2x \quad \text{and} \quad 2x = 2y$$

which we can rewrite as the following system of equations:

$$\begin{aligned} x - y &= 0 \\ 2x - 2y &= 0. \end{aligned}$$

From here we see that the 2-eigenspace for L is $\text{Span}\{(1, 1)\}$.

Our discussion above says that Span$\{(1,1)\}$ is invariant under $p(L)$ for any $p(t)$ in $\mathbb{R}[t]$. Grabbing a polynomial at random, let $p(t) = t^2 + 3t - 5$ so that

$$p(L)(1,1) = (L^2 + 3L - 5I)(1,1) = L^2(1,1) + 3L(1,1) - 5(1,1)$$
$$= 4(1,1) + 3(1,1) - (5,5) = (2,2)$$

which is certainly in Span$\{(1,1)\}$.

Our next theorem implies that the eigenvalue associated to an eigenvector is well-defined.

Theorem 8.25. Let L belong to $\mathcal{L}(V)$. If λ and μ are distinct L-eigenvalues, then $V_\lambda \cap V_\mu = \{\mathbf{0}\}$.

Proof. Let everything be as hypothesized. If $\mathbf{w}$ is in $V_\lambda \cap V_\mu$, then

$$L(\mathbf{w}) = \lambda \mathbf{w} = \mu \mathbf{w}$$

implying that $(\lambda - \mu)\mathbf{w} = \mathbf{0}$. Since $\lambda \neq \mu$, Lemma 1.12 ensures that $\mathbf{w} = \mathbf{0}_V$, which proves the theorem. □

If we were to say that $\mathbf{0}$ is an eigenvector, then since $L(\mathbf{0}) = \lambda \mathbf{0}$ for all λ in $\mathbb{F}$, $\mathbf{0}$ would not have a well-defined eigenvalue. This is the reason we alluded to for denying $\mathbf{0}$ the label "eigenvector."

Since the eigenspaces associated to distinct eigenvalues intersect trivially, the sum of such eigenspaces is direct. The following corollary is an important consequence of this.

Corollary 8.26. Let V be an arbitrary vector space and let L belong to $\mathcal{L}(V)$. Any set of L-eigenvectors associated to distinct L-eigenvalues is linearly independent.

Proof. Let V and L be as hypothesized. Let $\mathbf{v}_1, \ldots, \mathbf{v}_k$ be L-eigenvectors associated to distinct eigenvalues $\lambda_1, \ldots, \lambda_k$. Since the sum of the associated eigenspaces is direct and each $\mathbf{v}_i$ is nonzero, there is a basis for the sum of the eigenspaces for which $\{\mathbf{v}_i\}_{i=1}^k$ is a subset. This is enough to establish that $\{\mathbf{v}_i\}_{i=1}^k$ is linearly independent, which implies the corollary. □

The next theorem says that, like the minimal polynomial, eigenvalues and the associated geometric multiplicities are the same for all operators in a given similarity class. Its proof is virtually identical to the proof of Theorem 8.10. Note that this does not depend on finite-dimensionality.

Theorem 8.27. If V is a vector space over $\mathbb{F}$, then similar operators in $\mathcal{L}(V)$ have the same eigenvalues and the associated eigenspaces are isomorphic.

While the λ-eigenspaces for similar operators are isomorphic, they are not necessarily identical.

Finally, we connect the minimal polynomial to the eigenvalues of a linear operator.

Theorem 8.28. Let $V \cong \mathbb{F}^n$, and let L belong to $\mathcal{L}(V)$. The roots of the minimal polynomial for L over $\mathbb{F}$ are precisely the L-eigenvalues for V over $\mathbb{F}$.

Proof. Let L and V be as hypothesized. Let $q(t)$ be the minimal polynomial for L. Suppose c is a root of $q(t)$ in $\mathbb{F}$. Write $q(t) = (t-c)p(t)$ for some $p(t)$ in $\mathbb{F}[t]$, $\deg p(t) < \deg q(t)$. If there is no nonzero $\mathbf{v}$ in V so that $(L-cI)(\mathbf{v}) = \mathbf{0}_V$, then $\mathbf{v} \mapsto (L-cI)(\mathbf{v})$ is injective, thus, surjective, since $\dim V = n$. Since

$$q(L)(\mathbf{v}) = p(L)(L-cI)(\mathbf{v}) = \mathbf{0}_V,$$

we have that $p(L)(\mathbf{v}) = \mathbf{0}_V$ for all $\mathbf{v}$ in V. This violates the definition of the minimal polynomial, so we conclude that any root for $q(t)$ in $\mathbb{F}$ is an L-eigenvalue.

Now let λ be an L-eigenvalue in $\mathbb{F}$ and let $\mathbf{v}$ be a λ-eigenvector. Factoring over $\mathbb{F}$ if necessary, write the minimal polynomial for L as

$$q(t) = \prod_{i=1}^{k}(t-c_i)^{r_i}.$$

If λ is not among the c_is, then $L - c_i I$ is injective on $\mathrm{Span}\{\mathbf{v}\}$ for all i. As $p(L)$ is the product of injective mappings on $\mathrm{Span}\{\mathbf{v}\}$, $p(L)(\mathbf{v})$ must itself be nonzero. This violates the definition of the minimal polynomial, thus verifies that every L-eigenvalue is a root of the minimal polynomial. $\square$

We have been able to scratch out the minimal polynomial for several operators but have not addressed anything like an algorithm for doing so. In the next section, we describe another device that gives us direct access to the minimal polynomial for an operator on a finite-dimensional space. This, in turn, gives us the means to identify eigenvalues.

Exercises 8.3.

8.3.1. Prove Lemma 8.18.

8.3.2. Write a proof for Lemma 8.23.

8.3.3. Identify a circumstance in which there is exactly one L-eigenvector for a linear operator L in $\mathcal{L}(V)$, associated to a single eigenvalue.

8.3.4. Show that the shear given in Example 8.19 has minimal polynomial $p(t) = t^2 - 2t + 1$.

8.3.5. Prove that the set $\{e^{\lambda t} \mid \lambda \in \mathbb{R}\}$ is linearly independent in $C^\infty(\mathbb{R})$. (This is the more elegant proof we promised in Exercise 5.4.4.)

8.3.6. Find three linearly independent eigenvectors, none of which is an exponential function, for D^2 in $\mathcal{L}(C^\infty(\mathbb{R}))$.

8.3.7. Prove that $\{\cos(\lambda t) \mid \lambda \in \mathbb{R}\} \subseteq C^\infty(\mathbb{R})$ is a linearly independent set.

8.3.8. In Example 8.24 we found that $\mathrm{Span}\{(1,1)\}$ was an L-eigenspace for L in $\mathcal{L}(\mathbb{R}^2)$ given by $L(x,y) = (3x-y, 2x)$.
   (a) Without using calculations, argue that there can be at most one other L-eigenspace.
   (b) Identify the second L-eigenspace.

8.3.9. Write out a proof of Theorem 8.27.

8.3.10. Let R_φ in $\mathcal{L}(\mathbb{R}^2)$ be the rotation through φ radians. Argue that unless $\varphi = k\pi$, for some k in $\mathbb{Z}_{\geq 0}$, R_φ has no eigenvectors.

8.3.11. Let $A = \begin{bmatrix} 3 & -2 \\ 4 & -3 \end{bmatrix}$ in $M_2(\mathbb{R})$. Check that $A^2 = I_2$. Since $A \neq \pm I_2$, its minimal polynomial must be $p(t) = t^2 - 1$. Show that $\mathbf{v} \mapsto A\mathbf{v}$ is a reflection on $\mathbb{R}^2$.

8.3.12. Define $L : \mathbb{R}^2 \to \mathbb{R}^2$ by $L(\mathbf{v}) = A\mathbf{v}$ where $A = \begin{bmatrix} -1/3 & -4/3 \\ 1/3 & 4/3 \end{bmatrix}$. Since $A^2 = A$, $L^2 = L$ so L is a projection. What is the associated direct sum decomposition of $\mathbb{R}^2$?

8.3.13. Let L be a linear operator on V and let λ be an L-eigenvalue in the underlying field $\mathbb{F}$. Show that $\mathrm{Ker}(L - \lambda I)$ and $(L - \lambda I)[V]$ are L-invariant, thus, $p(L)$ invariant for any polynomial $p(t)$ in $\mathbb{F}[t]$.

8.3.14. Let L be a linear operator on V. Show that if λ and μ are distinct L-eigenvalues, then $L - \mu I$ is injective on $V_\lambda = \mathrm{Ker}(L - \lambda I)$.

8.4. The Characteristic Polynomial

While eigenvalues and eigenvectors can be associated to any linear operator, a minimal polynomial is only guaranteed when the underlying vector space is finite-dimensional. Here we consider the case of a linear operator on an arbitrary finite-dimensional vector space. The determinant can guide us to important insights in this setting. Our first result is preliminary.

Lemma 8.29. Let $V \cong \mathbb{F}^n$. If L is in $\mathcal{L}(V)$ and t is an indeterminate, then $\det(L - tI)$ is a degree n polynomial in $\mathbb{F}[t]$.

Proof. Let everything be as hypothesized so that if $A = [a_{ij}] = [\![L]\!]_\mathcal{B}$, for some ordered basis $\mathcal{B}$ in V,

$$\det A = \sum_{\sigma \in \mathcal{S}_n} \mathrm{sgn}\, \sigma\, a_{1\sigma(1)} \cdots a_{n\sigma(n)}.$$

By Observation 6.45, $\det L = \det A$.

If t is an indeterminate, the diagonal entries in $A - tI_n$ are $a_{ii} - t$. When σ is the identity permutation in $\mathcal{S}_n$, the associated product in the alternating sum that is $\det(A - tI_n)$ has the form

$$(a_{11} - t) \cdots (a_{nn} - t).$$

This is a degree n polynomial in t. Every other product associated to a term in the alternating sum that is the determinant has $n - 1$ or fewer factors of the form $a_{ii} - t$, depending on σ. Certainly each of these is also a polynomial in t, of degree $n - 1$ or less. The result follows. □

A linear operator on a finite-dimensional vector space has a nontrivial kernel if and only if its determinant is zero. This informs our next definition.

Definition 8.30. Let $V \cong \mathbb{F}^n$ and let L belong to $\mathcal{L}(V)$. The **characteristic polynomial** for L is $p(t) = \det(L - tI)$. The **characteristic equation** for L is $p(t) = 0$.

8.4. The Characteristic Polynomial

The next theorem clarifies why we identify the characteristic polynomial for a linear operator.

Theorem 8.31. Let $V \cong \mathbb{F}^n$. If L belongs to $\mathcal{L}(V)$, then its eigenvalues are the zeroes of its characteristic polynomial.

Proof. Let L and V be as hypothesized. By definition, a scalar λ is an eigenvalue for L if and only if $L - \lambda I$ has nontrivial kernel which is the case if and only if $\det(L - \lambda I) = 0$. $\square$

The characteristic polynomial for L is often defined to be $\det(tI - L)$. Frequently, λ does double duty as an eigenvalue and an indeterminate so it is also common to see the characteristic polynomial for L written $\det(L - \lambda I)$ or $\det(\lambda I - L)$.

Theorem 8.31 is enough to prove our next corollary and the next theorem, which is one of the stars of linear algebra.

Corollary 8.32. If L belongs to $\mathcal{L}(V)$ where $V \cong \mathbb{F}^n$, then the minimal polynomial for L divides its characteristic polynomial.

Theorem 8.33 (Cayley-Hamilton). If L is a linear operator with characteristic polynomial $p(t)$, then $p(L) = O$.

We can leverage the definition of the characteristic polynomial to find the minimal polynomial for a linear operator on a finite-dimensional space.

Example 8.34. Let L in $\mathcal{L}(\mathbb{R}^3)$ be defined by

$$L(x, y, z) = (x, y + z, z).$$

It is easy to check that the characteristic polynomial for L is

$$p(t) = (1 - t)^3.$$

Since the minimal polynomial divides the characteristic polynomial, the possibilities for the minimal polynomial for L are $t - 1$, $(t-1)^2$, and $(t-1)^3$. (We are sticking to the convention that the minimal polynomial is monic.) Inspecting powers of $L - I$ we have

$$(L - I)(x, y, z) = (0, z, 0) \quad \text{and} \quad (L - I)^2 = (0, 0, 0).$$

We conclude that the minimal polynomial for L is $q(t) = (t - 1)^2$.

The characteristic polynomial is a feature of the similarity class associated to a linear operator on a finite-dimensional space, but it does not characterize a similarity class uniquely: Different similarity classes may have the same characteristic polynomial. For instance, the characteristic polynomials for $L_1(x, y) = (2x, 2y)$ and $L_2(x, y) = (2x + y, 2y)$ are identical. It is easy to check, though, that L_1 and L_2 have different minimal polynomials. This is enough to establish that L_1 and L_2 belong to different similarity classes in $\mathcal{L}(\mathbb{R}^2)$.

When V is defined over $\mathbb{F}$ and L belongs to $\mathcal{L}(V)$, we use $\mathcal{Z}(L)$ to designate the set of zeroes of the characteristic polynomial for L in $\overline{\mathbb{F}}$. For example, if L in $\mathcal{L}(\mathbb{R}^2)$ has characteristic polynomial $t^2 + 1$, then $\mathcal{Z}(L) = \{i, -i\} \subseteq \mathbb{C}$.

Definition 8.35. Let L belong to $\mathcal{L}(V)$ where $V \cong \mathbb{F}^n$. The **algebraic multiplicity** of λ in $\mathcal{Z}(L)$ is the multiplicity of λ as a root of the characteristic polynomial for L.

When $V \cong \mathbb{F}^n$ and $\mathbb{F}$ is algebraically closed, Theorem 8.31 ensures that every L in $\mathcal{L}(V)$ has n eigenvalues, counting algebraic multiplicity.

Since there are two types of multiplicity for an eigenvalue, there must be cases in which they differ.

Example 8.36. Let $L(x,y) = (x+y+z, y+z, z)$ belong to $\mathcal{L}(\mathbb{R}^3)$. We can calculate the characteristic polynomial for L using its standard matrix representation

$$L = \begin{bmatrix} 1 & 1 & 1 \\ 0 & 1 & 1 \\ 0 & 0 & 1 \end{bmatrix}.$$

We have

$$L - tI = \begin{bmatrix} 1 & 1 & 1 \\ 0 & 1 & 1 \\ 0 & 0 & 1 \end{bmatrix} - \begin{bmatrix} t & 0 & 0 \\ 0 & t & 0 \\ 0 & 0 & t \end{bmatrix} = \begin{bmatrix} 1-t & 1 & 1 \\ 0 & 1-t & 1 \\ 0 & 0 & 1-t \end{bmatrix}.$$

We see then that the characteristic polynomial for L is $p(t) = (1-t)^3$. The only L-eigenvalue is 1 with algebraic multiplicity 3.

The 1-eigenspace for L is the null space of

$$L - I = \begin{bmatrix} 0 & 1 & 1 \\ 0 & 0 & 1 \\ 0 & 0 & 0 \end{bmatrix}.$$

That subspace of $\mathbb{R}^3$ is the solution to the system of equations

$$\begin{aligned} y + z &= 0 \\ z &= 0 \end{aligned}$$

in $\mathbb{R}^3$. The solution to the system is $\{(x, 0, 0) \mid x \in \mathbb{R}\}$, which is 1-dimensional. We conclude that the geometric multiplicity of 1 as an L-eigenvalue is 1.

When a vector space has a basis of eigenvectors for a linear operator, we can understand the operator as scaling in each direction represented by a basis vector. This gives us a good conceptual grasp of the nature of the operator and simplifies calculations. In the next section, we study the conditions under which we can expect to find a basis of eigenvectors for a linear operator on a finite-dimensional space.

Exercises 8.4.

8.4.1. Let $L_1(x,y) = (2x, 2y)$ and $L_2(x,y) = (2x+y, 2y)$ be in $\mathcal{L}(\mathbb{R}^2)$. Show that L_1 and L_2 have the same characteristic polynomial, but different minimal polynomials.

8.4.2. Define L in $\mathcal{L}(\mathbb{R}^2)$ by $L(x,y) = (x+y, -x+y)$.
 (a) Verify that the characteristic polynomial for L is $p(t) = t^2 - 2t + 2$ and that $\mathcal{Z}(L) = \{1+i, 1-i\} \subseteq \mathbb{C}$.
 (b) Find the L-eigenspaces if we view L as belonging to $\mathcal{L}(\mathbb{C}^2)$.

8.4.3. Find L in $\mathcal{L}(\mathbb{R}^2)$ that has an eigenvalue for which the algebraic and geometric multiplicities are both 2.

8.4.4. Let R_φ in $\mathcal{L}(\mathbb{R}^2)$ be the rotation through φ radians. Show that unless $\varphi = k\pi$, for some k in $\mathbb{Z}_{\geq 0}$, the characteristic polynomial for R_φ is irreducible over $\mathbb{R}$.

8.4.5. Verify that the characteristic polynomial given in Example 8.34 is correct.

8.4.6. Find the minimal polynomial for $A = \begin{bmatrix} 2 & 2 & -5 \\ 3 & 7 & -15 \\ 1 & 2 & -4 \end{bmatrix}$, given that its characteristic polynomial is $p(t) = (1-t)^2(3-t)$.

8.4.7. Find the characteristic polynomial, the minimal polynomial, any eigenvalues in the underlying field, and a basis for the associated eigenspace, for each of the following matrices.

(a) $\begin{bmatrix} 4 & 2 \\ 3 & -1 \end{bmatrix}$ over $\mathbb{R}$

(b) $\begin{bmatrix} 5 & -1 \\ 1 & 3 \end{bmatrix}$ over $\mathbb{R}$

(c) $\begin{bmatrix} 1 & -2 \\ 4 & 5 \end{bmatrix}$ over $\mathbb{R}$

(d) $\begin{bmatrix} 1 & -2 \\ 4 & 5 \end{bmatrix}$ over $\mathbb{C}$

(e) $\begin{bmatrix} 0 & 1 & 1 \\ 1 & 0 & 1 \\ 1 & 1 & 0 \end{bmatrix}$ over $\mathbb{R}$

(f) $\begin{bmatrix} 0 & -1 \\ 1 & 0 \end{bmatrix}$ over $\mathbb{R}$

(g) $\begin{bmatrix} 0 & -1 \\ 1 & 0 \end{bmatrix}$ over $\mathbb{C}$

8.4.8. Find the characteristic polynomial, eigenvalues, and associated eigenvectors for $A = \begin{bmatrix} 4 & 1 & -1 \\ 2 & 5 & -2 \\ 1 & 1 & 2 \end{bmatrix}$ over $\mathbb{R}$.

8.4.9. Find the minimal polynomial for $A = \begin{bmatrix} 0 & 0 & 1 \\ 1 & 0 & 2 \\ 0 & -1 & 4 \end{bmatrix}$.

8.4.10. For each of the following real matrices, find the characteristic polynomial, the eigenvalues, a basis for each eigenspace, and the minimal polynomial. Which matrices are similar?

(a) $\begin{bmatrix} 2 & 1 & 0 \\ 0 & 2 & 1 \\ 0 & 0 & 2 \end{bmatrix}$

(b) $\begin{bmatrix} 2 & 1 & 0 \\ 0 & 2 & 0 \\ 0 & 0 & 2 \end{bmatrix}$

(c) $\begin{bmatrix} 2 & 0 & 0 \\ 0 & 2 & 0 \\ 0 & 0 & 2 \end{bmatrix}$

8.5. Diagonalizability

If L in $\mathcal{L}(V)$ is a reflection or a projection, V has a basis of L-eigenvectors. When V is finite-dimensional, this means that L has a representation as a diagonal matrix.

Definition 8.37. Let $V \cong \mathbb{F}^n$. A mapping L in $\mathcal{L}(V)$ is **diagonalizable** or **diagonalizable over $\mathbb{F}$** or **$\mathbb{F}$-diagonalizable** provided there is a matrix representation for L with the form $\mathrm{diag}(a_1, \ldots, a_n)$.

The mapping in $\mathcal{L}(\mathbb{R}^2)$ given by $L(x, y) = (-y, x)$ is not diagonalizable. If we apply the same rule to define a mapping in $\mathcal{L}(\mathbb{C}^2)$, it is diagonalizable. To emphasize that things may change when we extend the underlying field, we say that L is not $\mathbb{R}$-diagonalizable but is $\mathbb{C}$-diagonalizable.

A diagonalizable linear operator scales each vector in some basis for the underlying space. The following theorem is a list of equivalent criteria that can help us think about what else we know when we know that a linear operator is diagonalizable. The rest of this section will be about adding to this list.

Theorem 8.38. Let $V \cong \mathbb{F}^n$ and let L belong to $\mathcal{L}(V)$. The following conditions are logically equivalent.

(a) L is $\mathbb{F}$-diagonalizable.

(b) V has a basis of L-eigenvectors.

(c) The geometric multiplicities of the L-eigenvalues add up to n.

(d) V is the direct sum of its L-eigenspaces.

Proof. Let V and L be as hypothesized.

The equivalence of the first two conditions is immediate by the definition of diagonalizability.

Assume that V has a basis of L-eigenvectors, $\mathcal{B}$. For each L-eigenvalue λ, let
$$\mathcal{B}_\lambda = \mathcal{B} \cap V_\lambda.$$
We leave it as an exercise to show that $\mathcal{B}_\lambda$ is a basis for V_λ and that
$$\{\mathcal{B}_\lambda \mid \lambda \in \mathcal{Z}(L)\}$$

8.5. Diagonalizability

is a partition on $\mathcal{B}$. It follows that

$$\sum_{\lambda \in \mathcal{Z}(L)} |\mathcal{B}_\lambda| = n,$$

which is to say that the sum of the geometric multiplicities of the L-eigenvalues is n. This proves that (b) implies (c).

Next assume that (c) is true. For each λ in $\mathcal{Z}(L)$, let $\mathcal{B}_\lambda$ be a basis for V_λ. By our assumption,

$$\sum_{\lambda \in \mathcal{Z}(L)} |\mathcal{B}_\lambda| = n,$$

which implies that $\sum_{\lambda \in \mathcal{Z}(L)} V_\lambda = V$. Since the sum is direct, (c) implies (d).

Finally, assume that $\bigoplus_{\lambda \in \mathcal{Z}(L)} V_\lambda = V$. Let

$$\mathcal{B} = \bigcup_{\lambda \in \mathcal{Z}(L)} \mathcal{B}_\lambda$$

where $\mathcal{B}_\lambda$ is a basis for V_λ. Since the L-eigenspaces intersect trivially, $\mathcal{B}$ is linearly independent. Since every element in V is a sum of L-eigenvectors, $\mathcal{B}$ is a basis for V. Since the elements in $\mathcal{B} = \{\mathbf{b}_i\}$ are L-eigenvectors, $[\![L]\!]_\mathcal{B} = \operatorname{diag}(a_1, \ldots, a_n)$, where $L(\mathbf{b}_i) = a_i \mathbf{b}_i$, for a_i in $\mathbb{F}$. This shows that condition (d) implies condition (a) and thus completes the proof. $\square$

Example 8.39. Let $L = \begin{bmatrix} 1 & 1 & 1 \\ 1 & 1 & 1 \\ 1 & 1 & 1 \end{bmatrix}$ belong to $\mathcal{L}(\mathbb{R}^3)$. We would like to determine whether L is diagonalizable.

Since $\operatorname{rank} L = 1$, L has a 2-dimensional kernel. Since

$$L \stackrel{\sim}{r} \begin{bmatrix} 1 & 1 & 1 \\ 0 & 0 & 0 \\ 0 & 0 & 0 \end{bmatrix},$$

$\operatorname{Ker} L$ is the plane in $\mathbb{R}^3$ given by $x + y + z = 0$. Let $\mathcal{B}_0 = \{(1, 0, -1), (1, -1, 0)\}$. This is a basis for $\operatorname{Ker} L$.

The next thing we can see without any calculations is that $L[V]$ has basis $\{(1, 1, 1)\}$. Indeed,

$$L(1, 1, 1) = (3, 3, 3) = 3(1, 1, 1)$$

so $(1, 1, 1)$ is a 3-eigenvector for L. Taking $\mathcal{B}_3 = \{(1, 1, 1)\}$, we see that

$$\mathcal{B} = \mathcal{B}_0 \cup \mathcal{B}_3$$

is a basis for $\mathbb{R}^3$. Since $\mathbb{R}^3$ has a basis of L-eigenvectors, L is diagonalizable. In particular, treating $\mathcal{B}$ as an ordered basis, we have

$$[\![L]\!]_\mathcal{B} = \begin{bmatrix} 0 & 0 & 0 \\ 0 & 0 & 0 \\ 0 & 0 & 3 \end{bmatrix}.$$

If L is in $\mathcal{L}(V)$ and W is an L-invariant subspace, then $L[W] \subseteq W$ does not mean that L or $p(L)$ cannot map vectors from outside of W into W. In Example 8.36, for instance, the 1-eigenspace for $L: \mathbb{R}^3 \to \mathbb{R}^3$ is spanned by $\{(1,0,0)\}$, yet

$$(L-I)(0,1,0) = \begin{bmatrix} 0 & 1 & 1 \\ 0 & 0 & 1 \\ 0 & 0 & 0 \end{bmatrix} \begin{bmatrix} 0 \\ 1 \\ 0 \end{bmatrix} = (1,0,0).$$

In other words, $L - I$ maps $(0,1,0)$ into the 1-eigenspace. We will see that this is exactly what cannot happen when L is diagonalizable.

Lemma 8.40. Let L belong to $\mathcal{L}(V)$ where $V \cong \mathbb{F}^n$. If λ is an L-eigenvalue in $\mathbb{F}$, then

$$\operatorname{Ker}(L - \lambda I) \oplus (L - \lambda I)[V] = V$$

if and only if λ has multiplicity 1 as a factor of the minimal polynomial for L.

Proof. Let everything be as hypothesized. Write $V_\lambda = \operatorname{Ker}(L - \lambda I)$. By the Rank Theorem,

$$\dim V_\lambda + \dim(L - \lambda I)[V] = n$$

so the content of the lemma is about when there is a nontrivial intersection between V_λ and $(L - \lambda I)[V]$.

Let $q(t)$ be the minimal polynomial for L. We can see that $(t - \lambda)^r$ divides $q(t)$ for $r > 1$ if and only if there is $\mathbf{v}$ in V so that $(L - \lambda I)^{r-1}(\mathbf{v}) \neq \mathbf{0}$ and $(L - \lambda I)^r(\mathbf{v}) = \mathbf{0}$. Note that if $r > 1$, then $r - 1 \geq 1$, in which case $(L - \lambda I)^{r-1}(\mathbf{v})$ is in $(L - \lambda I)[V] \cap V_\lambda$. The result follows immediately. $\square$

Proof of the next theorem follows Lemma 8.40.

Theorem 8.41. Let L belong to $\mathcal{L}(V)$ where $V \cong \mathbb{F}^n$. If $\mathcal{Z}(L) \subseteq \mathbb{F}$, then the minimal polynomial for L is the product of distinct degree one factors $t - \lambda$, λ in $\mathcal{Z}(L)$, if and only if

$$\operatorname{Ker}(L - \lambda I) \oplus (L - \lambda I)[V] = V$$

for all λ in $\mathcal{Z}(L)$.

One more lemma assists in the proof of our main theorem on diagonalizability.

Lemma 8.42. Let $V \cong \mathbb{F}^n$ and let L belong to $\mathcal{L}(V)$. Suppose $\mathcal{Z}(L) \subseteq \mathbb{F}$ and that λ is an L-eigenvalue with multiplicity 1 as a root of the minimal polynomial for L. If $W = (L - \lambda I)[V]$, then $\mathcal{Z}(L|_W) = \mathcal{Z}(L) \setminus \{\lambda\}$.

Proof. Let everything be as hypothesized with $V_\lambda = \operatorname{Ker}(L - \lambda I)$. Let μ belong to $\mathcal{Z}(L) \setminus \{\lambda\}$ and suppose that $\mathbf{v}$ is a μ-eigenvector in V. Lemma 8.40 gives us

$$V = W \oplus V_\lambda$$

so we can write $\mathbf{v} = \mathbf{w} + \mathbf{v}_\lambda$ for $\mathbf{w}$ in W and $\mathbf{v}_\lambda$ in V_λ. From there,

$$L(\mathbf{v}) = L(\mathbf{w}) + \lambda \mathbf{v}_\lambda = \mu \mathbf{w} + \mu \mathbf{v}_\lambda.$$

Since W is L-invariant, $L(\mathbf{w})$ belongs W, implying that

$$(\lambda - \mu)\mathbf{v}_\lambda = \mu \mathbf{w} - L(\mathbf{w}) \in W.$$

8.5. Diagonalizability

Since $\lambda - \mu$ is nonzero and $V_\lambda \cap W = \{\mathbf{0}_V\}$, $\mathbf{v}_\lambda = \mathbf{0}$. It follows that $V_\mu \subseteq W$, thus, that
$$\mathcal{Z}(L) \setminus \{\lambda\} \subseteq \mathcal{Z}(L|_W).$$

Now suppose $\mathbf{w}$ is nonzero in W with $L(\mathbf{w}) = \mu \mathbf{w}$ for some scalar μ. Since $\mathbf{w}$ is in V, μ is in $\mathcal{Z}(L) \setminus \{\lambda\}$, so
$$\mathcal{Z}(L|_W) \subseteq \mathcal{Z}(L) \setminus \{\lambda\}.$$

This is enough to prove the lemma. □

The following corollary drops out of the first part of the proof of the lemma.

Corollary 8.43. Let $V \cong \mathbb{F}^n$, and suppose L is in $\mathcal{L}(V)$. If λ is an L-eigenvalue in $\mathbb{F}$ with multiplicity 1 as a root of the minimal polynomial for L, then for every μ in $\mathcal{Z}(L) \setminus \{\lambda\}$, $\mathrm{Ker}(L - \mu I) \subseteq (L - \lambda I)[V]$.

We have done most of the work to prove the following.

Theorem 8.44. Let $V \cong \mathbb{F}^n$. Let L belong to $\mathcal{L}(V)$ and let $\mathcal{Z}(L) = \{\lambda_i\}_{i=1}^k \subseteq \mathbb{F}$. The operator L is diagonalizable if and only if the minimal polynomial for L is $q(t) = \prod_{i=1}^k (t - \lambda_i)$.

Proof. Suppose everything is as hypothesized. Assume L is diagonalizable, with $\mathcal{B} = \{\mathbf{b}_i\}_{i=1}^n$ a basis of L-eigenvectors for V. Since each $\mathbf{b}_i$ in $\mathcal{B}$ is in an L-eigenspace, we have
$$\prod_{\lambda \in \mathcal{Z}(L)} (L - \lambda I)(\mathbf{b}_i) = \mathbf{0}_V$$
for i in $\{1, \ldots, n\}$. Since λ is a root of the minimal polynomial for L if and only if λ is in $\mathcal{Z}(L)$, the minimal polynomial for L must be
$$q(t) = \prod_{i=1}^k (t - \lambda_i).$$

Suppose now that L has minimal polynomial $q(t) = \prod_{i=1}^k (t - \lambda_i)$ where $\mathcal{Z}(L) = \{\lambda_i\}_{i=1}^k$. We prove that L has a diagonal representation using induction on k. When $k = 1$, L is a scalar mapping, $\mathbf{v} \mapsto \lambda \mathbf{v}$ for all $\mathbf{v}$ in V. Any basis for V is then a basis of L-eigenvectors so the theorem is true in the base case.

Assume now that $k > 1$. Take μ in $\mathcal{Z}(L)$ and let $W = (L - \mu I)[V]$. By Lemma 8.42, $\mathcal{Z}(L|_W) = \mathcal{Z}(L) \setminus \{\mu\}$ so
$$q(t) = \prod_{\substack{\lambda \in \mathcal{Z}(L) \\ \lambda \neq \mu}} t - \lambda$$
must be the minimal polynomial for $L|_W$. By the induction hypothesis, $L|_W$ is diagonalizable. Let $\mathcal{B}_W$ be a basis of $L|_W$-eigenvectors, and let $\mathcal{B}_\mu$ be a basis for $\mathrm{Ker}(L - \mu I)$. The union $\mathcal{B}_W \cup \mathcal{B}_\mu$ is then a basis of L-eigenvectors for V, which implies the result in this direction. □

We can say at this point that diagonalizability of L in $\mathcal{L}(V)$, in the finite-dimensional setting, comes down to whether the minimal polynomial for L is a product of distinct degree one factors, or equivalently, whether

$$V = \operatorname{Ker}(L - \lambda I) \oplus (L - \lambda I)[V]$$

for each L-eigenvalue, λ. All of this is under the assumption that $\mathcal{Z}(L) \subseteq \mathbb{F}$, when $V \cong \mathbb{F}^n$. We can finesse the last condition, by expanding the scalar field, but we cannot finesse the multiplicity of a root of the minimal polynomial.

Exercises 8.5.

8.5.1. The L-eigenvalues for L in Example 8.39 are $\{0, 3\}$. Check directly that $\operatorname{Ker} L \cap L[V] = \{\mathbf{0}\}$ and that $\operatorname{Ker}(L - 3I) \cap (L - 3I)[V] = \{\mathbf{0}\}$.

8.5.2. What is the characteristic polynomial for the linear operator given in Example 8.39?

8.5.3. Let L in $\mathcal{L}(\mathbb{F}^3)$ be given by $L(x, y, z) = (y, z, 0)$. Find $\operatorname{Ker} L$ and $L[V]$. Are they complementary subspaces in $\mathbb{F}^3$?

8.5.4. Suppose there are no L-eigenvalues for L in $\mathcal{L}(\mathbb{R}^2)$. Show that L is $\mathbb{C}$-diagonalizable.

8.5.5. Argue geometrically that $L(x, y) = (-y, x)$ is not diagonalizable over $\mathbb{R}$.

8.5.6. Diagonalize $L(x, y) = (-y, x)$ over $\mathbb{C}$.

8.5.7. Does a diagonalizable operator on V scale every vector in V?

8.5.8. Suppose L in $\mathcal{L}(\mathbb{R}^2)$ has a real eigenvalue but that L is not $\mathbb{R}$-diagonalizable. Show that L cannot be $\mathbb{C}$-diagonalizable.

8.5.9. Suppose L in $\mathcal{L}(\mathbb{R}^2)$ is not diagonalizable but that it has a real eigenvalue. What form does the characteristic polynomial for L have to have?

8.5.10. Suppose the characteristic polynomial for L in $\mathcal{L}(\mathbb{R}^3)$ has two distinct degree one factors. Prove that L is $\mathbb{R}$-diagonalizable.

8.5.11. Suppose L is diagonalizable. When does this mean that the minimal and the characteristic polynomials for L are identical?

8.5.12. Give an example of a diagonalizable linear operator for which the characteristic polynomial and the minimal polynomial are different.

8.5.13. Give an example of a diagonalizable linear operator for which the characteristic polynomial and the minimal polynomial are the same.

8.5.14. Give an example of a nondiagonalizable linear operator for which all eigenvalues are in the scalar field and for which the characteristic polynomial and the minimal polynomial are the same.

8.5.15. Give an example of a nondiagonalizable linear operator for which all the eigenvalues are in the scalar field and for which the characteristic polynomial and the minimal polynomial are different.

8.5.16. Referring to the proof of Theorem 8.38, show that $\mathcal{B}_\lambda$ is a basis for V_λ and that $\{\mathcal{B}_\lambda \mid \lambda \in \mathcal{Z}(L)\}$ is a partition on $\mathcal{B}$.

8.5.17. Let L in $\mathcal{L}(\mathbb{R}^3)$ be given by $L(x,y,z) = (x+y, y, z)$.
   (a) Find the characteristic polynomial and the minimal polynomial for L.
   (b) Find any eigenvalues and associated eigenvectors for L.
   (c) Is L diagonalizable?

8.5.18. Let $V \cong \mathbb{F}^n$ and let L belong to $\mathcal{L}(V)$. Suppose the dimensions of the L-eigenspaces add up to n. Is every nonzero vector in V then an eigenvector?

8.5.19. Let $V \cong \mathbb{F}^n$. Prove the following lemma: If L in $\mathcal{L}(V)$ has n distinct eigenvalues, then L is diagonalizable.

8.5.20. Show that the converse of the lemma in the last exercise is false.

8.5.21. Define L in $\mathcal{L}(\mathbb{R}^3)$ by $L(x,y,z) = (y, z, x)$.
   (a) Use the standard matrix representation to find the characteristic polynomial for L.
   (b) Determine that L is not diagonalizable over $\mathbb{R}$.
   (c) Show that if we view L as belonging to $\mathcal{L}(\mathbb{C}^3)$, then L is diagonalizable.
   (d) Find a basis of L-eigenvectors in $\mathbb{C}^3$.

8.6. Self-Adjoint Matrices Are Diagonalizable

Let L belong to $\mathcal{L}(V)$. Typically, we can expect a fair amount of work ahead if we want to determine whether or not V has a basis of L-eigenvectors. In a finite-dimensional inner product space, though, there is a single criterion that guarantees diagonalizability of a linear operator. Since an inner product space for us is always real or complex, in this section, $\mathbb{F}$ can mean $\mathbb{R}$ or $\mathbb{C}$.

If (V, σ) is an inner product space, then we have seen that an operator L in $\mathcal{L}(V)$ is self-adjoint provided
$$\sigma(L(\mathbf{v}), \mathbf{w}) = \sigma(\mathbf{v}, L(\mathbf{w})) \,\forall\, \mathbf{v}, \mathbf{w} \in V.$$
A self-adjoint operator over $\mathbb{R}$ is called a symmetric operator. A self-adjoint matrix is A such that $A = A^\dagger = \overline{A^T}$.

Theorem 8.45. *The eigenvalues of a self-adjoint operator are real.*

Proof. Let (V, σ) be an inner product space and let L in $\mathcal{L}(V)$ be self-adjoint. If $\mathbf{v}$ is an L-eigenvector with eigenvalue λ, then
$$\sigma(L(\mathbf{v}), \mathbf{v}) = \sigma(\lambda \mathbf{v}, \mathbf{v}) = \bar{\lambda} \sigma(\mathbf{v}, \mathbf{v}).$$
We also have
$$\sigma(\mathbf{v}, L(\mathbf{v})) = \sigma(\mathbf{v}, \lambda \mathbf{v}) = \lambda \sigma(\mathbf{v}, \mathbf{v}).$$
Since $\sigma(L(\mathbf{v}), \mathbf{v}) = \sigma(\mathbf{v}, L(\mathbf{v}))$ and $\sigma(\mathbf{v}, \mathbf{v}) > 0$, it must be the case that $\bar{\lambda} = \lambda$, in other words, λ is real. $\square$

Example 8.46. Consider the following self-adjoint matrix in $M_2(\mathbb{C})$:
$$A = \begin{bmatrix} 1 & 1-i \\ 1+i & 2 \end{bmatrix}.$$

The characteristic polynomial for A is
$$(1-t)(2-t) - (1-i)(1+i) = t^2 - 3t = t(t-3).$$
Since its eigenvalues are distinct, A is similar to $\begin{bmatrix} 0 & 0 \\ 0 & 3 \end{bmatrix}$. To find an associated basis of eigenvectors for $\mathbb{C}^2$, we look for the null spaces of A and $A - 3I_2$, respectively. Without too much difficulty, we find that $(-2, 1+i)$ spans $\operatorname{Nul} A$ and that $(1, 1+i)$ spans the 3-eigenspace.

Note that in the infinite-dimensional setting, a self-adjoint operator need not have any eigenvectors.

Example 8.47. Consider $\mathscr{C}([a,b])$ for any real interval $[a,b] \subseteq \mathbb{R}$, with the inner product
$$\sigma(f,g) = \int_a^b \bar{f}(t)g(t)\,dt.$$
Exercise 7.8.9 implies that the mapping $f \mapsto tf$ is self-adjoint. We leave it as an exercise to verify that this mapping has no eigenvalues.

The basis of eigenvectors that we found for $\mathbb{C}^2$ in Example 8.46 was
$$\mathcal{B} = \{(-2, 1+i), (1, 1+i)\}.$$
Notice that this is an orthogonal set. Where linear independence shows up in an arbitrary vector space, we often see orthogonality in an inner product space.

Lemma 8.48. Let (V, σ) be an inner product space. If L in $\mathcal{L}(V)$ is self-adjoint, then any set of L-eigenvectors associated to distinct eigenvalues is orthogonal.

Proof. Suppose (V, σ) and L are as hypothesized. Let $\mathbf{v}$ and $\mathbf{x}$ be L-eigenvectors associated to distinct eigenvalues. Say $L(\mathbf{v}) = \lambda_1 \mathbf{v}$ and $L(\mathbf{x}) = \lambda_2 \mathbf{x}$. Since λ_1 and λ_2 are real, we have
$$\sigma(L(\mathbf{v}), \mathbf{x}) = \overline{\lambda_1}\sigma(\mathbf{v}, \mathbf{x}) = \lambda_1\sigma(\mathbf{v}, \mathbf{x}) = \sigma(\mathbf{v}, L(\mathbf{x})) = \lambda_2 \sigma(\mathbf{v}, \mathbf{x}).$$
Since $\lambda_1 \neq \lambda_2$, $\sigma(\mathbf{v}, \mathbf{x}) = 0$. $\square$

Let (V, σ) be an inner product space. Suppose L in $\mathcal{L}(V)$ is self-adjoint. If V contains an L-eigenvector, it contains a nontrivial L-invariant subspace, W. Consider $\mathbf{v}$ in $W^\perp$ and $\mathbf{w}$ in W. We have
$$\sigma(L(\mathbf{v}), \mathbf{w}) = \sigma(\mathbf{v}, L(\mathbf{w})) = 0.$$
This shows that if W is L-invariant, then $W^\perp$ is also L-invariant, leading us to one of the crown jewels of the theory.

Theorem 8.49 (Finite-Dimensional Spectral Theorem). There is an orthonormal basis of eigenvectors associated to any self-adjoint operator on a finite-dimensional inner product space.

Proof. Let (V, σ) be a finite-dimensional inner product space and let L in $\mathcal{L}(V)$ be self-adjoint. Since all L-eigenvalues are in $\mathbb{R}$, there is an L-eigenvector in V. Let $\mathbf{v}$ be an L-eigenvector and take $W = \operatorname{Span}\{\mathbf{v}\}$. Since $W^\perp$ is L-invariant, $L|_{W^\perp}$ is self-adjoint in $\mathcal{L}(W^\perp)$. Since σ restricted to $W^\perp$ is an inner product and $\dim W^\perp < \dim V$, the proof follows by induction on $\dim V$. $\square$

The spectral theorem suggests a definition.

Definition 8.50. Matrices A and B in $M_n(\mathbb{F})$ are **unitarily equivalent** provided there is a unitary matrix P in $M_n(\mathbb{F})$ such that $B = PAP^\dagger$.

When the underlying field is $\mathbb{R}$, we usually replace "unitarily" with "orthogonally," that is, real matrices A and B are **orthogonally equivalent** when there is a real orthogonal matrix P such that $B = PAP^T$.

The following is immediate by the spectral theorem.

Corollary 8.51. Any self-adjoint matrix in $M_n(\mathbb{C})$ is unitarily equivalent to a diagonal matrix with real entries.

The next is a corollary to the corollary.

Corollary 8.52. Any symmetric matrix in $M_n(\mathbb{R})$ is orthogonally equivalent to a diagonal matrix.

The proof of the Finite-Dimensional Spectral Theorem hinges, essentially, on the existence of a single eigenvector for a linear transformation. We see this idea exploited to an extreme in the next chapter. Before we go there, we consider some important mappings that do not succumb nicely to analysis via eigenvectors.

Exercises 8.6.

8.6.1. Let A belong to $M_n(\mathbb{F})$. Show that $A^\dagger A$ is unitarily equivalent to a diagonal matrix with real entries.

8.6.2. A matrix A in $M_n(\mathbb{F})$ is **normal** provided $AA^\dagger = A^\dagger A$. Let A be a normal matrix.
  (a) Show that for any $\mathbf{v}$ in $\mathbb{F}^n$, $\|A\mathbf{v}\|^2 = \|A^\dagger \mathbf{v}\|^2$.
  (b) Show that if λ is any scalar in $\mathbb{F}$, then $A - \lambda I$ is normal.
  (c) Conclude that $\mathbf{v}$ is a λ-eigenvector for A if and only if $\mathbf{v}$ is a $\bar{\lambda}$-eigenvector for $A^\dagger$.
  (d) Prove Theorem 8.45, substituting "normal matrix" for self-adjoint operator.
  (e) Prove that any orthogonal matrix is normal.

8.6.3. For the ambitious: Prove that the Finite-Dimensional Spectral Theorem applies to all normal matrices.

8.6.4. Show that the determinant of a self-adjoint matrix is a real number.

8.6.5. Verify that the mapping in Example 8.47 has no eigenvectors.

8.7. Rotations and Translations

An in-depth study of rotations can become a lifelong project so we restrict our attention to rotations in $\mathbb{R}^2$ and $\mathbb{R}^3$. By a rotation, we always mean a linear mapping, that is, one for which the origin is fixed.

We saw in Example 5.44 that $O_2(\mathbb{R})$ has two classes of operators, one of which is comprised of matrices that can be written

$$(8.3) \qquad R_\varphi = \begin{bmatrix} \cos\varphi & -\sin\varphi \\ \sin\varphi & \cos\varphi \end{bmatrix}.$$

In Exercise 5.5.16, you verified that $\mathbf{v} \mapsto R_\varphi(\mathbf{v})$ is the rotation through φ in $\mathbb{R}^2$. We call a matrix of the form given in (8.3) a **rotation matrix**.

Via trigonometric identities, we can verify that
$$R_{\varphi_2} R_{\varphi_1} = R_{\varphi_1 + \varphi_2}.$$
Rotations on $\mathbb{R}^2$ are commutative.

We treat I_2 as a rotation through 0 radians. With that, we can say that the rotations on $\mathbb{R}^2$ are the mappings in $O_2(\mathbb{R})$ with determinant 1. Since $\det AB = \det A \det B$ for matrices A and B when the product AB is defined, and since $\det A^{-1} = 1/\det A$ when $\det A \neq 0$, we see that the rotations on $\mathbb{R}^2$ form a subgroup of $O_2(\mathbb{R})$. That subgroup is denoted $SO_2(\mathbb{R})$ and is called the **special orthogonal group** on $\mathbb{R}^2$.

An eigenvector for a linear operator L is mapped to a multiple of itself under L. When we scale a vector in $\mathbb{R}^n$, we stretch it, shrink it, and/or reverse its direction. Unless φ is an integer multiple of π, a rotation in $\mathbb{R}^2$ through φ has no real eigenvalues. It does have eigenvectors in $\mathbb{C}^2$, though.

The characteristic polynomial for the rotation on $\mathbb{R}^2$ through φ is
$$\det \begin{bmatrix} \cos\varphi - t & -\sin\varphi \\ \sin\varphi & \cos\varphi - t \end{bmatrix} = t^2 - (2\cos\varphi)t + 1.$$
Its eigenvalues are thus $\lambda = \cos\varphi + i\sin\varphi$ and $\bar{\lambda}$. Notice that $|\lambda|^2 = 1$. This shows that scaling a vector in $\mathbb{C}^2$ by a complex number with modulus 1 effects a rotation.

Example 8.53. We have $R_{\pi/3} = \begin{bmatrix} 1/2 & -\sqrt{3}/2 \\ \sqrt{3}/2 & 1/2 \end{bmatrix}$, which has eigenvalues $(1 \pm i\sqrt{3})/2$. The eigenspace associated to $(1 + i\sqrt{3})/2$ is the null space of
$$\begin{bmatrix} -i\sqrt{3}/2 & -\sqrt{3}/2 \\ \sqrt{3}/2 & -i\sqrt{3}/2 \end{bmatrix}$$
which has basis $\{(i, 1)\}$. We leave it as an exercise to find a basis for the other eigenspace.

Though not necessarily the case in other applications, in our setting we define a mapping entirely by where each element in the domain is mapped. We do not concern ourselves with the route it took to get to its destination. Whether we rotate $\mathbb{R}^2$ through $\pi/4$ or $-7\pi/4$ or $9\pi/4$, the image of a given vector remains the same. For any angle φ, $R_\varphi = R_{\varphi + 2k\pi}$ for all k in $\mathbb{Z}$.

Rotation matrices have disguises that may be uncovered by application of trigonometric identities. For instance,
$$A = \begin{bmatrix} \cos\varphi & \sin\varphi \\ -\sin\varphi & \cos\varphi \end{bmatrix} = \begin{bmatrix} \cos\theta & -\sin\theta \\ \sin\theta & \cos\theta \end{bmatrix}$$
where $\theta = -\varphi$, $A = R_{-\varphi}$.

Since R_φ has such a nice representation with respect to the standard basis, it is hard to imagine a reason to look for other matrix representations of the same rotation. Yet something interesting arises when we consider what we get when we do this.

Take $\mathcal{B} = \{(1,0), (1,-1)\}$ to be an ordered basis for $\mathbb{R}^2$. We have

$$[R_\varphi]_\mathcal{B} = \begin{bmatrix} 1 & 1 \\ 0 & -1 \end{bmatrix} R_\varphi \begin{bmatrix} 1 & 1 \\ 0 & -1 \end{bmatrix} = \begin{bmatrix} \cos\varphi + \sin\varphi & 2\sin\varphi \\ -\sin\varphi & \cos\varphi - \sin\varphi \end{bmatrix}.$$

Unless $\varphi = k\pi$, for $k \in \mathbb{Z}$, this is not a rotation matrix. In other words, there is no angle ξ such that

$$[R_\varphi]_\mathcal{B} = \begin{bmatrix} \cos\xi & -\sin\xi \\ \sin\xi & \cos\xi \end{bmatrix}.$$

How could a rotation be represented by a matrix that is not a rotation matrix?

We saw in Section 7.7 that an orthogonal operator has an orthogonal matrix representation with respect to any orthonormal basis for the underlying space. Indeed, change of basis is itself an automorphism on $\mathbb{R}^2$. If $P\mathbf{v} = [\mathbf{v}]_\mathcal{B}$ for some ordered basis $\mathcal{B}$, then the matrix product $PR_\varphi P^{-1}$ corresponds to a composition of three mappings: (1) the inverse of $[\ ]_\mathcal{B}$, (2) the rotation, and (3) $[\ ]_\mathcal{B}$. In general, $[\ ]_\mathcal{B}$ and its inverse cannot be expected to preserve the lengths and relative positions of vectors, as a rotation does. To guarantee that they do, we must take $\mathcal{B}$ to be an orthonormal basis.

Theorem 8.54. Let R_φ be the rotation of $\mathbb{R}^2$ about the origin, through the angle φ. If $\mathcal{B}$ is an orthonormal basis, then $[R_\varphi]_\mathcal{B}$ is a rotation matrix.

Proving Theorem 8.54 with coordinates requires several steps that we outline in the exercises.

Another way to state Theorem 8.54 follows.

Corollary 8.55. Every rotation matrix is orthogonally similar to one of the form

$$\begin{bmatrix} \cos\varphi & -\sin\varphi \\ \sin\varphi & \cos\varphi \end{bmatrix}.$$

What of rotations in $\mathbb{R}^3$? Before embarking on this part of the journey, we note that a *right-handed coordinate system* for $\mathbb{R}^3$ is an orthonormal basis of the form $\mathcal{B} = \{\mathbf{u}_1, \mathbf{u}_2, \mathbf{u}_3\}$, where $\mathbf{u}_i \times \mathbf{u}_{i+1} = \mathbf{u}_{i+2}$, taking i in $\{1, 2, 3\}$ and doing the arithmetic mod 3. Section B.6 contains a more detailed discussion.

We only needed an angle to specify a rotation in $\mathbb{R}^2$, but in $\mathbb{R}^3$, we need a second piece of data to identify a rotation: an axis.

Let $R_{\ell,\varphi} = R_{\mathbf{v},\varphi}$ designate the rotation in $\mathbb{R}^3$ through the angle φ about the line $\ell = \text{Span}\{\mathbf{v}\}$. Say that $\mathbf{v} = (a, b, c)$. Since $\text{Span}\{\mathbf{v}\}$ is a line, $\mathbf{v}$ is nonzero, so a, b, and c cannot all be zero. Consider the plane α defined by $ax + by + cz = 0$. This is a 2-dimensional subspace of $\mathbb{R}^3$ in which the rotation behaves exactly as it does in $\mathbb{R}^2$. In other words, if α is your entire world and you identify α with $\mathbb{R}^2$, then $R_{\ell,\varphi}$ looks exactly like R_φ on $\mathbb{R}^2$. Since a rotation is linear, we can understand $R_{\ell,\varphi}$ by how it behaves on α and on ℓ.

We construct an orthonormal basis for $\mathbb{R}^3$, $\mathcal{B} = \{\mathbf{u}_1, \mathbf{u}_2, \mathbf{u}_3\}$ so that $\mathbf{u}_3$ is in the direction of ℓ. This is enough to guarantee that $\mathcal{B}_\alpha = \{\mathbf{u}_1, \mathbf{u}_2\}$ is a basis for α.

The matrix representation of $R_{\ell,\varphi}$ restricted to α is
$$\begin{bmatrix} \cos\theta & -\sin\theta \\ \sin\theta & \cos\theta \end{bmatrix}$$
for some angle θ. If $\mathcal{B}$ is a right-handed coordinate system for $\mathbb{R}^3$, then $\theta = \varphi$.

Let $\mathbf{u}_1$ be an arbitrary unit vector in α, and take $\mathbf{u}_2$ in α orthogonal to $\mathbf{u}_1$. Letting $\mathbf{u}_3 = \mathbf{u}_1 \times \mathbf{u}_2$, we get
$$[\![R_{\ell,\varphi}]\!]_\mathcal{B} = \left[\begin{array}{cc|c} \cos\theta & -\sin\theta & 0 \\ \sin\theta & \cos\theta & 0 \\ \hline 0 & 0 & 1 \end{array}\right].$$

Example 8.56. We would like a matrix representation of $R_{\ell,\varphi}$ where $\varphi = \pi/3$ and $\ell = \text{Span}\{(1,-1,1)\}$. To construct an orthonormal basis for $\alpha = \ell^\perp$, we start by noticing that α is given by the equation $x - y + z = 0$. Taking $\mathbf{v} = (1,1,0)$ in α and normalizing, we get $\mathbf{u}_1 = (1/\sqrt{2}, 1/\sqrt{2}, 0)$. Our choice for $\mathbf{u}_2$ must be in α and orthogonal to $\mathbf{u}_1$. This means $\mathbf{u}_2$ is either $(1/\sqrt{6}, -1/\sqrt{6}, -2/\sqrt{6})$ or $(-1/\sqrt{6}, 1/\sqrt{6}, 2/\sqrt{6})$. The latter is the vector we get by traveling $\pi/2$ radians in the counterclockwise direction in α, starting from $\mathbf{u}_1$. Taking
$$\mathbf{u}_2 = (-1/\sqrt{6}, 1/\sqrt{6}, 2/\sqrt{6}),$$
we let $\mathbf{u}_3 = \mathbf{u}_1 \times \mathbf{u}_2$. This gives us an orthonormal basis for $\mathbb{R}^3$,
$$\mathcal{B} = \{(1/\sqrt{2}, 1/\sqrt{2}, 0), (-1/\sqrt{6}, 1/\sqrt{6}, 2/\sqrt{6}), (1/\sqrt{3}, -1/\sqrt{3}, 1/\sqrt{3})\},$$
so that
$$A = [\![R_{\ell,\pi/3}]\!]_\mathcal{B} = \left[\begin{array}{cc|c} 1/2 & -\sqrt{3}/2 & 0 \\ \sqrt{3}/2 & 1/2 & 0 \\ \hline 0 & 0 & 1 \end{array}\right].$$

We can find $R_{\ell,\varphi} = [\![R_{\ell,\varphi}]\!]_\mathcal{E}$ by calculating PAP^{-1}, where P is the $\mathcal{B}$ to $\mathcal{E}$ change of basis matrix. We have
$$P = \begin{bmatrix} 1/\sqrt{2} & -1/\sqrt{6} & 1/\sqrt{3} \\ 1/\sqrt{2} & 1/\sqrt{6} & -1/\sqrt{3} \\ 0 & 2/\sqrt{6} & 1/\sqrt{3} \end{bmatrix}.$$
Since P is orthogonal, $P^{-1} = P^T$. Notice for instance that
$$[\mathbf{e}_1]_\mathcal{B} = (1/\sqrt{2}, -1/\sqrt{6}, 1/\sqrt{3}).$$

Relegating the calculations to a machine, we find
$$R_{\ell,\pi/3} = [\![R_{\ell,\pi/3}]\!]_\mathcal{E} = \begin{bmatrix} 2/3 & -2/3 & -1/3 \\ 1/3 & 2/3 & -2/3 \\ 2/3 & 1/3 & 2/3 \end{bmatrix}.$$
We then have that $R_{\ell,\pi/3}(\mathbf{e}_1) = (2/3, 1/3, 2/3)$, for instance.

We turn next to translations, which are anomalous here because they are not linear. They do, however, play an important role in applications, most notably, computer graphics. Programmers love matrices and rely heavily on the support of the theory of linear algebra. How can we use linear algebra to understand these important affine mappings?

8.7. Rotations and Translations

The trick to working with a translation arithmetically is to embed the vector space we care about into a vector space of higher dimension. We can do this with any finite-dimensional vector space because any finite-dimensional vector space is isomorphic to $\mathbb{F}^n$, for some field $\mathbb{F}$, and geometrically, we can view a hyperplane in $\mathbb{F}^{n+1}$ as a copy of $\mathbb{F}^n$. In applying our trick, we identity $\mathbb{F}^n$ with a hyperplane in $\mathbb{F}^{n+1}$ that is not a hyperspace. This allows us to treat a translation in the hyperplane as a linear transformation on the ambient space. Our job here is to turn this idea into mathematics.

Map $\mathbb{F}^n \to \mathbb{F}^{n+1}$ by

$$\mathbf{x} = (x_1, \ldots, x_n) \mapsto (x_1, \ldots, x_n, 1) = (\mathbf{x}; 1).$$

Whenever we write $(\mathbf{x}; 1)$ going forward, we are indicating the image of $\mathbf{x}$ in $\mathbb{F}^n$ under this mapping. Notice that this mapping is an injection.

Suppose now that $\mathbf{w} = (w_1, w_2, \ldots, w_n)$ is nonzero in $\mathbb{F}^n$ and that we wish to map $\mathbb{F}^n$ to itself by $t_\mathbf{w}$, that is, translation by $\mathbf{w}$. Define $L : \mathbb{F}^{n+1} \to \mathbb{F}^{n+1}$ by

$$[\![L]\!]_\mathcal{E} = \left[\begin{array}{c|c} I_n & \begin{array}{c} w_1 \\ \vdots \\ w_n \end{array} \\ \hline 0 \cdots 0 & 1 \end{array}\right].$$

We have $L(\mathbf{e}_i) = \mathbf{e}_i$ for $i = 1, \ldots, n$, and $L(\mathbf{e}_{n+1}) = (\mathbf{w}; 1)$. Given $\mathbf{x} = (x_1, \ldots, x_n)$ in $\mathbb{F}^n$, we have

$$L(\mathbf{x}; 1) = x_1 \mathbf{e}_1 + \cdots + x_n \mathbf{e}_n + (\mathbf{w}; 1)$$
$$= (\mathbf{x}; 0) + (\mathbf{w}; 1) = (\mathbf{x} + \mathbf{w}; 1) = (t_\mathbf{w}(\mathbf{x}); 1).$$

We can exploit the same idea to compose translation by $\mathbf{w} = (w_1, \ldots, w_n)$ on $\mathbb{F}^n$ with a linear operator T on $\mathbb{F}^n$. Let

$$A = \begin{bmatrix} \mathbf{a}_1 & \cdots & \mathbf{a}_n \end{bmatrix}$$

be the matrix representation of T with respect to the standard basis for $\mathbb{F}^n$. Let L be the linear operator on $\mathbb{F}^{n+1}$ with standard matrix representation

$$(8.4) \qquad \left[\begin{array}{c|c} A & \begin{array}{c} w_1 \\ \vdots \\ w_n \end{array} \\ \hline 0 \cdots 0 & 1 \end{array}\right].$$

For $\mathbf{v} = (c_1, \ldots, c_{n+1})$ in $\mathbb{F}^{n+1}$, we have

$$L(\mathbf{v}) = c_1(\mathbf{a}_1; 0) + \cdots + c_n(\mathbf{a}_n; 0) + c_{n+1}(\mathbf{w}; 1).$$

Notice in particular that for $\mathbf{v} = (\mathbf{x}; 1)$ in $\mathbb{F}^{n+1}$,

$$L(\mathbf{v}) = (A\mathbf{x} + \mathbf{w}; 1) = (T(\mathbf{x}) + \mathbf{w}; 1).$$

The order of operations is important here. The matrix in (8.4) effects the mapping $t_\mathbf{w} \circ T$. We leave it as an exercise to find the matrix that effects $T \circ t_\mathbf{w}$.

Exercises 8.7.

8.7.1. Prove that a rotation matrix R_φ preserves the dot product on $\mathbb{R}^2$.

8.7.2. Regardless of the value of n in $\mathbb{Z}_+$, we can define the rotations on $\mathbb{R}^n$ as the operators in $O_n(\mathbb{R})$ with determinant 1. The argument that $SO_n(\mathbb{R}) = \{A \in O_n(\mathbb{R}) \mid \det A = 1\}$ is a subgroup of $O_n(\mathbb{R})$ is identical to the argument that $SO_2(\mathbb{R})$ is a subgroup of $O_2(\mathbb{R})$. Does the set of elements in $O_n(\mathbb{R})$ with determinant different from 1 comprise a subgroup of $O_n(\mathbb{R})$?

8.7.3. Use matrices and trigonometric identities to prove that in $\mathbb{R}^2$,
$$R_{\varphi_2} R_{\varphi_1} = R_{\varphi_1 + \varphi_2}.$$
What are the obstacles in addressing the same matter in $\mathbb{R}^3$? Are you sure it is true in $\mathbb{R}^3$? Why or why not?

8.7.4. Determine whether each of the following is a rotation matrix. For each that is, determine an angle θ so that the matrix effects the rotation R_θ.

(a) $\begin{bmatrix} \cos\varphi & \sin\varphi \\ -\sin\varphi & \cos\varphi \end{bmatrix}$

(b) $\begin{bmatrix} \sin\varphi & \cos\varphi \\ \cos\varphi & -\sin\varphi \end{bmatrix}$

(c) $\begin{bmatrix} -\sin\varphi & -\cos\varphi \\ \cos\varphi & -\sin\varphi \end{bmatrix}$

8.7.5. Write the matrix representation of R_φ with respect to each of the following ordered, orthonormal bases.
  (a) $\mathcal{B} = \{(1,0), (0,-1)\}$
  (b) $\mathcal{B} = \{(0,1), (1,0)\}$
  (c) $\mathcal{B} = \{(1,0), (0,-1)\}$
  (d) $\mathcal{B} = \{(1/\sqrt{2}, 1/\sqrt{2}), (-1/\sqrt{2}, 1/\sqrt{2})\}$
  (e) $\mathcal{B} = \{(1/\sqrt{2}, 1/\sqrt{2}), (1/\sqrt{2}, -1/\sqrt{2})\}$

8.7.6. (a) Show that every (ordered) orthonormal basis of $\mathbb{R}^2$ has the form either
$$\mathcal{B}_1 = \{(\cos\psi, \sin\psi), (-\sin\psi, \cos\psi)\} \text{ or}$$
$$\mathcal{B}_2 = \{(\cos\psi, \sin\psi), (\sin\psi, -\cos\psi)\}.$$

  (b) Show that the coordinate mapping $L_1(\mathbf{v}) = [\mathbf{v}]_{\mathcal{B}_1}$ is a rotation and that the coordinate mapping $L_2(\mathbf{v}) = [\mathbf{v}]_{\mathcal{B}_2}$ is not a rotation.
  (c) Describe the mapping on $\mathbb{R}^2$ effected by L_2.
  (d) Conclude that $L_1 R_\varphi L_1^{-1}$ is a rotation matrix. What is the angle of rotation?
  (e) Use matrices and trigonometric identities to find $R_\varphi L_2^{-1}$. Do the same to see that $L_2 R_\varphi L_2^{-1}$ is a rotation matrix. What is the angle of rotation?
  (f) Conclude that for any ordered orthonormal basis $\mathcal{B}$, $[\![R_\varphi]\!]_\mathcal{B}$ is a rotation matrix.

8.7.7. Let ℓ be the line in $\mathbb{R}^3$ orthogonal to the plane $x + y = 0$. Find the matrix representation of $R_{\ell, \pi/4}$ with respect to a right-handed coordinate system for $\mathbb{R}^3$.

8.7. Rotations and Translations

8.7.8. Check that the standard matrix representation of $R_{\pi/3,\ell}$ as given in Example 8.56 is an orthogonal matrix.

8.7.9. Check that you get the standard matrix representation of $R_{\pi/3,\ell}$ as given in Example 8.56 if you construct a right-handed coordinate system for $\mathbb{R}^3$ starting with $\mathbf{v} = (1,0,-1)$ in α.

8.7.10. Let $\mathbf{w} = (1,1)$ in $\mathbb{F}^2$.
  (a) Describe $t_w : R^2 \to R^2$ geometrically.
  (b) What effect does $t_{\mathbf{w}}$ have on $\mathbb{F}_2{}^2$?
  (c) What effect does $t_{\mathbf{w}}$ have on $\mathbb{F}_3{}^2$?

8.7.11. Let $\mathbf{w}$ belong to $\mathbb{F}^n$. Prove that $\mathbf{w} \mapsto (\mathbf{w};1)$ is an injection of $\mathbb{F}^n$ into $\mathbb{F}^{n+1}$.

8.7.12. Identify $\mathbf{v}$ in $\mathbb{R}^2$ with the point $(\mathbf{v};1)$ in $\mathbb{R}^3$. Let $\mathbf{w} = (-2,3)$ in $\mathbb{R}^2$.
  (a) Let $L : \mathbb{R}^3 \to \mathbb{R}^3$ be the linear transformation that first rotates $\mathbb{R}^3$ through $\pi/6$ about the z-axis and then maps $(\mathbf{v};1)$ to $(\mathbf{v}+\mathbf{w};1)$. Find the standard matrix representation of L.
  (b) Let $T : \mathbb{R}^3 \to \mathbb{R}^3$ be the linear transformation that first maps $(\mathbf{v};1)$ to $(\mathbf{v}+\mathbf{w};1)$ and then rotates $\mathbb{R}^3$ through $\pi/6$ about the z-axis. Find the standard matrix representation of T.

8.7.13. Recall from Section 2.3, that if L belongs to $\mathcal{L}(\mathbb{F}^n)$ and $\mathbf{w}$ belongs to $\mathbb{F}^n$, then the affine mapping
$$A(L,\mathbf{w}) : \mathbb{F}^n \to \mathbb{F}^n$$
is defined by
$$A(L,\mathbf{w})(\mathbf{v}) = L(\mathbf{v}) + \mathbf{w},$$
for all $\mathbf{v}$ in $\mathbb{F}^n$.
  (a) If L_1 and L_2 belong to $\mathcal{L}(\mathbb{F}^n)$ and $\mathbf{w}_1, \mathbf{w}_2$ are vectors in $\mathbb{F}^n$, show that
  $$A(L_1,\mathbf{w}_1) \circ A(L_2,\mathbf{w}_2)$$
  is an affine mapping on $\mathbb{F}^n$.
  (b) Suppose L is invertible and $\mathbf{w}$ is arbitrary in $\mathbb{F}^n$. Show that $A(L,\mathbf{w})$ is an affine transformation by identifying its inverse.

8.7.14. Suppose T belongs to $\mathcal{L}(\mathbb{F}^n)$ and that $\mathbf{w}$ is nonzero in $\mathbb{F}^n$. Find the matrix representation for the linear operator on $\mathbb{F}^{n+1}$ that effects $T \circ t_{\mathbf{w}}$ on $\mathbb{F}^n$, if we identify $\mathbb{F}^n$ with the plane $x_{n+1} = 1$ in $\mathbb{F}^{n+1}$.

Chapter 9

Similarity

An equivalence relation is a device for grouping different objects that are, in some sense, "the same." Different equivalence relations pick up different "samenesses." To this point, we have defined several different equivalence relations on linear transformations and matrices: row equivalence, rank equivalence, and similarity. Row equivalent matrices represent linear transformations with the same kernels and ranges. A row equivalence class is characterized by its reduced row echelon form. Rank equivalent matrices represent linear transformations for which the kernels and the ranges have the same dimensions. Rank equivalence is characterized by a matrix of the form given in Theorem 3.37. Similar matrices represent identical linear transformations. Here, we pursue a matrix form that characterizes a similarity class.

Whenever possible, our discussion applies to all vector spaces, but the classification theorem applies to finite-dimensional vector spaces over algebraically closed fields.

9.1. Triangularization

Let $V \cong \mathbb{F}^n$ and let L belong to $\mathcal{L}(V)$. Based on our work to this point, we know that both the minimal and the characteristic polynomials for L can be written[1]

$$\prod_{i=1}^{k}(t-\lambda_i)^{r_i}$$

where $\lambda_1, \ldots, \lambda_k$ are the distinct eigenvalues for L in $\overline{\mathbb{F}}$. We also know that if $\mathcal{Z}(L)$ is not contained in $\mathbb{F}$, L is not diagonalizable over $\mathbb{F}$. As Example 8.36 suggests, there are other ways things can go wrong. Sometimes, there is no diagonal matrix representation for a linear operator.

---

[1] Because we defined the characteristic polynomial to be $\det(L - tI)$, it is not necessarily monic but since we care only about its zeroes, we are happy to take the characteristic polynomial to be either $\pm \det(L - tI)$, as convenient.

If we cannot have a diagonal matrix representation, can we find a matrix representation that is "nearly" diagonal? What form would "nearly" diagonal take?

The matrix in Example 8.36 is an upper triangular matrix. It turns out that this type of representation is rather easy to come by, as long as the eigenvalues are in the scalar field. Our next task is proving that. We start by thinking about what a triangular representation says about the underlying linear operator.

It does not matter much whether we deal with upper or lower triangular matrices here. We pick one type and stay with it. Our results will all be stated and proved for lower triangular matrices but everything can be adjusted to apply to upper triangular matrices.

Recall that $A = [a_{ij}]$ in $M_n(\mathbb{F})$ is lower triangular provided $a_{ij} = 0$ whenever $i < j$.

Lemma 9.1. Let $V \cong \mathbb{F}^n$. An operator L in $\mathcal{L}(V)$ has a lower triangular matrix representation if and only if V has a set of L-invariant subspaces, H_j, such that
$$V = H_1 \supseteq H_2 \supseteq \cdots \supseteq H_n,$$
where $\dim H_j = n - j + 1$.

Proof. Let V be as hypothesized and let L belong to $\mathcal{L}(V)$.

Suppose L has a lower triangular representation, $A = [a_{ij}] = [\![L]\!]_\mathcal{B}$. Let $\mathcal{B} = \{\mathbf{b}_i\}_{i=1}^n$. For each j in $\{1, \ldots, n\}$, let $H_j = \mathrm{Span}\{\mathbf{b}_j, \ldots, \mathbf{b}_n\}$. Notice that
$$L(\mathbf{b}_j) = a_{jj}\mathbf{b}_j + \cdots + a_{nj}\mathbf{b}_n \in H_j.$$
This establishes that H_j is L-invariant. Since
$$V = H_1 \supseteq H_2 \supseteq \cdots \supseteq H_n,$$
this proves the lemma in one direction.

We leave proof of the other direction as an exercise. $\square$

The lesson of Section 8.5 is that any difficulty in dealing with linear operators on finite-dimensional spaces arises out of eigenvalues that, as roots of the minimal polynomial, have multiplicity greater than 1. If λ is an L-eigenvalue with multiplicity $\ell > 1$ as a root of the minimal polynomial for L, we do not have an obvious L-decomposition of V, as we did when $\ell = 1$, nor can we switch between L and $L - \lambda I$ to analyze $\mathrm{Ker}(L - \lambda I)^\ell$, as we were able to do when $\ell = 1$. There is something we can exploit here, though. Consider a linear operator L on $V \cong \mathbb{F}^n$ with minimal polynomial $q(t) = (t - \lambda)^\ell$ for λ in $\mathbb{F}$ and some positive integer $\ell > 1$. Note that
$$V = \mathrm{Ker}(L - \lambda I)^\ell \supseteq \mathrm{Ker}(L - \lambda I)^{\ell-1} \supseteq \cdots \supseteq \mathrm{Ker}(L - \lambda I).$$

This leads us to a decomposition of the form given in Lemma 9.1. From there, we can tease out a lower triangular representation for L. This idea will bubble up again as we make our way through this chapter. In the meantime, we can prove triangularizability in the general case without much trouble.

Theorem 9.2. Let $V \cong \mathbb{F}^n$ and let L belong to $\mathcal{L}(V)$. If $\mathcal{Z}(L) \subseteq \mathbb{F}$, then L has a lower triangular representation.

9.1. Triangularization

Proof. Assume V, L, and $\mathcal{Z}(L)$ are as hypothesized. We prove the theorem by induction on $\dim V$, the case $\dim V = 1$ being evident. Assume that $n = \dim V > 1$. Let $\mathbf{v}$ be a λ-eigenvector for L, for some λ in $\mathcal{Z}(L)$. Consider the quotient space $W = V/\operatorname{Span}\{\mathbf{v}\}$, and notice that $\dim W = n - 1$. Since $\operatorname{Span}\{\mathbf{v}\}$ is L-invariant, L induces a linear operator on W, $\tilde{L}$. Let $q(t)$ be the minimal polynomial for L and let $\tilde{q}(t)$ be the minimal polynomial for $\tilde{L}$. Since $q(L)(\mathbf{x}) = \mathbf{0}$ for all $\mathbf{x}$ in V, $q(t)$ must also annihilate $\tilde{L}$. It follows that $\tilde{q}(t)$ divides $q(t)$, proving that $\mathcal{Z}(\tilde{L}) \subseteq \mathbb{F}$. This allows us to apply the induction hypothesis to W. Let $\mathcal{B}_W = \{\tilde{\mathbf{b}}_i\}_{i=1}^{n-1}$ be an ordered basis for W so that $A = [\![\tilde{L}]\!]_{\mathcal{B}_W}$ is lower triangular.

Choose a coset representative, $\mathbf{b}_i$, for each $\tilde{\mathbf{b}}_i$ in $\mathcal{B}_W$. By Exercise 2.2.7, $\{\mathbf{b}_i\}_{i=1}^{n-1}$ is linearly independent in V.

Our choice of $\mathcal{B}_W$ implies that
$$\tilde{L}(\tilde{\mathbf{b}}_i) \in \operatorname{Span}\{\tilde{\mathbf{b}}_i, \tilde{\mathbf{b}}_{i+1}, \ldots, \tilde{\mathbf{b}}_{n-1}\}.$$
This means that for each $i \in \{1, \ldots, n-1\}$,
$$L(\mathbf{b}_i + \mathbf{v}) = L(\mathbf{b}_i) + \lambda \mathbf{v} = \sum_{j=i}^{n-1} a_j \mathbf{b}_j + \lambda \mathbf{v}$$
for some scalars a_j. It follows that $L(\mathbf{b}_i)$ is in $\operatorname{Span}\{\mathbf{b}_j\}_{j=i}^{n-1} \cup \{\mathbf{v}\}$. Letting
$$\mathcal{B} = \{\mathbf{b}_1, \ldots, \mathbf{b}_{n-1}, \mathbf{v}\}$$
we see that
$$[\![L]\!]_{\mathcal{B}} = \left[\begin{array}{c|c} A & \begin{matrix} 0 \\ \vdots \\ 0 \end{matrix} \\ \hline * \cdots * & \lambda \end{array}\right],$$
for $A = [\![\tilde{L}]\!]_{\mathcal{B}_W}$, which is itself lower triangular. This is enough to prove the theorem. $\square$

Triangular matrices present certain advantages, but a given similarity class of operators may have many triangular forms. Our goal for the rest of this chapter is to refine Theorem 9.2. This is significantly more involved than refining row echelon form to reduced row echelon form. Not only does it require a more developed theory, it also requires navigating what at times may feel like thick jungle undergrowth. The next leg of our journey will be to establish conditions under which a vector space V has an L-decomposition for L in $\mathcal{L}(V)$.

Exercises 9.1.

9.1.1. Discuss some of the advantages of having a triangular representation for a linear operator.

9.1.2. Finish the proof of Lemma 9.1.

9.1.3. State and prove a version of Lemma 9.1 that would apply to upper triangular matrix representations.

9.1.4. Prove that if $V \cong \mathbb{F}^n$, then L in $\mathcal{L}(V)$ has an upper triangular matrix representation if and only if it has a lower triangular matrix representation.

9.1.5. Let $V \cong \mathbb{F}^n$ and let L belong to $\mathcal{L}(V)$. Suppose $\mathcal{Z}(L) \subseteq \mathbb{F}$. Prove that the diagonal entries in any upper or lower triangular representation of L always comprise the same collection of numbers.

9.1.6. Let $V \cong \mathbb{F}^n$ and suppose L is in $\mathcal{L}(V)$. If L has an upper triangular matrix representation $A = [a_{ij}]$, prove that the characteristic polynomial for L is $\prod_{i=1}^n (a_{ii} - t)$.

9.1.7. Find an upper triangular form for each of the following matrices over $\mathbb{C}$. Specify the associated basis.

(a) $\begin{bmatrix} 1 & 1 \\ 1 & 1 \end{bmatrix}$

(b) $\begin{bmatrix} 0 & 1 & 0 \\ 0 & 0 & 1 \\ 0 & 1 & 0 \end{bmatrix}$

(c) $\begin{bmatrix} 1 & 0 & 0 \\ 1 & 1 & 0 \\ 1 & 1 & 1 \end{bmatrix}$

9.1.8. Let $V \cong \mathbb{F}^n$ and let L belong to $\mathcal{L}(V)$. Prove that if $\mathcal{Z}(L) = \{\lambda_i\}_{i=1}^k$ is contained in $\mathbb{F}$ and the characteristic polynomial of L is $\prod_{i=1}^k (t - \lambda_i)^{r_i}$, then $\det L = \prod_{i=1}^k \lambda_i^{r_i}$.

9.2. The Primary Decomposition

Our proximate goal is to see a linear operator L on a finite-dimensional space V as the direct sum of operators of certain types. Saying that we want to see L as a direct sum of operators is just another way of saying that we want to see V as an L-sum, or that we want to see L as having a diagonal square-block matrix representation. The following result suggests why we might want these things. This is a corollary to Theorem 6.46.

Corollary 9.3. If A in $M_n(\mathbb{F})$ has diagonal square-block form
$$A = \text{diag}(A_1, \ldots, A_k),$$
then the characteristic polynomial for A is the product of the characteristic polynomials for $A_1, \ldots, A_k$. Moreover, the set of roots of the characteristic polynomial for A is the union of the sets of roots of the characteristic polynomials for the A_is.

Proof. Suppose everything is as hypothesized. Let A_j belong to $M_{n_j}(\mathbb{F})$. If the diagonal entries in A_j are $(a_{1_j}, \ldots, a_{n_j})$, then the diagonal entries in A are
$$(a_{1_1}, \ldots, a_{n_1}, \ldots, a_{1_k}, \ldots, a_{n_k}).$$
The diagonal entries in $A - tI$ are thus
$$(a_{1_1} - t, \ldots, a_{n_1} - t, \ldots, a_{1_k} - t, \ldots, a_{n_k} - t).$$
On the other hand, the diagonal entries in $A_j - tI$ are
$$(a_{1_j} - t, \ldots, a_{n_j} - t).$$

9.2. The Primary Decomposition

Letting $I = I_{n_j}$, we can thus write

$$A - tI = \begin{bmatrix} A_1 - tI & 0 & \cdots & \cdots & 0 \\ 0 & A_2 - tI & 0 & \cdots & 0 \\ & & \cdots & & \\ 0 & \cdots & \cdots & 0 & A_k - tI \end{bmatrix}.$$

The first part of the result now follows Theorem 6.46 and the definition of the characteristic polynomial. The second part of the result follows immediately from the first. □

We now move on to our proximate goal. The first new idea we run across is *nilpotence*.

Definition 9.4. A linear operator L in $\mathcal{L}(V)$ is **nilpotent** provided there is a positive integer, ℓ, such that L^ℓ is identically zero. The least positive integer ℓ such that $L^\ell = O$ is the **nilpotence index of** L or, when L is understood, simply the **nilpotence index**.

The zero mapping is nilpotent with nilpotence index one.

Nilpotence implies noninvertibility but a noninvertible operator need not be nilpotent. For example, if $L_1 = \begin{bmatrix} 1 & 0 \\ 0 & 0 \end{bmatrix}$, then $L_1^j = L_1$, for all j in $\mathbb{Z}_+$, regardless of the underlying field: L_1 is singular but not nilpotent. On the other hand, if $L_2 = \begin{bmatrix} 0 & 1 \\ 0 & 0 \end{bmatrix}$, we have $L_2^2 = O$, regardless of the underlying field, thus, L_2 is nilpotent with index two.

Let L belong to $\mathcal{L}(V)$ where $V \cong \mathbb{F}^n$ and suppose $\mathcal{Z}(L) \subseteq \mathbb{F}$. If λ is in $\mathcal{Z}(L)$, $L - \lambda I$ is nilpotent on $\text{Ker}(L - \lambda I)^j$, for any positive integer j. In itself, this observation is trivial. Once we establish that for some j, $\text{Ker}(L - \lambda I)^j$ has an L-complement in V, this trivial observation becomes essential to our analysis.

Consider again the operator L in $\mathcal{L}(\mathbb{R}^3)$ given by

$$L(x, y, z) = (x + y + z, y + z, z).$$

The only L-eigenvalue is 1. Since

$$(L - I)(x, y, z) = (y + z, z, 0),$$

we see that

$$(L - I)^2(x, y, z) = (L - I)(y + z, z, 0) = (z, 0, 0)$$

so that

$$(L - I)^3(x, y, z) = (L - I)(z, 0, 0) = (0, 0, 0).$$

It follows that $L - I$ is nilpotent on $\mathbb{R}^3$ with nilpotence index three.

Next consider L from Example 8.39 given by

$$L(x, y, z) = (x + y + z, x + y + z, x + y + z).$$

There we noticed that the L-eigenvalues are 0 and 3. It is easy to see that $L - 0I = L$ is not nilpotent. We leave it as an exercise to find a formula for $L^j(x, y, z)$ for positive integer j and to determine whether $L - 3I$ is nilpotent.

Note that the concept of a nilpotent operator L is defined essentially in terms of its minimal polynomial, $q(t) = t^\ell$, where ℓ is the nilpotence index for L. Since similar linear operators have the same minimal polynomials, all linear operators in a given similarity class have the same nilpotence index.

Our next theorem looks like a technical lemma but it is at the heart of our story.

Theorem 9.5. Let $V \cong \mathbb{F}^n$ and let L belong to $\mathcal{L}(V)$. If $K_i = \operatorname{Ker} L^i$ for any positive integer i, there is then ℓ in $\mathbb{Z}_+$ such that
$$K_1 \subseteq K_2 \subseteq \cdots \subseteq K_\ell = K_{\ell+1} = \cdots = K_{\ell+j}$$
for all j in $\{2, 3, \ldots\}$.

Proof. Let everything be as hypothesized. If $\mathbf{x}$ belongs to $\operatorname{Ker} L^j$ for j in $\mathbb{Z}_+$, then $\mathbf{x}$ is in $\operatorname{Ker} L^{j+1}$:
$$L^{j+1}(\mathbf{x}) = L\left(L^j(\mathbf{x})\right) = L(\mathbf{0}_V) = \mathbf{0}_V.$$
This proves that
$$(9.1) \qquad K_1 \subseteq \cdots \subseteq K_j \subseteq K_{j+1} \subseteq \cdots.$$
Since V is finite-dimensional and K_j is a subspace of V, $\dim V$ is an upper bound on $\dim K_j$. The implication is that there is ℓ in $\mathbb{Z}_+$ so that $K_\ell = K_{\ell+1}$. Once we find ℓ with this property, we have to show that $K_\ell = K_{\ell+j}$ for all j in $\mathbb{Z}_+$. This ensures that the kernels cannot start growing again once they have stabilized.

Take ℓ in $\mathbb{Z}_+$ minimal so that $K_\ell = K_{\ell+1}$. For any j in $\{2, 3, \ldots\}$ and $\mathbf{x}$ in $K_{\ell+j}$, we have
$$\mathbf{0}_V = L^{\ell+j}(\mathbf{x}) = L^{\ell+1}\left(L^{j-1}(\mathbf{x})\right)$$
$$= L^\ell\left(L^{j-1}(\mathbf{x})\right) = L^{\ell+j-1}(\mathbf{x}).$$

We can conclude that $K_{\ell+j} \subseteq K_{\ell+j-1}$. Since we established that $K_{\ell+j-1} \subseteq K_{\ell+j}$, this proves that $K_\ell = K_{\ell+j}$ for all j in $\mathbb{Z}_+$ and thus completes the proof of the theorem. $\square$

Consider the extreme cases addressed in Theorem 9.5. First, suppose L is identically zero on V. For all j in $\mathbb{Z}_+$, we then have $\operatorname{Ker} L^j = V$. In this case, the least positive integer j so that
$$\operatorname{Ker} L^j = \operatorname{Ker} L^{j+1} = V$$
is $j = 1$. Second, suppose L is nonsingular. When it is defined, the product of nonsingular mappings is nonsingular so for all $j = 2, 3, \ldots$, $\operatorname{Ker} L^j = \{\mathbf{0}_V\}$. In this case as well, the least positive integer j so that
$$\operatorname{Ker} L^j = \operatorname{Ker} L^{j+1} = \{\mathbf{0}_V\}$$
is $j = 1$.

Generally, the kernels for L^j may grow with j, but in the extreme cases, they are stable and equal either to all of V or to $\{\mathbf{0}_V\}$.

Theorem 9.5 suggests the following definition.

9.2. The Primary Decomposition

Definition 9.6. Let L belong to $\mathcal{L}(V)$ where $V \cong \mathbb{F}^n$. The **kernel index of** L is the least ℓ in $\mathbb{Z}_+$ so that $\operatorname{Ker} L^\ell = \operatorname{Ker} L^{\ell+1}$.

The examples of the zero mapping and a nonsingular operator make it evident that the kernel index and the nilpotence index for a linear operator are different measurements. Nilpotence index is only defined for nilpotent operators while kernel index is defined for all operators in the finite-dimensional setting.

Example 9.7. Let $A = \begin{bmatrix} 0 & 1 & 0 \\ 0 & 0 & 1 \\ 0 & 0 & 0 \end{bmatrix}$ and let $L = L_A$ on $\mathbb{F}^3$. We have $\operatorname{Ker} L = \operatorname{Span}\{\mathbf{e}_1\}$. Since $A^2 = \begin{bmatrix} 0 & 0 & 1 \\ 0 & 0 & 0 \\ 0 & 0 & 0 \end{bmatrix}$, $\operatorname{Ker} L^2 = \operatorname{Span}\{\mathbf{e}_1, \mathbf{e}_2\}$. Finally $A^3 = O$ so $\operatorname{Ker} L^3 = \mathbb{F}^3$. We see then that L has both nilpotence index and kernel index equal to 3.

We are ready for our first L-decomposition of V. Our only hypothesis here is that V is finite-dimensional.

Theorem 9.8. Let $V \cong \mathbb{F}^n$. If L is in $\mathcal{L}(V)$, then

(9.2) $$V = \operatorname{Ker} L^j \oplus L^j[V]$$

if and only if $\operatorname{Ker} L^j = \operatorname{Ker} L^\ell$ and $L^j[V] = L^\ell[V]$ where ℓ is the kernel index of L.

Proof. Assume V and L are as hypothesized. Let ℓ be the kernel index of L. To prove that
$$V = \operatorname{Ker} L^\ell \oplus L^\ell[V],$$
we have to show that
$$\operatorname{Ker} L^\ell \cap L^\ell[V] = \{\mathbf{0}_V\}.$$

If $\mathbf{x}$ belongs to $\operatorname{Ker} L^\ell \cap L^\ell[V]$, then $L^\ell(\mathbf{x}) = \mathbf{0}_V$ and for some $\mathbf{y}$ in V, $L^\ell(\mathbf{y}) = \mathbf{x}$. We then have
$$\mathbf{0}_V = L^\ell(\mathbf{x}) = L^{2\ell}(\mathbf{y}).$$
By our choice of ℓ, $L^{2\ell}(\mathbf{y}) = \mathbf{0}_V$ implies that $\mathbf{y}$ belongs to $\operatorname{Ker} L^\ell$. This gives us
$$L^\ell(\mathbf{y}) = \mathbf{x} = \mathbf{0}_V,$$
proving the result in one direction.

Suppose now that (9.2) holds for some j in $\mathbb{Z}_+$. If $j < \ell$, then there is $\mathbf{y}$ in V so that $L^j(\mathbf{y}) \neq \mathbf{0}_V$ but $L^{j+1}(\mathbf{y}) = \mathbf{0}_V$. We then have
$$L^{j+1}(\mathbf{y}) = L^j(L(\mathbf{y})) = \mathbf{0}_V.$$
Since $L^j(\mathbf{y}) \neq \mathbf{0}_V$ implies $L(\mathbf{y}) \neq \mathbf{0}_V$, we have a nonzero element, $L(\mathbf{y})$ in both $\operatorname{Ker} L^j$ and $L^j[V]$. This contradicts our assumption that V is the direct sum of $\operatorname{Ker} L^j$ and $L^j[V]$. The contradiction implies that $j \geq \ell$.

Suppose $j > \ell$. By Theorem 9.5, $\operatorname{Ker} L^j = \operatorname{Ker} L^\ell$. Since
$$\dim V = \dim \operatorname{Ker} L^j + \dim L^j[V],$$
$\dim L^j[V] = \dim L^\ell[V]$. Note further that if $j = \ell + i$, then
$$L^j(\mathbf{v}) = L^\ell\left(L^i(\mathbf{v})\right).$$

This proves that $L^j[V] \subseteq L^\ell[V]$. Since $L^j[V]$ and $L^\ell[V]$ are finite-dimensional, we must conclude that $L^j[V] = L^\ell[V]$. This proves the result in the other direction and thus completes the proof of the theorem. □

We can check the theorem against the linear transformation in Example 9.7. Consider the kernel and image of L given by $A = \begin{bmatrix} 0 & 1 & 0 \\ 0 & 0 & 1 \\ 0 & 0 & 0 \end{bmatrix}$ on $\mathbb{F}^3$. We have $\operatorname{Ker} L = \operatorname{Span}\{\mathbf{e}_1\}$ and $L[V] = \operatorname{Span}\{\mathbf{e}_1, \mathbf{e}_2\}$, so the two are not complements. Next, we found that $\operatorname{Ker} L^2 = \operatorname{Span}\{\mathbf{e}_1, \mathbf{e}_2\}$, while $L^2[V] = \operatorname{Span}\{\mathbf{e}_1\}$, so these two are not complements, either. As guaranteed by the theorem, we have to get to L^3 before the associated kernel and image are complements in $\mathbb{F}^3$.

Our application of Theorem 9.8 will be to $L - \lambda I$ when λ is an L-eigenvalue. We refine Theorem 9.2 by peeling certain L-invariant subspaces off $\operatorname{Ker}(L - \lambda I)^\ell$ when ℓ is the $(L - \lambda I)$-kernel index. The restriction of L to each of these subspaces has a simple normal form. The next lemma suggests how to corral those subspaces, though we will not see this until the next section. We bring the lemma into our narrative here because it has other applications in the sequel.

Proof of the lemma is by induction, which comes as no surprise. What may be a surprise is that in the base case, the hypothesis of the lemma does not hold. We say that the statement of the lemma is *vacuously true* in this case: If the hypothesis cannot hold, then when it does hold, the conclusion must as well. "If pigs had wings, they'd fly," is an example of a statement that is vacuously true.

Lemma 9.9. Let V be a vector space over $\mathbb{F}$ and let L in $\mathcal{L}(V)$ be nilpotent with nilpotence index, ℓ. If $L^{\ell-1}(\mathbf{x})$ is not zero, then

$$\{\mathbf{x}, L(\mathbf{x}), \ldots, L^{\ell-1}(\mathbf{x})\}$$

is linearly independent in V. In this case, $\dim \operatorname{Ker} L^\ell \geq \ell$.

Proof. Suppose V and L are as hypothesized. We prove the lemma by induction on ℓ.

If $\ell = 1$, then L is identically zero so L^0 is not defined. Since there can be no $\mathbf{x}$ in V such that $L^{\ell-1}(\mathbf{x}) \neq \mathbf{0}_V$, the lemma is vacuously true in the base case.

Assume the lemma is true for any linear operator on V with nilpotence index greater than one.

Since the nilpotence index of L is ℓ, there is $\mathbf{x}$ in V not in $\operatorname{Ker} L^{\ell-1}$. It follows that $\mathbf{x}$ and $L^j(\mathbf{x})$ are nonzero for $j \in \{1, \ldots, \ell-1\}$. Suppose there are scalars a_i such that

(9.3) $$a_1 \mathbf{x} + a_2 L(\mathbf{x}) + \cdots + a_\ell L^{\ell-1}(\mathbf{x}) = \mathbf{0}_V.$$

Evaluating L on both sides of (9.3), we see that what remains when $L(a_\ell L^{\ell-1}(\mathbf{x}))$ drops out is

(9.4) $$a_1 L(\mathbf{x}) + a_2 L^2(\mathbf{x}) + \cdots + a_{\ell-1} L^{\ell-1}(\mathbf{x}) = \mathbf{0}_V.$$

Let $R = L[V] \subseteq V$. We have established that R is L-invariant. Note also that L is nilpotent on R with nilpotence index $\ell - 1$. By our choice of $\mathbf{x}$, $L(\mathbf{x})$ is nonzero in

9.2. The Primary Decomposition

R with
$$L^{\ell-2}(L(\mathbf{x})) = L^{\ell-1}(\mathbf{x}) \neq \mathbf{0}_V.$$
The induction hypothesis thus applies to give us that
$$\{L(\mathbf{x}), \ldots, L^{\ell-1}(\mathbf{x})\}$$
is linearly independent in R. From there is follows that the coefficients a_i in (9.4) must all be zero. Applying this to (9.3), we have
$$a_\ell L^{\ell-1}(\mathbf{x}) = \mathbf{0}_V.$$
Since $L^{\ell-1}(\mathbf{x})$ is nonzero, it follows that $a_\ell = 0$.

We have shown, then, that if (9.3) holds, $a_i = 0$ for all i. This implies the result. $\square$

Recall that if $V = W \oplus U$ is an L-decomposition for some L in $\mathcal{L}(V)$, we write $L_W = L|_W$ and we write $L = L_W \oplus L_U$. In this case, we say that L is a direct sum.

The next result realizes our proximate goal.

Theorem 9.10. Let $V \cong \mathbb{F}^n$. Let L belong to $\mathcal{L}(V)$. If the kernel index of L is ℓ, then for $K = \operatorname{Ker} L^\ell$ and $R = L^\ell[V]$, L is the direct sum, $L = L_K \oplus L_R$, where L_K is nilpotent and L_R is invertible.

Proof. Let everything be as hypothesized.

Theorem 9.8 established that $V = K \oplus R$ and we have long established that K and R are L-invariant. We thus have $L = L_K \oplus L_R$.

That L_K is nilpotent with index ℓ is immediate by Definition 9.4. We have left to show that L_R is invertible.

Let $\mathbf{x}$ belong to $\operatorname{Ker} L_R$. Since $\mathbf{x}$ belongs to $R = L^\ell[V]$, we can write $\mathbf{x} = L^\ell(\mathbf{y})$ for some $\mathbf{y}$ in V. This gives us
$$L(\mathbf{x}) = L^{\ell+1}(\mathbf{y}) = \mathbf{0}_V,$$
implying that $\mathbf{y}$ belongs to $\operatorname{Ker} L^{\ell+1} = \operatorname{Ker} L^\ell$. This proves that
$$\mathbf{x} = L^\ell(\mathbf{y}) = \mathbf{0}_V,$$
thus, that $\operatorname{Ker} L_R$ is trivial. It follows that L_R is invertible. $\square$

The next theorem follows Theorem 9.10, giving us a finer L-decomposition of V. Here, the underlying field must contain all of $\mathcal{Z}(L)$.

We have subjected the minimal polynomial for a linear operator on a finite-dimensional vector space to a pretty good workout on this journey. Theorem 9.10 may be disturbing because it is stated in term of the kernel index, which is not terribly transparent, instead of maybe the exponents for the linear factors in the minimal polynomial for L. Our forbearance will soon be rewarded, however. In the meantime, the characteristic polynomial still has secrets to reveal.

Recall that the algebraic multiplicity of λ in $\mathcal{Z}(L)$ is the multiplicity of λ as a root of the characteristic polynomial for L.

Theorem 9.11 (Primary Decomposition). Let $V \cong \mathbb{F}^n$. Let L belong to $\mathcal{L}(V)$. Assume that $\mathcal{Z}(L) = \{\lambda_i\}_{i=1}^k \subseteq \mathbb{F}$. If ℓ_i is the $(L - \lambda_i I)$-kernel index, then V has L-decomposition

$$V = \bigoplus_{i=1}^k \operatorname{Ker}(L - \lambda_i I)^{\ell_i}.$$

Moreover, the dimension of $\operatorname{Ker}(L - \lambda_i I)^{\ell_i}$ is the algebraic multiplicity of λ_i.

Proof. Let everything be as hypothesized. We prove the result by induction on k, the number of L-eigenvalues.

If $\mathcal{Z}(L) = \{\lambda\}$, the characteristic polynomial for L must be $(\lambda - t)^n$ so the algebraic multiplicity of λ is n. Applying Theorem 9.8 to $L - \lambda I$, we have the L-decomposition of V

$$V = \operatorname{Ker}(L - \lambda I)^\ell \oplus (L - \lambda I)^\ell[V]$$

where ℓ is the $(L - \lambda I)$-kernel index. Let $K = \operatorname{Ker}(L - \lambda I)^\ell$ and $R = (L - \lambda I)^\ell[V]$. Applying Theorem 9.10 to $L - \lambda I$, we see that $(L - \lambda I)_R$ is invertible: There are no L-eigenvectors in R. Since

(9.5) $$\mathcal{Z}(L) = \{\lambda\} = \mathcal{Z}(L_K) \cup \mathcal{Z}(L_R)$$

and since $\mathcal{Z}(L) \subseteq \mathbb{F}$ by assumption, it must be the case that $\mathcal{Z}(L_R)$ is empty and that $R = \{\mathbf{0}_V\}$. We conclude that $V = \operatorname{Ker}(L - \lambda I)^\ell$ so that $\dim \operatorname{Ker}(L - \lambda I)^\ell = n$. This proves the result in the base case.

Now suppose $p(t) = \prod_{i=1}^k (\lambda_i - t)^{r_i}$ is the characteristic polynomial for L in $\mathcal{L}(V)$. Let $K = \operatorname{Ker}(L - \lambda_1 I)^{\ell_1}$ and $R = (L - \lambda_1 I)^{\ell_1}[V]$. Since $(L - \lambda_1 I)_R$ is invertible, there are no λ_1-eigenvectors in R. By (9.5), $|\mathcal{Z}(L_R)| < k$. The induction hypothesis thus applies to R and L_R. Again invoking (9.5), we can write

$$R = \bigoplus_{i=2}^k \operatorname{Ker}(L - \lambda_i I)^{\ell_i},$$

where $\prod_{i=2}^k (\lambda_i - t)^{r_i}$ is the characteristic polynomial for L_R and where $\dim \operatorname{Ker}(L - \lambda_i I)^{\ell_i} = r_i$ for $i = 2, \ldots, k$. Notice that $\dim R = \sum_{i=2}^k r_i$.

Corollary 9.3 implies that

$$p(t) = \det(L - tI)_K \det(L - tI)_R = \det(L - tI)_K \prod_{i=2}^k (\lambda_i - t)^{r_i}.$$

It follows that

$$\det(L - tI)_K = (\lambda_1 - t)^{r_1}.$$

Since $n = \dim K + \dim R$ and $\dim R = \sum_{i=2}^k r_i$, we see that

$$\dim K = n - \sum_{i=2}^k r_i = r_1.$$

9.2. The Primary Decomposition

Putting everything together, we have V as the L-sum,

$$V = \bigoplus_{i=1}^{k} \operatorname{Ker}(L - \lambda_i I)^{\ell_i},$$

with $\dim \operatorname{Ker}(L - \lambda_i I)^{\ell_i} = r_i$ for $i = 1, \ldots, k$. This proves the theorem. $\square$

We refer to

$$V = \bigoplus_{i=1}^{r} \operatorname{Ker}(L - \lambda_i I)^{\ell_i}$$

as the **primary L-decomposition of** V. When L is understood, we call it the primary decomposition of V.

Proof of the Primary Decomposition Theorem bears more fruit.

Corollary 9.12. Let $V \cong \mathbb{F}^n$ and let L belong to $\mathcal{L}(V)$. Suppose $\mathcal{Z}(L) \subseteq \mathbb{F}$. The primary L-decomposition of V is

$$V = \bigoplus_{i=1}^{k} \operatorname{Ker}(L - \lambda_i I)^{\ell_i}$$

if and only if the minimal polynomial for L is

$$\prod_{i=1}^{k}(t - \lambda_i)^{\ell_i}.$$

Proof. Let everything be as hypothesized. Assume that

$$V = \bigoplus_{i=1}^{k} \operatorname{Ker}(L - \lambda_i I)^{\ell_i}.$$

Suppose the minimal polynomial for L is $q(t) = \prod_{i=1}^{k}(t - \lambda_i)^{r_i}$ and that for some i, $r_i < \ell_i$. We lose no generality by assuming $i = 1$.

The definition of the kernel index guarantees that there is $\mathbf{x}$ in $\operatorname{Ker}(L - \lambda_1 I)^{\ell_1}$ satisfying

$$(L - \lambda_1 I)^{\ell_1 - 1}(\mathbf{x}) \neq \mathbf{0}_V.$$

Since $L - \lambda_i I$ is invertible on $\operatorname{Ker}(L - \lambda_1 I)^{r_1}$, so is $\prod_{i=2}^{k}(L - \lambda_i I)^{r_i}$. It follows that $q(L)(\mathbf{x})$ is nonzero. This shows that $r_i \geq \ell_i$ for $i = 1, \ldots, k$. Since $\prod_{i=1}^{k}(t - \lambda_i)^{\ell_i}$ annihilates L, $\ell_i \geq r_i$. This is enough to prove the corollary. $\square$

Exercises 9.2.

9.2.1. Consider the statement, "An empty set with one element is filled with gold." Restate it as a logical implication. Is the statement true?

9.2.2. Find a nonzero, noninvertible linear operator L with kernel index of L equal to one.

9.2.3. Referring to Lemma 9.9, find the matrix representation of the restriction of L to the subspace of V with basis $\{\mathbf{x}, \ldots, L^{\ell-1}(\mathbf{x})\}$.

9.2.4. Let L in $\mathcal{L}(\mathbb{R}^3)$ be defined by $L(x,y,z) = (x+y+z, y+z, z)$.
   (a) Find the standard matrix representation for L.
   (b) Find the standard matrix representation for $L - I$.
   (c) Use the matrix for $L - I$ to show that it is nilpotent with index three.

9.2.5. Let L in $\mathcal{L}(\mathbb{R}^3)$ be given by $L(x,y,z) = (x+y+z, x+y+z, x+y+z)$.
   (a) Find a general expression for $L^j(x,y,z)$ when j is a positive integer.
   (b) We noted that 3 is an L-eigenvalue. Is $L - 3I$ nilpotent on $\mathbb{R}^3$?

9.2.6. Let L in $\mathcal{L}(\mathbb{R}^3)$ be given by $L(x,y,z) = (y,z,x)$. Without using calculations, find an L-eigenvalue λ in $\mathbb{R}$. Determine whether $L - \lambda I$ is nilpotent on $\mathbb{R}^3$.

9.2.7. Let $V \cong \mathbb{F}^n$ and let λ be an eigenvalue for L belonging to $\mathcal{L}(V)$. Let ℓ be the $(L - \lambda I)$-kernel index. We know that $L = L_K \oplus L_R$, if we take $K = \mathrm{Ker}(L - \lambda I)^\ell$ and $R = (L - \lambda_i I)[V]$. Is L_K nilpotent? Is L_R invertible? Discuss.

9.2.8. Suppose L belonging to $\mathcal{L}(\mathbb{R}^7)$ has minimal polynomial
$$q(t) = (t-2)^2(t+3)^2(t-5)^3.$$
   (a) Find the primary L-decomposition of $\mathbb{R}^7$.
   (b) What can you say about the characteristic polynomial for L?
   (c) How many L-eigenspaces are there? What are their dimensions?

9.2.9. Repeat the last problem except this time say that the minimal polynomial for L is $q(t) = (t-2)(t+3)^2(t-5)^3$.

9.3. Nilpotent Operators, Part I

A *normal form* for a linear operator on a finite-dimensional vector space is a type of matrix that characterizes the similarity class of the operator. Different normal forms are associated to different conditions restricting the operators under consideration. Our condition is that $\mathcal{Z}(L)$ must be contained in the scalar field. In this section, we find a normal form for nilpotent operators. Much of our approach is modeled after [8].

The background to our study in this section is finite-dimensional
$$V = \bigoplus_{i=1}^{r} \mathrm{Ker}(L - \lambda_i I)^{\ell_i},$$
in its primary decomposition with respect to L in $\mathcal{L}(V)$. Here we focus entirely on one component in the primary decomposition, $K = \mathrm{Ker}(L - \lambda I)^\ell$. Notice that $(L - \lambda I)_K$ has nilpotence index ℓ.

Since we stop thinking about the role of λ and about the role of K as an L-invariant subspace of V, we streamline the discussion and focus on L in $\mathcal{L}(V)$ with nilpotence index ℓ on V.

The only result in this section is a technical lemma. The proof is busy enough to warrant its own section.

9.3. Nilpotent Operators, Part I

Lemma 9.13. Let V be finite-dimensional. Let L in $\mathcal{L}(V)$ be nilpotent with nilpotence index ℓ. Given any $\mathbf{x}$ in V for which $L^{\ell-1}(\mathbf{x})$ is nonzero,

(9.6) $$K = \text{Span}\{\mathbf{x}, L(\mathbf{x}), \ldots, L^{\ell-1}(\mathbf{x})\}$$

has an L-complement in V.

Proof. Let everything be as hypothesized. We prove the lemma by induction on ℓ, the nilpotence index of L.

If $\ell = 1$, then L is identically zero so the result is vacuously true. This proves the lemma in the base case.

Assume that $\ell > 1$. If we let $R = L[V]$, then R is L-invariant. Evidently, L_R is nilpotent. We verify that its nilpotence index is $\ell - 1$.

When $\mathbf{x}$ is as hypothesized, Lemma 9.9 ensures that
$$\{\mathbf{x}, L(\mathbf{x}), \ldots, L^{\ell-1}(\mathbf{x})\}$$
is a basis for $K = \text{Span}\{\mathbf{x}, L(\mathbf{x}), \ldots, L^{\ell-1}(\mathbf{x})\}$. Since $\mathbf{x}$ is not in $\text{Ker}\, L^{\ell-1}$, $L(\mathbf{x})$ is not in the kernel of $L^{\ell-2}(\mathbf{x})$. This verifies that R contains an element that is not mapped to zero by $L^{\ell-2}$ so the nilpotence index of L_R must be $\ell - 1$.

Now we can apply the induction hypothesis to R and L_R, guaranteeing an L_R-complement to $K_0 = L[K]$, H_0, giving us an L_R-decomposition of R,
$$R = K_0 \oplus H_0.$$

Let $H_1 = L^{-1}[H_0] \subseteq V$. Since H_0 is L_R-invariant, we have
$$L_R[H_0] = L[H_0] \subseteq H_0,$$
implying that $H_0 \subseteq H_1$.

We claim that $V = K + H_1$.

Let $\mathbf{v}$ belong to V. Since $L(\mathbf{v})$ is in R, there are unique $\mathbf{k}_0$ in K_0 and $\mathbf{h}_0$ in H_0 such that
$$L(\mathbf{v}) = \mathbf{k}_0 + \mathbf{h}_0.$$
Since $K_0 = L[K]$, we can take $\mathbf{k}$ in K so that $L(\mathbf{k}) = \mathbf{k}_0$. We then have
$$L(\mathbf{v}) - \mathbf{k}_0 = L(\mathbf{v} - \mathbf{k}) = \mathbf{h}_0.$$
Since $\mathbf{v} - \mathbf{k}$ is mapped into H_0, it belongs to H_1. This proves that
$$\mathbf{v} = \mathbf{k} + \mathbf{v} - \mathbf{k} = \mathbf{k} + \mathbf{h}_1,$$
for some $\mathbf{h}_1$ in H_1, thus that $V \subseteq K + H_1$. Since containment in the other direction is given, our claim follows.

Since $L(L^{\ell-1}(\mathbf{x})) = \mathbf{0}$ in H_0, we see that $L^{\ell-1}(\mathbf{x})$ belongs to H_1. Since $L^{\ell-1}(\mathbf{x})$ is also in K, $K \cap H_1$ is nontrivial. This shows that the sum $K + H_1$ is not direct: H_1 is too big to be a direct complement to K in V.

Our quarry now is a subspace of H_1 that forms an L-complement to K. Towards this end, we claim that
$$H_0 \cap (K \cap H_1) = \{\mathbf{0}_V\}.$$
Proving this will establish that, while $H_0 \oplus (K \cap H_1)$ may not be all of K, we can at least form the direct sum of H_0 and $K \cap H_1$ inside K.

Since $H_0 \subseteq H_1$, we see that $H_0 \cap (K \cap H_1) = H_0 \cap K$. Let $\mathbf{v}$ belong to $H_0 \cap K$.

Since $\mathbf{v}$ belongs to K, $L(\mathbf{v})$ belongs to K_0. Since H_0 is L-invariant, $L(\mathbf{v})$ is also in H_0. It follows that $L(\mathbf{v}) = \mathbf{0}$.

As an element in K, $\mathbf{v}$ can be written
$$\mathbf{v} = a_1 \mathbf{x} + \cdots + a_\ell L^{\ell-1}(\mathbf{x}),$$
for some scalars a_i. Applying L to $\mathbf{v}$ we get
$$(9.7) \qquad L(\mathbf{v}) = a_1 L(\mathbf{x}) + \cdots + a_{\ell-1} L^{\ell-1}(\mathbf{x}).$$

As a subset of a linearly independent set, $\{L(\mathbf{x}), \ldots, L^{\ell-1}(\mathbf{x})\}$ is itself linearly idependent. By (9.7), then, $a_i = 0$ for $i = 1, \ldots, \ell - 1$. It follows that if $\mathbf{v}$ belongs to $H_0 \cap K$, then
$$(9.8) \qquad \mathbf{v} = a_\ell L^{\ell-1}(\mathbf{x})$$
for some scalar a_ℓ. In particular, $\mathbf{v}$ is in K_0.

We have now shown that if $\mathbf{v}$ is in $H_0 \cap K$, then $\mathbf{v}$ is in $H_0 \cap K_0$, in other words,
$$H_0 \cap K \subseteq H_0 \cap K_0.$$
Since
$$H_0 \cap K_0 = \{\mathbf{0}_V\},$$
$H_0 \cap K = \{\mathbf{0}\}$. Our claim that $H_0 \cap (K \cap H_1) = \{\mathbf{0}\}$ follows.

We have now established that we can write
$$(9.9) \qquad H_1 = H_0 \oplus (K \cap H_1) \oplus U,$$
where U is any direct complement to $H_0 \oplus (K \cap H_1)$ in H_1.

We saw that H_1 was too big to be a direct complement to K in V. Now we throw $K \cap H_1$ out of H_1 and show that what is left over is a direct complement to K in V. To be precise, we claim that $H_0 \oplus U$ is an L-complement to K in V.

First, we argue that $V = K \oplus (H_0 \oplus U)$.

We have seen that $V = K + H_1$. We can put that together with (9.9) to say that if $\mathbf{v}$ belongs to V, then there are $\mathbf{k}$ in K, $\mathbf{h}_0$ in H_0, $\mathbf{k}'$ in $K \cap H_1$, and $\mathbf{u}$ in U so that
$$\mathbf{v} = \mathbf{k} + (\mathbf{h}_0 + \mathbf{k}' + \mathbf{u}) = (\mathbf{k} + \mathbf{k}') + \mathbf{h}_0 + \mathbf{u}.$$
Since K is a subspace, $\mathbf{k} + \mathbf{k}'$ is in K. This proves that $V \subseteq K \oplus H_0 \oplus U$. Since inclusion in the other direction is given, we have established that $V = K \oplus (H_0 \oplus U)$.

The last step of the proof is to show that $H_0 \oplus U$ is L-invariant.

As U is a subset of H_1, and $H_1 = L^{-1}[H_0]$, $L[U] \subseteq H_0$. Since H_0 is L-invariant, we also know that $L[H_0] \subseteq H_0$. It follows that any element in $H_0 \oplus U$ is mapped into H_0 by L. This proves that $H_0 \oplus U$ is L-invariant. From here, we see that by taking $H = H_0 \oplus U$,
$$V = K \oplus H$$
is an L-decomposition. $\square$

Next, we train our sights on K, as given in (9.6), and on L_K.

Exercises 9.3.

9.3.1. Let V be any finite-dimensional vector space. If there is a matrix representation of L in $\mathcal{L}(V)$ with a nonzero entry on the main diagonal, show that L cannot be nilpotent.

9.3.2. Let $V \cong \mathbb{F}^n$ where $\mathbb{F}$ is algebraically closed. Let L belong to $\mathcal{L}(V)$. Show that L is nilpotent if and only if there is no triangular matrix in the L-similarity class with a nonzero entry on its diagonal.

9.3.3. Let $\dim V = n$. What is the maximum nilpotence index for L in $\mathcal{L}(V)$? Prove your assertion.

9.3.4. Let $L = \begin{bmatrix} 0 & 0 & 0 \\ 1 & 0 & 0 \\ 1 & 1 & 0 \end{bmatrix}$ belong to $\mathbb{F}^3$ for any field $\mathbb{F}$.
  (a) Find all the distinct positive powers of L.
  (b) What is the nilpotence index of L?
  (c) Find $\mathbf{x}$ in $\mathbb{F}^3$ so that $L^{\ell-1}(\mathbf{x}) \neq \mathbf{0}$ where ℓ is the nilpotence index of L.
  (d) Find $\mathcal{B} = \{\mathbf{x}, L(\mathbf{x}), \ldots, L^{\ell-1}(\mathbf{x})\}$.
  (e) Let $\mathbf{b}_i = L^{i-1}(\mathbf{x})$ for $i = 1, \ldots, \ell$. Find the $\mathcal{B}$-coordinate vector for each $L(\mathbf{b}_i)$.

9.3.5. Taking K and H_1 as in the proof of Lemma 9.13, show that $K \cap H_1 = \mathrm{Span}\{L^{\ell-1}(\mathbf{x})\}$.

9.4. Nilpotent Operators, Part II

We are still working under the assumptions of Lemma 9.13. Let $V \cong \mathbb{F}^n$ and let L in $\mathcal{L}(V)$ be nilpotent with nilpotence index ℓ. For $\mathbf{x}$ in V with $L^{\ell-1}(\mathbf{x}) \neq \mathbf{0}_V$, let

$$K = \mathrm{Span}\{\mathbf{x}, L(\mathbf{x}), \ldots, L^{\ell-1}(\mathbf{x})\}.$$

We treat $\mathcal{B} = \{\mathbf{x}, L(\mathbf{x}), \ldots, L^{\ell-1}(\mathbf{x})\}$ as an ordered basis for K. Write $\mathbf{b}_1 = \mathbf{x}$ and $\mathbf{b}_i = L^{i-1}(\mathbf{x})$, for $i = 2, \ldots, \ell$, so that

$$L(\mathbf{b}_i) = \mathbf{b}_{i+1},$$

for i in $\{1, \ldots, \ell - 1\}$, while $L(\mathbf{b}_\ell) = \mathbf{0}_V$. This gives us

$$[L(\mathbf{b}_i)]_\mathcal{B} = \mathbf{e}_{i+1},$$

for i in $\{2, \ldots, \ell - 1\}$ and $[L(\mathbf{b}_\ell)]_\mathcal{B} = \mathbf{0}_n$. The matrix representation of L_K with respect to $\mathcal{B}$ is then

$$[\![L_K]\!]_\mathcal{B} = (\mathbf{e}_2, \ldots, \mathbf{e}_\ell, \mathbf{0}).$$

More explicitly, $[\![L_K]\!]_\mathcal{B}$ is the $\ell \times \ell$ matrix

$$\begin{bmatrix} 0 & \cdots & & \cdots & 0 \\ 1 & 0 & & \cdots & 0 \\ 0 & 1 & 0 & \cdots & 0 \\ & & \ddots & & \\ 0 & 0 & \cdots & 1 & 0 \end{bmatrix}.$$

Definition 9.14. The **subdiagonal** of an $n \times n$ matrix $A = [a_{ij}]$ is the $(n-1)$-tuple
$$(a_{21}, \ldots, a_{n\,n-1}).$$
The **super-diagonal** of A is the $(n-1)$-tuple
$$(a_{12}, \ldots, a_{n-1\,n}).$$
A **sub-1 matrix** is an $n \times n$ matrix A with subdiagonal
$$(1, \ldots, 1)$$
and all other entries equal to zero.

A **super-1 matrix** is an $n \times n$ matrix A with super-diagonal
$$(1, \ldots, 1)$$
and all other entries equal to zero.

Notice that $A = [a_{ij}]$ is a sub-1 matrix provided $a_{ij} = \delta(i, j+1)$ and that A is a super-1 matrix provided $a_{ij} = \delta(i, j-1)$.

Whatever we say about sub-1 matrices can be adjusted to apply to super-1 matrices.

Notice that $[\![L_K]\!]_\mathcal{B}$ is the $\ell \times \ell$ sub-1 matrix. In particular, there is no 1×1 sub-1 matrix, and for k in $\mathbb{Z}_{\geq 2}$, there is exactly one $k \times k$ sub-1 matrix.

The 2×2 sub-1 matrix is
$$A = \begin{bmatrix} 0 & 0 \\ 1 & 0 \end{bmatrix}.$$
The 3×3 sub-1 matrix is
$$\begin{bmatrix} 0 & 0 & 0 \\ 1 & 0 & 0 \\ 0 & 1 & 0 \end{bmatrix}.$$
While
$$B = \begin{bmatrix} 0 & 0 & 0 \\ 1 & 0 & 0 \\ \hline 0 & 0 & 0 \end{bmatrix}$$
is nilpotent, it is not a sub-1 matrix. It is, however, a diagonal square-block matrix $\mathrm{diag}(A, O_1)$, where A is the 2×2 sub-1 matrix and O_1 is the 1×1 zero matrix. With this observation, we see a normal form for nilpotent operators taking shape.

We have one more lemma before our theorem.

Lemma 9.15. Let $V \cong \mathbb{F}^n$ and let L in $\mathcal{L}(V)$ be nilpotent with nilpotence index $\ell > 1$. The L-decomposition $V = K \oplus H$, where
$$K = \mathrm{Span}\{\mathbf{x}, L(\mathbf{x}), \ldots, L^{\ell-1}(\mathbf{x})\}$$
for $\mathbf{x}$ in V satisfying $L^{\ell-1}(\mathbf{x}) \neq \mathbf{0}$, is unique up to isomorphism.

Proof. Let everything be as hypothesized. Suppose there is $\mathbf{y}$ in V, so that $L^{\ell-1}(\mathbf{y})$ is nonzero. For
$$K' = \mathrm{Span}\{\mathbf{y}, \ldots, L^{\ell-1}(\mathbf{y})\}$$
and H' an L-complement to K', $\dim K = \dim K' = \ell$ implies $\dim H = \dim H' = n - \ell$. This proves that $K \cong K'$ and $H \cong H'$, which is what we had to show. $\square$

9.4. Nilpotent Operators, Part II

Notice that the lemma says that the dimensions of K and H are uniquely determined by L, not that K and H themselves are uniquely determined by L.

We can now prove our normal form theorem for nilpotent operators.

Theorem 9.16. Let $V \cong \mathbb{F}^n$. If L in $\mathcal{L}(V)$ is nilpotent, then V has a basis, $\mathcal{B}$, such that $[\![L]\!]_\mathcal{B} = \mathrm{diag}(A_1, \ldots, A_k)$, where each square-block matrix A_i is either a sub-1 matrix or a zero matrix. If there is another basis for V, $\mathcal{C}$, such that $[\![L]\!]_\mathcal{C} = \mathrm{diag}(B_1, \ldots, B_r)$, where each square-block B_i is either a sub-1 matrix or a zero matrix, then $r = k$, and for some permutation σ in S_k, $B_{\sigma(i)} = A_i$, for $i = 1, \ldots, k$.

Proof. Let L and V be as hypothesized. We proceed by induction on $\dim V$.

If $\dim V = 1$, then $L = O$ and its only matrix representation is $[0]$. This proves the theorem in the base case.

Assume $\dim V > 1$. Let ℓ be the nilpotence index of L.

If there is no $\mathbf{x}$ in V so that $L^{\ell-1}(\mathbf{x}) \neq \mathbf{0}_V$, then $\ell = 1$ so $L = O$. In this case, the only matrix representation of L is O_n. This proves the lemma when $\ell = 1$.

We go forward assuming $\ell > 1$.

Choose $\mathbf{x}$ in V so that $L^{\ell-1}(\mathbf{x}) \neq \mathbf{0}$. Let

$$\mathcal{B}_1 = \{\mathbf{x}, L(\mathbf{x}), \ldots, L^{\ell-1}(\mathbf{x})\}$$

and let $K = \mathrm{Span}\,\mathcal{B}_1$ so $\dim K = \ell$. By Lemma 9.13, V has an L-decomposition

$$V = K \oplus H$$

so that $\dim H = n - \ell < n$. Since L_H is nilpotent, the induction hypothesis ensures that H has a basis with respect to which L_H has the form

$$\mathrm{diag}(A_2, \ldots, A_k),$$

each A_i a sub-1 matrix or a zero matrix. Treating $\mathcal{B}_1$ as an ordered basis for K, we have

$$A_1 = [\![L_K]\!]_{\mathcal{B}_1} = \begin{bmatrix} 0 & \cdots & & \cdots & 0 \\ 1 & 0 & & \cdots & 0 \\ 0 & 1 & 0 & \cdots & 0 \\ & & \ddots & & \\ 0 & 0 & \cdots & 1 & 0 \end{bmatrix},$$

the $\ell \times \ell$ sub-1 matrix. It follows that L has matrix representation

$$\mathrm{diag}(A_1, \ldots, A_k),$$

in which each A_i is a sub-1 matrix or a zero matrix. This proves the existence part of the theorem.

Since each square-block is associated to an L-invariant subspace K as defined in (9.6), the uniqueness of the L-decomposition $V = K \oplus H$, up to isomorphism, implies uniqueness of the square-blocks up to permutation. This proves the theorem. $\square$

Exercises 9.4.

9.4.1. Show that the transpose of a sub-1 matrix is a super-1 matrix.

9.4.2. Suppose L in $\mathcal{L}(V)$ has a sub-1 matrix representation with respect to the ordered basis $\mathcal{B} = \{\mathbf{b}_1, \ldots, \mathbf{b}_n\}$. Find a basis $\mathcal{C}$ such that $[\![L]\!]_{\mathcal{C}}$ is a super-1 matrix.

9.4.3. Let $A = \text{diag}(A_1, \ldots, A_k)$ be a diagonal square-block matrix. Show that $A^T = \text{diag}(A_1^T, \ldots, A_k^T)$.

9.4.4. Let
$$A = \begin{bmatrix} 0 & 0 & 0 & 0 \\ 1 & 0 & 0 & 0 \\ 0 & 0 & 0 & 0 \\ 0 & 0 & 1 & 0 \end{bmatrix}.$$

Let
$$B = \begin{bmatrix} 0 & 0 & 0 & 0 \\ 1 & 0 & 0 & 0 \\ 0 & 1 & 0 & 0 \\ 0 & 0 & 1 & 0 \end{bmatrix}.$$

(a) Identify the sub-1 and zero square-blocks for each matrix.
(b) Find the nilpotence indices for A and B.
(c) Find the minimal polynomials and the characteristic polynomials for A and B.
(d) Are A and B similar?

9.4.5. Let $A = [a_{ij}]$, where $a_{ij} = 0$ or 1.
(a) List all the 2×2 versions of A that are nilpotent.
(b) Repeat the exercise for 3×3 matrices. Group the matrices on the list by similarity class. Find the diagonal square-block form given in Theorem 9.16 for each similarity class.

9.5. Jordan Canonical Form

We start with a definition. Note that a scalar matrix is one of the form cI_n for some scalar c.

Definition 9.17. A **Jordan block** matrix is either a 1×1 matrix or the sum of a scalar matrix and a sub-1 matrix.

Any sub-1 matrix is a Jordan block.

Example 9.18. Since $A = \begin{bmatrix} 2 & 0 & 0 \\ 1 & 2 & 0 \\ 0 & 1 & 2 \end{bmatrix}$ is the sum of $2I_2$ and the 3×3 sub-1 matrix, A is a Jordan block matrix. Note that $B = \begin{bmatrix} 2 & 0 & 0 \\ 1 & 1 & 0 \\ 0 & 1 & 2 \end{bmatrix}$ is not a Jordan block matrix.

9.5. Jordan Canonical Form

As in the last section, everything we do here can be recast in terms of super-1 matrices, starting with the definition of Jordan block matrix. We made the choice to stick with sub-1 matrices.

Note that a Jordan block matrix is associated to exactly one eigenvalue and that the associated eigenspace is 1-dimensional. An $n \times n$ diagonal matrix is the sum of n 1×1 Jordan blocks.

Definition 9.19. A matrix in diagonal square-block form $J = \text{diag}(J_1, \ldots, J_k)$ is in **Jordan canonical form** or **Jordan form** provided each J_i is a Jordan block matrix.

There are two types of 2×2 Jordan form matrices over a field $\mathbb{F}$. The first has the form
$$\begin{bmatrix} \lambda_1 & 0 \\ 0 & \lambda_2 \end{bmatrix}$$
for scalars λ_1, λ_2 that are not necessarily distinct. This is a block diagonal matrix with two Jordan blocks. The second type has the form
$$\begin{bmatrix} \lambda & 0 \\ 1 & \lambda \end{bmatrix}$$
for a scalar λ. This is a single Jordan block, equal to the sum
$$\begin{bmatrix} \lambda & 0 \\ 0 & \lambda \end{bmatrix} + \begin{bmatrix} 0 & 0 \\ 1 & 0 \end{bmatrix}.$$

The theorem we have been working towards is this.

Theorem 9.20 (Jordan Canonical Form). *Let $V \cong \mathbb{F}^n$ and let L belong to $\mathcal{L}(V)$. If $\mathcal{Z}(L) \subseteq \mathbb{F}$, then L has a matrix representation in Jordan form. The Jordan form for an operator is unique up to permutation of its Jordan blocks.*

The proof is behind us. We just need some discussion to put everything together.

Assume throughout that we are working under the hypotheses of Theorem 9.20.

Let ℓ be the $(L - \lambda I)$-kernel index for λ in $\mathcal{Z}(L)$. Let $\mathcal{B}_\lambda$ be any basis for $K_\lambda = \text{Ker}(L - \lambda I)^\ell$. The Primary Decomposition Theorem implies that the disjoint union

(9.10) $$\mathcal{B} = \bigcup_{\lambda \in \mathcal{Z}(L)} \mathcal{B}_\lambda$$

is a basis for V. Moreover, if $\mathcal{Z}(L) = \{\lambda_1, \ldots, \lambda_r\}$ and we order $\mathcal{B}$ in the obvious fashion (we leave the details to an exercise), then

(9.11) $$[\![L]\!]_\mathcal{B} = \text{diag}(A_1, \ldots, A_r),$$

where each A_i has the form $[\![L_{K_\lambda}]\!]_{\mathcal{B}_\lambda}$ for a distinct element λ in $\mathcal{Z}(L)$.

The primary decomposition allows us to deal with one L-eigenvalue at a time. Fix λ in $\mathcal{Z}(L)$ and let $A_\lambda = [\![L_{K_\lambda}]\!]_{\mathcal{B}_\lambda}$. Our job now is to argue that we can choose $\mathcal{B}_\lambda$ so that A_λ is in Jordan form. We have already made the argument. This will be evident once we put two more observations together.

The first observation is that $L \mapsto [\![L]\!]_\mathcal{B}$ is a linear transformation from $\mathcal{L}(V)$ to $M_n(\mathbb{F})$. Since we can write
$$L = (L - \lambda I) + \lambda I,$$
we have
$$[\![L]\!]_\mathcal{B} = [\![L - \lambda I]\!]_\mathcal{B} + [\![\lambda I]\!]_\mathcal{B}.$$
Since $[\![\lambda I]\!]_\mathcal{B} = \lambda I_n$, we have
$$[\![L]\!]_\mathcal{B} = [\![L - \lambda I]\!]_\mathcal{B} + \lambda I_n.$$
The application here is to A_λ, for a given choice of $\mathcal{B}_\lambda$:

(9.12) $$A_\lambda = [\![L_{K_\lambda}]\!]_{\mathcal{B}_\lambda} = [\![(L - \lambda I)_{K_\lambda}]\!]_{\mathcal{B}_\lambda} + \lambda I_{m_\lambda},$$
where
$$m_\lambda = \dim K_\lambda = \dim \mathrm{Ker}(L - \lambda I)^\ell.$$

Note that the sum $(L - \lambda I) + \lambda I$ is a sum of elements in $\mathcal{L}(K_\lambda)$. Likewise, the sum in (9.12) is the sum of two square matrices.

If $\ell = 1$, then $\mathrm{Ker}(L - \lambda I) = K_\lambda$ is the λ-eigenspace in V. In this case,
$$[\![(L - \lambda I)_{K_\lambda}]\!]_{\mathcal{B}_\lambda} = O_{m_\lambda}$$
so the sum in (9.12) is the scalar matrix, λI_{m_λ}. As noted above, λI_{m_λ} is a sum of m_λ 1×1 Jordan blocks of the form $[\ \lambda\ ]$. We proceed with the discussion then assuming that $\ell > 1$.

The second observation snaps into focus when we look at Lemma 9.9 and Lemma 9.13 together. From Lemma 9.9 and the definition of the $(L - \lambda I)$-kernel index, we know that there is $\mathbf{x}$ in K_λ such that

(9.13) $$\mathcal{B}_\mathbf{x} = \{\mathbf{x}, (L - \lambda I)(\mathbf{x}), \ldots, (L - \lambda I)^{\ell-1}(\mathbf{x})\}$$

is linearly independent in K_λ. Let $K_\mathbf{x}$ be the span of $\mathcal{B}_\mathbf{x}$. From Lemma 9.13, we know that $K_\mathbf{x}$ has an L-complement, $H_\mathbf{x}$, in K_λ. Take $\mathcal{B}_\mathbf{x}$ to be an ordered basis and $\mathcal{C}$ to be any basis for $H_\mathbf{x}$. Let
$$\mathcal{B}_\lambda = \mathcal{B}_\mathbf{x} \cup \mathcal{C}.$$
If we order $\mathcal{B}_\lambda$ appropriately, we have
$$A_\lambda = [\![L_{K_\lambda}]\!]_{\mathcal{B}_\lambda} = \mathrm{diag}(B_1, B_2),$$
where
$$B_1 = [\![L_{K_\mathbf{x}}]\!]_{\mathcal{B}_\mathbf{x}} = [\![(L - \lambda I)_{K_\mathbf{x}}]\!]_{\mathcal{B}_\mathbf{x}} + \lambda I_\ell$$
and $B_2 = [\![L_{H_\mathbf{x}}]\!]_\mathcal{C}$. Since $[\![(L - \lambda I)_{K_\mathbf{x}}]\!]_{\mathcal{B}_\mathbf{x}}$ is the $\ell \times \ell$ sub-1 matrix, B_1 is in Jordan canonical form. This gives us
$$A_\lambda = \mathrm{diag}(B_1, B_1'),$$
where B_1 is in Jordan form.

Since B_1 is a matrix representation of L restricted to a subspace of K_λ, we are not done with L_{K_λ}, unless $K_\mathbf{x}$ is all of K_λ. In this case, $H_\mathbf{x}$ is nontrivial, and we move on to evaluate the $(L - \lambda I)_{H_\mathbf{x}}$-kernel index, $\ell_\mathbf{x}$. Note that $\ell_\mathbf{x} \leq \ell$. If $\ell_\mathbf{x} = 1$, $L_{H_\mathbf{x}} = \lambda I$ in which case
$$A_\lambda = \mathrm{diag}(B_1, \lambda I_j),$$

9.5. Jordan Canonical Form

where $j = \dim H_\mathbf{x}$. If $\ell_\mathbf{x} > 1$, we produce a linearly independent ordered set
$$\mathcal{B}_\mathbf{y} = \{\mathbf{y}, (L - \lambda I)(\mathbf{y}), \ldots, (L - \lambda I)^{\ell_\mathbf{x}-1}(\mathbf{y})\} \subseteq H_\mathbf{x}$$
and apply our handiwork now to $K_\mathbf{y} = \operatorname{Span} \mathcal{B}_\mathbf{y}$. This puts us back to (9.13). Following through, we find
$$B_1' = \operatorname{diag}(B_2, B_2'),$$
where B_2 is a Jordan block.

Repeating this process as many times as necessary, we get
$$A_\lambda = \operatorname{diag}(B_1, B_2, \ldots, B_t),$$
where each B_i is a Jordan block.

Having taken care of A_λ, we return to the remaining pool of L-eigenvalues and start all over again.

Since V is finite-dimensional, we finish eventually with $[\![L]\!]_\mathcal{B}$ in Jordan canonical form.

The Jordan blocks themselves are completely determined by the $(L-\lambda I)$-kernel indices and the subspace structure of each K_λ, in particular, the dimensions of the respective subspaces
$$\operatorname{Ker}(L - \lambda I) \subseteq \cdots \subseteq \operatorname{Ker}(L - \lambda I)^\ell.$$

Those dimensions determine, for instance, whether our very first $K_\mathbf{x}$ exhausts all of K_λ or not. The ordering of the Jordan blocks may vary. This is what it means when we say that the Jordan canonical form is unique up to permutations on the Jordan blocks.

Example 9.21. Let $L(x, y, z) = (2x + z, 2y + z, z)$ on $V = \mathbb{R}^3$. We leave it as an easy exercise to verify that $\mathcal{Z}(L) = \{2, 1\}$.

We have $(L - 2I)(x, y, z) = (z, z, -z)$ so that
$$\operatorname{Ker}(L - 2I) = \operatorname{Span}\{(1, 0, 0), (0, 1, 0)\}.$$
This is a subspace of $K_2 = \operatorname{Ker}(L - 2I)^\ell$, where ℓ is the $(L - 2I)$-kernel index. Since $\dim V = 3$ and there is another L-eigenvalue to go, we see that $\ell = 1$. Double-checking, we calculate
$$(L - 2I)^2(x, y, z) = (-z, -z, z)$$
so that $\operatorname{Ker}(L - 2I)^2 = \operatorname{Ker}(L - 2I)$ as we suspected.

Notice that $\ell = 1$ is the exponent for $t - 2$ in the minimal polynomial for L, while $m_2 = 2 = \dim \operatorname{Ker}(L - 2I)$ is the exponent for $2 - t$ in the characteristic polynomial for L. Since $\ell = 1$, the Jordan blocks associated to $\lambda = 2$ look like $\begin{bmatrix} 2 \end{bmatrix}$. How many are there? There must be two, since $\dim \operatorname{Ker}(L - 2I) = 2$.

We are done with $\lambda = 2$, so we move on to $\lambda = 1$. Checking that
$$(L - I)(x, y, z) = (x + z, y + z, 0),$$
we see that we can take $\{(-1, -1, 1)\}$ as a basis for $\operatorname{Ker}(L - I)$. Again, there is no room in V for the $(L - I)$-kernel index to be any bigger than one but we check anyway to see that
$$(L - I)^2(x, y, z) = (x + z, y + z, 0),$$

so that $\text{Ker}(L - I) = \text{Ker}(L - I)^2$, as we knew it had to be. The associated Jordan block is $\begin{bmatrix} 1 \end{bmatrix}$.

The exponent for $t - 1$ in both the minimal and characteristic polynomials for L is 1.

Taking $\mathcal{B} = \{(1, 0, 0), (0, 1, 0), (-1, -1, 1)\}$, we have $[\![L]\!]_\mathcal{B}$, the Jordan canonical form for L, given by

$$\begin{bmatrix} 2 & 0 & 0 \\ 0 & 2 & 0 \\ 0 & 0 & 1 \end{bmatrix}.$$

We marked off the Jordan blocks just to emphasize that there are three.

It is nice to see how all these subspaces and bases play out in a simple example. At the same time, the example suggests that we can come by the Jordan form for a linear transformation from a couple of different angles.

Suppose, for instance, that all we know about L is that its characteristic polynomial is $p(t) = (2-t)^2(1-t)$ and that its minimal polynomial is $q(t) = (t-2)(t-1)$. First off, since $\deg p(t) = 3$, we are looking at a linear operator on $V \cong \mathbb{F}^3$, for some field $\mathbb{F}$. Presumably, $\text{char}\,\mathbb{F} \neq 2$. Next, Theorem 8.44 tells us that L is diagonalizable. The exponent for $\lambda - t$ in the characteristic polynomial for L is the dimension of $\text{Ker}(L - \lambda I)^\ell$ so we know that $\dim \text{Ker}(L - 2I) = 2$ and $\dim \text{Ker}(L - I) = 1$. This is enough. We know that for some ordered basis $\mathcal{B}$ for V,

$$[\![L]\!]_\mathcal{B} = \begin{bmatrix} 2 & 0 & 0 \\ 0 & 2 & 0 \\ 0 & 0 & 1 \end{bmatrix}.$$

Exercises 9.5.

9.5.1. Show that the eigenspace associated to a single Jordan block is 1-dimensional.

9.5.2. Prove that the union in (9.10) is disjoint.

9.5.3. Describe how $\mathcal{B}$ must be ordered in (9.10) so that $[\![L]\!]_\mathcal{B}$ is in diagonal square-block form.

9.5.4. Find the Jordan canonical form for $\begin{bmatrix} 2 & 1 \\ 0 & 3 \end{bmatrix}$. How many Jordan blocks does it have?

9.5.5. List the types of 3×3 Jordan form matrices, the way we listed the types of 2×2 Jordan form matrices. How many Jordan blocks does each type have?

9.5.6. Let $L(x, y, z) = (2x + z, 2y + z, x + y + z)$ on $\mathbb{R}^3$.
(a) Find the characteristic polynomial for L.
(b) Find the Jordan canonical form for L.

9.5.7. Let $L(x, y, z, w) = (2x, x + 2y, 2z, 3w)$ on $\mathbb{R}^4$.
(a) What is the standard matrix representation of L?
(b) Reorder the natural basis so that the matrix representation is in Jordan form with sub-1 matrices.

(c) Find the $(L-2I)$-kernel index directly, that is, by calculating $\text{Ker}(L-2I)^j$ as often as necessary.

(d) Write down the characteristic polynomial and the minimal polynomial without doing any more calculations.

9.5.8. Suppose that all you know about L in $\mathcal{L}(V)$ is that its characteristic polynomial is $p(t) = (2-t)^4(1+t)^3(5-t)^2$ and its minimal polynomial is $q(t) = (t-2)^3(t+1)^2(t-5)$. What can you say about its Jordan canonical form?

Chapter 10

$GL_n(\mathbb{F})$ and Friends

The general linear group associated to a vector space V is $GL(V)$, the full group of invertible linear operators on V. When V is n-dimensional over $\mathbb{F}$, we identify $GL(V)$ with $GL_n(\mathbb{F})$, the invertible $n \times n$ matrices over $\mathbb{F}$. The **classical groups** are subgroups of $GL_n(\mathbb{F})$, when $\mathbb{F}$ is the real or complex numbers. This chapter is a brief survey of the classical groups and a few other associated structures.

Studies of the classical groups sometimes include matrix groups over the quaternions. This requires more algebra than we are prepared to tackle. Nonetheless, the quaternions are an irresistible corner in the world of vector spaces so we introduce them in this chapter, along with some of the associated geometry.

One way to view $GL(V)$ is as the set of linear operators that permute the set of all ordered bases for V. In particular, $GL_n(\mathbb{F})$ is the group of matrices that, in their action on $\mathbb{F}^n$, leave the collection of ordered bases for $\mathbb{F}^n$ invariant. The idea of invariance is a theme in the study of the classical groups: Each of the classical groups can be associated to some linear structure that it respects, that is, that it leaves invariant. The orthogonal group of real $n \times n$ matrices, $O_n(\mathbb{R})$, leaves the dot product on $\mathbb{R}^n$ invariant and, in so doing, leaves the collection of ordered orthonormal bases for $\mathbb{R}^n$ invariant. One of our projects in this chapter is to identify classical groups in these sorts of terms.

There are several instances in the sequel in which we reference *unit spheres*. In this chapter, the locution *unit n-sphere* and the notation S^n refer to the subset of $\mathbb{R}^{n+1}$ comprised of vectors of length 1:

$$S^n := \{\mathbf{v} \in \mathbb{R}^{n+1} \mid \|\mathbf{v}\| = 1\}.$$

When $n = 1$, S^1 is the unit circle, that is, the set $\{(x, y) \mid x^2 + y^2 = 1\}$.

While this is the end of our study of linear operators, it is the beginning of an inquiry into a venerable subject that informs many different areas of mathematics and applications, among them algebraic geometry, number theory, topology, and physics. As a farewell tour of sorts, we keep our treatment brief and our ambition in check. The aim of the discussion is to whet the appetite for deeper study.

10.1. More about Groups

We need a bit more group theory before we start digging into subgroups of $GL_n(\mathbb{F})$.

Recall from Definition 5.38 that a group is a set G with an associative binary operation, indicated by juxtaposition, an identity element e, and an inverse g^{-1} for every element g in G.

When the binary operation on a group is commutative, the group is abelian. When discussing abelian groups abstractly, we use additive notation. Recall that every vector space is an abelian group under addition.

Our first lemma collects some facts about groups. The proofs were either exercises in earlier parts of the text or imitations of the analogous proofs for fields or $\mathbb{F}$-algebras.

Lemma 10.1. *The following results hold in any group G.*

(a) The identity element in G is unique.

(b) The inverse of any element in G is unique.

(c) If g is in G, then $(g^{-1})^{-1} = g$.

(d) If g and h are in G, then $(gh)^{-1} = h^{-1}g^{-1}$.

The next lemma helps streamline some of our discussion. Problems in Exercises 5.5 addressed the proof.

Lemma 10.2. *Let G be a group with identity element e. Suppose g and h belong to G.*

(a) If $gh = e$, then $h = g^{-1}$.

(b) If $gh = g$, then $h = e$.

There is a version of the next theorem that applies to vector spaces. Articulating it is an exercise.

Lemma 10.3. *If G is a group and H is a nonempty subset of G, then H is a subgroup of G if and only if hk^{-1} is in H for all h and k in H.*

Proof. Let G be a group with nonempty subset H.

Suppose H is a subgroup of G. Let h and k belong to H. Since H is closed under inverses, k^{-1} belongs to H. Since H is closed under the binary operation, hk^{-1} is also in H. This proves the lemma in one direction.

Suppose now that H is a nonempty subset of G such that hk^{-1} belongs to H for all h and k in H. Since H is nonempty, there is h in H so that $hh^{-1} = e$ is in H. Given any h in H, we have $eh^{-1} = h^{-1}$ in H so H is closed under inverses. If h and k are in H, then $h(k^{-1})^{-1} = hk$ is in H, so H is closed under the binary operation. Since the binary operation is associative on all of G, it is associative on H. The proves that H is a subgroup of G. $\square$

Corollary 10.4. *Let G be a group with identity element e. If H is a subgroup of G, then H contains e.*

10.1. More about Groups

Every field is an abelian group under addition and the nonzero elements in a field comprise an abelian group under multiplication. In Exercise 1.1.14, we alluded to the additive analog to exponential notation. In general, if $(G, +)$ is an abelian group, the identity element is 0_G. If 0 is the integer, we define $0g := 0_G$. For n in $\mathbb{Z}_+$,
$$ng := \overbrace{g + \cdots + g}^{n \text{ summands}}$$
and
$$(-n)g := \overbrace{(-g) + \cdots + (-g)}^{n \text{ summands}}.$$

Example 10.5. It is easy to see that $\mathbb{Z}$ is an abelian group under addition. Consider $2\mathbb{Z} = \{\ldots, -4, -2, 0, 2, 4, 6, \ldots\}$ and notice that if $2a$ and $2b$ belong to $2\mathbb{Z}$, then
$$2a - 2b = 2(a - b) \in 2\mathbb{Z}.$$
By Lemma 10.3, $2\mathbb{Z}$ is a subgroup of $\mathbb{Z}$.

A cornerstone of group theory is Cayley's Theorem, which says that every group is a subgroup of the group of permutations on some set. We can see $GL(V)$ as the group of linear permutations on the set of all ordered bases of V. We listed all the elements of $GL_2(\mathbb{F}_2)$ in Example 5.40. Notice the bijective correspondence between ordered bases for $\mathbb{F}_2{}^2$ and matrices on that list.

If g belongs to a group G, the set $\{g^n \mid n \in \mathbb{Z}\}$ is the **group generated by** g. We saw this idea play out in $\mathcal{L}(V)$ when we looked at the polynomials in a linear operator L. In a group, there is only one binary operation, and every element is invertible so instead of polynomials in a single element, the group generated by a single element is the collection of all powers of that element.

Let $G = GL_2(\mathbb{F}_2)$. Taking $g = \begin{bmatrix} 1 & 0 \\ 1 & 1 \end{bmatrix}$ in G, we have $g^2 = \begin{bmatrix} 1 & 0 \\ 0 & 1 \end{bmatrix}$. This means $g = g^{-1}$, so for all integers n, $g^n = g$ or e. We say that the **order of** g is two and we write $|g| = 2$. Since
$$H_1 = \{e, g\}$$
is closed under multiplication and contains all of its elements' inverses, H_1 is an order two subgroup of G. Since H_1 is generated by a single element, g, it is **cyclic** with generator g. We indicate this by writing $H_1 = \langle g \rangle$.

If we take $h = \begin{bmatrix} 0 & 1 \\ 1 & 1 \end{bmatrix}$ in G, we find that $h^2 = \begin{bmatrix} 1 & 1 \\ 1 & 0 \end{bmatrix}$ and that $h^3 = e$. From here it is easy to verify that $H_2 = \{e, h, h^2\}$ is another subgroup of G. Both H_2 and h have order three.

The conjugation mapping, $x \mapsto axa^{-1}$, is defined on a multiplicative group for any element a in the group. When H is a subgroup of G and a is an element in G, the set
$$aHa^{-1} = \{aha^{-1} \mid h \in H\}$$
is called a **conjugate of** H. A conjugate of a subgroup H is also a subgroup in G. A conjugate of H may or may not be equal to H itself.

Still working in $G = GL_2(\mathbb{F}_2)$ with g and h defined as above, we have
$$gH_2g^{-1} = \{e, ghg^{-1}, gh^2g^{-1}\}.$$
Following through on the matrix products, we find that
$$gH_2g^{-1} = H_2.$$
Does $gH_2g^{-1} = H_2$ for every g in G? We leave that as an exercise.

What about conjugates of H_1?
$$hH_1h^{-1} = \{e, hgh^{-1}\} = \left\{ \begin{bmatrix} 1 & 0 \\ 0 & 1 \end{bmatrix}, \begin{bmatrix} 0 & 1 \\ 1 & 1 \end{bmatrix} \right\}.$$
We leave it as an exercise to show that hH_1h^{-1} is another subgroup of $GL_2(\mathbb{F}_2)$ that we have not yet considered.

Exercises 10.1.

10.1.1. In this problem, we consider two *dihedral groups*, the groups of symmetries on regular polygons. In each case, we have a regular n-gon with its vertices labeled $1, 2, \ldots, n$. There is a corresponding set of points in the xy-plane, also labeled. Think of that set of points in the xy-plane as parking places for the vertices of the n-gon. When we meet the n-gon, it is parked so that the label on each vertex matches the label of its parking spot. A symmetry of the n-gon is any positioning of the n-gon so that each vertex lies in a parking place. To repark the n-gon, you are allowed to rotate it and to reflect it across its axes of symmetry so that each vertex winds up in a parking spot. The symmetry group for the n-gon is the full set of different parking configurations of the n-gon. Figure 10.1 shows four

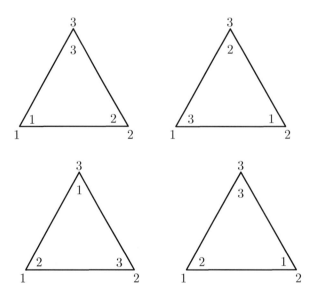

Figure 10.1. Four symmetries of an equilateral triangle.

different symmetries of an equilateral triangle. The axes of symmetry for an equilateral triangle are its medians.
(a) Show that the symmetry group of an equilateral triangle is the same as the permutation group on $\{1,2,3\}$, $\mathcal{S}_3$.
(b) Determine the symmetry group of a square. Is it the same as $\mathcal{S}_4$, the group of permutations on $\{1,2,3,4\}$?

10.1.2. Formulate a version of Lemma 10.3 that applies to a subspace of a vector space.

10.1.3. Show that for any k in $\mathbb{Z}$, $k\mathbb{Z} = \{ka \,|\, a \in \mathbb{Z}\}$ is a subgroup of $\mathbb{Z}$.

10.1.4. Show that if G is a group with subgroups H and K, then $H \cap K$ is a subgroup of G.

10.1.5. A **permutation matrix** in $GL_n(\mathbb{F})$ is a product of type-1 elementary matrices.
(a) Show that the permutation matrices form a subgroup of $GL_n(\mathbb{F})$.
(b) Let P in $GL_n(\mathbb{F})$ be a permutation matrix. Describe $P[\mathcal{E}]$, where $\mathcal{E}$ is the standard ordered basis for $\mathbb{F}^n$.

10.1.6. Show that $GL_2(\mathbb{R})$ is a subgroup of $GL_2(\mathbb{C})$.

10.1.7. Is $GL_2(\mathbb{F}_2)$ a subgroup of $GL_2(\mathbb{F}_3)$?

10.1.8. Find two matrices in $GL_2(\mathbb{C})$ that are not in $GL_2(\mathbb{R})$.

10.1.9. Find all subgroups of $GL_2(\mathbb{F}_2)$.

10.1.10. Find the subgroup of $GL_3(\mathbb{F}_2)$ that maps the standard basis to itself, possibly with a different order.

10.1.11. Find the subgroup of $GL_3(\mathbb{R})$ that maps the standard basis to itself, possibly with a different order.

10.1.12. Let G be a group and let g belong to G. Show that
$$\langle g \rangle = \{g^n \,|\, n \in \mathbb{Z}\}$$
is a subgroup of G. Show that if g belongs to any subgroup H in G, then H contains $\langle g \rangle$.

10.1.13. Let G be a group with subgroup H. Let a belong to G. Prove that aHa^{-1} is a subgroup of G.

10.2. Homomorphisms and Normal Subgroups

Just as we saw connections between linear transformations on a vector space and its subspaces, here we will see connections between homomorphisms on a group and its so-called *normal subgroups*.

Properties of group homomorphisms have analogs in the properties of linear transformations. A group homomorphism $\varphi : G \to G'$ maps the identity element e in G to the identity element e' in G', and the inverse of g to the inverse of $\varphi(g)$. These are easy consequences of the group homomorphism property and Lemma 10.2. Since
$$\varphi(e)\varphi(g) = \varphi(eg) = \varphi(g),$$

we have $\varphi(e) = e'$. Since
$$\varphi(gg^{-1}) = \varphi(g)\varphi(g^{-1}) = e',$$
we have $\varphi(g^{-1}) = \varphi(g)^{-1}$.

Recall that groups G and G' are isomorphic provided there is a bijective group homomorphism — a group isomorphism — $G \to G'$. In this case, we write $G \cong G'$.

Theorem 10.6. If V and W are isomorphic vector spaces over a field $\mathbb{F}$, then $GL(V)$ and $GL(W)$ are isomorphic groups.

Proof. Let V and W be as hypothesized. Fix an isomorphism $L : V \to W$. Given T in $GL(V)$, define T' in $\mathcal{L}(W)$ by
$$T' = L \circ T \circ L^{-1}.$$
Since L, T, and L^{-1} are all vector space isomorphisms, T' is in $GL(W)$. Define $\varphi : GL(V) \to GL(W)$ by $\varphi(T) = L \circ T \circ L^{-1}$. We will show that φ is a group isomorphism, which is enough to prove the theorem.

Suppose $\varphi(T_1) = \varphi(T_2)$ for some T_1, T_2 in $GL(V)$. We then have
$$L \circ T_1 \circ L^{-1} = L \circ T_2 \circ L^{-1}.$$
Multiply on the right by L and on the left by L^{-1} to conclude that $T_1 = T_2$. This proves that φ is injective.

Given any T' in $GL(W)$, define $T = L^{-1} \circ T' \circ L$. By the argument above, T is in $GL(V)$. Since $\varphi(T) = T'$, φ is surjective.

Finally, consider that for T_1 and T_2 in $GL(V)$,
$$\varphi(T_1 \circ T_2) = L \circ T_1 \circ T_2 \circ L^{-1} = L \circ T_1 \circ L^{-1} \circ L \circ T_2 \circ L^{-1} = \varphi(T_1) \circ \varphi(T_2).$$
This shows that φ is a group homomorphism, thus, an isomorphism of groups. $\square$

Theorem 10.6 picks up on a theme that has played repeatedly through this course. In the finite-dimensional setting, there is a mechanism to swap abstract mappings associated to vector spaces for matrices. The theorem tells us that when $V \cong \mathbb{F}^n$, we can understand $GL(V)$ as a group by studying $GL_n(\mathbb{F})$.

A linear transformation L in $\mathcal{L}(V, W)$ is a group homomorphism from the additive group of V to the additive group of W. We saw a version of the following as it applies to linear transformations.

Theorem 10.7. Let G and G' be groups. If $\varphi : G \to G'$ is a group homomorphism, then

(a) $\varphi[H]$ is a subgroup of G' for any subgroup $H \subseteq G$ and

(b) $\varphi^{-1}[H']$ is a subgroup of G for any subgroup $H' \subseteq G'$.

Proof. Let G, G', and φ be as hypothesized. Let $H \subseteq G$ be a subgroup, with h and k in H. Since e is in H, $\varphi(e)$ is in $\varphi[H]$, so $\varphi[H]$ is nonempty. Moreover, since hk^{-1} is in H,
$$\varphi(h)\varphi(k)^{-1} = \varphi(hk^{-1}) \in \varphi[H].$$
It follows by Lemma 10.3 that $\varphi[H]$ is a subgroup of G'.

10.2. Homomorphisms and Normal Subgroups

Now let $H' \subseteq G'$ be a subgroup. Since the identity element in G is mapped to the identity element in G' and since H' contains the identity element in G', $\varphi^{-1}[H']$ is nonempty. Consider h and k in $\varphi^{-1}[H']$. We have
$$\varphi(h)\varphi(k)^{-1} = \varphi(hk^{-1})$$
in H', which implies that hk^{-1} belongs to $\varphi^{-1}[H']$. It follows again by Lemma 10.3 that $\varphi^{-1}[H']$ is a subgroup of G. □

When G is any group with identity element e, it has subgroups $\{e\}$ and G. If $\varphi : G \to G'$ is a group homomorphism, then $\varphi[G]$ is a subgroup of G' and $\varphi^{-1}[\{e'\}]$ is a subgroup of G.

Definition 10.8. The **kernel of a group homomorphism** is the inverse image of the identity element in the codomain.

We saw that conjugation was an automorphism on any $\mathbb{F}$-algebra. Likewise, $x \mapsto gxg^{-1}$ is a group automorphism. It follows that if G is a group with subgroup H and element g, then gHg^{-1} is a subgroup of G and $|H| = |gHg^{-1}|$.

We can streamline some of our proofs by noting the following. Proofs of two of the three statements were problems in Exercises 6.4.

Lemma 10.9. If G is a group and g is in G, the mappings $x \mapsto gx$ and $x \mapsto xg$ are bijections. The mapping $x \mapsto x^{-1}$ is also a bijection on any group.

Theorem 10.10. If G is a group and $H = \operatorname{Ker}\varphi$ for some group homomorphism φ on G, then for any g in G, $gHg^{-1} = H$.

Proof. Let everything be as hypothesized. For any h in $H = \operatorname{Ker}\varphi$,
$$\varphi(ghg^{-1}) = \varphi(g)\varphi(g^{-1}) = \varphi(gg^{-1}) = \varphi(e).$$
This proves that for each h in H, ghg^{-1} is in H. In other words, $gHg^{-1} \subseteq H$ for all g in G. From here, Lemma 10.9 implies that $gH \subseteq Hg$ and $H \subseteq g^{-1}Hg$. Since $H = \{h^{-1} \,|\, h \in H\}$ and since $g \mapsto g^{-1}$ is a bijection, from here we have $H \subseteq gHg^{-1}$, which proves the theorem. □

If G is a group with subgroup H, recall that we define the left H-coset containing g to be
$$gH = \{gh \,|\, h \in H\}.$$
The right H-coset containing g is
$$Hg = \{hg \,|\, h \in H\}.$$

We have seen that left H-cosets in a group intersect if and only if they are the same coset. In particular, the left H-cosets partition the ambient group. Proof of the following is an exercise.

Lemma 10.11. If G is a group with subgroup H, then $gH \cap g'H$ is nonempty if and only if $g^{-1}g'$ is in H, in which case $gH = g'H$.

Recall that when V is a vector space with subspace W, V/W is also a vector space. That idea generalizes to apply in a group, but there are conditions.

Theorem 10.12. If G is a group with subgroup H satisfying $gHg^{-1} = H$ for all g in G, then the collection of left H-cosets is a group under the binary operation given by

(10.1) $$g_1 H \circ g_2 H = g_1 g_2 H.$$

Proof. Assume that everything is as hypothesized. We must show that the binary operation given in (10.1) is well-defined.

Suppose $g_1 H = g_1' H$ and $g_2 H = g_2' H$. We have $g_1^{-1} g_1'$ and $g_2^{-1} g_2'$ in H. To show that
$$g_1 g_2 H = g_1' g_2' H$$
we must show that $(g_1 g_2)^{-1} g_1' g_2'$ is in H.

We have $(g_1 g_2)^{-1} g_1' g_2' = g_2^{-1} g_1^{-1} g_1' g_2'$. Let $g_1^{-1} g_1' = h$, to remind us that this element is in H. Then we have
$$(g_1 g_2)^{-1} g_1' g_2' = g_2^{-1} h g_2.$$
Since $g_2^{-1} H g_2 \subseteq H$, $(g_1 g_2)^{-1} g_1' g_2'$ is in H. This shows that the binary operation on the left H-cosets is well-defined.

Since the binary operation on G is associative, the binary operation on the left H-cosets is as well.

The identity element among the left H-cosets is $eH = H$.

The inverse of gH is $g^{-1} H$. This is enough to prove the theorem. □

The theorem suggests a definition.

Definition 10.13. A **normal subgroup** of a group G is a subgroup H with the property that $gHg^{-1} = H$ for all g in G.

When H is a normal subgroup in G, the group of left H-cosets is denoted G/H, read G mod H. We call G/H a **quotient group**.

We have seen that when H is a subgroup in G, the left H-coset partition and the right H-coset partition may not be identical. The next theorem establishes that the two partitions are identical precisely when H is normal in G.

Theorem 10.14. Let G be a group with subgroup H. Every left H-coset is a right H-coset if and only if H is normal in G.

Proof. Let G be a group with subgroup H. Suppose every left H-coset is a right H-coset. Given g in G, there is then g' in G so that $gH = Hg'$. Since this implies that g belongs to Hg', Theorem 6.30 implies $Hg = Hg'$. It follows that $gH = Hg$, thus, that $gHg^{-1} = H$.

Conversely, if $gHg^{-1} = H$, then $gH = Hg$. □

We saw that in a vector space, the kernel of a mapping was not a special subspace. This changes in the context of groups.

Theorem 10.15. A subgroup of a group is normal if and only if it is the kernel of a homomorphism.

10.2. Homomorphisms and Normal Subgroups

Proof. Let G be a group with subgroup H.

Suppose H is normal in G so that G/H is a group. Define
$$\varphi : G \to G/H$$
by $\varphi(g) = gH$. Since
$$\varphi(gg') = gg'H = (gH)(g'H) = \varphi(g)\varphi(g'),$$
φ is a group homomorphism. Since the identity element in G/H is H,
$$\operatorname{Ker} \varphi = H.$$
This proves the result in one direction.

Theorem 10.10 is the result in the other direction. $\square$

Example 10.16. When $\mathbb{F}$ is a field, its multiplicative group of nonzero elements is often denoted $\mathbb{F}^*$. One of our favorite group homomorphisms is the determinant mapping $\det : GL_n(\mathbb{F}) \to \mathbb{F}^*$. The identity element in $\mathbb{F}^*$ is 1, so
$$\operatorname{Ker} \det = \{A \in GL_n(\mathbb{F}) \mid \det A = 1\}.$$
Matrices with determinant equal to 1 form a normal subgroup of $GL_n(\mathbb{F})$, called the **special linear group** of $n \times n$ matrices over $\mathbb{F}$.

Exercises 10.2.

10.2.1. Show that isomorphism is an equivalence relation on groups.

10.2.2. Show that $\mathbb{Z}$ and $2\mathbb{Z}$ are isomorphic as groups.

10.2.3. Is the additive group of $\mathbb{F}_p$ isomorphic to the multiplicative group $\mathbb{F}_p^*$ when p is prime?

10.2.4. Is the additive group of $\mathbb{R}$ isomorphic to the multiplicative group $\mathbb{R}^*$?

10.2.5. Finish the proof of Lemma 10.9.

10.2.6. Let G be a group and let g belong to G. Show that conjugation by g is an automorphism on G.

10.2.7. Let G be a group with subgroup H. Let g and g' belong to G. Show that $gH \cap g'H$ is nonempty if and only if $g^{-1}g'$ is in H.

10.2.8. Let G be a group with subgroup H. Let g and g' in G. Prove or disprove each of the following statements.
 (a) $gH \subseteq g'H$ implies $gH = g'H$.
 (b) $gHg^{-1} = g^{-1}Hg$.
 (c) $gHg^{-1} \subseteq H$ implies $gHg^{-1} = H$.
 (d) $gH = Hg'$ implies $gHg'^{-1} = g^{-1}Hg' = H$.

10.2.9. Let $H_2 = \{e, h, h^2\}$ in $G = GL_2(\mathbb{F}_2)$, where $h = \begin{bmatrix} 0 & 1 \\ 1 & 1 \end{bmatrix}$. Show that H_2 is normal in G.

10.2.10. Let $g = \begin{bmatrix} 1 & 0 \\ 1 & 1 \end{bmatrix}$ belong to $G = GL_2(\mathbb{F}_2)$ and let $H_1 = \langle g \rangle$.
   (a) List the elements in each of the left H_1-cosets in G.
   (b) List the elements in each of the right H_1-cosets in G.
   (c) Is any left H_1-coset also a right H_1-coset? Is H_1 normal in G?

10.2.11. Verify that $A \mapsto \det A$ is a homomorphism from $GL_n(\mathbb{F})$ to $\mathbb{F}^*$.

10.3. The Quaternions

The quaternions were discovered by William Rowan Hamilton in 1843, well before the theory of vector spaces had been fully developed. Hamilton was an Irish physicist and mathematician then at the height of his renown. Early on, he understood that the complex numbers served as a model for 2-dimensional real space, a setting for elementary physics. Three-space is an even better setting for physics so Hamilton set about finding a model analogous to $\mathbb{C}$ that would provide an algebraic framework for modeling 3-space. After a years-long search, he discovered the quaternions, designated $\mathbb{H}$ in Hamilton's honor.

As a vector space over $\mathbb{R}$, $\mathbb{H}$ is 4-dimensional, which is one reason Hamilton had trouble finding it. He was looking for something 3-dimensional.

During the course of the nineteenth century, linear algebra and tensor analysis were developed. Anything $\mathbb{H}$ could do in service to the models that most physicists needed, tensors could do just as well if not better. By the twentieth century, $\mathbb{H}$ had fallen into obscurity, largely unknown except among specialists. Late in the twentieth century, the quaternions enjoyed a renaissance. At that point, people came to understand that the quaternions could be used to solve a problem that plagued computer programmers: $\mathbb{H}$ is perfect for coding rotations for graphics applications. Interest in $\mathbb{H}$ has since surged.

One way to see $\mathbb{H}$ is as a generalization of the complex numbers.

We coordinatize the complex plane with a horizontal coordinate axis corresponding to the real numbers and a vertical axis corresponding to the pure imaginary numbers. The **unit complex numbers** are the elements in $\mathbb{C}$ with modulus one. We denote this set by $\mathcal{U}(\mathbb{C})$, so that

$$\mathcal{U}(\mathbb{C}) = \{z \in \mathbb{C} \mid |z| = 1\}.$$

Note, for instance, that 1 is in $\mathcal{U}(\mathbb{C})$ so $\mathcal{U}(\mathbb{C})$ is nonempty.

Let z and w belong to $\mathcal{U}(\mathbb{C})$. Since

$$|zw| = |z||w| = 1,$$

$\mathcal{U}(\mathbb{C})$ is closed under multiplication. Since any z in $\mathcal{U}(\mathbb{C})$ is nonzero, we also have

$$|1/z| = \left|\frac{\bar{z}}{z\bar{z}}\right| = |\bar{z}|\,|1/z\bar{z}| = |z| = 1.$$

This proves the following theorem.

Lemma 10.17. $\mathcal{U}(\mathbb{C})$ is a group under multiplication.

10.3. The Quaternions

Recall the exponential form of a complex number $z = a+bi$. If θ is the argument for (a, b) — the angle when we write it in polar form in $\mathbb{R}^2$ — then
$$z = |z|e^{i\theta}.$$
Notice in particular that $e^{i\theta} = \cos\theta + i\sin\theta$ is in $\mathcal{U}(\mathbb{C})$. Conversely, if $z = a + bi$ is in $\mathcal{U}(\mathbb{C})$, there is θ in $\mathbb{R}$ so that $a = \cos\theta$ and $b = \sin\theta$. This shows that
$$\mathcal{U}(\mathbb{C}) = \{e^{i\theta} \mid \theta \in \mathbb{R}\} \subseteq \mathbb{C}.$$

There is an obvious bijective correspondence between the elements in S^1, the points on the unit circle in $\mathbb{R}^2$, and the elements in $\mathcal{U}(\mathbb{C})$:
$$(\cos\theta, \sin\theta) \mapsto e^{i\theta}.$$
Since we can identify $\mathcal{U}(\mathbb{C})$ and S^1 point for point, we can use $\mathcal{U}(\mathbb{C})$ to define a multiplication on S^1:
$$(\cos\theta_1, \sin\theta_1)(\cos\theta_2, \sin\theta_2) := (\cos(\theta_1 + \theta_2), \sin(\theta_1 + \theta_2)).$$

This is an important idea that we see all over mathematics. In this instance, we employ the bijection to make S^1 a group, called **the circle group**.

Proof of the following is an exercise.

Theorem 10.18. If we define multiplication on S^1 by
$$(\cos\theta_1, \sin\theta_1)(\cos\theta_2, \sin\theta_2) = (\cos(\theta_1 + \theta_2), \sin(\theta_1 + \theta_2)),$$
then S^1 is a group with $S^1 \cong \mathcal{U}(\mathbb{C})$.

We are going to some trouble here to distinguish $\mathbb{R}^2$ from $\mathbb{C}$. Is the difference not purely notational? In some contexts, the difference between the two is blurred or even ignored. Here, we maintain the distinction because while the real plane and the complex plane are isomorphic as real vector spaces, they are crucially different as algebraic objects. The complex numbers form a field. A field has two binary operations. We could apply the definition of multiplication in $\mathbb{C}$ to multiplication of vectors in $\mathbb{R}^2$ — as we used $\mathcal{U}(\mathbb{C})$ to define multiplication on S^1 — but to do so would be to step outside the structure we associate to $\mathbb{R}^2$ as a vector space or an inner product space.

Definition 10.19. The **quaternions** are a 4-dimensional vector space over $\mathbb{R}$ with basis $\{1, i, j, k\}$ where 1 is in $\mathbb{R}$ and i, j, k are symbols with the property
(10.2) $$i^2 = j^2 = k^2 = ijk = -1.$$

Understanding that $\{1, i, j, k\}$ is a linearly independent set with the property given in (10.2), we can write
$$\mathbb{H} = \{a + bi + cj + dk \mid a, b, c, d \in \mathbb{R}\}.$$
As is the case with complex numbers, whenever we write q in $\mathbb{H}$ as
$$q = a + bi + cj + dk,$$
it is to be understood that a, b, c, d are all real. The **real part** of $q = a+bi+cj+dk$ in $\mathbb{H}$ is a and the **pure quaternion part** is $bi + cj + dk$. If $a = 0$, then q is a **pure quaternion**.

Though $\mathbb{H}$ is a 4-dimensional real vector space, we do not identify it with $\mathbb{R}^4$. Like $\mathbb{C}$ and $\mathbb{R}^2$, $\mathbb{H}$ and $\mathbb{R}^4$ are quite different, apart from linear structure. However, we will see that multiplication in the quaternions meshes perfectly with the structure of $\mathbb{R} \oplus \mathbb{R}^3$ as a vector space. Because of that, we will later blur the distinction between the quaternions and $\mathbb{R} \oplus \mathbb{R}^3$.

We define multiplication in $\mathbb{H}$ using (10.2), the distributive law, and the usual rules of arithmetic in $\mathbb{R}$. Letting the chips fall where they may from there, we find that multiplication in $\mathbb{H}$ is more exciting than multiplication in $\mathbb{C}$. Postmultiplying both sides of $ijk = -1$ by k, we get $-ij = -k$, or $ij = k$. Premultiplying $ijk = -1$ by i, we get $-jk = -i$, or $jk = i$. Premultiplying again by j gets us $-k = ji$. In this fashion we can find out what all six products are when we draw from $\{i, j, k\}$ taking two elements at a time in either order.

Note that among i, j, k, multiplication is *anticommutative*:

$$ij = k, \quad ji = -k, \quad jk = i, \quad kj = -i, \quad ki = j, \quad ik = -j.$$

This recipe is captured in Figure 10.2. Taking the product of two successive elements in the clockwise direction, we get the third element. When going counterclockwise, we introduce a minus sign.

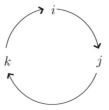

Figure 10.2. Multiplication rule for quaternions.

We have for example that

$$(1 - 2i + 3k)(2 + 3i - 4j - k)$$
$$= (2 + 6 + 3) + (3 - 4 + 12)i + (-4 - 2 + 9)j + (-1 + 6 + 8)k$$
$$= 11 + 11i + 3j + 13k.$$

We calculate $11i$, for instance, by

$$(1)(3i) + (-2i)(2) + (3k)(-4j) = 3i - 4i + 12i = 11i,$$

because $kj = -i$.

Multiplication in $\mathbb{H}$ is associative. Since its multiplication is not commutative, $\mathbb{H}$ is not a field.

As in $\mathbb{C}$, the multiplicative inverse of nonzero $q = a + bi + cj + dk$ in $\mathbb{H}$ employs the **quaternion conjugate**

$$\bar{q} = a - bi - cj - dk.$$

We leave it to the reader to formulate and prove the analog to Lemma B.2 for quaternion conjugates.

Since $q\bar{q} = a^2 + b^2 + c^2 + d^2$ is a nonnegative real number, we can define the modulus of q in $\mathbb{H}$ by

$$|q| = \sqrt{q\bar{q}}.$$

If $q = a + bi + cj + dk$ is in $\mathbb{H}$, then $|q|^2 = a^2 + b^2 + c^2 + d^2$ is the square of the distance between the origin and the point (a, b, c, d) in $\mathbb{R}^4$.

Given nonzero q in $\mathbb{H}$, we see that

$$q\frac{\bar{q}}{q\bar{q}} = 1$$

so $\bar{q}/(q\bar{q}) = 1/q$, the multiplicative inverse of q.

Certainly, we have the inclusion $\mathbb{R} \subseteq \mathbb{C} \subseteq \mathbb{H}$. As a set with two binary operations that satisfy all the field axioms except commutativity of multiplication, $\mathbb{H}$ is a **skew-field**. The nonzero elements in $\mathbb{H}$ comprise a nonabelian group, the *multiplicative group of* $\mathbb{H}$ which we often denote by $\mathbb{H}^*$.

The **unit quaternions** are given by

$$\mathcal{U}(\mathbb{H}) = \{q \in \mathbb{H} \mid |q| = 1\}.$$

Since $|1/q| = 1/|q|$ and $|q_1 q_2| = |q_1||q_2|$, for all q, q_1, q_2 in $\mathbb{H}$, it is easy to prove that $\mathcal{U}(\mathbb{H})$ is a subgroup of $\mathbb{H}^*$.

Notice that $q = a + bi + cj + dk$ belongs to $\mathcal{U}(\mathbb{H})$ if and only if

$$|q|^2 = a^2 + b^2 + c^2 + d^2 = 1.$$

We thus have a bijection between $\mathcal{U}(\mathbb{H})$ and the unit 3-sphere

$$S^3 = \{(a, b, c, d) \in \mathbb{R}^4 \mid a^2 + b^2 + c^2 + d^2 = 1\}.$$

Since $\mathcal{U}(\mathbb{H})$ is a multiplicative group, we again use the bijection to define multiplication on S^3. This makes S^3 into a group that is isomorphic to $\mathcal{U}(\mathbb{H})$.

We have seen now that we can make S^1, the unit circle, and S^3, the unit 3-sphere, into multiplicative groups. What about S^2? Hamilton's difficulties in finding $\mathbb{H}$ had to do with the fact that he could not find a way to define multiplication — with an identity element and inverses — on what we would call 3-vectors, that is, elements in $\mathbb{R}^3$. This was not a failure of Hamilton's. It was later proved to be impossible. We get the group structure on $S^1 \subseteq \mathbb{R}^2$ using $\mathbb{C}$, a field that is isomorphic to $\mathbb{R}^2$ as a real vector space. We get the group structure on $S^3 \subseteq \mathbb{R}^4$ using $\mathbb{H}$, a skew-field that is isomorphic to $\mathbb{R}^4$ as a real vector space. Following this template to define a group structure on S^2, we would look for a field or skew-field that is isomorphic to $\mathbb{R}^3$. As Hamilton effectively discovered, though, there is no field — skew or otherwise — that forms a 3-dimensional vector space over $\mathbb{R}$. Digging further into that is a worthy project, but one that goes beyond the scope of this text.

Proof of the following important theorem is an exercise.

Theorem 10.20. If $f : \mathbb{H} \to \mathbb{R} \oplus \mathbb{R}^3$ is given by

$$f(a + bi + cj + dk) = (a; \mathbf{v}), \text{ where } \mathbf{v} = (b, c, d) \in \mathbb{R}^3,$$

then f is an isomorphism of vector spaces. Moreover, if q_1, q_2 are in $\mathbb{H}$ with $f(q_1) = (a_1; \mathbf{v}_1)$ and $f(q_2) = (a_2; \mathbf{v}_2)$, then

$$f(q_1 q_2) = (a_1 a_2 - (\mathbf{v}_1 \cdot \mathbf{v}_2); a_1 \mathbf{v}_2 + a_2 \mathbf{v}_1 + (\mathbf{v}_1 \times \mathbf{v}_2))$$

where $\cdot$ is the Euclidean dot product on $\mathbb{R}^3$ and $\times$ is the cross-product on $\mathbb{R}^3$.

Exploiting Theorem 10.20, we identify a **vector part** for each q in $\mathbb{H}$, $\mathbf{v}$ in $\mathbb{R}^3$ so that $f(q) = (a; \mathbf{v})$. The pure quaternion part of q is then

$$(\mathbf{v} \cdot \mathbf{i})i + (\mathbf{v} \cdot \mathbf{j})j + (\mathbf{v} \cdot \mathbf{k})k.$$

It is because multiplication in $\mathbb{H}$ is on such friendly terms with the dot product and the cross-product in $\mathbb{R}^3$ that we frequently identify q in $\mathbb{H}$ with $f(q)$ in $\mathbb{R} \oplus \mathbb{R}^3$.

If $q = (a; \mathbf{v})$, then $\bar{q} = (a; -\mathbf{v})$. In this case, since $\mathbf{v} \times \mathbf{v} = \mathbf{0}$,

$$|q|^2 = q\bar{q} = (a^2 + \|\mathbf{v}\|^2; \mathbf{0}).$$

Usually, we just write $|q|^2 = a^2 + \|\mathbf{v}\|^2$. When $q = (0; \mathbf{v})$, $|q|^2 = \|\mathbf{v}\|^2$. We may thus identify the **unit pure quaternions**, $\{q = (0; \mathbf{v}) \mid \|\mathbf{v}\| = 1\}$, with S^2. Since S^2 is not a multiplicative group, neither are the unit pure quaternions. This is easy to verify via an example.

Since $\mathbb{H}$ is not a field, $\mathbb{H}^2$ is not a vector space. In treatments of matrix groups for mature audiences, though, matrices over $\mathbb{H}$ are embraced as part of the family of linear operators. Though we will not delve into this, it is worth a moment to look at an actual 2×2 matrix over $\mathbb{H}$ to see what transpires when our scalars come from $\mathbb{H}$.

Let

$$A = \begin{bmatrix} i & j \\ k & 1 \end{bmatrix}.$$

If we calculate $\det A$, we either get

$$i - jk = i - i = 0$$

or

$$i - kj = i + i = 2i \neq 0.$$

Clearly, we cannot treat $\mathbb{H}$ as just another field. A workable definition of determinant that applies to $\mathbb{H}$ is sorted out in [**2**], for instance. It involves taking a certain quotient group of the multiplicative subgroup of $\mathbb{H}$. In the exercises, you get a chance to explore how the invertibility of 2×2 matrices over $\mathbb{H}$ is more delicate than it is over a field. All this is to say that for any postitive integer n, there is a general linear group over $\mathbb{H}$, $GL_n(\mathbb{H})$, the set of nonsingular matrices with entries in $\mathbb{H}$. We will not say anything more about $GL_n(\mathbb{H})$ and its subgroups, but in Section 10.7, we revisit $\mathbb{H}$ as a device to work with rotations in $\mathbb{R}^3$.

Exercises 10.3.

10.3.1. In this problem we discuss the unit complex numbers $\mathcal{U}(\mathbb{C}) \subseteq \mathbb{C}$.
  (a) Show that $e^{i\theta}e^{i\varphi} = e^{i(\theta+\varphi)}$.
  (b) Show that multiplication in $\mathcal{U}(\mathbb{C})$ is well-defined: If $e^{i\theta} = e^{i\eta}$, then $e^{i\theta}e^{i\varphi} = e^{i\eta}e^{i\varphi}$.
  (c) What is the identity element in $\mathcal{U}(\mathbb{C})$?
  (d) What is the inverse of $e^{i\theta}$ in $\mathcal{U}(\mathbb{C})$?
  (e) Show that the set of all real multiples of 2π comprises a subgroup of the additive group of real numbers, $\mathbb{R}$. Designate that subgroup by $\langle 2\pi \rangle$.
  (f) Show that the multiplicative group $\mathcal{U}(\mathbb{C})$ is isomorphic to the additive group $\mathbb{R}/\langle 2\pi \rangle$.

10.3.2. Prove Theorem 10.18.

10.3.3. State and prove the analog to Lemma B.2 for $\mathbb{H}$.

10.3.4. Prove that $\mathcal{U}(\mathbb{H})$ is a multiplicative subgroup of $\mathbb{H}^*$. What is the identity element?

10.3.5. Prove that the set $\{i, j, k\}$ generates a multiplicative subgroup of $\mathcal{U}(\mathbb{H})$ of order eight.

10.3.6. Show that if q in $\mathbb{H}$ satisfies $qq' = q'q$ for all q' in $\mathbb{H}$, then q is a real number.

10.3.7. Show that q in $\mathbb{H}$ is a pure quaternion if and only if q^2 is a nonpositive real number.

10.3.8. Prove Theorem 10.20.

10.3.9. Verify that if $q = (a; \mathbf{v})$ belongs to $\mathbb{H}$, then $q\bar{q} = (a^2 + \|\mathbf{v}\|^2; \mathbf{0})$.

10.3.10. Show that if $q = (a; \mathbf{v})$ is in $\mathcal{U}(\mathbb{H})$, then $q^{-1} = (a; -\mathbf{v})$.

10.3.11. Why do the unit pure quaternions fail to form a group under multiplication?

10.3.12. The unit coordinate vectors in $\mathbb{R}^3$ — the elements in the standard ordered basis — are often called $\mathbf{i}, \mathbf{j}, \mathbf{k}$.
  (a) Check the six cross-products that we get by taking ordered pairs from $\{\mathbf{i}, \mathbf{j}, \mathbf{k}\}$.
  (b) What is to stop us from identifying $\mathbf{i}, \mathbf{j}, \mathbf{k}$ with i, j, k in $\mathbb{H}$?

10.3.13. Show that the pure quaternions are invariant under conjugation by any unit quaternion, that is, if q is a pure quaternion and x is in $\mathcal{U}(\mathbb{H})$, then xqx^{-1} is a pure quaternion.

10.3.14. A matrix A in $M_n(\mathbb{H})$ is invertible provided it has a left inverse, that is, B in $M_n(\mathbb{H})$ so that BA is the identity matrix. We then define $GL_n(\mathbb{H})$ to be the invertible matrices in $M_n(\mathbb{H})$. Determine whether the following matrices belong to $GL_2(\mathbb{H})$. For those that do, find their left inverse.
  (a) $A_1 = \begin{bmatrix} i & j \\ k & 1 \end{bmatrix}$.
  (b) A_1^T.

(c) $A_2 = \begin{bmatrix} j & 1 \\ i & k \end{bmatrix}$.

(d) A_2^T.

10.4. The Special Linear Group

The special linear group, which we designate $SL_n(\mathbb{F})$, is the normal subgroup of matrices with determinant 1 in $GL_n(\mathbb{F})$. In this section, our efforts go towards seeing elements in $SL_n(\mathbb{F})$ geometrically.

A symmetry is a bijective mapping on a set that preserves some feature of the set. The symmetric group $\mathcal{S}_n$ is the group of symmetries on a set, $\{a_1, \ldots, a_n\}$. It preserves the cardinality of the set. The general linear group $GL_n(\mathbb{F})$ is the group of symmetries of coordinate systems for $\mathbb{F}^n$. It maps one coordinate system to another, and given any two coordinate systems for $\mathbb{F}^n$, there is an element in $GL_n(\mathbb{F})$ that maps one to the other. We can understand $SL_n(\mathbb{F})$ as a group of symmetries if we think about how we use coordinate systems in $\mathbb{R}^2$ and $\mathbb{R}^3$ to study geometry.

A parallelogram in $\mathbb{R}^2$ is determined by two linearly independent vectors, $\mathbf{v}$ and $\mathbf{w}$. To calculate the area of the region enclosed by this parallelogram, we have to pirouette around the minus signs that bubble up in our use of coordinates. By analytic geometry, we know that the parallelogram has area

$$\|\mathbf{v}\|\|\mathbf{w}\| \sin \theta$$

where θ in the interval $(0, \pi)$ is the angle between $\mathbf{v}$ and $\mathbf{w}$. If $\mathbf{v} = (v_1, v_2)$ and $\mathbf{w} = (w_1 w_2)$, then

$$\|\mathbf{v}\|\|\mathbf{w}\| \sin \theta = |v_1 w_2 - v_2 w_1|.$$

If we embrace the signs, getting rid of all the absolute values, we get a *signed area* or the area of an *oriented parallelogram*. An oriented parallelogram $\square \mathcal{B} \subseteq \mathbb{R}^2$ is associated to an ordered basis for $\mathbb{R}^2$, $\mathcal{B} = \{\mathbf{v}, \mathbf{w}\}$. To make sense of what this means geometrically, we take φ in the interval $(-\pi, \pi)$ to be the angle measured from $\mathbf{v}$ to $\mathbf{w}$. This is a directed angle: We take the shortest circular path starting at $\mathbf{v}$ and ending at $\mathbf{w}$. When the shortest arc from $\mathbf{v}$ to $\mathbf{w}$ is along a counterclockwise path, φ is positive and $\square \mathcal{B}$ is positively oriented. When the shortest arc from $\mathbf{v}$ to $\mathbf{w}$ is along a clockwise path, φ is negative and $\square \mathcal{B}$ is negatively oriented. For φ in $(0, \pi)$, $\sin \varphi$ and the area of $\square \mathcal{B}$ are positive. For φ in $(-\pi, 0)$, $\sin \varphi$ and the area of $\square \mathcal{B}$ are negative.

It is easy to check that if $\mathbf{v} = (v_1, v_2)$ and $\mathbf{w} = (w_1 w_2)$, then

$$\|\mathbf{v}\|\|\mathbf{w}\| \sin \varphi = v_1 w_2 - v_2 w_1.$$

Notice then that the area of the oriented parallelogram associated to $\mathcal{B} = \{\mathbf{v}, \mathbf{w}\}$ is

$$\det \begin{bmatrix} \mathbf{v} & \mathbf{w} \end{bmatrix}.$$

When we use coordinates to do geometry in $\mathbb{R}^2$ or $\mathbb{R}^3$, we fight to suppress negative numbers that might otherwise wind up measuring length, volume, or area. When we consider measurements associated to oriented objects, we put down our

10.4. The Special Linear Group

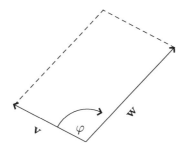

Figure 10.3. A negatively oriented parallelogram in $\mathbb{R}^2$.

weapons and go along with the coordinates, even when they are negative. This is not out of step with practice in other contexts. We do this when we study trigonometry in the real plane, accounting for coterminal angles, negative angles, and, when we extend that story to polar coordinates, negative radii. We also do this when we interpret definite integrals in terms of area.

A parallelepiped in $\mathbb{R}^3$ is a box determined by three linearly independent vectors $\mathbf{v}_1, \mathbf{v}_2, \mathbf{v}_3$. Its volume is usually given in terms of the so-called *scalar triple product*

$$|(\mathbf{v}_1 \times \mathbf{v}_2) \cdot \mathbf{v}_3| = |\det \begin{bmatrix} \mathbf{v}_1 & \mathbf{v}_2 & \mathbf{v}_3 \end{bmatrix}|.$$

As in the 2-dimensional case, an oriented parallelepiped is determined by an ordered basis for $\mathbb{R}^3$, $\mathcal{B} = \{\mathbf{v}_1, \mathbf{v}_2, \mathbf{v}_2\}$. The parallelepiped is negatively oriented if its signed volume

$$\det \begin{bmatrix} \mathbf{v}_1 & \mathbf{v}_2 & \mathbf{v}_3 \end{bmatrix}$$

is negative.

Having freed ourselves of absolute values, as well as sines and angles — none of which make sense over arbitrary fields — we arrive at the notion of an *oriented parallelotope*. A parallelotope is the generalization of a parallelepiped to higher dimensions. An oriented parallelotope is the object determined by an ordered basis for $\mathbb{F}^n$, $\{\mathbf{v}_1, \ldots, \mathbf{v}_n\}$. The volume of such an object is then defined to be

$$\det \begin{bmatrix} \mathbf{v}_1 & \mathbf{v}_2 & \cdots & \mathbf{v}_n \end{bmatrix}.$$

Theorem 10.21. Let $\mathbb{F}$ be any field and let n be a positive integer. The special linear group, $SL_n(\mathbb{F})$, is the symmetry group for oriented parallelotopes in $\mathbb{F}^n$. In other words, $SL_n(\mathbb{F})$ maps any oriented parallelotope in $\mathbb{F}^n$ to another oriented parallelotope with the same volume. Moreover, if $\square \mathcal{B}$ is the oriented parallelotope associated to the ordered basis $\mathcal{B} = \{\mathbf{b}_1, \ldots, \mathbf{b}_n\}$ for $\mathbb{F}^n$ and $\square \mathcal{C}$ is the oriented parallelotope associated to the ordered basis $\mathcal{C} = \{\mathbf{c}_1, \ldots, \mathbf{c}_n\}$ for $\mathbb{F}^n$, then when $\square \mathcal{B}$ and $\square \mathcal{C}$ have equal volume there is an element in $SL_n(\mathbb{F})$ that maps $\square \mathcal{B}$ to $\square \mathcal{C}$.

Proof. Let everything be as hypothesized. Let
$$B = \begin{bmatrix} \mathbf{b}_1 & \cdots & \mathbf{b}_n \end{bmatrix}.$$
If A is any matrix in $SL_n(\mathbb{F})$, then $\det A = 1$ so
$$\det AB = \det A \det B = \det B.$$
This means that for $A\mathcal{B} = \{A\mathbf{b}_i\}_{i=1}^n$, $\square A\mathcal{B}$ is an oriented parallelotope with the same volume as $\square \mathcal{B}$. That proves the first statement in the theorem.

Next suppose that $\square \mathcal{B}$ and $\square \mathcal{C}$ have equal volume. Let $C = \begin{bmatrix} \mathbf{c}_1 & \cdots & \mathbf{c}_n \end{bmatrix}$. Since $\det C = \det B$ is nonzero, we also have
$$\det C^{-1} = \det B^{-1} = 1/\det B.$$
It follows that $\det CB^{-1} = 1$, thus, that CB^{-1} is in $SL_n(\mathbb{F})$. Since $CB^{-1}\mathcal{B} = \mathcal{C}$, $\square CB^{-1}\mathcal{B} = \square\mathcal{C}$, which completes the proof of the theorem. $\square$

We discussed shears briefly in Section 8.3. If $\mathcal{B} = \{\mathbf{v}, \mathbf{w}\}$ is an arbitrary orthonormal basis for $\mathbb{R}^2$, a *shear in the $\mathbf{v}$-direction* is a linear operator that leaves $\mathbf{v}$ fixed and adds a multiple of $\mathbf{v}$ to $\mathbf{w}$. Figure 10.4 shows the effect of a shear in the x-direction on the unit square determined by $\mathcal{E}$. We leave it as an exercise to show that a shear leaves the area of $\square \mathcal{B}$ unchanged.

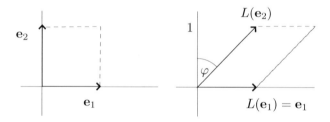

Figure 10.4. A shear on $\mathbb{R}^2$ in the x-direction.

The matrix representation for a shear with respect to the basis $\mathcal{B}$ is
$$\begin{bmatrix} 1 & c \\ 0 & 1 \end{bmatrix}.$$
Since its determinant is one, a shear must belong to $SL_2(\mathbb{R})$.

Is every element in $SL_2(\mathbb{R})$ a shear? We leave that question as an exercise.

We have noted that the special linear group is not just a subgroup of the general linear group, but a normal subgroup. This means that $GL_n(\mathbb{F})/SL_n(\mathbb{F})$ is also a group. What is its story?

Theorem 10.22. $GL_n(\mathbb{F})/SL_n(\mathbb{F})$ is isomorphic to the multiplicative group $\mathbb{F}^*$.

10.4. The Special Linear Group

Proof. Let $G = GL_n(\mathbb{F})$ and let $H = SL_n(\mathbb{F})$. For g_i in G, we have $g_1 H = g_2 H$ if and only if $g_1^{-1} g_2$ is in H, that is, if and only if $\det(g_1^{-1} g_2) = 1$. We have

$$1 = \det(g_1^{-1} g_2) = \det(g_1^{-1}) \det g_2 = (\det g_1)^{-1} \det g_2.$$

It follows that $g_1 H = g_2 H$ if and only if $\det g_1 = \det g_2$. From there, we see that $\varphi : G/H \to \mathbb{F}^*$ given by $\varphi(g) = \det g$ is well-defined and injective. That φ is a group homomorphism is immediate by the fact that the determinant mapping on G is a group homomorphism. We leave it as an easy exercise to show that φ is surjective. This proves it is an isomorphism of groups, thus, that $G/H \cong \mathbb{F}^*$. $\square$

Quotient groups can seem mysterious but Theorem 10.22 tells a straightforward tale. The partition on $GL_n(\mathbb{F})$ induced by the left (or right) $SL_n(\mathbb{F})$-cosets is according to determinant: Two matrices are in the same $SL_n(\mathbb{F})$-coset if and only if they have the same determinant. The thing distinguishing one element from another in $GL_n(\mathbb{F})/SL_n(\mathbb{F})$ is a nonzero element in the field.

Exercises 10.4.

10.4.1. Show that $SL_n(\mathbb{F}_2) = GL_n(\mathbb{F}_2)$.

10.4.2. List the elements in $SL_2(\mathbb{F}_3)$.

10.4.3. Let $\mathbf{v} = (v_1, v_2)$ and $\mathbf{w} = (w_1, w_2)$ belong to $\mathbb{R}^2$. Suppose θ in the interval $(0, \pi)$ is the angle between $\mathbf{v}$ and $\mathbf{w}$.
  (a) Show that $\|\mathbf{v}\|\|\mathbf{w}\| \sin \theta = |v_1 w_2 - v_2 w_1|$.
  (b) Identify $\mathbf{v}$ with $(v_1, v_2, 0)$ and $\mathbf{w}$ with $(w_1, w_2, 0)$ in $\mathbb{R}^3$. Show that the area of the parallelogram determined by $\mathbf{v}$ and $\mathbf{w}$ is $\|\mathbf{v} \times \mathbf{w}\|$.
  (c) When $\mathbf{x} = (r \cos \theta, r \sin \theta)$ is in $\mathbb{R}^2$, we write $\mathbf{x} = \text{Pol}(r, \theta)$. Let $\mathbf{v} = \text{Pol}(r_1, \theta_1)$ and $\mathbf{w} = \text{Pol}(r_2, \theta_2)$. Assume that θ_i is in $[0, 2\pi)$. Suppose that φ is the angle between $-\pi$ and π as measured from $\mathbf{v}$ to $\mathbf{w}$. Show that $\varphi = \theta_2 - \theta_1$ if θ_1 is a polar angle associated to $\mathbf{v}$ and θ_2 is a polar angle associated to $\mathbf{w}$.

10.4.4. Let $\mathbf{v} = (v_1, v_2, v_3)$, $\mathbf{w} = (w_1, w_2, w_3)$, and $\mathbf{u} = (u_1, u_2, u_3)$ belong to $\mathbb{R}^3$. If $\mathcal{B} = \{\mathbf{v}, \mathbf{w}, \mathbf{u}\}$ is linearly independent, then the volume of the parallelepiped determined by $\mathcal{B}$ is the area of the base — that is, any parallelogram determined by two vectors in $\mathcal{B}$ — times the height of the parallelepiped, which is the distance from the base to the oppposite side. Show that this volume is

$$|(\mathbf{v} \times \mathbf{w}) \cdot \mathbf{u}| = |(\mathbf{w} \times \mathbf{u}) \cdot \mathbf{v}| = |(\mathbf{u} \times \mathbf{v}) \cdot \mathbf{w}|.$$

10.4.5. Given f in $C'(\mathbb{R})$ and a and b in $\mathbb{R}$, we can interpret $\int_a^b f(x)\,dx$ in terms of areas of regions in the xy-plane. We break the region bounded by the x-axis and $y = f(x)$ into oriented regions, each of which has a signed area. The total of all those signed areas is then $\int_a^b f(x)\,dx$. Using the picture in Figure 10.5 as a hint, describe a rule that determines the orientation of a region bounded by points x_1 and x_2 on the x-axis and the curve $y = f(x)$. Here x_1 is not necessarily less than x_2, but the curve $y = f(x)$ does not cross the x-axis over the interval between x_1 and x_2.

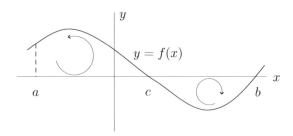

Figure 10.5. $\int_a^b f(x)\,dx = \int_a^c f(x)\,dx + \int_c^b f(x)\,dx.$

10.4.6. Let $G = GL_n(\mathbb{F})$ and $H = SL_n(\mathbb{F})$. Show that $\varphi : G/H \to \mathbb{F}^*$ given by $\varphi(gH) = \det g$ is surjective.

10.4.7. Is every operator in $SL_2(\mathbb{R})$ a shear? If so, prove it. If not, find an example to support your argument.

10.4.8. Recall that a scalar matrix is one of the form cI_n, where c is in $\mathbb{F}$ and I_n is the $n \times n$ identity matrix.
 (a) What scalar matrices belong to $SL_n(\mathbb{R})$?
 (b) What diagonal matrices belong to $SL_n(\mathbb{R})$?
 (c) What scalar matrices belong to $SL_n(\mathbb{C})$?
 (d) What diagonal matrices belong to $SL_n(\mathbb{C})$?

10.4.9. Show that the scalar matrices in $GL_n(\mathbb{F})$ form a normal subgroup of $GL_n(\mathbb{F})$.

10.4.10. Show that the diagonal matrices in $GL_n(\mathbb{F})$ form a normal subgroup of $GL_n(\mathbb{F})$.

10.5. The Projective Group

We start here by considering an important subgroup in any group.

Definition 10.23. The **center** of a group G is the set
$$Z(G) = \{g \in G \,|\, gx = xg \,\forall x \in G\}.$$

If G is abelian, it is its own center.

If G is any group, $Z(G)$ contains the identity element. Given g and g' in $Z(G)$, we have, for any x in G,
$$(g'g^{-1})x = g'(x^{-1}g)^{-1} = g'(gx^{-1})^{-1} = g'xg^{-1} = x(g'g^{-1}).$$
It follows that $g'g^{-1}$ also belongs to $Z(G)$. This proves that $Z(G)$ is a subgroup of G. Fix g in G. Since
$$gZ(G) = \{gh \,|\, h \in Z(G)\} = \{hg \,|\, h \in Z(G)\} = Z(G)g,$$
we have proved the following.

Theorem 10.24. If G is a group, then $Z(G)$ is a normal subgroup in G.

Theorem 10.25. The center of $GL_n(\mathbb{F})$ is the set of scalar matrices in $GL_n(\mathbb{F})$.

Proof. Let $C = [c_{ij}] = \begin{bmatrix} \mathbf{c}_1 & \cdots & \mathbf{c}_n \end{bmatrix}$ belong to $Z(GL_n(\mathbb{F}))$. We start by arguing that C is a diagonal matrix.

Let E_{ij} be the $n \times n$ type-3 elementary matrix that adds row j to row i on premultiplication and adds column j to column i on postmultiplication. For k in $\{1, \ldots, n\}$, the ik-entry in $E_{ij}C$ is $c_{ik} + c_{jk}$ while for $k \neq j$, the ik-entry in $CE_{ij} = c_{ik}$. We conclude that when $k \neq j$, $c_{jk} = 0$. In other words, C is a diagonal matrix, as claimed.

Write $C = \text{diag}(c_1, \ldots, c_n)$, for c_i in $\mathbb{F}$. Notice that when $j > 1$, the $1j$-entry of $E_{1j}C$ is c_j while the $1j$-entry of CE_{1j} is c_1. It follows that C belongs to $Z(GL_n(\mathbb{F}))$ if and only if $c_1 = c_j$ for $j = 2, \ldots, n$. This means that C is a scalar matrix. □

The quotient group is designated
$$PGL_n(\mathbb{F}) = GL_n(\mathbb{F})/Z(GL_n(\mathbb{F})).$$
How can we understand what this is?

Keep in mind that the center of $GL_n(\mathbb{F})$ is the identity element in $PGL_n(\mathbb{F})$. If $\mathbf{v}$ in $\mathbb{F}^n$ is nonzero, then the set
$$Z(GL_n(\mathbb{F}))[\{\mathbf{v}\}] = \{c\mathbf{v} \mid c \in \mathbb{F}\}$$
is the line in $\mathbb{F}^n$ determined by $\mathbf{v}$. To $PGL_n(\mathbb{F})$, $\mathbf{v}$ and $c\mathbf{v}$ are indistinguishable. This suggests that we should look at $PGL_n(\mathbb{F})$, not as acting on individual vectors, but as acting on the 1-dimensional subspaces in $\mathbb{F}^n$, lines through the origin.

Our job now is to identify $PGL_n(\mathbb{F})$ as a group of symmetries, presumably on an object comprised of subspaces in a vector space.

10.5.1. Projective Planes. Projective geometry is a type of incidence geometry that is used all over mathematics. We start our discussion with a set of axioms for the projective plane.[1]

Definition 10.26. A **projective plane** is a collection of points and lines satisfying the following three axioms.

Axiom 1. Any two points lie on a unique line.

Axiom 2. Any two lines intersect in a unique point.

Axiom 3. There are four points, no three of which are collinear.

Every projective plane enjoys a feature known as the *principle of duality*. The *dual* of any statement about a projective plane is the result of switching the words "point" and "line" and switching the ideas of *concurrence* or intersection, which applies to lines, and *collinearity*, which applies to points. (Recall that three or more lines are concurrent provided they have a common point of intersection. Dual to the notion of three concurrent lines is three collinear points, that is, three points that lie on a single line.) The principle of duality is that a statement about a

---
[1] Authors vary in their preferences for the axioms defining a projective plane, often based on a particular context.

projective plane is true if and only if the dual statement is also true. To prove that the principle of duality applies to projective planes, we must show that the axioms in Definition 10.26 hold if and only if the dual axioms hold. Since Axioms 1 and 2 are dual statements, we need only show that when we assume Axioms 1 and 2 are satisfied, Axiom 3 is equivalent to its dual.

Denote the line determined by points P_i and P_j by P_iP_j. Denote the point of intersection of lines ℓ_i and ℓ_j by $P_{ij} = \ell_i \cap \ell_j$.

Lemma 10.27. Every projective plane contains four lines, no three of which are concurrent.

Proof. Let P_1, P_2, P_3, P_4 be points in a projective plane such that no three of the points are collinear. By taking any two of these points, P_i and P_j, we get a line P_iP_j containing neither of the other two points.

Consider P_1P_2, P_2P_3, P_3P_4, and P_4P_1. We claim that no three of these lines are concurrent. Suppose, for instance, that P_1P_2, P_2P_3, and P_3P_4 were concurrent. Since $P_2 = P_1P_2 \cap P_2P_3$, the point of concurrency must be P_2. Since $P_2P_3 \cap P_3P_4 = P_3$, the point of concurrency must be P_3, which is absurd. We leave it as an exercise to check the three remaining triplets of lines to verify that they cannot be concurrent. From there, the lemma follows. □

Now we establish that if the dual of Axiom 3 holds, so does Axiom 3.

Lemma 10.28. Let α be a set of points and lines that satisfy Axioms 1 and 2 of Definition 10.26. If α contains four lines, no three of which are concurrent, then α satisfies Axiom 3 of Definition 10.26.

Proof. Let α be as hypothesized. Say that ℓ_i, $i = 1, 2, 3, 4$, are four lines, no three of which are concurrent. Let P_{ij} be the point of intersection of ℓ_i and ℓ_j.

Consider P_{12}, P_{23}, P_{34}, and P_{41}. We claim that no three of these points are collinear. Suppose, for instance, that P_{12}, P_{23}, and P_{34} were collinear. Since $P_{12}P_{23} = \ell_2$, then the line containing the three points is ℓ_2. Since $P_{23}P_{34} = \ell_3$, the line containing the three points must be ℓ_3, which is absurd. We leave it as an exercise to check the three remaining triplets of points to verify that they cannot be collinear. From there, the lemma follows. □

The two lemmas together imply that Axiom 3 and its dual are equivalent when Axioms 1 and 2 hold.

The following is now immediate.

Theorem 10.29 (Principle of Duality). The dual of any true statement about a projective plane is also a true statement.

The proof of Lemma 10.27 suggests the following.

Lemma 10.30. Every projective plane has at least seven points and seven lines.

10.5. The Projective Group

Proof. Let $S = \{P_1, P_2, P_3, P_4\}$ be points in a projective plane. Assume that no three of the points in S are collinear. We then have

$$\binom{4}{2} = 6$$

distinct lines in the plane, $P_i P_j = P_j P_i$, $i \neq j$.

While $P_1 P_2$ and $P_3 P_4$ must intersect, they cannot intersect in any of P_1, P_2, P_3, or P_4. If they did, it would mean that three points in S were collinear. The same can be said of $P_1 P_3$ and $P_2 P_4$. Let $P_1 P_2 \cap P_3 P_4 = P_5$ and let $P_1 P_3 \cap P_2 P_4 = P_6$. We leave it as an exercise to verify that $P_5 P_6$ is a seventh line in the plane.

By the principle of duality, every projective plane must also have seven points. $\square$

A **triangle** in a projective plane is a set of three noncollinear points and the $\binom{3}{2} = 3$ lines they determine. The three points are the *vertices* of the triangle and the lines are the *sides* of the triangle. There are no line segments in projective geometry. The sides of a projective triangle are entire projective lines.

A triangle is a self-dual configuration: We get the same object if we define it to be three nonconcurrent lines and the points they determine.

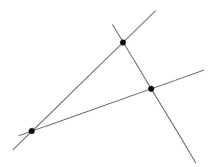

Figure 10.6. The sides of a projective triangle are lines, not line segments.

A **complete quadrangle** in a projective plane is determined by a set of four points, no three of which are collinear. These are the *vertices* of the quadrangle. By Lemma 10.30, we know that where there are four projective points, no three of which are collinear, there must actually be seven projective points. Those three additional points are the *diagonal points* of the complete quadrangle. The *sides* of a complete quadrangle are the $\binom{4}{2} = 6$ lines determined by the vertices. Two sides are *opposite sides* if they intersect in a diagonal point. The set of vertices, diagonal points, and sides is the complete quadrangle.

Since the number of lines in a complete quadrangle is not the same as the number of points, a complete quadrangle is not a self-dual configuration. We leave it as an exercise to define a **complete quadrilateral**, the dual of a complete quadrangle.

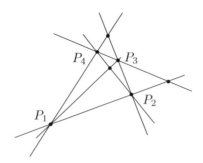

Figure 10.7. The complete quadrangle determined by vertex set $\{P_1, P_2, P_3, P_4\}$.

The axioms in Definition 10.26 define a projective plane abstractly. There are many types of projective planes — some are quite mysterious and remain the subject of active research — but the projective planes we study are based on vector spaces. With linear algebra as a boost, these are quite accessible.

We leave it as an easy exercise to verify that if $\mathbb{F}$ is any field, then the 1-dimensional subspaces of $\mathbb{F}^3$ behave as projective points while the 2-dimensional subspaces behave as projective lines. This will legitimize the following definition.

Definition 10.31. The **projective plane over** $\mathbb{F}$, $\mathbb{P}^2\mathbb{F}$, is comprised of **projective points**, which are the 1-dimensional subspaces in $\mathbb{F}^3$, and **projective lines**, which are the 2-dimensional subspaces in $\mathbb{F}^3$. Two projective objects in $\mathbb{P}^2\mathbb{F}$ intersect provided the underlying subspaces in $\mathbb{F}^3$ intersect nontrivially.

Lemma 10.30 says that a projective plane cannot have fewer than seven points. The following example shows that the smallest projective plane is one with exactly seven points.

Example 10.32. The **Fano plane** is the projective plane with seven points and seven lines. It can be modeled on $\mathbb{F}_2^3$. The points in $\mathbb{F}_2^3$ are

$$(10.3) \quad \{(0,0,0),(1,0,0),(0,1,0),(0,0,1),(0,1,1),(1,0,1),(1,1,0),(1,1,1)\}.$$

A point in $\mathbb{P}^2\mathbb{F}_2$ is a 1-dimensional subspace in $\mathbb{F}_2^3$, each of which is comprised of two points: $(0,0,0)$ and a nonzero point. Since it has seven nonzero points, $\mathbb{F}_2^3$ has seven 1-dimensional subspaces. Since each 1-dimensional subspace of $\mathbb{F}_2^3$ determines a single projective point, $\mathbb{P}^2\mathbb{F}_2$ has exactly seven points.

Lines in $\mathbb{P}^2\mathbb{F}_2$ are the 2-dimensional subspaces of $\mathbb{F}_2^3$. Each 2-dimensional subspace of $\mathbb{F}_2^3$ is comprised of $(0,0,0)$, along with two distinct nonzero points and their sum. The following diagram makes it easy to count the seven different lines in $\mathbb{P}^2\mathbb{F}_2$. Each is comprised of three points. The triplets that make up lines are represented in the diagram as lying on a line segment or lying on the circle. The line segments and the circular arcs in the diagram are not actually in the projective plane. They just show us which triplets of points comprise lines.

The definition of a projective plane over $\mathbb{F}$ suggests the following.

10.5. The Projective Group

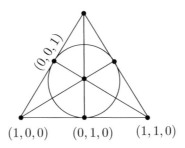

(1,0,0) (0,1,0) (1,1,0)

(0,0,1)

Figure 10.8. The Fano plane can be modeled using $\mathbb{F}_2{}^3$.

Definition 10.33. The **projective group** associated to $\mathbb{F}^n$ is $PGL_n(\mathbb{F})$.

We will find that $PGL_3(\mathbb{F})$ is the group of symmetries on $\mathbb{P}^2\mathbb{F}$.

10.5.2. Symmetries on a Projective Plane. Next we consider how $PGL_3(\mathbb{F})$ acts on the projective plane $\mathbb{P}^2\mathbb{F}$. For this discussion, $V = \mathbb{F}^3$ and we identify the subspaces of V with points and lines in $\mathbb{P}^2\mathbb{F}$.

Let $G = GL_3(\mathbb{F})$ and $H = Z(G)$. An element in $PGL_3(\mathbb{F})$ is an H-coset in G. For g in G, we use the notation
$$\bar{g} = gH \in PGL_3(\mathbb{F}).$$
Since H is the set of scalar matrices in G, $\bar{g} = \overline{cg}$ for any c in $\mathbb{F}^*$.

Let $P_{\mathbf{v}} = \text{Span}\{\mathbf{v}\}$ in $\mathbb{P}^2\mathbb{F}$. Notice that $P_{\mathbf{v}} = P_{c\mathbf{v}}$ for any c in $\mathbb{F}^*$. If we say
$$\bar{g}(P_{\mathbf{v}}) := P_{g(\mathbf{v})},$$
then $\bar{g} : \mathbb{P}^2\mathbb{F} \to \mathbb{P}^2\mathbb{F}$ is well-defined for each $\bar{g}$ in $PGL_3(\mathbb{F})$.

If $\bar{g}(P_{\mathbf{v}}) = \bar{g}(P_{\mathbf{w}})$, then $\text{Span}\{g(\mathbf{v})\} = \text{Span}\{g(\mathbf{w})\}$, which implies that $P_{\mathbf{v}} = P_{\mathbf{w}}$. It follows that $\bar{g}$ defines an injective mapping on $\mathbb{P}^2\mathbb{F}$.

Now suppose we are given arbitrary $P_{\mathbf{w}}$ in $\mathbb{P}^2\mathbb{F}$. Taking $\mathbf{v} = g^{-1}(\mathbf{w})$, we have $\bar{g}(P_{\mathbf{v}}) = P_{\mathbf{w}}$. This establishes that $\bar{g}$ is surjective on $\mathbb{P}^2\mathbb{F}$, thus, a bijection.

What does $PGL_3(\mathbb{F})$ do to the lines in $\mathbb{P}^2\mathbb{F}$? Suppose
$$\ell = \text{Span}\{\mathbf{v}, \mathbf{w}\} \in \mathbb{P}^2\mathbb{F}$$
where $\{\mathbf{v}, \mathbf{w}\}$ is linearly independent in V. For $\bar{g}$ in $PGL_3(\mathbb{F})$, we have
$$\bar{g}(\ell) = \text{Span}\{g(\mathbf{v}), g(\mathbf{w})\}.$$
Since g in G respects linear independence in V, elements in $PGL_3(\mathbb{F})$ send projective lines to projective lines.

We would like to know what $PGL_3(\mathbb{F})$ does to the triangles and quadrangles in $\mathbb{P}^2\mathbb{F}$, since those are fundamental structures in any projective plane.

Suppose $S = \{P_1, P_2, P_3\} \subseteq \mathbb{P}^2\mathbb{F}$ is a set of noncollinear projective points. Identify S with the triangle it determines. Fix $\mathbf{v}_i$ in V so that $\text{Span}\{\mathbf{v}_i\} = P_i$. If $\{\mathbf{v}_1, \mathbf{v}_2, \mathbf{v}_3\}$ is linearly dependent, then some $\mathbf{v}_i$ is in the span of the other two vectors. If that is the case, then P_i is on the projective line determined by the other two projective points, which is absurd. It follows that S determines a basis for V. Since every element in G permutes bases for V, we see that every element in $PGL_3(\mathbb{F})$ maps triangles to triangles in $\mathbb{P}^2\mathbb{F}$.

Now let $Q = \{P_1, P_2, P_3, P_4\}$ be the set of vertices that determine a quadrangle in $\mathbb{P}^2\mathbb{F}$. By omitting any one point P_i from Q, we get a set that determines a projective triangle, thus a basis, $\mathcal{B}_i$, for V. For g in G, $g[\mathcal{B}_i]$ is a basis for V from which it follows that $\bar{g}[Q]$ is a set of vertices that determine a quadrangle in $\mathbb{P}^2\mathbb{F}$. We conclude that $PGL_3(\mathbb{F})$ maps complete quadrangles to complete quadrangles.

Exercises 10.5.

10.5.1. Show that any two $n \times n$ symmetric matrices commute with one another.

10.5.2. Show that the symmetric matrices in $GL_n(\mathbb{F})$ form a subgroup of $GL_n(\mathbb{F})$.

10.5.3. Determine whether the symmetric matrices in $GL_n(\mathbb{F})$ form a normal subgroup.

10.5.4. Give an example of a symmetric matrix in $GL_2(\mathbb{R})$ that does not commute with $A = \begin{bmatrix} 1 & -1 \\ 0 & 1 \end{bmatrix}$.

10.5.5. Give an example of a diagonal matrix in $GL_2(\mathbb{R})$ that does not commute with $A = \begin{bmatrix} 0 & 1 \\ 1 & 0 \end{bmatrix}$.

Projective Planes

10.5.6. Finish the proof of Lemma 10.27.

10.5.7. Finish the proof of Lemma 10.28.

10.5.8. Finish the proof of Lemma 10.30.

10.5.9. Verify that the 1- and 2-dimensional subspaces in a 3-dimensional vector space satisfy the axioms for a projective plane.

10.5.10. Copy the diagram for the Fano plane in Figure 10.8 and finish labeling it.

10.5.11. Are there other ways to label the Fano plane? Can you pick any three points in the diagram and label them with any 3-tuples from (10.3)?

10.5.12. How many lines go through each point in the Fano plane? How many points lie on each line in the Fano plane?

10.5.13. How many triangles are in the Fano plane? Identify three of them.

10.5.14. How many vertices and diagonal points lie on each side of a complete quadrangle? How many sides go through each vertex of a complete quadrangle? How many sides go through each diagonal point of a complete quadrangle?

10.5.15. How many complete quadrangles are there in the Fano plane? Identify two of them.

10.5.16. Define a complete quadrilateral. Be sure to dualize the vertices, the diagonal points, and the sides of a complete quadrangle when you write your definition.

10.5.17. How many sides and diagonals go through each vertex of a complete quadrilateral? How many vertices lie on each side of a complete quadrangle? How many vertices lie on each diagonal of a complete quadrilateral?

10.5.18. How many complete quadrilaterals are there in the Fano plane? Identify two of them.

10.5.19. Let S be the projective triangle in $\mathbb{P}^2\mathbb{R}$ determined by the standard ordered basis for $\mathbb{R}^3$. Find a different set of basis vectors that determines S.

10.5.20. Let $S = \{(1,0,0),(0,1,0),(0,0,1),(1,1,1)\} \subseteq \mathbb{R}^3$.
   (a) Show that S determines a complete quadrangle, $Q \subseteq \mathbb{P}^2\mathbb{R}$.
   (b) Find a different set of vectors in $\mathbb{R}^3$ that determines the same complete quadrangle.
   (c) Find the sides and the diagonal points of Q.
   (d) Find the pairs of opposite sides of the complete quadrangle.

Symmetries on a Projective Plane

10.5.21. Find $PGL_3(\mathbb{F}_2)$, the symmetry group for the Fano plane.

10.5.22. Let ℓ be the projective line in $\mathbb{P}^2\mathbb{F}$ determined by a linearly independent set $\{\mathbf{v},\mathbf{w}\} \subseteq \mathbb{F}^3$. Show that if ℓ' is any other line in $\mathbb{P}^2\mathbb{F}$, then there is $\bar{g}$ in $PGL_3(\mathbb{F})$ so that $\bar{g}(\ell) = \ell'$.

10.5.23. Let $S = \{P_1, P_2, P_3\}$ be a triangle in $\mathbb{P}^2\mathbb{F}$. Find three nontrivial elements in $PGL_3(\mathbb{F})$ that map S to itself.

10.5.24. Let Q be the complete quadrangle in $\mathbb{P}^2\mathbb{R}$ determined by
$$S = \{(1,0,0),(0,1,0),(0,0,1),(1,1,1)\} \subseteq \mathbb{R}^3.$$
Find three nontrivial elements in $PGL_3(\mathbb{R})$ that map Q to itself.

10.5.25. Let $\ell_1 = \mathrm{Span}\{(1,0,0),(0,1,0)\}$, $\ell_2 = \mathrm{Span}\{(1,0,0),(0,0,1)\}$,
$$\ell_3 = \mathrm{Span}\{(0,1,0),(1,1,1)\},$$
and $\ell_4 = \mathrm{Span}\{(0,0,1),(1,1,1)\}$, in $\mathbb{R}^3$. Let $S = \{\ell_1, \ell_2, \ell_3, \ell_4\}$.
   (a) Show that S determines a complete quadrilateral $Q \subseteq \mathbb{P}^2\mathbb{R}$.
   (b) Find the vertices and diagonals of Q.
   (c) Find the pairs of opposite vertices of Q.
   (d) Find three nontrivial elements in $PGL_3(\mathbb{R})$ that map Q to itself.

10.5.26. Let S_1 and S_2 be triangles in some projective plane $\mathbb{P}^2\mathbb{F}$. Show that there is an element in $PGL_3(\mathbb{F})$ that maps S_1 to S_2. Can we play the same game with two complete quadrangles?

10.6. The Orthogonal Group

When (V, σ) is a real, n-dimensional inner product space, its orthogonal group can be identified with the orthogonal matrices in $M_n(\mathbb{R})$,
$$O_n(\mathbb{R}) = \{A \in M_n(\mathbb{R}) \mid AA^T = I_n\}.$$

This is the group that preserves the dot product on $\mathbb{R}^n$. It is the subject of our study in this section.

We have established that if A is in $O_n(\mathbb{R})$, then $\det A = \pm 1$. This means that some orthogonal matrices also belong to $SL_n(\mathbb{R})$. By Exercise 10.1.4 the intersection of $SL_n(\mathbb{R})$ and $O_n(\mathbb{R})$ is a group.

Definition 10.34. The **special orthogonal group** of $n \times n$ matrices over $\mathbb{R}$ is
$$SO_n(\mathbb{R}) = O_n(\mathbb{R}) \cap SL_n(\mathbb{R}).$$

We have already seen that rotations are orthogonal transformations on $\mathbb{R}^2$ and $\mathbb{R}^3$. What else might we run across in $O_n(\mathbb{R})$ and in $SO_n(\mathbb{R})$? We will satisfy ourselves with $n \leq 3$.

First consider A in $O_1(\mathbb{R})$. Any linear operator on $\mathbb{R}$ has the form
$$x \mapsto cx$$
for some scalar c. The dot product on $\mathbb{R}$ is multiplication. It follows that $O_1(\mathbb{R}) = \{1, -1\}$ and that $SO_1(\mathbb{R}) = \{1\}$.

We established in Example 5.44 that A in $O_2(\mathbb{R})$ can always be written in one of the following two forms:
$$A = \begin{bmatrix} \cos \varphi & \sin \varphi \\ \sin \varphi & -\cos \varphi \end{bmatrix},$$
in which case $\det A = -1$, and
$$A = \begin{bmatrix} \cos \varphi & \sin \varphi \\ -\sin \varphi & \cos \varphi \end{bmatrix},$$
in which case $\det A = 1$. The latter is the case in which A defines a rotation on $\mathbb{R}^2$. It follows that $SO_2(\mathbb{R})$ is the set of rotations on $\mathbb{R}^2$,
$$SO_2(\mathbb{R}) = \left\{ A = \begin{bmatrix} \cos \varphi & \sin \varphi \\ -\sin \varphi & \cos \varphi \end{bmatrix} \mid \varphi \in \mathbb{R} \right\}.$$

What about the other type of mapping in $O_2(\mathbb{R})$? If
$$A = \begin{bmatrix} \cos \theta & \sin \theta \\ \sin \theta & -\cos \theta \end{bmatrix},$$
its characteristic polynomial is
$$\det(A - tI_2) = (\cos \theta - t)(-\cos \theta - t) - \sin^2 \theta = -1 + t^2.$$
Since A has real eigenvalues 1 and -1, we now know that A effects an orthogonal reflection on $\mathbb{R}^2$.

We have proved the following theorem.

Theorem 10.35. An orthogonal operator on $\mathbb{R}^2$ is either an orthogonal reflection or a rotation. In particular, $SO_2(\mathbb{R})$ is the set of rotations on $\mathbb{R}^2$ centered at $\mathbf{0}$.

The fact that an element in $SO_2(\mathbb{R})$ is uniquely determined by an equivalence class of coterminal angles $\{\theta + 2n\pi \mid n \in \mathbb{Z}\}$ suggests a connection to $\mathcal{U}(\mathbb{C}) = \{e^{i\theta} \mid \theta \in \mathbb{R}\}$. Indeed, it is easy to verify that
$$\begin{bmatrix} \cos \theta & -\sin \theta \\ \sin \theta & \cos \theta \end{bmatrix} \mapsto e^{i\theta}$$
is a bijective mapping. Since $e^{i\theta} e^{i\varphi} = e^{i(\theta + \varphi)}$ and the composition of the rotations in $SO_2(\mathbb{R})$ through the angles θ and φ gives us the rotation through $\theta + \varphi$, the mapping is an isomorphism of groups.

This is enough to prove the following theorem.

10.6. The Orthogonal Group

Theorem 10.36. As groups, $SO_2(\mathbb{R}) \cong \mathcal{U}(\mathbb{C}) \cong S^1$.

Suppose next that A belongs to $O_3(\mathbb{R})$. Since the only irreducible polynomials over $\mathbb{R}$ are those of degrees one and two, and since the characteristic polynomial for A has degree three, A must have an eigenvalue in $\mathbb{R}$, thus, an eigenvector in $\mathbb{R}^3$. In fact, since an orthogonal matrix is normal — A in $O_n(\mathbb{R})$ implies $AA^T = A^TA$ — it is diagonalizable, by a generalization to the Finite-Dimensional Spectral Theorem. Since we left the proof of that as an exercise in Section 8.6, we state a more modest version in the following lemma. The (much easier) proof of this is an exercise, as well.

Lemma 10.37. Every orthogonal operator A in $O_3(\mathbb{R})$ has an eigenspace. Moreover, if there are two linearly independent A-eigenvectors in $\mathbb{R}^3$, then $\mathbb{R}^3$ has a basis of A-eigenvectors.

Suppose $\mathbf{u}$ is a unit eigenvector for A in $O_3(\mathbb{R})$. If the $\mathbf{u}$-eigenvalue is λ, then since A is nonsingular, $\lambda \neq 0$. Take $\mathbf{u}$ to be the third vector in an ordered, orthonormal basis for $\mathbb{R}^3$, $\mathcal{B} = \{\mathbf{u}_1, \mathbf{u}_2, \mathbf{u}\}$. Since A respects orthogonality,

$$A\mathbf{u} \cdot A\mathbf{u}_i = \lambda \mathbf{u} \cdot A\mathbf{u}_i = 0,$$

so that neither $A\mathbf{u}_1$ nor $A\mathbf{u}_2$ has a component in the $\mathbf{u}$-direction. For some A' in $M_2(\mathbb{R})$, we then have that

$$(10.4) \qquad [\![A]\!]_\mathcal{B} = \left[\begin{array}{cc|c} & & 0 \\ \multicolumn{2}{c|}{A'} & 0 \\ \hline 0 & 0 & \lambda \end{array}\right].$$

Letting $\mathbf{v} = v_1\mathbf{u}_1 + v_2\mathbf{u}_2$ and $\mathbf{w} = w_1\mathbf{u}_1 + w_2\mathbf{u}_2$, we see that

$$A\mathbf{v} \cdot A\mathbf{w} = A'\mathbf{v}' \cdot A'\mathbf{w}'$$

where $\mathbf{v}' = (v_1, v_2)$ and $\mathbf{w}' = (w_1, w_2)$ both belong to $\mathbb{R}^2$. Since

$$\mathbf{v} \cdot \mathbf{w} = \mathbf{v}' \cdot \mathbf{w}',$$

we conclude that A' belongs to $O_2(\mathbb{R})$. In this sense, understanding orthogonal operators on $\mathbb{R}^3$ reduces to understanding orthogonal operators on $\mathbb{R}^2$.

Notice next that if A in $O_3(\mathbb{R})$ is as given in (10.4), then $\det A = \lambda \det A'$. We have established that $\det A' = \pm 1$ so $\lambda = \pm 1$. By relabeling our coordinate system if necessary, we can interpret A as having a λ-eigenspace that coincides with the z-axis, and A' as an orthogonal operator on the xy-plane.

Suppose A' effects the orthogonal reflection on the xy-plane that fixes some nonzero $\mathbf{v} = (v_1, v_2, 0)$. We leave it as an exercise to show that, depending on the value of λ, either A effects the orthogonal reflection on $\mathbb{R}^3$ that fixes the plane Span$\{\mathbf{v}, \mathbf{e}_3\}$ or A effects the rotation on $\mathbb{R}^3$ with axis Span$\{\mathbf{v}\}$ through an angle coterminal with π.

Suppose next that A' effects the rotation on the xy-plane through the angle ψ. If $\lambda = 1$, then A effects the rotation on $\mathbb{R}^3$ through the angle ψ about the z-axis. If $\lambda = -1$, then A effects a composition of mappings: a rotation followed by an orthogonal reflection through the plane perpendicular to the axis of rotation. To think more carefully about this, we introduce the use of spherical coordinates, as shown in Figure 10.9.

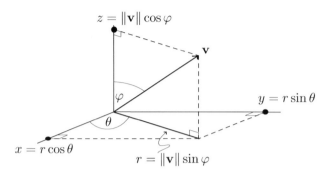

Figure 10.9. The angle between **v** in $\mathbb{R}^3$ and the z-axis is φ. The angle between the orthogonal projection of **v** onto the xy-plane and the x-axis is θ.

If $\mathbf{v} = (x, y, z)$ is in $\mathbb{R}^3$ and φ is the angle formed by **v** and the z-axis, then
$$(x, y, z) = \|\mathbf{v}\|(\sin\varphi\cos\theta, \sin\varphi\sin\theta, \cos\varphi),$$
where θ is the angle formed by the x-axis and the orthogonal projection of **v** onto the xy-plane. We then have
$$A\mathbf{v} = \begin{bmatrix} \cos\psi & -\sin\psi & 0 \\ \sin\psi & \cos\psi & 0 \\ 0 & 0 & \lambda \end{bmatrix} \|\mathbf{v}\| \begin{bmatrix} \sin\varphi\cos\theta \\ \sin\varphi\sin\theta \\ \cos\varphi \end{bmatrix}.$$
Using trigonometric identities, we get
$$A\mathbf{v} = \|\mathbf{v}\|(\sin\varphi\cos(\theta+\psi), \sin\varphi\sin(\theta+\psi), \lambda\cos\varphi).$$
Since
$$A = \begin{bmatrix} \cos\psi & -\sin\psi & 0 \\ \sin\psi & \cos\psi & 0 \\ 0 & 0 & 1 \end{bmatrix} \begin{bmatrix} 1 & 0 & 0 \\ 0 & 1 & 0 \\ 0 & 0 & \lambda \end{bmatrix}$$
$$= \begin{bmatrix} 1 & 0 & 0 \\ 0 & 1 & 0 \\ 0 & 0 & \lambda \end{bmatrix} \begin{bmatrix} \cos\psi & -\sin\psi & 0 \\ \sin\psi & \cos\psi & 0 \\ 0 & 0 & 1 \end{bmatrix},$$
we can now see that A either effects a rotation or a rotation followed by — or preceded by — orthogonal reflection through the xy-plane. These composition mappings are sometimes called *rotoreflections* or *improper rotations* to distinguish them from *proper rotations*, which have determinant one in any dimension.

Theorem 10.38. *The elements in $SO_3(\mathbb{R})$ are the proper rotations on $\mathbb{R}^3$. The elements in $O_3(\mathbb{R})$ are the proper rotations, the improper rotations, and the orthogonal reflections on $\mathbb{R}^3$.*

Exercises 10.6.

10.6.1. Prove Lemma 10.37 by showing that if A in $O_3(\mathbb{R})$ has two linearly independent eigenvectors, then $\mathbb{R}^3$ must have a basis of A-eigenvectors.

10.6.2. Suppose A has the form given in (10.4) and that A' effects a reflection on $\mathbb{R}^2$. Show that A effects either a reflection on $\mathbb{R}^3$ or a rotation through π on $\mathbb{R}^3$.

10.6.3. Suppose $[\![A]\!]_\mathcal{B}$ is as given in (10.4) and $[\![B]\!]_\mathcal{C}$ is as given in (10.4) where $\mathcal{B}$ and $\mathcal{C}$ are ordered orthonormal bases for $\mathbb{R}^3$. Can we assume $\mathcal{B} = \mathcal{E}$ in this case? If so, what is the implication for $\mathcal{C}$?

10.6.4. Describe the effect of the following operator on $\mathbb{R}^3$:

$$A = \begin{bmatrix} \cos\psi & 0 & -\sin\psi \\ 0 & 1 & 0 \\ \sin\psi & 0 & \cos\psi \end{bmatrix}.$$

10.6.5. Describe the effect of AB on $\mathbb{R}^3$ if A is as given in the previous exercise and

$$B = \begin{bmatrix} -1 & 0 & 0 \\ 0 & 1 & 0 \\ 0 & 0 & 1 \end{bmatrix}.$$

10.7. The Unitary Group

Just as we can identify the orthogonal mappings on a finite-dimensional real inner product space with the group of orthogonal matrices over $\mathbb{R}$, we can identify unitary mappings on a finite-dimensional complex inner product space with the set of unitary matrices,

$$U_n = \{A \in M_n(\mathbb{C}) \,|\, A^\dagger = A^{-1}\}.$$

We first met the unitary group $U_n = U_n(\mathbb{C})$ in Chapter 5. It was an exercise in Chapter 6 to show that if A is unitary, then $\det A$ is a complex number with modulus 1. This is consistent with $|\det A| = 1$ for an orthogonal matrix, A.

Suppose A belongs to U_n and that $\mathbf{v}$ is an A-eigenvector with eigenvalue λ. We then have

$$A\mathbf{v} \cdot A\mathbf{v} = \lambda \mathbf{v} \cdot \lambda \mathbf{v} = \bar{\lambda}\lambda \|\mathbf{v}\|^2$$
$$= |\lambda|^2 \|\mathbf{v}\|^2 = \mathbf{v} \cdot \mathbf{v} = \|\mathbf{v}\|^2.$$

Since $\mathbf{v}$ is nonzero, $\|\mathbf{v}\|^2$ is a positive real number. It follows that $|\lambda|^2 = 1$, from which we may deduce that $|\lambda| = 1$: Any A-eigenvalue is in $\mathcal{U}(\mathbb{C})$.

The mapping $\mathbf{v} \mapsto e^{i\theta}\mathbf{v}$ is just scaling in $\mathbb{C}^n$. Since $e^{i\theta}$ is in $\mathcal{U}(\mathbb{C})$, scaling by $e^{i\theta}$ is a unitary mapping on $\mathbb{C}^n$. On $\mathbb{C}$, these are the only unitary mappings.

Theorem 10.39. *As groups, $U_1 \cong \mathcal{U}(\mathbb{C}) \cong S^1 \cong SO_2(\mathbb{R})$.*

Every element in $\mathbb{C}$ can be written

$$z = |z|e^{i\varphi}$$

for some φ in $\mathbb{R}$ so for $e^{i\theta}$ in U_1, we have

$$e^{i\theta}z = |z|e^{i\theta}e^{i\varphi} = |z|e^{i(\theta+\varphi)}.$$

When we model $\mathbb{C}$ using the coordinate plane, $z \mapsto e^{i\theta}z$ is a rotation.

Taking $a, b, c,$ and d in $\mathbb{C}$, consider
$$A = \begin{bmatrix} a & b \\ c & d \end{bmatrix} \in U_2.$$
We have two ways to express A^{-1}. The formula for the inverse of a 2×2 matrix gives us
$$A^{-1} = \frac{1}{ad - bc} \begin{bmatrix} d & -b \\ -c & a \end{bmatrix}.$$
Since the determinant of a matrix in U_n is in $\mathcal{U}(\mathbb{C})$,
$$\frac{1}{ad - bc} = e^{-i\theta}$$
for some real number θ. For A in U_2, then,
$$A^{-1} = e^{-i\theta} \begin{bmatrix} d & -b \\ -c & a \end{bmatrix}.$$
At the same time, $A^{-1} = A^\dagger$ where
$$A^\dagger = \begin{bmatrix} \bar{a} & \bar{c} \\ \bar{b} & \bar{d} \end{bmatrix}.$$
Putting the two expressions for A^{-1} together we get
$$e^{-i\theta} d = \bar{a} \implies d = e^{i\theta} \bar{a}$$
and
$$-e^{-i\theta} c = \bar{b} \implies c = -e^{i\theta} \bar{b}.$$
This lets us write
$$A = \begin{bmatrix} a & b \\ -e^{i\theta}\bar{b} & e^{i\theta}\bar{a} \end{bmatrix},$$
where $e^{i\theta}(|a|^2 + |b|^2)$ is in $\mathcal{U}(\mathbb{C})$. Since $e^{i\theta}$ is in $\mathcal{U}(\mathbb{C})$ and $|a|^2 + |b|^2$ is in $\mathbb{R}_+$, we see that $|a|^2 + |b|^2 = 1$.

The **special unitary group** is $SU_n = U_n \cap SL_n(\mathbb{C})$. Since $\det z = z$, for z in $\mathbb{C}$, $SU_1 = \{1\}$. If
$$A = \begin{bmatrix} a & b \\ -e^{i\theta}\bar{b} & e^{i\theta}\bar{a} \end{bmatrix} \in SU_2,$$
then
$$e^{i\theta}(|a|^2 + |b|^2) = |a|^2 + |b|^2 = 1$$
implies that $e^{i\theta} = 1$. This gives us
$$A = \begin{bmatrix} a & b \\ -\bar{b} & \bar{a} \end{bmatrix},$$
where $|a|^2 + |b|^2 = 1$ and
$$A^\dagger = A^{-1} = \begin{bmatrix} \bar{a} & -b \\ \bar{b} & a \end{bmatrix}.$$
Suppose we take A in SU_2 given by
$$A = \begin{bmatrix} a_1 + ib_1 & a_2 + ib_2 \\ -a_2 + ib_2 & a_1 - ib_1 \end{bmatrix}$$

where a_1, a_2, b_1, and b_2 are real numbers. Since $\det A = 1$, we have
$$a_1^2 + b_1^2 + a_2^2 + b_2^2 = 1.$$
Indeed, we have a bijective mapping SU_2 to $\mathcal{U}(\mathbb{H})$ given by

(10.5) $\quad \begin{bmatrix} a_1 + ib_1 & a_2 + ib_2 \\ -a_2 + ib_2 & a_1 - ib_1 \end{bmatrix} \mapsto a_1 + b_1 i + a_2 j + b_2 k.$

To see that this is a group homomorphism, we have to verify that the mapping is compatible with multiplication in SU_2 and in $\mathcal{U}(\mathbb{H})$. For instance, if A is as given above and
$$A' = \begin{bmatrix} a_1' + ib_1' & a_2' + ib_2' \\ -a_2' + ib_2' & a_1' - ib_1' \end{bmatrix},$$
then the real part of the $(1, 2)$-entry in AA' is
$$a_1 a_2' - b_1 b_2' + a_2 a_1' + b_2 b_1'.$$
If our mapping is a homomorphism, this is the coefficient of j in the product
$$(a_1 + b_1 i + a_2 j + b_2 k)(a_1' + b_1' i + a_2' j + b_2' k)$$
which is
$$a_1 a_2' + a_2 a_1' + b_2 b_1' - b_1 b_2',$$
as we had hoped. We leave the rest of the verification to the reader. That will prove the following.

Theorem 10.40. As groups, $SU_2 \cong \mathcal{U}(\mathbb{H}) \cong S^3$.

We can "see" S^1 and S^2, but as a 3-dimensional object in $\mathbb{R}^4$, S^3 is difficult to appreciate spatially.[2] Theorem 10.40 can help us understand it in a different way. Towards that end, we take a closer look at $\mathcal{U}(\mathbb{H})$ and its connection to rotations in $\mathbb{R}^3$.

The connection between S^1 and rotations in $\mathbb{R}^2$, $SO_2(\mathbb{R})$, could not be more natural. A rotation in the plane is determined completely by an angle, and the points in S^1 correspond to all the angles. In $\mathbb{R}^3$, though, a rotation has an angle and an axis. When we follow one rotation by another, what is the axis of the resulting rotation? A few calculations will reveal that $\mathcal{U}(\mathbb{H})$ provides a device for describing rotations in $\mathbb{R}^3$ that takes care of this axis problem with dispatch.

We start by noticing that the pure quaternions are mapped to themselves under conjugation by any nonzero quaternion. To prove this and for the rest of our discussion, we identify $\mathbb{H}$ with $\mathbb{R} \oplus \mathbb{R}^3$. If $q = (a; \mathbf{v})$, then $|q|^2 = a^2 + \|\mathbf{v}\|^2$. The product of $q_1 = (a_1; \mathbf{v}_1)$ and $q_2 = (a_2; \mathbf{v}_2)$ is
$$q_1 q_2 = (a_1 a_2 - (\mathbf{v}_1 \cdot \mathbf{v}_2); a_1 \mathbf{v}_2 + a_2 \mathbf{v}_1 + (\mathbf{v}_1 \times \mathbf{v}_2)).$$
If q is a pure quaternion, then $q = (0; \mathbf{v})$. Exercise 10.3.10 was to show that if $q = (a; \mathbf{v})$ is in $\mathcal{U}(\mathbb{H})$, then $q^{-1} = (a; -\mathbf{v})$.

Lemma 10.41. Let q be nonzero in $\mathbb{H}$. If x is any pure quaternion, then $x \mapsto qxq^{-1}$ is a pure quaternion.

---

[2] Though not a vector space, S^3 does have a defined dimension. For us, that can mean that we have three degrees of freedom when we choose a point in S^3. Most importantly, S^3 is not a surface like S^2, which is 2-dimensional.

Proof. Let $q = (a; \mathbf{v})$ and $x = (0; \mathbf{w})$ be as hypothesized. An easy exercise verifies that if q is not a unit quaternion, then it is a scalar multiple of one. Since the pure quaternions are invariant under multiplication by a scalar, we can assume q is a unit quaternion. This allows us to take advantage of the formula for q^{-1}. We can write
$$qxq^{-1} = (a; \mathbf{v})(0; \mathbf{w})(-a; \mathbf{v}) = (-\mathbf{v} \cdot \mathbf{w}; a\mathbf{w} + \mathbf{v} \times \mathbf{w})(-a; \mathbf{v}).$$
Continuing with the calculations, we find the real part of qxq^{-1} to be given by
$$a\mathbf{v} \cdot \mathbf{w} - a\mathbf{w} \cdot \mathbf{v} - (\mathbf{v} \times \mathbf{w}) \cdot \mathbf{v}.$$
We leave it as an exercise to verify that
$$(\mathbf{v} \times \mathbf{w}) \cdot \mathbf{v} = 0.$$
This is enough to prove the lemma. $\square$

Given a rotation $R_{\ell,\varphi} : \mathbb{R}^3 \to \mathbb{R}^3$, we can find a right-handed coordinate system for $\mathbb{R}^3$, $\mathcal{B} = \{\mathbf{u}, \mathbf{u}_1, \mathbf{u}_2\}$, so that

(10.6) $$[\![R_{\ell,\varphi}]\!]_\mathcal{B} = \begin{bmatrix} 1 & 0 & 0 \\ 0 & \cos\varphi & -\sin\varphi \\ 0 & \sin\varphi & \cos\varphi \end{bmatrix}.$$

Theorem 10.42. Let $\mathbf{u}$ be a unit pure quaternion and let $q = (\cos(\varphi/2); \sin(\varphi/2)\mathbf{u})$ where φ is in $\mathbb{R}$. If we define $Q_{\mathbf{u},\varphi}$ on the pure quaternions by
$$Q_{\mathbf{u},\varphi}(x) = qxq^{-1},$$
then the effect of $Q_{\mathbf{u},\varphi}$ is identical to the rotation $R_{\mathbf{u},\varphi} = R_{\ell,\varphi}$ on $\mathbb{R}^3$, where $\ell = \text{Span}\{\mathbf{u}\}$.

Proof. Let everything be as hypothesized. Evidently, q belongs to $\mathcal{U}(\mathbb{H})$. Identifying the pure quaternions with $\mathbb{R}^3$, we have $Q_{\mathbf{u},\varphi}[\mathbb{R}^3] \subseteq \mathbb{R}^3$ by Lemma 10.41.

By the rules of arithmetic in $\mathbb{H}$, $Q_{\mathbf{u},\varphi}$ is linear so we just have to show that the effect of $Q_{\mathbf{u},\varphi}$ on a basis for $\mathbb{R}^3$ is the same as the effect of $R_{\mathbf{u},\varphi}$ on the same basis.

Let $\mathcal{B} = \{\mathbf{u}, \mathbf{u}_1, \mathbf{u}_2\}$ be an ordered right-handed coordinate system for $\mathbb{R}^3$ so that (10.6) holds when we take $\ell = \text{Span}\{\mathbf{u}\}$ in $\mathbb{R}^3$. We have
$$Q_{\mathbf{u},\varphi}(\mathbf{u}) = (\cos(\varphi/2); \sin(\varphi/2)\mathbf{u})(0; \mathbf{u})(\cos(\varphi/2); -\sin(\varphi/2)\mathbf{u})$$
$$= (-\sin(\varphi/2); \cos(\varphi/2)\mathbf{u})(\cos(\varphi/2); -\sin(\varphi/2)\mathbf{u})$$
$$= (0; (\sin^2(\varphi/2) + \cos^2(\varphi/2))\mathbf{u}) = \mathbf{u}.$$
Invoking the rules for cross-products among $\mathbf{u}, \mathbf{u}_1, \mathbf{u}_2$, we also have
$$Q_{\mathbf{u},\varphi}(\mathbf{u}_1) = (\cos(\varphi/2); \sin(\varphi/2)\mathbf{u})(0; \mathbf{u}_1)(\cos(\varphi/2); -\sin(\varphi/2)\mathbf{u})$$
$$= (0; \cos(\varphi/2)\mathbf{u}_1 + \sin(\varphi/2)\mathbf{u}_2)(\cos(\varphi/2); -\sin(\varphi/2)\mathbf{u})$$
$$= (0; \cos^2(\varphi/2)\mathbf{u}_1 + 2\cos(\varphi/2)\sin(\varphi/2)\mathbf{u}_2 - \sin^2(\varphi/2)\mathbf{u}_1)$$
$$= (0; \cos\varphi\,\mathbf{u}_1 + \sin\varphi\,\mathbf{u}_2) = R_{\mathbf{u},\varphi}(\mathbf{u}_1).$$
The same process reveals that
$$Q_{\mathbf{u},\varphi}(\mathbf{u}_2) = (0; -\sin\varphi\,\mathbf{u}_1 + \cos\varphi\,\mathbf{u}_2) = R_{\mathbf{u},\varphi}(\mathbf{u}_2),$$
which proves the theorem. $\square$

10.7. The Unitary Group

If we follow one rotation by another, we can use Theorem 10.42 to calculate the axis and angle of the resulting rotation.

Example 10.43. Suppose $\mathbf{v}$ and $\mathbf{w}$ are orthogonal unit vectors in $\mathbb{R}^3$. Let
$$R_{\ell,\eta} = R_{\mathbf{v},\pi/3} \circ R_{\mathbf{w},\pi/2}.$$
We see what it takes to find ℓ and η.

Identify $\mathbf{x}$ in $\mathbb{R}^3$ with the pure quaternion $x = (0; \mathbf{x})$. We then have
$$R_{\mathbf{v},\pi/3}(\mathbf{x}) = q_1 x q_1^{-1}$$
where
$$q_1 = (\cos(\pi/6); \sin(\pi/6)\mathbf{v})$$
and
$$R_{\mathbf{w},\pi/2}(\mathbf{x}) = q_2 x q_2^{-1}$$
where
$$q_2 = (\cos(\pi/4); \sin(\pi/4)\mathbf{w}).$$
The real part of $q_1 q_2$ is $\cos(\eta/2)$. Since $\mathbf{v}$ and $\mathbf{w}$ are orthogonal,
$$\cos(\eta/2) = \cos(\pi/6)\cos(\pi/4).$$
We then have
$$\cos(\pi/6)\cos(\pi/4) = \frac{\sqrt{3}}{2}\frac{\sqrt{2}}{2} = \sqrt{6}/4 = \cos(\eta/2).$$
Since
$$\cos^2(\eta/2) = \frac{1+\cos\eta}{2},$$
we get
$$6/16 = 3/8 = \frac{1+\cos\eta}{2},$$
so that $1 + \cos\eta = 3/4$. We conclude that $\eta = \cos^{-1}(-1/4)$, which is a little bit less than 0.6π.

The vector part of $q_1 q_2$ gives us ℓ, the axis of rotation of the composition of the two rotations. The vector part of $q_1 q_2$ is
$$\mathbf{y} = \cos(\pi/6)\sin(\pi/4)\mathbf{w} + \cos(\pi/4)\sin(\pi/6)\mathbf{v} + \sin(\pi/6)\sin(\pi/4)(\mathbf{v}\times\mathbf{w}).$$
Following through on the calculations, we have
$$\mathbf{y} = \sqrt{6}/4\,\mathbf{w} + \sqrt{2}/4\,\mathbf{v} + \sqrt{2}/4(\mathbf{v}\times\mathbf{w}),$$
so the axis of the product rotation is $\ell = \text{Span}\{\mathbf{y}\}$.

Exercises 10.7.

10.7.1. Verify that the mapping $\mathbf{v} \mapsto e^{i\theta}\mathbf{v}$ is unitary on $\mathbb{C}^n$.

10.7.2. Find an example of a unitary operator L in $\mathcal{L}(\mathbb{C}^2)$ and an ordered basis $\mathcal{B} \subseteq \mathbb{C}^2$ such that $[\![L]\!]_{\mathcal{B}}$ is not a unitary matrix.

10.7.3. Show that any $\mathbb{C}$-algebra containing the Pauli spin matrices is a subalgebra of $M_2(\mathbb{C})$.

10.7.4. Finish verifying that (10.5) is a group homomorphism to complete the proof that $SU_2 \cong \mathcal{U}(\mathbb{C}) \cong S^3$.

10.7.5. Verify all the calculations in the proof of Lemma 10.41.

10.7.6. Work through the calculations in the proof of Theorem 10.42 to show that
$$Q_{\mathbf{u},\varphi}(\mathbf{u}_2) = -\sin\varphi \mathbf{u}_1 + \cos\varphi \mathbf{u}_2.$$

10.7.7. Redo Example 10.43 switching the order of the rotations to see whether they commute.

10.8. The Symplectic Group

The term *classical groups* appears to be due to Hermann Weyl whose book with that title was first published in 1939. (It is worth mentioning that Weyl never defines the groups in the title of his book.) While it had certainly been studied before Weyl's book was published, the symplectic group, which we consider here, was at the time called the complex line group. Weyl himself seems to have coined the term *symplectic*, based on the ancient Greek, with meaning the same as *complex*, which is based on Latin.

Definition 10.44. A **symplectic form** is a nondegenerate bilinear form σ defined on a vector space V over a field $\mathbb{F}$ such that
$$\sigma(\mathbf{v}, \mathbf{v}) = 0$$
for all $\mathbf{v}$ in V.

When σ is a symplectic form on V, (V, σ) is a **symplectic space**.

Suppose (V, σ) is a symplectic space and that $\mathbf{v}$ and $\mathbf{w}$ belong to V. We then have
$$\sigma(\mathbf{v} + \mathbf{w}, \mathbf{v} + \mathbf{w}) = \sigma(\mathbf{v}, \mathbf{w}) + \sigma(\mathbf{w}, \mathbf{v}) = 0.$$
Over a field $\mathbb{F}$ with $\operatorname{char}\mathbb{F} \neq 2$, a symplectic form is skew-symmetric:
$$\sigma(\mathbf{v}, \mathbf{w}) = -\sigma(\mathbf{w}, \mathbf{v})\, \forall \mathbf{v}, \mathbf{w} \in V.$$

Symplectic spaces arise as naturally as real and complex inner product spaces. To see how, consider $R : \mathbb{C}^n \times \mathbb{C}^n \to \mathbb{R}$ defined by
$$R(\mathbf{z}_1, \mathbf{z}_2) = \operatorname{Re}(\mathbf{z}_1 \cdot \mathbf{z}_2).$$
Now define $C : \mathbb{C}^n \times \mathbb{C}^n \to \mathbb{R}$ by
$$C(\mathbf{z}_1, \mathbf{z}_2) = \operatorname{Im}(\mathbf{z}_1 \cdot \mathbf{z}_2).$$
This way we have
$$\mathbf{z}_1 \cdot \mathbf{z}_2 = R(\mathbf{z}_1, \mathbf{z}_2) + iC(\mathbf{z}_1, \mathbf{z}_2).$$
For example, for $\mathbf{z}_1 = (1 + 2i, 3 - i)$ and $\mathbf{z}_2 = (2 + i, -1 + 4i)$,
$$\mathbf{z}_1 \cdot \mathbf{z}_2 = (1+2i)(2-i) + (3-i)(-1-4i) = 2+2-3-4+i(4-1+1-12) = -3-8i.$$
This gives us $R(\mathbf{z}_1, \mathbf{z}_2) = -3$ and $C(\mathbf{z}_1, \mathbf{z}_2) = -8$.

We can write any $\mathbf{z}$ in $\mathbb{C}^n$ as $\mathbf{z} = \mathbf{x} + i\mathbf{y}$, where $\mathbf{x}$ and $\mathbf{y}$ belong to $\mathbb{R}^n$. Doing so, let $\mathbf{z}_1 = \mathbf{x}_1 + i\mathbf{y}_1$ and $\mathbf{z}_2 = \mathbf{x}_2 + i\mathbf{y}_2$. We have
$$R(\mathbf{z}_1, \mathbf{z}_2) = \operatorname{Re}\left((\mathbf{x}_1 + i\mathbf{y}_1)^\dagger (\mathbf{x}_2 + i\mathbf{y}_2)\right) = \mathbf{x}_1^T \mathbf{x}_2 + \mathbf{y}_1^T \mathbf{y}_2.$$

10.8. The Symplectic Group

This shows that we can view R as the dot product on $\mathbb{R}^{2n}$, when we identify $\mathbf{z} = \mathbf{x} + i\mathbf{y}$ in $\mathbb{C}^n$ with $(\mathbf{x};\mathbf{y})$ in $\mathbb{R}^{2n}$. Abusing notation, we write

$$R((\mathbf{x}_1;\mathbf{y}_1),(\mathbf{x}_2;\mathbf{y}_2)) = (\mathbf{x}_1 \cdot \mathbf{x}_2) + (\mathbf{y}_1 \cdot \mathbf{y}_2)$$

for $\mathbf{x}_j$ and $\mathbf{y}_j$ in $\mathbb{R}^n$.

Similarly,

$$C(\mathbf{z}_1, \mathbf{z}_2) = \operatorname{Im}\left((\mathbf{x}_1 + i\mathbf{y}_1)^\dagger (\mathbf{x}_2 + i\mathbf{y}_2)\right) = -\mathbf{y}_1^T \mathbf{x}_2 + \mathbf{x}_1^T \mathbf{y}_2.$$

Abusing notation once again, we write

(10.7) $$C((\mathbf{x}_1;\mathbf{y}_1),(\mathbf{x}_2;\mathbf{y}_2)) = (\mathbf{x}_1 \cdot \mathbf{y}_2) - (\mathbf{y}_1 \cdot \mathbf{x}_2)$$

for $\mathbf{x}_j$ and $\mathbf{y}_j$ in $\mathbb{R}^n$. It is easy to verify that C is bilinear and that

$$C((\mathbf{x};\mathbf{y}),(\mathbf{x};\mathbf{y})) = (\mathbf{x} \cdot \mathbf{y}) - (\mathbf{y} \cdot \mathbf{x}) = 0.$$

This shows that C defines a symplectic form on $\mathbb{R}^{2n}$.

Since every finite-dimensional complex inner product space is isomorphic to $(\mathbb{C}^n, \cdot)$, every complex inner product on a finite-dimensional space can be interpreted as the sum of a real inner product and a real symplectic form.

Let (V, σ) be an n-dimensional symplectic space with basis $\mathcal{B} = \{\mathbf{b}_1, \ldots, \mathbf{b}_n\}$. By means of a now familiar process, we can identify the matrix representation of σ with respect to $\mathcal{B}$ by

$$[\sigma]_\mathcal{B} = A = [a_{ij}]$$

where $a_{ij} = \sigma(\mathbf{b}_i, \mathbf{b}_j)$. Since σ is symplectic, A is skew-symmetric: $a_{ij} = -a_{ji}$. We then have

(10.8) $$\sigma(\mathbf{v}, \mathbf{w}) = [\mathbf{v}]_\mathcal{B}^T A [\mathbf{w}]_\mathcal{B}.$$

There is no 1-dimensional symplectic space, because a symplectic form is defined to be nondegenerate. In fact, a finite-dimensional symplectic space must have even dimension. We sketch a proof of this, using a variation on the Gram-Schmidt process. This will also yield the symplectic analog to an orthonormal basis.

Given a finite-dimensional symplectic space (V, σ), let $\mathbf{b}_1$ be any nonzero vector in V. Identify $\mathbf{w}$ in V so that $\sigma(\mathbf{b}_1, \mathbf{w}) \neq 0$. Scale $\mathbf{w}$ to get $\mathbf{b}_2$ that satisfies

$$\sigma(\mathbf{b}_1, \mathbf{b}_2) = 1.$$

Either $\{\mathbf{b}_1, \mathbf{b}_2\}$ is a basis for V or there is $\mathbf{x}$ in V so that $\{\mathbf{b}_1, \mathbf{b}_2, \mathbf{x}\}$ is linearly independent. In the latter case, let

$$\mathbf{b}_3 = \mathbf{x} + \sigma(\mathbf{b}_2, \mathbf{x})\mathbf{b}_1 - \sigma(\mathbf{b}_1, \mathbf{x})\mathbf{b}_2.$$

We then have

$$\sigma(\mathbf{b}_1, \mathbf{b}_3) = \sigma(\mathbf{b}_1, \mathbf{x}) + \sigma(\mathbf{b}_2, \mathbf{x})\sigma(\mathbf{b}_1, \mathbf{b}_1) - \sigma(\mathbf{b}_1, \mathbf{x})\sigma(\mathbf{b}_1, \mathbf{b}_2)$$
$$= \sigma(\mathbf{b}_1, \mathbf{x}) - \sigma(\mathbf{b}_1, \mathbf{x})\sigma(\mathbf{b}_1, \mathbf{b}_2) = \sigma(\mathbf{b}_1, \mathbf{x}) - \sigma(\mathbf{b}_1, \mathbf{x}) = 0.$$

Likewise, we can verify that $\sigma(\mathbf{b}_2, \mathbf{b}_3) = 0$.

Since $\mathbf{b}_3$ is orthogonal to every vector in $\operatorname{Span}\{\mathbf{b}_1, \mathbf{b}_2, \mathbf{b}_3\}$, there must be $\mathbf{y}$ in V so that $\sigma(\mathbf{b}_3, \mathbf{y}) \neq 0$. Let

$$\mathbf{b}_4' = \mathbf{y} + \sigma(\mathbf{b}_2, \mathbf{y})\mathbf{b}_1 - \sigma(\mathbf{b}_1, \mathbf{y})\mathbf{b}_2.$$

The calculations reveal that $\mathbf{b}_4'$ is orthogonal to $\mathbf{b}_1$ and $\mathbf{b}_2$, but not to $\mathbf{b}_3$. Scale $\mathbf{b}_4'$ to get $\mathbf{b}_4$ so that $\sigma(\mathbf{b}_3, \mathbf{b}_4) = 1$.

An induction argument can be brought to bear to finish a proof that V has a basis
$$\mathcal{B} = \{\mathbf{b}_1, \ldots, \mathbf{b}_{2n}\}$$
where, for $1 \leq i < j = 2, \ldots, 2n$, $\sigma(\mathbf{b}_i, \mathbf{b}_j) = \delta(i+1, j)$. $\mathcal{B}$ is a **symplectic basis** for V.

Theorem 10.45. A finite-dimensional symplectic space (V, σ) has a symplectic basis $\mathcal{B} = \{\mathbf{b}_1, \ldots, \mathbf{b}_{2n}\}$. With respect to $\mathcal{B}$, the matrix representation of σ has diagonal block matrix form
$$A = \mathrm{diag}(H_1, \ldots, H_n)$$
where $H_i = \begin{bmatrix} 0 & 1 \\ -1 & 0 \end{bmatrix}$.

The matrix representation of a symplectic form with respect to a symplectic basis is often given by a matrix in block form

(10.9) $$U = \left[\begin{array}{c|c} O_n & I_n \\ \hline -I_n & O_n \end{array}\right].$$

It is not difficult to see that we can get (10.9) by rearranging the vectors in what we defined above as a symplectic basis.

We can define a symplectic space using $V = \mathbb{F}^{2n}$ by
$$\sigma(\mathbf{v}, \mathbf{w}) := \mathbf{v}^T U \mathbf{w}$$
where U is given in (10.9).

The next definition is obvious.

Definition 10.46. Let (V, σ) be a finite-dimensional symplectic space over $\mathbb{F}$. The **symplectic group** $Sp_{2n}(\mathbb{F})$ is given by
$$Sp_{2n}(\mathbb{F}) = \{L \in GL_{2n}(\mathbb{F}) \mid \sigma(L(\mathbf{v}), L(\mathbf{w})) = \sigma(\mathbf{v}, \mathbf{w})\}$$
for all $\mathbf{v}$ and $\mathbf{w}$ in V. An element in $Sp_{2n}(\mathbb{F})$ is a **symplectic matrix**.

Using block-matrix arithmetic, we can see the advantage of using the form of a symplectic matrix given in (10.9). With U as given in (10.9), it is easy to check that M in $GL_{2n}(\mathbb{F})$ is in $Sp_{2n}(\mathbb{F})$ provided
$$M^T U M = U.$$

We follow through on this to see what a matrix in the symplectic group looks like.

Let

(10.10) $$M = \left[\begin{array}{c|c} A & B \\ \hline C & D \end{array}\right]$$

10.8. The Symplectic Group

belong to $Sp_{2n}(\mathbb{F})$. Using block-matrix arithmetic and U as given in (10.9), we have

$$M^T U M = \begin{bmatrix} A^T & C^T \\ \hline B^T & D^T \end{bmatrix} \begin{bmatrix} O_n & I_n \\ \hline -I_n & O_n \end{bmatrix} \begin{bmatrix} A & B \\ \hline C & D \end{bmatrix}$$

$$= \begin{bmatrix} -C^T & A^T \\ \hline -D^T & B^T \end{bmatrix} \begin{bmatrix} A & B \\ \hline C & D \end{bmatrix} = U.$$

This means

$$-C^T A + A^T C = -D^T B + B^T D = O$$

and

$$-C^T B + A^T D = I_n = -(-D^T A + B^T C).$$

Notice that

$$(D^T A - B^T C)^T = A^T D - C^T B.$$

We can now say that M as given in (10.10) has to satisfy the following conditions in order to belong to $Sp_{2n}(\mathbb{F})$:

(10.11) $$A^T C - C^T A = B^T D - D^T B = O$$

and

(10.12) $$A^T D - C^T B = I_n.$$

These generalize conditions that a matrix of the form (10.9) satisfies. In other words, a matrix that can be used to define a symplectic form itself defines an element in $Sp_{2n}(\mathbb{F})$.

We have already seen that some of the classical groups coincide in low dimensions. In the 2-dimensional setting, we have $\begin{bmatrix} a & b \\ c & d \end{bmatrix}$ in $Sp_2(\mathbb{F})$ provided $ad - bc = 1$. This shows that $Sp_2(\mathbb{F}) = SL_2(\mathbb{F})$.

Any matrix of the form (10.10) where, for instance, $C = O$, $A = D = I_n$, and $B = B^T$ satisfies (10.11) and (10.12). For example,

$$\begin{bmatrix} 1 & 0 & 0 & i \\ 0 & 1 & i & 0 \\ 0 & 0 & 1 & 0 \\ 0 & 0 & 0 & 1 \end{bmatrix}$$

belongs to $Sp_4(\mathbb{C})$.

Again using (10.10) as a template, we get M in $Sp_{2n}(\mathbb{F})$ if we take A in $GL_n(\mathbb{F})$, $D = (A^{-1})^T$, and $B = C = O$. For instance, for any real θ,

$$\begin{bmatrix} \cos\theta & -\sin\theta & 0 & 0 \\ \sin\theta & \cos\theta & 0 & 0 \\ 0 & 0 & \cos\theta & -\sin\theta \\ 0 & 0 & \sin\theta & \cos\theta \end{bmatrix}$$

is in $Sp_4(\mathbb{R})$.

As we have noted, this chapter just skims the surface of an ocean of mathematics fed by the classical groups. Further studies, taking you beyond our work here, might include a closer look at relationships among the classical groups themselves, as well as applications in physics and geometry. A first step in that direction would be an algebra course, that is, a course in groups, rings, and fields.

Exercises 10.8.

10.8.1. Let (V, σ) be a finite-dimensional symplectic space. Describe how to order a symplectic basis $\mathcal{B}$ so that $[\sigma]_\mathcal{B}$ looks like the matrix in (10.9).

10.8.2. Let (V, σ) be a finite-dimensional symplectic space with an ordered basis $\mathcal{B}$. Show that (10.8) holds when $A = [\sigma]_\mathcal{B}$.

10.8.3. Define $\sigma : \mathbb{C}^2 \to \mathbb{C}^2$ by
$$\sigma(\mathbf{v}, \mathbf{w}) = \mathbf{v}^T \begin{bmatrix} 0 & i \\ -i & 0 \end{bmatrix} \mathbf{w}.$$
Show that σ defines a symplectic form on $\mathbb{C}^2$.

10.8.4. Let σ be any bilinear form on a vector space, V. Assume that the underlying field has characteristic different from 2. Show that
$$\tau_1(\mathbf{v}, \mathbf{w}) := \frac{1}{2}(\sigma(\mathbf{v}, \mathbf{w}) + \sigma(\mathbf{w}, \mathbf{v}))$$
is symmetric and that
$$\tau_2(\mathbf{v}, \mathbf{w}) := \frac{1}{2}(\sigma(\mathbf{v}, \mathbf{w}) - \sigma(\mathbf{w}, \mathbf{v}))$$
is skew-symmetric.

10.8.5. Show that C as given in (10.7) is bilinear.

10.8.6. Show that M in $GL_{2n}(\mathbb{F})$ belongs to $Sp_{2n}(\mathbb{F})$, if and only if
$$M^T U M = U$$
where U is given in (10.9).

10.8.7. Using type-2 elementary row operations, argue that, for U as given in (10.9), $\det U = 1$.

10.8.8. Show that $\det M = \pm 1$, for any M in $Sp_{2n}(\mathbb{F})$. (As it happens, $Sp_{2n}(\mathbb{F}) \subseteq SL_{2n}(\mathbb{F})$ but this is much harder to prove.)

10.8.9. Show that $U^{-1} = -U$ for U as given in (10.9).

10.8.10. Show that if A belongs to $GL_n(\mathbb{F})$, then
$$M = \left[\begin{array}{c|c} A & O \\ \hline O & (A^T)^{-1} \end{array}\right]$$
is in $Sp_{2n}(\mathbb{F})$.

10.8.11. Show that if B is symmetric in $M_n(\mathbb{F})$, then
$$M = \left[\begin{array}{c|c} I_n & B \\ \hline O & I_n \end{array}\right]$$
is in $Sp_{2n}(\mathbb{F})$.

10.8.12. Find a formula for the inverse of
$$M = \left[\begin{array}{c|c} A & B \\ \hline C & D \end{array}\right]$$
in $Sp_{2n}(\mathbb{F})$.

10.8.13. Find two matrices in $M_4(\mathbb{C})$ with the form (10.10) where $C = O$, $A = D = I_n$, and $B = B^T$.

10.8.14. Find two matrices in $M_4(\mathbb{C})$ with the form (10.10) where A is in $GL_n(\mathbb{F})$, $D = (A^{-1})^T$, and $B = C = O$.

Appendix A

Background Review

This is a brief review of the foundations in general mathematics that we need to understand linear algebra. Consult it as needed while going through the course.

A.1. Logic and Proof

Most theorems in mathematics are *logical statements*, that is, statements that may be formulated, "If condition A is true, then condition B is true." This formulation is often shortened to read, "A implies B" or "$A \implies B$." Regardless of whether A includes one or several conditions, A is the *hypothesis*. The theorem only has relevance when A is true. B may also include several conditions. In any case, B is the *conclusion*.

A given theorem may not be stated explicitly in the form "If A is true, then B is true," or "If A, then B." In such a case, the reader must discern the hypothesis and the conclusion. For example, the theorem that says that a differentiable function is continuous can be stated, "If a function is differentiable, then it is continuous." The theorem that says that $\sqrt{2}$ is irrational can be stated, "If $x = \sqrt{2}$, then x is irrational."

Many theorems have the structure, "Condition A is true if and only if condition B is true." This is really two theorems: $A \implies B$ and $B \implies A$. For these, we may write $A \iff B$, which is read, "A if and only if B."

When A is some condition, $\sim A$ means "not A," in other words, that condition that A does not hold. For example, consider the theorem that says that the sum of two even integers is an even integer. If we think of the theorem as having the form $A \implies B$, then A is the condition that integers x and y are even and B is the condition that $x + y$ is even. Here $\sim A$ is the condition that we have two integers that are not both even — one could be even, the other odd, or both could be odd — while $\sim B$ is the condition that the sum of two integers is odd.

Every logical statement, $A \implies B$, is associated to three other logical statements: its **converse**, $B \implies A$; its **contrapositive**, $\sim B \implies \sim A$; and its inverse, $\sim A \implies \sim B$. Two statements are **logically equivalent** provided one is true if and only if the other is true. A statement and its contrapositive are logically equivalent. The converse and the inverse of a statement are logically equivalent. Consider the theorem that says the sum of two even integers is even. Its converse is, "If the sum of two integers is even, then both integers are even." The contrapositive of the theorem is, "If the sum of two integers is odd, then the integers are not both even." The inverse of the theorem is, "If two integers are not both even, then their sum is odd." Since $1 + 3 = 4$, we can see that the converse and the inverse of the theorem in this case are both false.

Understanding how to capture negation is important in logic and proof. If we are talking only about $\mathbb{Z}_+$, then the negation of "even integer" is "odd integer." Things are a bit trickier if there are more qualifiers in a statement. For instance, think about the statement, "For all $x \in \mathbb{R}$, condition A is true." Its negation is, "For some $x \in \mathbb{R}$, condition A is false." "For some x" can be expressed "there exists x." Using symbols, we write, "$\forall x \in \mathbb{R}$, A is true," and its negation, "$\exists x \in \mathbb{R}$, so that A is false." The symbol $\forall$ is read "for all" and $\exists$ is read "there exists" or "there exist."

Beginners often confuse a statement with its converse. Since a statement and its converse are not logically equivalent, statement/converse confusion must be eradicated posthaste. For example, suppose a friend who lives out of state says to you, "If it rains Saturday, I will go to the movies." Suppose further that you talk to your friend on Sunday and the friend mentions going to the movies the day before. If you surmise that it must have rained, you are confusing your friend's statement, "rain implies movies," with its the converse, "movies imply rain." Once your friend says "If it rains..." the only topic under discussion is what will ensue in the event of rain.

Not all theorems are best presented as logical statements. Some present more naturally as existence statements, for example, "There is a real-valued function that is everywhere continuous and nowhere differentiable." Since the proof of that theorem is the construction of a function that is everywhere continuous and nowhere differentiable, one might argue that it is not a theorem, but an example. There are existence theorems with nonconstructive proofs, for example, "Every vector space has a basis." This one does present naturally as a logical statement: If V is a vector space, then V has a basis.

When we encounter an existence statement, the question of uniqueness invariably arises. A vector space must have a basis, but is the basis unique? Every continuous function has an antiderivative, but is it unique? (The answer to both questions is "no.") *Existence and uniqueness* theorems abound in mathematics. The following is an example that arises in this course: Every $m \times n$ matrix A over $\mathbb{R}$ has a unique reduced row echelon form.

There are many different approaches to structuring an argument in mathematics but induction may be the method of proof that arises most often in a linear algebra course. We use it to prove statements that hold for all positive integers.

Let $P(n)$ be a statement about a positive integer n. For example, $P(n)$ could be that the sum of the first n positive integers is $n(n+1)/2$. The **principle of**

induction states that if $P(1)$ is true and $P(k)$ implies $P(k+1)$ for any positive integer k, then $P(n)$ is true for all positive integers n.

We take the principle of induction as an axiom: It is an assumption under which we conduct our studies.

There is some variability in induction proofs. Sometimes the statements in hand are better indexed by $\mathbb{Z}_{\geq 0}$ than $\mathbb{Z}_+$ or by $\{-1, 0, 1, 2, \ldots\}$ or by $\{2, 3, 4, \ldots\}$ or by some other infinite subset of $\mathbb{Z}$ with a least element. The **base case** for an induction proof is always $P(k_0)$, where k_0 is least in the index set.

The assumption that $P(k)$ is true is called the *induction hypothesis*. Sometimes the induction hypothesis is that $P(j)$ is true for all j between k_0 and k. This is slightly different from the induction hypothesis we described above and, surprisingly, the distinction between the two types of induction arguments does come up every now and again. We will not belabor this. The heart of an induction proof is proving $P(k+1)$ under "the" induction hypothesis (whichever is most convenient at the time), although sometimes the base case takes more work than one might expect.

Example A.1. Let $P(n)$ be the statement $1 + 2 + \cdots + n = n(n+1)/2$. We use induction to prove $P(n)$ for all $n \in \mathbb{Z}_+$.

We have $1 = (1 \cdot 2)/2$, so $P(1)$, the base case, is true.

The induction hypothesis is that for $k > 1$, $1 + 2 + \cdots + k = k(k+1)/2$. From here we have

$$1 + 2 + \cdots + (k+1) = \frac{k(k+1)}{2} + k + 1.$$

Adding fractions on the right-hand side of the last equation we get

$$\frac{k(k+1) + 2(k+1)}{2} = \frac{k^2 + k + 2k + 2}{2} = \frac{k^2 + 3k + 2}{2} = \frac{(k+1)(k+2)}{2}.$$

We conclude that if $P(k)$ is true, then $P(k+1)$ is true. By the principle of induction, $P(n)$ is true for all $n \in \mathbb{Z}_+$.

Another method of proof called *indirect proof* or *proof by contradiction* arises frequently throughout mathematics and we will see several examples in this course. It can be easy to confuse proof by contradiction with proof of the contrapositive of a theorem. For example, an important theorem in calculus states that if a series $\sum_{n=1}^{\infty} a_n$ converges, then the sequence of terms, $\{a_n\}_{n=1}^{\infty}$, converges to 0. The standard proof of that is actually a proof of the contrapositive: If $\{a_n\}_{n=1}^{\infty}$ does not converge to 0, then $\sum_{n=1}^{\infty} a_n$ diverges.

A proof by contradiction proceeds from the assumption that the conclusion of the statement of interest is false. The goal of such a proof is to reach a point at which it is evident that one of the hypotheses of the theorem, or a known truth — for instance that $0 \neq 1$ — is violated. In contrast, proof of the contrapositive proceeds from the assumption that the conclusion of a statement is false to a point at which it is evident that the negation of the hypothesis of the theorem is true. Clearly, there is overlap in the two approaches to proof. This is another point we will not press.

As an example of proof by contradiction, we cite a very old proof that $\sqrt{2}$ is irrational. Before we do this, however, we recall one of the building blocks of number theory.

Theorem A.2 (Fundamental Theorem of Arithmetic). If q is any positive integer greater than 1, then there are distinct prime numbers p_i such that

(A.1) $$q = p_1^{r_1} \cdots p_n^{r_n}$$

where each r_i is a positive integer. Moreover, the factorization in (A.1) is unique except for a possible reordering of the factors.

Proof of the Fundamental Theorem of Arithmetic requires preparation that would take us too far afield so we omit it here.

The factorization in (A.1) is called the **prime factorization** of q and the p_is are the **prime factors** of q. Notice that for any positive integer k,

$$q^k = p_1^{kr_1} \cdots p_n^{kr_n}.$$

By uniqueness, this must be the prime factorization of q^k. In particular, q and q^k have the same set of prime factors.

Theorem A.3. $\sqrt{2}$ is irrational.

Proof. Suppose by way of contradiction that we can write $\sqrt{2} = p/q$ where p and q are positive integers with no common factors. From $\sqrt{2} = p/q$, we have $2 = p^2/q^2$ implying that $2q^2 = p^2$. Since 2 is a prime factor of p^2, it must be a prime factor of p. Write $p = 2b$, for some positive integer b. Now we have $2 = 4b^2/q^2$ implying $2q^2 = 4b^2$ so $q^2 = 2b^2$. Since 2 is a prime factor of q^2, it must be a prime factor of q. This contradicts our assumption that p and q have no common factors. We conclude that $\sqrt{2}$ must be irrational. □

Exercises A.1.

A.1.1. Write the converse, the inverse, and the contrapositive of the following statement: If a car drives down the road, then Macky is back in town.

A.1.2. Recast the following as a logical statement and write its converse, inverse, and contrapositive: All's well that ends well.

A.1.3. Negate the following statements in cogent English without using the word "not."
  (a) All is fair in love and war.
  (b) Every day, in every way, I get better.
  (c) You can fool all of the people some of the time.

A.1.4. Show that the sum of the first n even positive integers is $n(n+1)$.

A.1.5. Use induction to prove that the sum of the squares of the first n positive integers is $n(n+1)(2n+1)/6$.

A.1.6. Write down the prime factorization for 360.

A.1.7. The number 108 is considered sacred in some ancient traditions. Find its prime factorization to see why this may be the case.

A.1.8. Prove that $\sqrt{p}$ is irrational for any prime number p.

A.2. Sets

A *set* is a certain type of collection. The things in a set could be students in a classroom, stairs going up the Empire State Building, cars in a parking lot, real numbers greater than 10, functions from $\mathbb{Z}$ to $\mathbb{R}$, lines through the origin in the xy-plane, or the pieces of furniture in a waiting room. A set may have nothing in it in which case we say it is empty and we call it *the empty set*. We use $\emptyset$ to designate the empty set. If a set is not empty, the things in it are called its *elements*. The fundamental nature of a set is in its elements, but the set itself is the aggregate of its elements. When x is an element in a set A, we say x *belongs* to A and we write $x \in A$. If we want to emphasize that an element does not belong to a set A, we may write $x \notin A$.

If a set A has at least one element, we say it is *nonempty* and we may write $A \neq \emptyset$. If a set A has exactly one element, we say A is a *singleton* or a *singleton set*.

Sometimes it is convenient to identify the elements in a set using a list enclosed by braces. The empty set may be designated $\{\,\}$. If we write $A = \{1, 2, 3\}$, we may also describe A as the set of the first three positive integers. Sometimes we identify elements in a set as objects that satisfy some rule. For example,

$$\mathbf{B}_0 = \{x \in \mathbb{R} \mid |x| < 1\}$$

is the collection of real numbers strictly between -1 and 1.

Different objects in a set must be distinguishable. For example, $(1,1)$ is not a set.

A set is independent of any arrangement of its elements. This means that, as sets, $\{a, b\}$ and $\{b, a\}$ are identical. If we choose to distinguish sets with the same elements but different arrangements, we refer to them as **ordered sets**. Thus, as sets, $\{a, b, c\}$ and $\{b, a, c\}$ are identical, but as ordered sets, they are not identical.

The question as to whether a given object is in a set must have an unambiguous answer. If it does not, the collection in question cannot be a set. For instance, let $\mathcal{S}$ be the collection containing every set that is not a member of itself. $\mathcal{S}$ is not empty because, for example, $A = \{1, 2, 3\}$ is in $\mathcal{S}$. If $\mathcal{S}$ is a set, it either belongs to itself or it does not. If $\mathcal{S}$ is a set that does not belong to itself, then $\mathcal{S}$ must be in $\mathcal{S}$. This, of course, is absurd. On the other hand, if $\mathcal{S}$ is a set that belongs to itself, then by virtue of its belonging to $\mathcal{S}$, it must be a set that does not belong to itself. This is also absurd. These dual absurdities are together known as *Russell's Paradox*. We resolve the paradox by concluding that while we may be able to describe and to talk about $\mathcal{S}$, it cannot be a set. Russell's Paradox is important for establishing that set theory cannot be used to analyze every collection.

As Russell's Paradox suggests, distinguishing sets from collections that are not sets can be delicate. Here we are more concerned with a basic understanding of subsets, intersections, and unions.

Let A and B be sets. When A and B have exactly the same elements, we say they are equal and we write $A = B$. If A and B are sets that are not equal, we write $A \neq B$. We say A is a **subset** of B provided each element in A is an element in B. In this case, we write $A \subseteq B$. If $A \subseteq B$ and there are elements in B not in A, then A is a **proper subset** of B. When A is a proper subset of B, we may write $A \subsetneq B$ or $A \subset B$. Notice that if $A \subseteq B$ and $B \subseteq C$, then $A \subseteq C$.

Our first lemma illustrates an approach that we use frequently to show that two apparently different sets are actually identical.

Lemma A.4. If A and B are sets with $A \subseteq B$ and $B \subseteq A$, then $A = B$.

Proof. Suppose A and B are sets that satisfy the two conditions as hypothesized. If $A \neq B$, then there must be either (1) $x \in A$ not in B or (2) $x \in B$ not in A. Since $A \subseteq B$, every element in A is in B so there can be no x satisfying the first condition. Likewise, since $B \subseteq A$, every element in B is in A so there can be no x satisfying the second condition. It follows that $A = B$. $\square$

Notice that the proof of the lemma is a strict application of the definition of subset.

It is critical to distinguish between the members of a set and the subsets of a set. Every set is a subset of itself. If A is a nonempty set, it has at least two subsets: $\emptyset$ and A itself. Though the empty set is a subset of every set, it is not an element in every set.

There are several natural ways to build new sets out of old sets.

Definition A.5. The **Cartesian product** of sets A and B, $A \times B$, is the set of all ordered pairs, (a,b), where $a \in A$ and $b \in B$.

When (a,b) and (a',b') are both in $A \times B$, $(a,b) = (a',b')$ if and only if $a = a'$ and $b = b'$. For example, if $A = B = \{1,2\}$, then $A \times B = \{(1,1),(1,2),(2,1),(2,2)\}$. In particular, $(1,2) \neq (2,1)$.

The set underlying the xy-plane is $\mathbb{R} \times \mathbb{R}$.

Lemma A.6. If A is any set, then $A \times \emptyset = \emptyset$.

Proof. Suppose $(a,b) \in A \times \emptyset$. By definition, $b \in \emptyset$, which is impossible. We conclude that $A \times \emptyset$ must itself be empty. $\square$

We frequently have reason to designate a family of sets using subscripts, as in A_1, A_2, etc. If A_i designates an arbitrary member of such a family, we say that i is an *index* and that the A_is form an **indexed family of sets**. The *index set*, I, is the set of all i where A_i is under consideration.

It is natural to extend the notion of the Cartesian product of two sets to the notion of the Cartesian product of a family of $n \in \mathbb{Z}_+$ sets or of a family of sets indexed by all of $\mathbb{Z}_+$. For example, if $A_1 = \{1,2\}$, $A_2 = \{2,3\}$, and $A_3 = \{3,4\}$, we can understand that $(2,2,3)$ is an element in $A_1 \times A_2 \times A_3$. Note for instance that we can understand the collection of all infinite sequences of real numbers as the elements in $\mathbb{R} \times \mathbb{R} \times \cdots$.

Definition A.7. Let A and B be sets. The **union** of A and B is the set

$$C = \{x \,|\, x \in A \text{ or } x \in B\}.$$

We write $C = A \cup B$.

A.2. Sets

When A and B are sets and we say, "x is in A or B," we mean $x \in A$, $x \in B$, or x is in both A and B. When "or" means "and/or" like this, it is called the **inclusive or**. This is the default meaning of "or" in mathematics.

When taking the union of an indexed family of sets, $\{A_i\}_{i \in I}$, we may write
$$\bigcup_{i \in I} A_i.$$
When the index set is finite, we may write
$$\bigcup_{i=1}^{n} A_i = A_1 \cup \cdots \cup A_n.$$

Definition A.8. Let A and B be sets. The **intersection** of A and B is the set
$$C = \{x \mid x \in A \text{ and } x \in B\}.$$
We write $C = A \cap B$. When $A \cap B = \emptyset$, we say that A and B are **disjoint** or **nonintersecting** sets.

When A and B are disjoint and $C = A \cup B$, we say that C is a **disjoint union**.

Let $A = \{0, 1\}$ and $B = \{0, 2\}$. Then $A \cup B = \{0, 1, 2\}$ and $A \cap B = \{0\}$. Note that the union of sets is the set of elements that belong to either set and the intersection is the set of elements that belong to both sets.

If $A = \{1, 2, 3\}$ and $B = \{0, 7\}$, then $A \cup B = \{0, 1, 2, 3, 7\}$ is a disjoint union.

We often designate the intersection of a family of sets, A_i, with index set I, by
$$\bigcap_{i \in I} A_i.$$
When the intersection is over a finite family of sets, we may write
$$\bigcap_{i=1}^{n} A_i = A_1 \cap \cdots \cap A_n.$$

Union and intersection are operations on sets, much as addition and multiplication are operations on numbers. Since sets are not ordered, $A \cup B = B \cup A$ and $A \cap B = B \cap A$, so we can say that union and intersection are commutative.

Lemma A.9. If A and B are sets, then

(a) $A \cap B \subseteq A$;

(b) $A \subseteq A \cup B$; and

(c) $A \cap B \subseteq A \cup B$.

Proof. (a) Since every element in $A \cap B$ is in A, (a) is true.

(b) If $x \in A$, then x is in A or B so $x \in A \cup B$. This shows $A \subseteq A \cup B$.

(c) Since $A \cap B \subseteq A$ and $A \subseteq A \cup B$, $A \cap B \subseteq A \cup B$. $\square$

Venn diagrams can be useful for thinking about sets. In Figure A.1, the region enclosed by the square represents the so-called **universal set**, that is, the largest set in the context at hand. For example, when we write $\mathbf{B}_0 = \{x \in \mathbb{R} \mid |x| < 1\}$,

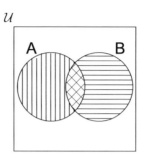

Figure A.1. A Venn diagram.

we are defining $\mathbf{B}_0$ as a subset of $\mathbb{R}$. In this context, we can think of $\mathcal{U}$ as $\mathbb{R}$. In another context, we might work with

$$\mathbf{B}_0 = \{(x,0) \mid |x| < 1\}.$$

In this case, we might take the universal set $\mathcal{U}$ to be $\mathbb{R}^2$.

Definition A.10. Let A be a set and let $\mathcal{U}$ be the universal set. The **complement of A** is designated A^c and is given by

$$A^c = \{x \mid x \notin A\} \subseteq \mathcal{U}.$$

The shaded region on the left in Figure A.1 is the set of elements in A not in B, that is $A \cap B^c$. The shaded region on the right is the set of elements in B not in A, that is $B \cap A^c$. The cross-hatched region is $A \cap B$. This helps us see that we can write A as the union of disjoint sets, $A = (A \cap B) \cup (A \cap B^c)$. Likewise, $B = (B \cap A) \cup (B \cap A^c)$.

If A is a set with $B \subsetneq A$, then $A \setminus B$ is the complement of B in A, that is,

$$A \setminus B = A \cap B^c.$$

We use this notation frequently throughout mathematics. For instance the set of nonzero real numbers, $\mathbb{R}^*$, may also be designated $\mathbb{R} \setminus \{0\}$.

Lemma A.11. Let A be a set in the universal set $\mathcal{U}$.

(a) $A^c \cup A = \mathcal{U}$ and

(b) $(A^c)^c = A$.

Proof. (a) Every element in $\mathcal{U}$ is either in A or not in A, that is, every element in $\mathcal{U}$ is in A or A^c. In other words, $\mathcal{U} = A \cup A^c$.

(b) Suppose $x \in \mathcal{U}$ is not in A^c. We just established that x must then be in A. This shows that $(A^c)^c \subseteq A$. On the other hand, if $x \in A$, x is not in A^c so it must be in $(A^c)^c$. This shows that $A \subseteq (A^c)^c$, which proves that $(A^c)^c = A$, as desired. □

Theorem A.12 (DeMorgan's Laws). If A and B are sets, then

(a) $(A \cap B)^c = A^c \cup B^c$ and

(b) $(A \cup B)^c = A^c \cap B^c$.

Proof. Let A and B be sets.

(a) If $x \in (A \cap B)^c$, then x is outside $A \cap B$ so it is not in both A and B. This means that either $x \notin A$ or $x \notin B$, that is, $x \in A^c \cup B^c$. This shows $(A \cap B)^c \subseteq A^c \cup B^c$.

If $x \in A^c \cup B^c$, then either x is in A^c or x is in B^c. If $x \in A^c$, then x cannot belong to $A \cap B \subseteq A$ so $x \in (A \cap B)^c$. The same argument applies to show that when $x \in B^c$, $x \in (A \cap B)^c$. This proves that $A^c \cup B^c \subseteq (A \cap B)^c$. Lemma A.4 thus ensures that $(A \cap B)^c = A^c \cup B^c$.

(b) If $x \in (A \cup B)^c$, then x is outside the set of elements belonging to either A or B. In other words, x is neither in A nor in B. This is to say $x \in A^c$ and $x \in B^c$, thus $x \in A^c \cap B^c$. This shows $(A \cup B)^c \subseteq A^c \cap B^c$.

If $x \in A^c \cap B^c$, then x is not in A and not in B. It follows that x is outside $A \cup B$ so $x \in (A \cup B)^c$. This shows $A^c \cap B^c \subseteq (A \cup B)^c$. By Lemma A.4, $(A \cup B)^c = A^c \cap B^c$. $\square$

Exercises A.2.

A.2.1. Let A be any set containing exactly four elements. How many singleton subsets does A have? How many two-element subsets does A have?

A.2.2. Write down all the elements in $A \times B \times C$ if $A = \{1\}$, $B = \{1, 2\}$, and $C = \{1, 2, 3\}$.

A.2.3. We have shown that the Cartesian product of the empty set with another set must be empty. Show that if any $A_i = \emptyset$, then $A_1 \times A_2 \times \cdots \times A_n = \emptyset$.

A.2.4. Give an example of a set A and a set B where $A \times B \neq B \times A$.

A.2.5. Let A_1, A_2, A_3 be nonempty sets. Discuss possible differences among $(A_1 \times A_2) \times A_3$, $A_1 \times (A_2 \times A_3)$, and $A_1 \times A_2 \times A_3$ or argue that they are all equal.

A.2.6. If A is any set, what is $A \cup \emptyset$?

A.2.7. If A is any set, what is $A \cap \emptyset$?

A.2.8. Let $A_0 = \emptyset$, $A_1 = \{\emptyset\}$, $A_2 = \{\emptyset, \{\emptyset\}\}$, $A_3 = \{\emptyset, \{\emptyset\}, \{\emptyset, \{\emptyset\}\}\}$, etc. Note that while A_0 contains no elements, A_1 contains exactly one element.
  (a) We can write $A_1 = \{A_0\}$: Every element in A_1 is an A_i for some $i < 1$. For $i = 2, 3, 4$, write A_i as a list of elements A_j for $j < i$. Generalize to A_n for n any positive integer greater than 4.
  (b) How many elements are in A_n?
  (c) What is $A_n \cup \{A_n\}$?
  (d) Find $A_1 \cup A_2$ and $A_1 \cap A_2$.
  (e) List the subsets of A_0, A_1, A_2, and A_3.
  (f) Note that $A_0 \subseteq A_i$ for all $i = 0, 1, 2, \ldots$ but also that $A_0 \in A_i$ for all $i = 1, 2, \ldots$. Is $A_1 \subseteq A_2$? Is $A_1 \in A_2$?
  (g) Is any A_i a member of itself?

A.2.9. Let A, B, C be sets.
  (a) Show that $(A \cap B) \cap C = A \cap (B \cap C)$ and that $(A \cup B) \cup C = A \cup (B \cup C)$.
  (b) Show that $A \cap (B \cup C) = (A \cap B) \cup (A \cap C)$.

A.2.10. Let $A = \{a,b,c,d\}$, $B = \{b,d,e,f\}$, $C = \{a,f,g\}$.
   (a) Find $A \cap B$, $A \cap C$, $B \cap C$, and $A \cap B \cap C$.
   (b) Find $A \cup B$, $A \cup C$, and $A \cup B \cup C$.
   (c) Find $A \cap (B \cup C)$ and $B \cap (A \cup C)$.
   (d) Write down all the elements in $A \times B$.

A.2.11. Show that for any sets A and B, $A = (A \cap B) \cup (A \cap B^c)$ and that $A \cap B$ and $A \cap B^c$ are disjoint. Sketch a Venn diagram to illustrate what goes on here.

A.2.12. Identify the regions in Figure A.1 that represent the sets discussed in DeMorgan's Laws.

A.3. Well-Definedness

It is difficult to get off the ground in any study of mathematics without the notion of *well-definedness*, a term that is effectively synonymous with *unambiguous*. Well-definedness is a question anytime there is more than one way to represent a mathematical entity, be it a number or something else. Students do not run across well-definedness knowingly while learning arithmetic and school algebra, but questions of well-definedness are central to elementary arithmetic. Every number in $\mathbb{Z}$ has a well-defined integer expression which makes addition and multiplication in $\mathbb{Z}$ straightforward: There is only one correct integer expression for 7×2. On the other hand, no number in $\mathbb{Q}$ has a well-defined rational expression: $1/1 = 2/2$ and $3/6 = 1/2 = 2/4$. In elementary school, children learn that for any noninteger in $\mathbb{Q}_+$, there is a well-defined rational expression in lowest terms, that is, for which numerator and denominator are relatively prime. Toggling between a fraction in lowest terms and other fractions that represent the same number is a skill students must master to add fractions. Its necessity is due to the fact that a fraction is not a well-defined expression for a rational number.

Another important rule of arithmetic that students learn early on is the stricture against division by zero. It is easy to see there is a problem if we attempt to assign a real value to $1/0$. If there is $a \in \mathbb{R}$ so that $1/0 = a$, then when we cross-mulitply we get $1 = a \cdot 0 = 0$. This is not a matter of $1/0$ not being well-defined. It is a matter of $1/0$ not representing a real number.

The subject of well-definedness as it relates to decimal expansions for numbers in $\mathbb{R}$ is frequently avoided entirely in school mathematics. Notice for instance that $1 = 0.99\ldots$ and $0.12 = 0.11999\ldots$. We can play that trick with any real number that has a terminating decimal expansion. The decimal expansion for an irrational number, though, is well-defined. For instance, 3 is the 17th digit in the decimal expansion of π. There is no other candidate. Maybe this should not be surprising: Irrational numbers are the reason we need decimal expansions in the first place.

Despite its centrality in the theory of limits, well-definedness often gets a glancing treatment in calculus courses. The analysis of left- and right-hand limits is typically applied to address the question as to whether a given limit is well-defined: A limit exists at a point provided its left- and right-hand limits exist and agree at that point. In this context, we can shift from viewing $1/0$ as not defined in $\mathbb{R}$

to viewing $1/0$ as bad shorthand for $\lim_{x \to 0} 1/x$, which itself is not well-defined, because the underlying right- and left-hand limits disagree.

L'Hôpital's Rule deals expressly with questions of well-definedness in limits of ratios of expressions that both approach zero or both approach something infinite. While we call $0/0$, ∞/∞, $0 \cdot \infty$ *indeterminate forms*, we could just as well say that these expressions are not well-defined as elements of the **extended real numbers**, $\mathbb{R} \cup \{\infty, -\infty\}$. Granted, indeterminate forms are more interesting than $1/0$ because $1/0$ in some sense stands for $\pm\infty$, while an indeterminate form could stand for something that has any value in $\mathbb{R} \cup \{\infty, -\infty\}$ or that is undefined.

The slope of a nonvertical line in the xy-plane is an artifact of the coordinate system. In Euclidean geometry, there is no definition of slope. Once we agree to fix a coordinate system, the slope of a nonvertical line in the real plane is well-defined. If (x_0, y_0) and (x_1, y_1) are any two points on the line, its slope is

$$\frac{y_0 - y_1}{x_0 - x_1} = \frac{y_1 - y_0}{x_1 - x_0}.$$

Well-definedness here is a consequence of the fact that the calculation yields the same number regardless of the two points chosen. The slope of a vertical line in $\mathbb{R}^2$ is not well-defined for the same reason that $\lim_{x \to 0} 1/x$ is not well-defined.

Another place in which well-definedness lies in the background is in discussions of functions and equations. A function is a rule that gives a single, unambiguous output for a given allowed input. This means that if f is a function with domain D, then for x in D, $y = f(x)$ is well-defined. An equation in x and y does not necessarily yield y as a function of x, though. For instance, if we take x from the interval $[-1, 1] \subset \mathbb{R}$, then y that satisfies $x^2 + y^2 = 1$ is not well-defined: $x = 0$ means $y = 1$ or $y = -1$, for instance. Here we say that y is not a function of x but we may just as well say that y is not well-defined in terms of x. The two statements mean the same thing in this context.

The equation describing a line in $\mathbb{R}^2$ is not well-defined. Indeed, $x + y = 1$, $y = 1 - x$, and $2x + 2y = 2$ are all different equations that describe the same line in $\mathbb{R}^2$. Any curve in $\mathbb{R}^2$ and any surface in $\mathbb{R}^3$ can be described using many different equations. The situation is even more ambiguous when it comes to describing curves and surfaces parametrically. Consider, for example,

$$x = \cos t, \quad y = \sin t, \quad t \in \mathbb{R},$$

a parametric description of the unit circle in $\mathbb{R}^2$. We can describe the same circle using

$$x = \sin(2t), \quad y = \cos(2t), \quad t \in \mathbb{R}.$$

The association of the unit circle to a parametrization is not well-defined.

Note that if we want to describe the path along a curve taken by some object — rather than the curve itself — a parametric description is often the solution. For example, suppose an object traverses the unit circle in the clockwise direction in the xy-plane, starting from $(0, 1)$ and moving with a uniform speed of 3 units per second. Taking t to stand for time measured in seconds, we can describe the path of the object using

$$x = \sin 3t, \quad y = \cos 3t.$$

We leave it as an exercise to find a parametrization for the unit circle that does not use sines and cosines.

Exercises A.3.

A.3.1. Find two different decimal expansions for 1.13 and for 2.1.

A.3.2. Which rational numbers have well-defined decimal expansions?

A.3.3. What is the connection between the slope of a vertical line in $\mathbb{R}^2$ and $\lim_{x \to 0} 1/x$? (Hint: Start by thinking about a vertical line given by $x = a$, for some fixed number $a \in \mathbb{R}$.)

A.3.4. Suppose $y = x^2$. Is y well-defined in terms of x? Is x well-defined in terms of y? Explain.

A.3.5. Suppose $y = x^3$. Is y well-defined in terms of x? Is x well-defined in terms of y? Explain.

A.3.6. Suppose $\sin x = 1/2$. Is x well-defined? If not, how do we solve an equation like $\sin x = 1/2$?

A.3.7. What does $\sin^{-1}(1/2) = \pi/6$ mean? How is this related to the question as to whether x is well-defined when we say $\sin x = 1/2$?

A.3.8. Sketch $x^2 + y^2 = 1$ in the xy-plane. Give a graphic illustration of what it means when we say that y is not well-defined in terms of x when $x^2 + y^2 = 1$.

A.3.9. Sketch $y = \sqrt{1 - x^2}$ in the xy-plane. How is this example related to the example in the previous problem? Is y well-defined in terms of x? Is x well-defined in terms of y?

A.3.10. The text cites two different ways to describe the unit circle parametrically. Find a third way that does not use sines and cosines.

A.4. Counting

A passing familiarity with certain rules of counting can be helpful in the study of linear algebra. We review some elementary principles here, starting with the most basic axiom of counting.

Axiom A.13. If we have n ways of choosing one thing and m ways of choosing a second thing, then the total number of ways we can choose a first thing and then a second thing is nm.

Axiom A.13 can be extended to apply to any process involving finitely many steps.

Example A.14. Suppose we have a class of twenty students and that we want to pick a president, a vice president, and a treasurer from the class. A student may fill at most one of the three positions, and each position must be filled. We can count how many ways to do this by thinking of making three choices. For president, there are 20 possible outcomes. Once we make a selection for president, we are left with 19 candidates for vice president. Once president and vice president are chosen, we have 18 possible outcomes in our choice of treasurer. This gives us

$20 \cdot 19 \cdot 18 = 6{,}840$ different configurations of students from the class serving as president, vice president, and treasurer.

Factorial notation is useful in any discussion of counting. Recall that $0! = 1$ and that for any $n \in \mathbb{Z}_+$, $n! = n \cdot (n-1)!$.[1] For example, $3! = 3 \cdot 2! = 3 \cdot 2 \cdot 1! = 3 \cdot 2 \cdot 1 \cdot 0! = 6$.

Example A.15. Consider the letters $AMDE$. Any arrangement of these four letters is called a *word* whether or not it is a word in any language. There are four choices for the first letter, three remaining choices for the second letter, two for the third letter, and one left for the fourth letter so there are $4! = 4 \cdot 3 \cdot 2 \cdot 1 = 24$ different words we can make from the letters $AMDE$, using each letter exactly once. We leave it as an exercise to write them all down.

We could have used factorial notation in Example A.14. In general, if we have n items and we are to choose $m < n$ of them, in a particular order, we have a total of
$$n(n-1) \cdots (n-m+1) = \frac{n!}{(n-m)!}$$
possible outcomes. This result comes up often enough that we frame it as a lemma.

Lemma A.16. *The number of ways of arranging m objects drawn from a collection of n distinct objects is $n \cdot (n-1) \cdots (n-m+1) = \frac{n!}{(n-m)!}$.*

Sometimes we need to count things without considering order or ranking. For example, in a certain card game, a player receives five cards drawn from a standard bridge deck, which has 52 different cards. These five cards are called a *hand*. How many different five-card hands are in such a deck?

We can think of this problem using the principles we have in place. Let x be our unknown, the number of ways to pick a five-card hand from a standard bridge deck. If we care about which card is dealt first, which second, etc., then there are
$$52 \cdot 51 \cdot 50 \cdot 49 \cdot 48$$
different ordered hands. Next, we can think of choosing an ordered hand as a two-step process: First choose a five-card hand, then choose an ordering for that hand. That two-step process has $x \cdot 5!$ different outcomes. This makes it clear that
$$x \cdot 5! = 52 \cdot 51 \cdot 50 \cdot 49 \cdot 48 = \frac{52!}{47!}$$
allowing us to solve for
$$x = \frac{52!}{47!5!}.$$
These sorts of problems come up so often that the expression
$$\frac{n!}{(n-m)!m!}$$
has its own name and symbol.

---

[1] $n!$ is read *n factorial*.

Definition A.17. Let n, m be positive integers with $m \leq n$. The **binomial coefficient**, $\binom{n}{m}$, is
$$\binom{n}{m} = \frac{n!}{(n-m)!m!}.$$
We read $\binom{n}{m}$ as "n choose m."

Note that $\binom{n}{m}$ is the number of ways we can choose m items from among n distinct items, regardless of the order in which the items are chosen. It may take a moment, but some thought should reveal that if $n \in \mathbb{Z}_+$ and $m \leq n$, $\binom{n}{m}$ is precisely the coefficient for $x^{n-m}y^m$ in the binomial expansion

(A.2) $(x+y)^n = x^n + nx^{n-1}y + \binom{n}{2}x^{n-2}y^2 + \cdots + \binom{n}{n-2}x^2y^{n-2} + nxy^{n-1} + y^n.$

This means, for instance, that
$$(x-1)^6 = x^6 - 6x^5 + 15x^4 - 20x^3 + 15x^2 - 6x + 1.$$
We leave the verification of this as an exercise.

Counting is the basis for an introduction to probability so we close this section with the following simple observation.

Axiom A.18. Suppose there are n equally likely events that may occur. The probability of a specific event occurring is $1/n$.

Example A.19. Black Jack is a card game for two or more players, played with a standard bridge deck. There is an initial deal, in which each player receives two cards. After the initial deal, there are rounds of betting and subsequent deals. That sequence continues until the game ends by various mechanisms that need not concern us. The goal of the game is to get 21, which is calculated by adding values associated to the cards. We would like to calculate the probability of getting 21 on the initial deal.

A bridge deck has four of each numbered cards that range from 2 to 10. Each of those cards is valued according to its number. The deck also contains four of each face card — jack, queen, king. Each face card is worth 10 points. There are four aces. An ace can be worth 1 point or 11 points, whichever works best for the player.

We start by counting how many different two-card hands there are in a bridge deck. We know this is
$$\binom{52}{2} = \frac{52 \cdot 51}{2} = 26 \cdot 51 = 1{,}326.$$

Next we count how many different two-card hands give us 21. There are not that many. To have 21, we need an ace and a face, or an ace and a ten. There are 4 aces and a total of 16 faces and tens, so there are $4 \cdot 16 = 48$ different hands that qualify. The probability of getting such a hand is then $48/1{,}326$ which, to four decimal places, is 0.0362. Some people prefer to express probability in terms of percentages, and they would say that you have a 3.62% chance of a getting 21 in the initial deal.

A.4. Counting

The probability of an event occurring should be a measure of the likelihood of that event occurring. Strictly speaking, an outcome with a probability of 1 must occur and an outcome with a probability of 0 can never occur. In life, though, events with a probability of 0 do actually occur — very rarely — and events with a probability of 1 can fail to occur, again, very rarely. This disconnect happens somewhere on the bridge we traverse between mathematics and the world. Mathematics is not a perfect match with the reality we experience, so sometimes we must modulate our assumptions to get a better reflection of what we are actually trying to understand about the world. There is more to it when we look at probabilities, though. For instance, suppose we were to select an integer at random. Since $\mathbb{Z}$ is infinite, the probability of choosing, say, 2,137 is $1/\infty = 0$. This is true, though, for every number in $\mathbb{Z}$, thus, no matter the outcome of the experiment, the probability of getting that particular outcome is zero. If this is disturbing, so much the better. Mathematics gets at truths that are not obvious, and this can be unsettling.

Exercises A.4.

A.4.1. Suppose you plan to order a hummus and cheese sandwich from a local delicatessen. They have three different kinds of hummus, four different bread choices, and six different cheese choices.
  (a) How many different sandwiches could you order by choosing one type of hummus, one type of bread, and one type of cheese?
  (b) How many different sandwiches could you order by choosing two types of hummus for your sandwich, one type of bread, and one type of cheese?

A.4.2. Write down all the words we can form from the letters $AMDE$.

A.4.3. A senior-level linear algebra class has 15 students and three must be chosen to present the solution to a problem at a local conference.
  (a) How many different choices of three students are available for the task?
  (b) Suppose that the students must be chosen so that one presents an introduction to the problem, one presents the solution to the problem, and one discusses applications. How many different ways can we choose three students to do this?

A.4.4. Verify that for $m, n \in \mathbb{Z}_+$, $n \cdot (n-1) \cdots (n-m+1) = \frac{n!}{(n-m)!}$ provided $n \geq m$.

A.4.5. Use the distributive law to expand $(x+y)^3$ and $(x+y)^4$. Verify that the coefficients are in fact the binomial coefficients; for example, verify that the coefficient for $x^2 y^2$ in $(x+y)^4$ is $\binom{4}{2}$.

A.4.6. Verify that $(x-1)^6 = x^6 - 6x^5 + 15x^4 - 20x^3 + 15x^2 - 6x + 1$ using (A.2).

A.4.7. Use the idea that $\binom{n}{k}$ counts the number of sets with k elements that we can choose from a set with n elements to explain briefly why $\binom{n}{k} = \binom{n}{n-k}$.

A.4.8. Use Definition A.17 to verify that $\binom{52}{2} = \frac{52 \cdot 51}{2}$.

A.4.9. Use Definition A.17 to verify that $\binom{n}{k} = \frac{n(n-1)\cdots(n-k+1)}{k!}$. How many factors are in the numerator?

A.4.10. Suppose you are to choose a number at random from $\mathbb{Z}_+$.
  (a) What is the probability of choosing an even number?
  (b) What is the probability of choosing a multiple of 3?

A.4.11. Suppose you ask each of five people to pick a positive integer and that two of them pick the same number. How can that happen, given that the probability of it happening is zero? Is there a way to look at this experiment so that it better aligns with what humans are likely to do in the face of having to choose a positive integer?

A.5. Equivalence Relations

The notion of a relation is quite general and, on some level, easily accessible to intuition. In everyday conversation, we may talk about our relations, in which case we usually mean people with whom we share a familial bond. Since we take the notion of relation beyond what we talk about in casual conversation, we start with a formal definition.

Definition A.20. A **relation** on a nonempty set A is a subset of $A \times A$.

As a subset of $\mathbb{R} \times \mathbb{R}$, a line in the xy-plane is a relation on $\mathbb{R}$. For example if $\ell = \{(x,y) \mid y = 2x, x \in \mathbb{R}\}$, then $(1,2) \in \ell$.

Frequently we define a relation on a set using natural relationships among its elements. For instance, one relation on $\mathbb{R}$ is $\{(a,b) \mid a \leq b\}$. In this setting, we write $a \leq b$ to indicate that (a,b) is in the relation.

When we say that two lines in the xy-plane are parallel, we are saying they are related by *parallelism*, a relation $\mathcal{R}$ on the lines in the xy-plane:

$$(\ell_1, \ell_2) \in \mathcal{R} \iff \ell_1 \| \ell_2.$$

Parallelism has properties that are not shared by all relations. For example, if $\ell_1 \| \ell_2$, then $\ell_2 \| \ell_1$. This is not the case in the child/parent relation, for example.

Definition A.21. A relation $\mathcal{R}$ is **reflexive** on a set A provided $(a,a) \in \mathcal{R}$ for all $a \in A$. A relation $\mathcal{R}$ is **symmetric** on A provided $(a,b) \in \mathcal{R}$ implies $(b,a) \in \mathcal{R}$ for $a,b \in A$. A relation $\mathcal{R}$ is **antisymmetric** on A provided $(a,b) \in \mathcal{R}$ and $(b,a) \in \mathcal{R}$ imply $a = b$ for $a,b \in A$. A relation $\mathcal{R}$ is **transitive** on A provided $(a,b) \in \mathcal{R}$ and $(b,c) \in \mathcal{R}$ together imply that $(a,c) \in \mathcal{R}$, for $a,b,c \in A$.

We noted that parallelism is symmetric. That parallelism is transitive is a special feature of the geometry of the Euclidean plane: If $\ell_1 \| \ell_2$ and $\ell_2 \| \ell_3$, then $\ell_1 \| \ell_3$. Frequently, we treat parallelism as reflexive, meaning that for a given line ℓ, $\ell \| \ell$. This allows us to classify parallelism as an *equivalence relation*.

Definition A.22. An **equivalence relation** on a set A is a relation on A that is reflexive, symmetric, and transitive.

Example A.23. Let $A = \{a,b,c\}$. Let $\mathcal{R} = \{(a,a),(b,b),(c,c)\}$. We leave it as an exercise to show that $\mathcal{R}$ is an equivalence relation on A.

An equivalence relation may be taken as a generalized form of equality. For example, let $\mathcal{R}$ be the relation on triangles in the xy-plane defined by congruence.

A.5. Equivalence Relations

Congruent triangles are identical geometrically. Another example is similarity defined on the triangles in the xy-plane. Except for scale, similar triangles look identical.

The next example is of a type that occurs frequently throughout the course.

Example A.24. Define a relation $\mathcal{R}$ on $\mathbb{Z}$ by
$$\mathcal{R} = \{(a,b) \mid a - b = 2k, \text{ for some } k \in \mathbb{Z}\}.$$

We call $\mathcal{R}$ **equivalence mod** 2. Note, for instance, that $(1,3)$ and $(0,2)$ are in $\mathcal{R}$ and that $(1,2)$ is not in $\mathcal{R}$. We usually indicate that $(a,b) \in \mathcal{R}$ by writing
$$a \equiv_2 b,$$
read, "a is equivalent to b mod 2."

Equivalence mod 2 is reflexive since $a - a = 0$ is even. If $a \equiv_2 b$, then $a - b = 2k$, for some integer k, so that $b - a = -2k = 2(-k)$ implying $b \equiv_2 a$. This shows that equivalence mod 2 is symmetric. If $a \equiv_2 b$ and $b \equiv_2 c$, then $a - b = 2k$ and $b - c = 2m$, for integers k and m, so that
$$a - c = (a - b) + (b - c) = 2k + 2m = 2(k + m).$$

The implication is that if $a \equiv_2 b$ and $b \equiv_2 c$, then $a \equiv_2 c$. It follows that equivalence mod 2 is transitive, thus, an equivalence relation on $\mathbb{Z}$.

As a generalization of equality, an equivalence relation on a set A enables us to treat elements in certain subsets of A as one object.

Definition A.25. When $\mathcal{R}$ is an equivalence relation on a set A,
$$\mathcal{E}_a = \{b \in A \mid (a,b) \in \mathcal{R}\}$$
is the **equivalence class** determined by a.

We apply Definition A.25 to equivalence mod 2 on $\mathbb{Z}$ in the next example.

Example A.26. Consider $\mathcal{E}_0 = \{a \in \mathbb{Z} \mid a \equiv_2 0\}$. If $a \equiv_2 0$, then $a - 0 = a = 2k$ for some integer k. It follows that $a \in \mathcal{E}_0$ if and only if a is even. Since 1 is odd, it is not in $\mathcal{E}_0$. Indeed, $a \in \mathcal{E}_1$ means $a - 1 = 2k$, so that $a = 2k + 1$, that is, $a \equiv_2 1$ if and only if a is odd. There are then two equivalence classes associated to equivalence mod 2 on $\mathbb{Z}$:
$$\mathcal{E}_0 = \{\ldots, -4, -2, 0, 2, 4, \ldots\} = \{2k \mid k \in \mathbb{Z}\}$$
and
$$\mathcal{E}_1 = \{\ldots, -3, -1, 1, 3, 5, \ldots\} = \{2k + 1 \mid k \in \mathbb{Z}\}.$$
Notice that every integer is either in $\mathcal{E}_0$ or $\mathcal{E}_1$.

Equivalence classes lead naturally to the notion of a *partition* on a set.

Definition A.27. A **partition** on a set A is a collection of nonempty, nonintersecting subsets of A, the union of which is all of A. The subsets in a partition are called **cells**.

Theorem A.28. (a) The equivalence classes associated to an equivalence relation on a set A are the cells in a partition on A.

(b) A partition on a set determines an equivalence relation on the set.

Proof. (a) Let A be a set. Fix an equivalence relation $\mathcal{R}$ on A. Let $\mathcal{E}_a$ be the equivalence class determined by an element $a \in A$. Since $a \in \mathcal{E}_a$, no $\mathcal{E}_a$ is empty and
$$\bigcup_{a \in A} \mathcal{E}_a = A.$$

Suppose there is $x \in \mathcal{E}_a \cap \mathcal{E}_b$ for some $a, b \in A$. We then have (a, x) and (x, b) in $\mathcal{R}$. By transitivity, $(a, b) \in \mathcal{R}$. Since anything related to a is related to b, $\mathcal{E}_a \subseteq \mathcal{E}_b$. By the same argument, $\mathcal{E}_b \subseteq \mathcal{E}_a$ so that $\mathcal{E}_a = \mathcal{E}_b$. This proves (a).

(b) Let Π be a partition on A. Define $\mathcal{R}$ on A by $(a, b) \in \mathcal{R}$ if and only if a and b are in the same cell in Π. Since the union of cells in Π is all of A, every $a \in A$ belongs to some cell. From there it is immediate that $\mathcal{R}$ is reflexive. Evidently, $\mathcal{R}$ is symmetric. Since cells are nonintersecting, $(a, b) \in \mathcal{R}$ and $(b, c) \in \mathcal{R}$ together imply that a, b, and c all belong to the same cell. From there it follows that $(a, c) \in \mathcal{R}$, thus, $\mathcal{R}$ is transitive. □

Equivalence relations and partitions come up repeatedly in linear algebra. The theorem says they are two different ways to look at the same thing.

Exercises A.5.

A.5.1. Let $A = \{a, b, c\}$.
   (a) Show that $\mathcal{R} = \{(a, a), (b, b), (c, c)\}$ is an equivalence relation on $A = \{a, b, c\}$. What are the equivalence classes?
   (b) Define another equivalence relation on $A = \{a, b, c\}$. Specify the equivalence classes.
   (c) Define a relation on A that is neither reflexive, symmetric, nor transitive.
   (d) Define a relation on A that is reflexive, but neither symmetric nor transitive.
   (e) Define a relation on A that is transitive, but neither reflexive nor symmetric.
   (f) Define a relation on A that is reflexive and symmetric but not transitive.
   (g) Define a relation on A that is reflexive and transitive but not symmetric.
   (h) Define a relation on A that is symmetric and transitive but not reflexive.

A.5.2. When using polar coordinates in the xy-plane, we measure an angle by taking the vertex at the origin and the initial side of the angle to coincide with the positive x-axis. If we measure counterclockwise from the initial side to the terminal side, we get a positive angle. If we measure clockwise from the initial side to the terminal side, we get a negative angle. Different angles with the same terminal side are *coterminal*. Define $\mathcal{R}$ on angle

A.6. Mappings

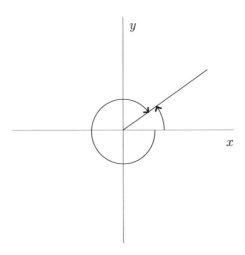

Figure A.2. Coterminal angles in $\mathbb{R}^2$.

measurements in the xy-plane so that $(\theta_1, \theta_2) \in \mathcal{R}$ provided θ_1 and θ_2 measure coterminal angles. Show that $\mathcal{R}$ is an equivalence relation.

A.5.3. Fix $n \in \mathbb{Z}_+$. Define a relation $\mathcal{R}_n$ on $\mathbb{Z}$ by $(a,b) \in \mathcal{R}_n$ provided $a - b = kn$ for some $k \in \mathbb{Z}$. If $(a,b) \in \mathcal{R}_n$, we write $a \equiv_n b$.
  (a) Show that $\mathcal{R}_n$ is an equivalence relation on $\mathbb{Z}$. It is called **equivalence mod** n.
  (b) How many cells are in the partition on $\mathbb{Z}$ determined by $\mathcal{R}_3$? What about $\mathcal{R}_n$?

A.5.4. Let A be the set of lines in the xy-plane. We can write the equation for any line in the xy-plane in the form $ax + by = c$, for some $a, b, c \in \mathbb{R}$, where a and b are not both zero. Define a relation $\mathcal{R}$ on A by $(\ell_1, \ell_2) \in \mathcal{R}$ provided $a_1 b_2 - a_2 b_1 = 0$ where ℓ_1 is given by $a_1 x + b_1 y = c_1$ and ℓ_2 is given by $a_2 x + b_2 y = c_2$.
  (a) Since there are many equations of the form $ax + by = c$ that describe a given line, it is not immediately clear that $\mathcal{R}$ is well-defined. Prove that it is.
  (b) Show that $\mathcal{R}$ is an equivalence relation on A.
  (c) Use words to describe the equivalence class of lines containing the line given by $x = 1$.
  (d) Describe the equivalence class of lines containing the line given by $y = 3 - 2x$.

A.6. Mappings

Functions on $\mathbb{R}$ are familiar to students who have completed high school mathematics. In this course, we consider functions on sets that are not necessarily subsets of $\mathbb{R}$.

Definition A.29. Let A and B be sets. A **mapping** or **function** $f : A \to B$ is a rule that associates to each element $x \in A$ exactly one element $f(x)$ in B. We say f **maps** A **into** B and we write $x \mapsto f(x)$ to indicate that x in A maps to $f(x)$ in B. The **domain** of f is A and the **codomain** of f is B. The set $\{y \in B \,|\, y = f(x) \text{ for some } x \in A\}$ is the **range** of f. If $C \subseteq A$, we write $f[C] = \{f(x) \,|\, x \in C\} \subseteq B$. We say that $f[C]$ is the **image** of C under f. In this case, the mapping from C into B is called **the restriction of** f **to** C and is designated $f|_C$.

If $U \subseteq B$, then $f^{-1}[U] = \{x \in A \,|\, f(x) \in U\}$. We say that $f^{-1}[U]$ is the **pre-image** or **inverse image** of U.

If $f[A] = B$, then f is **onto** or **surjective**. In this case, we also say that f is a **surjection**. If $f(x_1) = f(x_2)$ implies $x_1 = x_2$, for all $x_1, x_2 \in A$, f is **one-to-one** (often written 1-1) or **injective**. In this case, we also say f is an **injection**. If f is both 1-1 and onto, f is **bijective**. Bijective mappings are called **bijections**.

The question as to whether $f : A \to B$ is surjective is entirely separate from the question as to whether $f : A \to B$ is injective. Surjectivity is about whether everything in the codomain, B, is the image of one or more elements in A. Injectivity is about whether each element in the range, $f[A]$, is the image of *no more than* one element in A.

Every mapping is onto its range. This is something that comes up frequently all over mathematics. For example, consider $f : \mathbb{R} \to \mathbb{R}$ given by $f(x) = e^x$. As a strictly increasing function, f is an injection. However, because $e^x > 0$ for all $x \in \mathbb{R}$, f is not onto $\mathbb{R}$. It is, however, onto the set of positive real numbers, $\mathbb{R}_+ = f[\mathbb{R}]$. Likewise, $g : \mathbb{R} \to \mathbb{R}$ given by $g(x) = \sin x$ is not onto $\mathbb{R}$, but it is onto the closed interval $[-1, 1]$, its range. Indeed, this sort of "reverse engineering" of the codomain to force a mapping to be onto is always available, though it may not be appropriate in some applications. We may be able to manipulate a given mapping to associate it to one that is injective, as well, but in changing the domain of the original mapping, we often run into problems that are more serious.

Lemma A.30. Let A and B be sets and suppose $f : A \to B$ is a bijective mapping. If we define $f^{-1} : B \to A$ by

$$f^{-1}(b) = a \iff b = f(a),$$

then f^{-1} is also a bijective mapping.

Proof. Suppose f, A, and B are as hypothesized and that f^{-1} is defined as in the statement of the lemma. We must show that f^{-1} is well-defined and that it is bijective.

Suppose $b \in f[A]$. Since f is injective, there is a unique $a \in A$ such that $f(a) = b$. It follows that $f^{-1}(b) = a$ is unambiguous, thus f^{-1} is well-defined on $f[A]$.

Since f is surjective, $f[A] = B$ so f^{-1} is defined on all of B. This proves that $f^{-1} : B \to A$ is indeed a mapping.

A.6. Mappings

Next we show that f^{-1} is injective. Suppose $f^{-1}(b) = f^{-1}(b') = a$ for some $b, b' \in B$, $a \in A$. We then have $f(a) = b$ and $f(a) = b'$. Since f is well-defined, $b = b'$, which is to say that f^{-1} is injective.

Given $a \in A$, let $b = f(a)$. In this case $f^{-1}(b) = a$, so f^{-1} is surjective. □

The following is a slight generalization of Lemma A.30. It comes up frequently.

Corollary A.31. *If A and B are sets and $f : A \to B$ is an injective mapping, then $f^{-1} : f[A] \subseteq B \to A$ defined by*

$$f^{-1}(b) = a \iff b = f(a)$$

is a well-defined bijective mapping.

Proof. Let f and A be as hypothesized so that $f : A \to f[A]$ is bijective. The corollary follows now from the lemma. □

When $f : A \to B$ is bijective, f^{-1}, as defined in Lemma A.30, is called the **inverse** of f. Note the distinction between f^{-1}, the inverse of an invertible mapping, and $f^{-1}[U]$, the inverse image of a subset of the range of an arbitrary mapping. In the latter case, f may not be invertible at all.

Recall that the composition of mappings $f : A \to B$ and $g : B \to C$ is denoted $g \circ f$ where, for $x \in A$, $(g \circ f)(x) = g(f(x))$. It is easy to verify that in this case, $(g \circ f) : A \to C$ is another mapping.

Lemma A.32. *Let $f : A \to B$ and $g : B \to C$ be mappings of sets.*

(a) *If f and g are both injective, then $g \circ f$ is injective.*

(b) *If f and g are both surjective, then $g \circ f$ is surjective.*

(c) *If f and g are both bijective, then $g \circ f$ is bijective.*

Proof. Suppose $f : A \to B$ and $g : B \to C$ are injective and that $(g \circ f)(a_1) = (g \circ f)(a_2)$, for some a_1 and $a_2 \in A$. We can write $g(f(a_1)) = g(f(a_2))$. Since g is injective, $f(a_1) = f(a_2)$. Since f is injective, $a_1 = a_2$, which proves that $g \circ f$ is injective. We leave the rest of the proof as an exercise. □

Lemma A.32 applies to a bijective mapping $f : A \to B$ and its inverse $f^{-1} : B \to A$. We write $f^{-1} \circ f = \iota_A$, where $\iota_A : A \to A$ is the **identity mapping**: $\iota_A(a) = a$ for all $a \in A$. Likewise, $f \circ f^{-1} = \iota_B$ is the identity mapping on B.

Exercises A.6.

A.6.1. Let A and B be sets and let $f : A \to B$ be a mapping. Show that f is onto $f[A]$.

A.6.2. Let A and B be finite sets, each with n elements. Let $f : A \to B$ be a mapping.
(a) Show that if f is injective, then it must be surjective.
(b) Show that if f is surjective, then it must be injective.

A.6.3. Show that the inverse of a bijective mapping is unique.

A.6.4. Finish the proof of Lemma A.32.

A.6.5. Suppose $f : A \to B$ is a bijection. Show that $f \circ f^{-1}$ is the identity mapping on B and that $f^{-1} \circ f$ is the identity mapping on A.

A.6.6. Suppose $f : A \to B$ is a bijection. Show that the inverse of f^{-1} is f.

A.7. Binary Operations

Binary operations are an important class of mappings that arise in the study of algebra. The most familiar binary operations are addition and multiplication of numbers.

Definition A.33. A **binary operation** on a set A is a mapping $\odot : A \times A \to A$.

When $\odot$ is a binary operation on A, we write $a \odot b = c$ instead of $\odot(a, b) = c$.

If $\odot$ is a binary operation on A and $(a \odot b) \odot c = a \odot (b \odot c)$ for all $a, b, c \in A$, then $\odot$ is **associative**. If $a \odot b = b \odot a$ for all $a, b \in A$, then $\odot$ is **commutative**. If there is $e \in A$ so that $a \odot e = e \odot a = a$ for all $a \in A$, then we say that e is a $\odot$ **identity** in A. If e is a $\odot$ identity in A and if $a \odot b = b \odot a = e$, then b is a $\odot$ **inverse** for a in A.

When $\odot$ is a binary operation on a set A, we also say that A is **closed** under $\odot$.

We will use the following lemma repeatedly throughout the course.

Lemma A.34. Let $\odot$ be an associative binary operation on A. If e in A is a $\odot$ identity and a in A has inverse a', then

(a) e is the only $\odot$ identity element in A and

(b) a' is the only $\odot$ inverse for a.

Proof. Suppose A, e, a, and a' are as hypothesized.

If there is e' in A so that $e' \odot a = a \odot e' = a$ for all a in A, then
$$e = e \odot e' = e' \odot e = e',$$
so the $\odot$ identity is unique.

If there is a'' in A so that $a \odot a'' = a'' \odot a = e$, then
$$a' = a' \odot e = a' \odot (a \odot a'') = (a' \odot a) \odot a'' = e \odot a'' = a''.$$

This proves that the $\odot$ inverse of a is unique. □

Addition on $\mathbb{Z}$ is an associative, commutative binary operation with identity 0. Every element a in $\mathbb{Z}$ has an additive inverse $-a$.

Multiplication on $\mathbb{Z}$ is an associative, commutative binary operation with identity 1. The only elements in $\mathbb{Z}$ with multiplicative inverse are 1 and -1.

Division on the nonzero real numbers, $\mathbb{R}^*$, is a binary operation. It is not commutative — $1 \div 2 \neq 2 \div 1$, for example — neither is it associative: $1 \div (2 \div 3) = 3/2$ but $(1 \div 2) \div 3 = 1/6$, for example. For all $a \in \mathbb{R}^*$ we have $a \div 1 = a$ but since $1 \div a \neq a$ for all $a \in \mathbb{R}^*$, there is no $\div$ identity in $\mathbb{R}^*$.

A.7. Binary Operations

Composition on $\mathcal{F}(\mathbb{R})$, the collection of functions from $\mathbb{R}$ to $\mathbb{R}$, is a binary operation. It is associative: For $f, g, h \in \mathcal{F}$,

$$((f \circ g) \circ h)(x) = (f \circ g)(h(x)) = f(g(h(x))) = f((g \circ h)(x)) = (f \circ (g \circ h))(x),$$

for all $x \in \mathbb{R}$. Since a mapping is defined by what it does to elements in its domain, we see that for any mappings $f, g, h \in \mathcal{F}$, $(f \circ g) \circ h = f \circ (g \circ h)$.

Note that in showing that the composition of functions in $\mathcal{F}(\mathbb{R})$ is associative, we did not use properties of $\mathbb{R}$. Indeed, our argument proves the following important result.

Theorem A.35. *If A is any set and $\mathcal{F}(A)$ is the collection of functions $f : A \to A$, then function composition on $\mathcal{F}(A)$ is associative.*

Composition on $\mathcal{F}(\mathbb{R})$ is not commutative, but there is the identity element $\iota : \mathbb{R} \to \mathbb{R}$, given by $\iota(x) = x$ for all $x \in \mathbb{R}$.

Since the composition of continuous functions is continuous, composition on $C(\mathbb{R})$, the set of continuous functions from $\mathbb{R}$ to $\mathbb{R}$, is an associative binary operation with identity.

The next example may be less familiar.

Example A.36. Consider $\bar{0} = \mathcal{E}_0 = \{\ldots, -4, -2, 0, 2, 4, \ldots\}$, the even integers, and $\bar{1} = \mathcal{E}_1 = \{\ldots, -3, -1, 1, 3, 5, \ldots\}$, the odd integers. These are the equivalence classes associated to equivalence mod 2 on $\mathbb{Z}$. Let $\mathbb{Z}_2 = \{\bar{0}, \bar{1}\}$. Define $\oplus_2$ on $\mathbb{Z}_2$ according to the following addition table (this is an example of a *Cayley table*).

(A.3)
$$\begin{array}{c|cc} \oplus_2 & \bar{0} & \bar{1} \\ \hline \bar{0} & \bar{0} & \bar{1} \\ \bar{1} & \bar{1} & \bar{0} \end{array}$$

Notice the sense in which $\oplus_2$ corresponds to the actual addition of elements in the equivalence classes: The set of all sums, $a + b$, for $a, b \in \bar{0}$ is $\bar{0}$. Likewise, the set of all sums, $a + b$, for $a, b \in \bar{1}$ is $\bar{0}$. The set of all sums, $a + b$, for $a \in \bar{0}$ and $b \in \bar{1}$ is $\bar{1}$. We call $\oplus_2$ **addition mod 2.**

We define a second binary operation on $\mathbb{Z}_2$. This one we designate $\odot_2$. What follows is the Cayley table for $\odot_2$.

(A.4)
$$\begin{array}{c|cc} \odot_2 & \bar{0} & \bar{1} \\ \hline \bar{0} & \bar{0} & \bar{0} \\ \bar{1} & \bar{0} & \bar{1} \end{array}$$

Notice the sense in which $\odot_2$ corresponds to the actual multiplication of elements in the equivalence classes: The set of all products ab, for $a, b \in \bar{0}$, is all of $\bar{0}$. The set of all products ab, for $a \in \bar{0}$ and $b \in \bar{1}$, is all of $\bar{0}$. The set of all products ab, for $a, b \in \bar{1}$, is all of $\bar{1}$. We call $\odot_2$ **multiplication mod 2.**

We leave it as an exercise to show that addition and multiplication mod 2 are associative and commutative and that each has an identity element.

Exercises A.7.

A.7.1. Show that addition on $\mathcal{F}(\mathbb{R})$ is an associative, commutative, binary operation with identity and that every nonzero element in $\mathcal{F}(\mathbb{R})$ has an additive inverse.

A.7.2. Write out the set $2 + b$, where $b \in \bar{0} = \{\ldots, -6, -4, -2, 0, 2, 4, 6, 8, \ldots\}$. Write out the set $-4 + b$, $b \in \bar{0}$. Repeat the exercise adding 1 to each element in $\bar{0}$ and again adding -3 to each element in $\bar{0}$. Repeat it now adding 1 to each element in $\bar{1} = \{\ldots, -5, -3, -1, 1, 3, 5, 7, \ldots\}$ and again adding -5 to each element in $\bar{1}$.

A.7.3. Prove that addition and multiplication mod 2 are associative and commutative on $\mathbb{Z}_2$. Verify that there is an identity for $\oplus_2$ on $\mathbb{Z}_2$ and that each element in $\mathbb{Z}_2$ has an inverse relative to $\oplus_2$. Verify that there is a multiplicative identity for $\odot_2$ on $\mathbb{Z}_2$. What can you say about inverses relative to $\odot_2$?

A.7.4. Let $\mathbb{Z}_3$ be the set of equivalence classes mod 3 in $\mathbb{Z}$. Mimic Example A.36 to define $\oplus_3$ on $\mathbb{Z}_3$, a binary operation that corresponds to addition of elements in the equivalence classes. We call this *addition mod* 3. Verify that addition mod 3 is associative and commutative. Verify that there is an identity in $\mathbb{Z}_3$ for $\oplus_3$ and that each element in $\mathbb{Z}_3$ has an inverse with respect to $\oplus_3$. Now define $\odot_3$, multiplication mod 3, on $\mathbb{Z}_3$. Verify that $\odot_3$ is associative and commutative. Is there an identity element in $\mathbb{Z}_3$ for $\odot_3$? Discuss multiplicative inverses.

A.7.5. Repeat the last exercise on the set of equivalence classes mod 4 in $\mathbb{Z}$. What properties of $\odot_2$ and $\odot_3$ fail to carry over to $\odot_4$?

Appendix B

$\mathbb{R}^2$ and $\mathbb{R}^3$

This appendix treats properties of $\mathbb{R}^n$ that should be familiar to students from precalculus, calculus, and physics courses. Our specific objective is to review connections between geometry and algebra in $\mathbb{R}^2$ and $\mathbb{R}^3$. A review of complex numbers and their relationship to $\mathbb{R}^2$ is included. We start with a discussion of the word *vector* as it is used in elementary physics and in linear algebra.

B.1. Vectors

We learn in elementary physics courses that a vector is an entity determined by magnitude and direction. Examples of vectors in this sense are displacement, velocity, acceleration, and force. In physics problems, vectors are typically represented by arrows. An arrow extends from its tail to its head. The direction of the extension is the direction of the vector it represents. The magnitude of the vector is indicated by the length of the arrow.

We learn in linear algebra that a vector space is a type of set that is closed under scaling and addition. We usually refer to the elements in a vector space as points but they are also called vectors. Since the vectors we work with in elementary physics can be scaled and added, physics vectors are also vectors in the mathematical sense. In other words, vectors that arise in physics are points in vector spaces.

Elementary physics does not rely heavily on the theory of vector spaces. When we use vectors to analyze the forces on an object, for example, we want to define a coordinate system that facilitates our understanding of the various components of a given force. While we may do some of that type of analysis in linear algebra, the mathematics is mostly about properties of large sets of vectors, that is, vector spaces and subspaces.

A point $(a_1, \ldots, a_n)$ in $\mathbb{R}^n$ can be identified with the arrow that has its tail at the origin in $\mathbb{R}^n$ and its head at $(a_1, \ldots, a_n)$. This arrow can then represent a vector as we would use it in a physics problem. The magnitude or length of the arrow is the distance from the origin to $(a_1, \ldots, a_n)$. Angles formed by the arrow and the coordinate axes can be identified using the coordinates, $a_1, \ldots, a_n$. For instance, say $\mathbf{v} = (a, b)$ is in $\mathbb{R}^2$. Let φ be the angle formed by $\mathbf{v}$ and the positive x-axis. The magnitude of $\mathbf{v}$ is then $\|\mathbf{v}\| = \sqrt{a^2 + b^2}$, while $\sin\varphi = b/\sqrt{a^2 + b^2}$ and $\cos\varphi = a/\sqrt{a^2 + b^2}$. If $\mathbf{v}$ represents a force, then the component of the force in the x-direction has magnitude $|a|$ and its component in the y-direction has magnitude $|b|$.

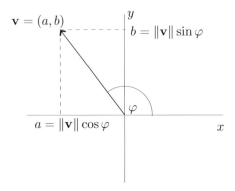

Figure B.1. Identify a point in $\mathbb{R}^2$ with an arrow.

We can depict a vector as emanating from any point in space, as long as its magnitude and direction are preserved. If $\mathbf{v}$ is the arrow in the xy-plane with tail at (a, b) and head at (c, d), we can identify $\mathbf{v}$ with the point $(c-a, d-b)$ in $\mathbb{R}^2$. An easy exercise verifies that $\mathbf{v}$ and the arrow starting at $(0, 0)$ and ending at $(c - a, d - b)$ have the same length and direction.

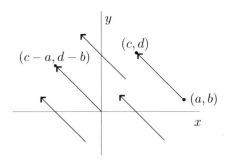

Figure B.2. A vector has many different representations as arrows in the xy-plane.

When we refer to a vector in $\mathbb{R}^n$ as a geometric entity, we are usually thinking of the vector as an arrow or a directed line segment. When we refer to a vector in $\mathbb{R}^n$ as an algebraic entity, we are usually thinking about the vector in terms of its coordinates.

B.1. Vectors

Suppose we are given two vectors $\mathbf{v} = (a_1, \ldots, a_n)$ and $\mathbf{w} = (b_1, \ldots, b_n)$ in $\mathbb{R}^n$. If we view $\mathbf{v}$ and $\mathbf{w}$ as arrows emanating from the origin, they determine three points: the origin, $(a_1, \ldots, a_n)$, and $(b_1, \ldots, b_n)$. In Euclidean geometry, three noncollinear points determine a plane. This means that the three points determined by the two vectors either lie on a line in $\mathbb{R}^n$ or they determine a plane in $\mathbb{R}^n$. In the first case, we can analyze our two vectors by treating them as elements in $\mathbb{R}$. In the second case, we can analyze the two vectors by treating them as lying in $\mathbb{R}^2$. These two scenarios account for most of the problems we see in elementary physics courses.

When we add vectors $\mathbf{v} = (a, b)$ and $\mathbf{w} = (c, d)$ geometrically, we apply the Parallelogram Rule. The given vectors, emanating from the origin, determine the sides of a parallelogram. The vector with tail at $(0, 0)$ and head at the opposite vertex of that parallelogram is the vector sum $\mathbf{v} + \mathbf{w}$. Algebraically, we add vectors componentwise:

$$\mathbf{v} + \mathbf{w} = (a, b) + (c, d) = (a + c, b + d).$$

We leave it as an exercise to show that the Parallelogram Rule and addition of components yield the same vector sum.

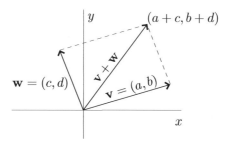

Figure B.3. Adding vectors by the Parallelogram Rule.

While every physics vector is an element in a vector space, not all vector spaces contain what we usually think of as vectors in physics. In particular, there are vector spaces in which the elements do not have magnitude or direction in the usual sense. It is possible to go through a first course in linear algebra never having seen such vector spaces but we will see them in this course.

Exercises B.1.

B.1.1. Let $\mathbf{v}$ be the arrow in $\mathbb{R}^2$ that starts at (a, b) and ends at (c, d). Show that the length and the direction of $\mathbf{v}$ are the same as the length and direction of the arrow that starts at $(0, 0)$ and ends at $(c - a, d - b)$.

B.1.2. A human lobs a ball across a field, giving it an initial velocity of 10 meters per second at an angle of 30 degrees to the horizontal. Sketch the initial velocity vector in the xy-plane and find its vertical and horizontal components.

B.1.3. Show that if $\mathbf{v} = (a,b)$ and $\mathbf{w} = (c,d)$, then $\mathbf{v} + \mathbf{w} = (a+c, b+d)$ is actually the vector describing the diagonal of the parallelogram with sides determined by $\mathbf{v}$ and $\mathbf{w}$, as shown in Figure B.3. This establishes that vector addition in $\mathbb{R}^2$ conforms to the so-called Parallelogram Rule.

B.2. The Real Plane

Most students' first encounter with $\mathbb{R}^2$ is not as a vector space, but as a setting for analytic geometry. Aside from its role in analytic geometry, $\mathbb{R}^2$ also serves as a canvas for the graphs we see in single variable calculus. In physics, calculus, and analytic geometry, we use $\mathbb{R}^2$ as a tool. In linear algebra, we study $\mathbb{R}^2$ itself. That study includes the geometry of points and lines in $\mathbb{R}^2$.

The setting for high school geometry is the Euclidean plane. The study of pure Euclidean geometry employs neither direction nor distance. Objects are only measured in comparison to one another. Two lines may be parallel. Two line segments may be congruent. One angle may exceed another.

Analytic geometry involves the use of coordinates to study Euclidean geometry. We take $\mathbb{R}^2$ as a model for the Euclidean plane. In defining a coordinate system in $\mathbb{R}^2$, we establish standards for direction and distance. *Up* in the xy-plane is the direction of increasing y-coordinates. *Right* is the direction of increasing x-coordinates. *Vertical* is parallel to the y-axis. *Horizontal* is parallel to the x-axis. Units of distance are determined by the scale we choose on the x- and y-axes. These units can be deployed to get a distance formula, based on the Pythagorean Theorem. Using the unit circle, we can define sines and cosines.

Once coordinates enter the story, many proofs in Euclidean geometry become simplified. For example, if we use $\mathbb{R}^2$ to model the Euclidean plane, we can assign coordinates to the vertices of a triangle. Via the distance formula and a definition of cosine, we then get an easy proof of the Law of Cosines.

Every tool comes at a price. One price of using the coordinate plane for Euclidean geometry is the apparently exceptional nature of vertical lines. Geometrically, there is nothing special about a vertical line. The concepts of slope and verticality are artifacts of the choice of a particular coordinate system.

The geometry of elementary linear algebra is called *affine geometry*. This is the geometry of points, lines, planes, and the incidence relations among them in vector spaces. This is in contrast to Euclidean geometry and analytic geometry, which treat angles, polygons, circles, and congruent objects. Especially in higher dimensions, affine geometry is surprisingly rich.

The following is a set of axioms that defines any affine plane, that is, the incidence plane based on the points and lines in a 2-dimensional vector space such as $\mathbb{R}^2$.

Axioms B.1 (Affine Plane). An affine plane contains *points* and *lines*. Every point *lies on* a line. Every line contains at least two points. Lines *intersect* when they have common points. Lines that do not intersect are *parallel*. There exist a point and a line not through that point.

Points and lines in an affine plane obey the following rules of incidence:

(1) Two points determine a unique line.
(2) Given a line and a point not on the line, there is exactly one line through the point parallel to the given line.

Another way to state Axiom B.1(1) is to say that two points lie on a unique line. In affine geometry, incidence is the only relationship objects have relative to one another: Two objects are either incident or not.

We use $\mathbb{R}^2$ as a model for the *real affine plane* by taking points to be ordered pairs (x, y). A line in $\mathbb{R}^2$ is a collection of points (x, y) that satisfy a *linear equation in x and y*, that is, an equation of the form

$$ax + by = c$$

where $a, b, c \in \mathbb{R}$, a and b not both zero. Based on long experience with the xy-plane, we know that the points and lines in $\mathbb{R}^2$ behave as expected, that is, that two points determine a unique line, for example. The algebraic verification of this is an important, though messy, task. We leave it to the exercises.

Exercises B.2.

B.2.1. The Law of Cosines says that if a triangle has sides with lengths x, y, z and if the angle formed by the sides with lengths x and y is θ, then $x^2 + y^2 - 2xy \cos \theta = z^2$. Prove the Law of Cosines using the triangle in Figure B.4, along with the distance formula and the definition of cosine.

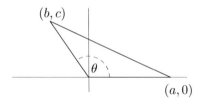

Figure B.4. Proving the Law of Cosines using $\mathbb{R}^2$.

B.2.2. Explain how the slope of a line in $\mathbb{R}^2$ is defined in terms of coordinates.

B.2.3. Consider points $(-2, 1)$ and $(1, 7)$ in $\mathbb{R}^2$.
   (a) Write down two equations in three unknowns that allow you to find a, b, c in $\mathbb{R}$ so that both of the given points satisfy $ax + by = c$.
   (b) Use the equations you found in part (a) to find a, b, c in $\mathbb{R}$ so that both of the given points satisfy $ax + by = c$. Let ℓ be the line given by $ax + by = c$.
   (c) Argue that any equation of the form $ax + by = c$ satisfied by points $(-2, 1)$ and $(1, 7)$ describes the same line, ℓ.
   (d) Use coordinates to argue that two points (x_0, y_0), (x_1, y_1) in $\mathbb{R}^2$ determine a unique line. (Hint: Treat the following two cases separately: the case in which $x_0 = x_1$, and the case in which $y_0 = y_1$.)

B.2.4. Consider the line ℓ in $\mathbb{R}^2$ given by $2x - 3y = 1$.
  (a) Is the point $(-1, 1)$ on the line?
  (b) Find an equation for the line through $(0,0)$ parallel to ℓ. Argue that this line is the only line through $(0,0)$ parallel to ℓ.

B.2.5. Consider an arbitrary line ℓ in $\mathbb{R}^2$ given by $ax + by = c$, $a, b, c \in \mathbb{R}$, a and b not both zero. Let (x_0, y_0) be a point not on ℓ. Find the line parallel to ℓ through (x_0, y_0). Argue that the line is unique.

B.3. The Complex Numbers and $\mathbb{R}^2$

One way to define the complex numbers, $\mathbb{C}$, is as the algebraic closure of $\mathbb{R}$. This means that $\mathbb{C}$ contains $\mathbb{R}$ and the roots of all polynomials over $\mathbb{R}$. Alternatively, we may define $\mathbb{C}$ as the 2-dimensional vector space over $\mathbb{R}$ with basis $\{1, i\}$ where i is a symbol with the property $i^2 = -1$. This means that we can write every complex number in the form $a + bi$, for a, b in $\mathbb{R}$; that $a + bi = 0$ only if a and b in $\mathbb{R}$ are both zero; and that if a, b, c, d and r are all real, then

$$(a + bi) + r(c + di) = (a + rc) + (b + rd)i.$$

Applying the distributive law and the rules of arithmetic in $\mathbb{R}$, we define multiplication in $\mathbb{C}$ by

$$(a + bi)(c + di) = (ac - bd) + (ad + bc)i.$$

From this we can see that multiplication is commutative and associative.

If $z = a + bi$ where a and b are in $\mathbb{R}$, then a is the **real part** of z and b is the **imaginary part** of z. We write $\operatorname{Re} z = a$ and $\operatorname{Im} z = b$. If z and w belong to $\mathbb{C}$, notice that $\operatorname{Re}(z + w) = \operatorname{Re} z + \operatorname{Re} z$ and $\operatorname{Im}(z + w) = \operatorname{Im} z + \operatorname{Im} w$. This is to say that addition in $\mathbb{C}$ is componentwise.

When we write z in $\mathbb{C}$ as either $z = a + bi$ or $z = x + iy$, it is to be understood that $a, b, x,$ and y are real numbers.

Where we use a line to model $\mathbb{R}$, we use a coordinate plane to model $\mathbb{C}$. Except for the names of the elements, $\mathbb{C}$ and $\mathbb{R}^2$ are identical as vector spaces. We capture this by saying they are *isomorphic vector spaces*. It is important to understand, though, that $\mathbb{C}$ and $\mathbb{R}^2$ are different sorts of objects. The complex numbers are closed under multiplication — $\mathbb{C}$ is an example of a *field* — while $\mathbb{R}^2$ is not. The real plane plays different roles in mathematics, and modeling $\mathbb{C}$ as a vector space is just one of them.

When we model $\mathbb{C}$ using a coordinate plane, we identify the vertical axis with the pure imaginary numbers, bi, $b \in \mathbb{R}$, and the horizontal axis with $\mathbb{R}$.

Complex conjugation is the mapping $\mathbb{C} \to \mathbb{C}$ defined by

$$a + bi \mapsto a - bi.$$

The complex conjugate of $z \in \mathbb{C}$ is denoted $\bar{z}$.

We leave proof of the following as an exercise.

Lemma B.2. If z and w belong to $\mathbb{C}$, then

(a) $\bar{\bar{z}} = z$,
(b) $\overline{z+w} = \bar{z} + \bar{w}$,
(c) $z + \bar{z} = 2\operatorname{Re} z$,
(d) $z - \bar{z} = 2i \operatorname{Im} z$,
(e) $\overline{zw} = \bar{z}\,\bar{w}$,
(f) $\bar{z} = z$ if and only if $z \in \mathbb{R}$,
(g) $\operatorname{Re} z = \frac{\bar{z}+z}{2}$, and
(h) $\operatorname{Im} z = \frac{z-\bar{z}}{2i}$.

The **modulus** of $z = a + bi$ is given by
$$|z| := \sqrt{z\bar{z}} = \sqrt{a^2 + b^2}.$$
Since the modulus on $\mathbb{C}$ reduces to the absolute value on $\mathbb{R} \subseteq \mathbb{C}$, we use the same symbol for both. Note also that if $z = a + bi$, then $|z|$ is the distance from the origin to (a, b) in $\mathbb{R}^2$, thus, the distance from 0 to $a + bi$ in $\mathbb{C}$. Certainly
$$|\bar{z}| = |z|.$$
Lemma B.2 implies that for any z, w in $\mathbb{C}$,
$$|zw| = |z||w|.$$
If z is nonzero in $\mathbb{C}$, then
$$\frac{z\bar{z}}{z\bar{z}} = 1$$
so
$$\frac{1}{z} = \frac{\bar{z}}{z\bar{z}} = \frac{\bar{z}}{|z|^2}.$$
The multiplicative inverse of nonzero $z = a + bi \in \mathbb{C}$ is given explicitly by
$$\frac{1}{z} = \frac{a}{a^2 + b^2} - \frac{b}{a^2 + b^2}i.$$
If $z = 1 - i$, for example, then $\bar{z} = 1 + i$, $|z| = \sqrt{2}$, and $1/z = 1/\sqrt{2} + i/\sqrt{2}$.

There are settings in which it is convenient to have complex numbers in exponential form, which itself derives from the polar form of a point in $\mathbb{R}^2$. Recall that if $\operatorname{Pol}(r, \theta)$ is the polar form of (a, b) in $\mathbb{R}^2$, then $a = r\cos\theta$, $b = r\sin\theta$, and
$$r = \sqrt{a^2 + b^2}.$$
This means that if $z = a + bi$ in $\mathbb{C}$, then
$$z = |z|(\cos\theta + i\sin\theta).$$
By Euler's formula,
$$e^{i\theta} = \cos\theta + i\sin\theta$$
for any θ in $\mathbb{R}$. The **exponential form** of z in $\mathbb{C}$ is then
$$z = |z|e^{i\theta}.$$

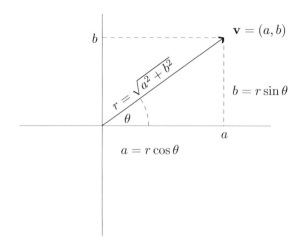

Figure B.5. Polar coordinates for **v** in $\mathbb{R}^2$ come from basic trigonometry.

Example B.3. Consider $z = 1 + i\sqrt{3}$ and $(1, \sqrt{3})$, the corresponding point in $\mathbb{R}^2$. The polar form of $(1, \sqrt{3})$ is $\text{Pol}(2, \pi/3)$ so the exponential form of z is
$$z = 2e^{i\pi/3}.$$

Multiplication and division in $\mathbb{C}$ are simplified when we use exponential form. It is easy to verify, for instance, that
$$|z|e^{i\theta}|w|e^{i\varphi} = |zw|e^{i(\theta+\varphi)}.$$
The algebra of complex exponentials mirrors that of the real-valued exponential function. We leave the verification as an exercise.

Exercises B.3.

B.3.1. Prove Lemma B.2.

B.3.2. Let i be the complex number with $i^2 = -1$. Show that for any $k \in \mathbb{Z}$, i^k is $1, -1, i$, or $-i$. Be sure to address both positive and negative integral powers.

B.3.3. Prove that $|\bar{z}| = |z|$ and that $|zw| = |z||w|$ for all z, w in $\mathbb{C}$.

B.3.4. Find $1/z$ if $z = 2 + 3i$.

B.3.5. Check that if $z = x + iy$ belongs to $\mathbb{C}$, then
$$\bar{z}^2 - i\bar{z} = (x^2 - y^2 - y) - i(x + 2xy).$$

B.3.6. Use Euler's formula to show that $e^{i\theta}e^{i\varphi} = e^{i(\theta+\varphi)}$ for all θ and φ in $\mathbb{R}$.

B.3.7. Write $1 - i\sqrt{2}$ in exponential form.

B.4. Real 3-Space

Real 3-space, $\mathbb{R}^3$, is a model for Euclidean 3-space, an idealization of the 3-dimensional space we experience day to day at the human scale on earth. This is also the

B.4. Real 3-Space

setting for 3-dimensional analytic geometry, which is often treated in a multivariable calculus course. In that context, for instance, we may use coordinates to study conic sections as intersections of a cone and a plane. As in the case of $\mathbb{R}^2$, we can think of $\mathbb{R}^3$ as Euclidean space with the additional notions of direction and distance built into it via a coordinate system.

The standard coordinatization of $\mathbb{R}^3$ is illustrated in Figure B.6.

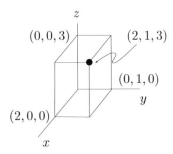

Figure B.6. The standard coordinatization of $\mathbb{R}^3$.

Direction in $\mathbb{R}^3$ is as indicated in the figure. The axes shown are all positive. The positive z-axis goes up. The positive y-axis extends to the right. The positive x-axis comes out of the page towards the viewer. These conventions are standard in mathematics, though other applications may use different conventions.

While $\mathbb{R}^2$ is not a subset of $\mathbb{R}^3$, there are many isomorphic copies of $\mathbb{R}^2$ in $\mathbb{R}^3$. For example, any two coordinate axes in $\mathbb{R}^3$ can play the role of coordinate axes in a copy of $\mathbb{R}^2$ sitting inside $\mathbb{R}^3$. These copies of $\mathbb{R}^2$ are the *coordinate planes* of $\mathbb{R}^3$. The xy-plane contains the x- and y-axes. The xz-plane contains the x- and z-axes. The yz-plane contains the y- and z-axes.

Just as the coordinate axes in $\mathbb{R}^2$ divide the real plane into four quadrants, the coordinate planes in $\mathbb{R}^3$ divide real 3-space into eight *octants*, four on the viewer's side of the page and four behind the page. While there is consensus about numbering the quadrants in $\mathbb{R}^2$, there appears to be no such consensus when it comes to numbering the octants in $\mathbb{R}^3$, with the exception of the first octant, which is framed by the positive halves of the coordinate axes. Algebraically, the first octant can be specified by $x > 0, y > 0, z > 0$. To describe any other octant, we use inequalities. Note for instance that $x < 0, y > 0, z < 0$ describes the octant behind the page on the lower right.

A linear equation in x, y, and z has the form $ax + by + cz = d$, where a, b, c, d are in $\mathbb{R}$ and the coefficients a, b, c are not all zero. If any of the coefficients is zero, we may be tempted to think that the equation describes a line, but in $\mathbb{R}^3$, a linear equation always describes a plane. Consider, for example, $x = 0$. We can view this as a description of all points in $\mathbb{R}^3$ that have the form $(0, y, z)$, where y and z can take any value. This is a description of the yz-plane.

There are a couple of ways to describe lines in $\mathbb{R}^3$. We can use two equations, each representing a plane that contains the line. A parametric description is another approach. In a parametric description of a line, we give each of the three coordinates of any point on the line in terms of a parameter, t. This parameter, or variable, ranges over all of $\mathbb{R}$. Specifically, a parametric description of the line through a point (x_0, y_0, z_0) with direction vector (a, b, c) is

$$x = x_0 + at, \quad y = y_0 + bt, \quad z = z_0 + ct.$$

The same idea can be applied to describe a line parametrically in $\mathbb{R}^2$, or $\mathbb{R}^n$ for $n > 3$, for that matter.

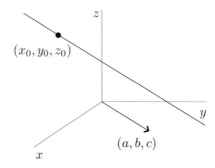

Figure B.7. A parametric description of a line in $\mathbb{R}^3$ depends on a point and a direction vector.

The points, lines, and planes in $\mathbb{R}^3$, along with their incidence relations, comprise the elements of *real affine 3-space*. More generally, affine 3-space is the incidence geometry associated to an arbitrary 3-dimensional vector space. It satisfies the following axioms.

Axioms B.4 (Affine 3-space). Affine 3-space contains points, lines, and planes. Every point lies on a line, every line lies in a plane, and every plane is an affine plane. There exist four noncoplanar points. Two planes that intersect do so in a line. Two nonintersecting planes are *parallel*. Nonintersecting coplanar lines are *parallel*. Nonintersecting noncoplanar lines are *skew*.

The following incidence relations hold.

(1) Two points determine a unique line.

(2) A point and a line not through the point lie in a unique plane.

(3) Given a plane and a point not in the plane, there is a unique plane containing the point and parallel to the given plane.

We leave it as an easy exercise to verify that Axiom B.4(2) can be replaced with Axiom 2′: Three noncollinear points determine a unique plane.

The affine geometry of $\mathbb{R}^3$ arises when we solve a system of linear equations in three unknowns. Each equation represents a plane in $\mathbb{R}^3$ and the solution set is the collection of points in $\mathbb{R}^3$ that lie in all the planes. Two equations in three unknowns represent two planes in 3-space. If one equation is a multiple of the other, the equations represent the same plane. In that case, every point in the plane is a solution to the system. When the planes are distinct, either they are parallel or they intersect in a line. If the planes are parallel, they have no points in common. In this case, the system has no solution. If the planes intersect, then since a line in $\mathbb{R}^3$ has infinitely many points, there are infinitely many solutions to the system.

The affine geometry of $\mathbb{R}^3$ can give us other insights into the algebra of a system of linear equations. For instance, suppose that all we know about a system of two linear equations in three unknowns is that the system has two solutions. Each solution is a point in affine 3-space and two points in affine 3-space determine a unique line. That line must lie in any plane containing those two points. Since a line in real affine 3-space contains infinitely many points, the line determined by the two known solutions represents a collection of infinitely many solutions to the system of equations. In other words, once we know that the system has two solutions, the geometry guarantees that the system must actually have infinitely many solutions.

Exercises B.4.

B.4.1. Suppose you stand up straight on level ground and that the coordinate axes of $\mathbb{R}^3$ run through your body, intersecting at your center of gravity. (Assume your center of gravity is at the height of your belly button and two inches inside your abdomen.) Consider the standard coordinatization of $\mathbb{R}^3$ with the axes labeled for use by someone who is facing you. Your right arm and fingers extend parallel to the ground to your right. Your left arm and fingers extend parallel to the ground directly in front of you: Your arms are at right angles to one another and to your torso, at shoulder height.
 (a) Which coordinate axis does your right arm run parallel to?
 (b) Would that coordinate axis actually run along your arm? Why or why not?
 (c) Are the fingers of your right hand pointing in the positive or the negative direction of the associated coordinate axis?
 (d) Answer the three questions above, this time about your left arm.
 (e) With your nose pointing in the direction of your left arm, in which octant is your right eye?
 (f) In which octant is your left big toe?

B.4.2. We have seen that $x = 0$ is an equation describing the yz-plane in $\mathbb{R}^3$. Describe the other two coordinate planes using equations.

B.4.3. Each coordinate axis in $\mathbb{R}^3$ is the intersection of two coordinate planes.
 (a) Which two coordinate planes intersect in the x-axis? Specify them using equations.
 (b) What are the equations for the two coordinate planes that intersect in the z-axis?

B.4.4. Using Axioms B.4, show that three noncollinear points determine a unique plane in affine 3-space.

B.4.5. Assume everything in Axioms B.4 except Axiom 2 and, instead, assume Axiom 2'. Show that for a point and a line not through the point, there is a unique plane containing the point and the line.

B.4.6. Consider the planes in $\mathbb{R}^3$ determined by the equations $2x - y + z = 1$ and $x + y - z = 0$.
   (a) Find a point of intersection of the two planes.
   (b) Find a parametric description of the line of intersection of the two planes.

B.4.7. Find an equation for a plane in $\mathbb{R}^3$ parallel to the plane given by
$$x + y - z = 1.$$

B.4.8. Find a, b, c, d so that $(1, -3, -3)$, $(0, -1, 1)$, and $(1, 1, 0)$ lie in the plane given by $ax + by + cz = d$ in $\mathbb{R}^3$.

B.4.9. Give a parametric description of the line in $\mathbb{R}^2$ given by $2x + 3y = 1$.

B.5. The Dot Product

The dot product allows us to define length and angle algebraically in $\mathbb{R}^n$. This gives us better access to Euclidean geometry, something that distinguishes $\mathbb{R}^n$ among vector spaces.

Definition B.5. Given $\mathbf{v} = (a_1, \ldots, a_n)$ and $\mathbf{w} = (b_1, \ldots, b_n)$ in $\mathbb{R}^n$, the **dot product** $\mathbf{v} \cdot \mathbf{w}$ is
$$\mathbf{v} \cdot \mathbf{w} = a_1 b_1 + \cdots + a_n b_n = \sum_{i=1}^n a_i b_i.$$

The dot product on $\mathbb{R}^n$ goes by several names, among them the **scalar product**, **inner product**, and **Euclidean inner product**. Notice that the dot product is not a binary operation, in general, since it maps into $\mathbb{R}$, not into $\mathbb{R}^n$.

Algebraic properties of the dot product follow. These are all consequences of the properties of arithmetic in $\mathbb{R}$. We leave details of the proofs as an exercise.

Lemma B.6. If $\mathbf{u}, \mathbf{v}, \mathbf{w}$ are vectors in $\mathbb{R}^n$ and $c \in \mathbb{R}$, then
   (a) $\mathbf{v} \cdot \mathbf{w} = \mathbf{w} \cdot \mathbf{v}$;
   (b) $\mathbf{v} \cdot \mathbf{v} \geq 0$ and $\mathbf{v} \cdot \mathbf{v} = 0$ if and only if $\mathbf{v} = (0, \ldots, 0)$;
   (c) $c(\mathbf{v} \cdot \mathbf{w}) = (c\mathbf{v}) \cdot \mathbf{w}$; and
   (d) $(\mathbf{u} + \mathbf{v}) \cdot \mathbf{w} = \mathbf{u} \cdot \mathbf{w} + \mathbf{v} \cdot \mathbf{w}$.

Property (a) of Lemma B.6 says that the dot product is *symmetric*. Property (b) says that the dot product is *positive definite*. Properties (c) and (d) say that the dot product is linear in its first variable. Property (a) then implies that it is also linear in its second variable. Since it is linear in each of its two variables, the dot product is *bilinear*.

The dot product of a vector in $\mathbb{R}^n$ with itself is the sum of the squares of its coordinates. Since the sum of squares of real numbers is always nonnegative, $\sqrt{\mathbf{v} \cdot \mathbf{v}}$ is defined in $\mathbb{R}$ for all $\mathbf{v} \in \mathbb{R}^n$. In particular, the sum of the squares of the

B.5. The Dot Product

coordinates of a point $\mathbf{v}$ in $\mathbb{R}^n$ is the square of the distance from the origin to $\mathbf{v}$. This is a consequence of the Pythagorean Theorem.

Definition B.7. The **length** of $\mathbf{v} \in \mathbb{R}^n$ is $\|\mathbf{v}\| = \sqrt{\mathbf{v} \cdot \mathbf{v}}$. The **distance** between $\mathbf{v}$ and $\mathbf{w}$ in $\mathbb{R}^n$ is $\|\mathbf{v} - \mathbf{w}\|$.

Like the modulus of a complex number, the length of a vector in $\mathbb{R}^n$ is an extension of the absolute value of an element in $\mathbb{R} = \mathbb{R}^1$: $|a| = \sqrt{a^2}$ for all $a \in \mathbb{R}$.

Proof of the following is an easy exercise.

Lemma B.8. *If $\mathbf{v}$ is in $\mathbb{R}^n$ and $c \in \mathbb{R}$, then $\|c\mathbf{v}\| = |c|\,\|\mathbf{v}\|$.*

The connection between the dot product of two vectors and the angle between those vectors in $\mathbb{R}^n$ is given by

$$\mathbf{v} \cdot \mathbf{w} = \|\mathbf{v}\|\|\mathbf{w}\| \cos \varphi. \tag{B.1}$$

We leave it as an exercise to verify (B.1) for $\mathbf{v}$ and $\mathbf{w}$ in $\mathbb{R}^2$.

Notice that nonzero vectors $\mathbf{v}$ and $\mathbf{w}$ in $\mathbb{R}^n$ are **perpendicular** or **orthogonal** if and only if $\mathbf{v} \cdot \mathbf{w} = 0$.

The dot product on $\mathbb{R}^n$ allows us to define the component of one vector in the direction of another vector. This idea comes up frequently in precalculus and high school physics.

Definition B.9. Let $\mathbf{v}$ be any vector in $\mathbb{R}^n$ and let $\mathbf{w}$ in $\mathbb{R}^n$ be nonzero. The **component of v in the direction of w**, or the **projection of v onto w**, is

$$\operatorname{proj}_{\mathbf{w}}(\mathbf{v}) := \frac{\mathbf{v} \cdot \mathbf{w}}{\|\mathbf{w}\|^2} \mathbf{w}.$$

Figure B.8 illustrates relationships among vectors $\mathbf{v}$, $\mathbf{w}$, and $\operatorname{proj}_{\mathbf{w}}(\mathbf{v})$ in $\mathbb{R}^n$.

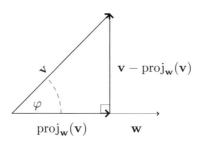

Figure B.8. The projection of $\mathbf{v}$ onto $\mathbf{w}$ in $\mathbb{R}^n$.

If $\mathbf{w}$ is nonzero in $\mathbb{R}^n$, then $(1/\|\mathbf{w}\|)\mathbf{w}$ is a **unit vector**, that is, its length is one unit. Scaling that unit vector by the factor $\frac{\mathbf{v} \cdot \mathbf{w}}{\|\mathbf{w}\|}$ gets us $\operatorname{proj}_{\mathbf{w}}(\mathbf{v})$. Moreover, $\mathbf{v} - \operatorname{proj}_{\mathbf{w}}(\mathbf{v})$ is perpendicular to $\mathbf{w}$. This means that

$$(\mathbf{v} - \operatorname{proj}_{\mathbf{w}}(\mathbf{v})) \cdot \mathbf{w} = 0.$$

We leave it as an exercise to check this algebraically.

Exercises B.5.

B.5.1. Prove Lemma B.6.

B.5.2. Prove that the formula given in (B.1) is true for **v** and **w** in $\mathbb{R}^2$.

B.5.3. Prove Lemma B.8.

B.5.4. Consider a block sitting on an inclined plane, as in Figure B.9. If the block does not slide, it stays in place because the force due to the friction of the inclined plane cancels the component of the force due to gravity on the block in the direction of the incline. Label the figure to illustrate this and explain how the notion of the projection of one vector onto another comes into the analysis.

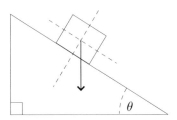

Figure B.9. The force due to gravity on the block is indicated by the arrow.

B.5.5. Use properties of the dot product to show that if **v** and **w** are in $\mathbb{R}^n$, then the component of **v** orthogonal to **w** has length $\|\mathbf{v}\| \sin \varphi$, where
$$\varphi = \cos^{-1}\left(\frac{\mathbf{v} \cdot \mathbf{w}}{\|\mathbf{v}\|\|\mathbf{w}\|}\right).$$

B.6. The Cross-Product

While the dot product is defined on $\mathbb{R}^n$ for all $n \in \mathbb{Z}_+$, the cross-product is a binary operation peculiar to $\mathbb{R}^3$. There are situations in which we apply it to $\mathbb{R}^2$ by identifying $\mathbb{R}^2$ with one of the coordinate planes in $\mathbb{R}^3$. We see an example of this below.

Definition B.10. The **cross-product** of $\mathbf{v} = (a_1, a_2, a_3)$ and $\mathbf{w} = (b_1, b_2, b_3)$ in $\mathbb{R}^3$ is
$$\mathbf{v} \times \mathbf{w} = (a_2 b_3 - a_3 b_2, -(a_1 b_3 - a_3 b_1), a_1 b_2 - a_2 b_1)$$
$$= (a_2 b_3 - a_3 b_2, a_3 b_1 - a_1 b_3, a_1 b_2 - a_2 b_1).$$

There is a mnemonic for remembering the cross-product. It uses the determinant of a 3×3 matrix and the **unit coordinate vectors** in $\mathbb{R}^3$, $\mathbf{i} = (1, 0, 0)$, $\mathbf{j} = (0, 1, 0)$, and $\mathbf{k} = (0, 0, 1)$:
$$\mathbf{v} \times \mathbf{w} = \begin{vmatrix} \mathbf{i} & \mathbf{j} & \mathbf{k} \\ a_1 & a_2 & a_3 \\ b_1 & b_2 & b_3 \end{vmatrix}$$
$$= (a_2 b_3 - a_3 b_2)\mathbf{i} - (a_1 b_3 - a_3 b_1)\mathbf{j} + (a_1 b_2 - a_2 b_1)\mathbf{k}.$$

B.6. The Cross-Product

Our first lemma concerning the cross-product is a list of its algebraic properties. The reader should verify these, either appealing to properties of the determinant or using coordinates and letting the fur fly.

The first property says that the cross-product is *anticommutative*. The last property says that the cross-product obeys the *Jacobi identity*. Verification of the Jacobi identity is laborious but straightforward.

Lemma B.11. The following properties hold for all $\mathbf{v}, \mathbf{w}, \mathbf{u}$ in $\mathbb{R}^3$ and $c \in \mathbb{R}$.

(a) $\mathbf{v} \times \mathbf{w} = -\mathbf{w} \times \mathbf{v} = -(\mathbf{w} \times \mathbf{v})$.

(b) $(\mathbf{v} + \mathbf{w}) \times \mathbf{u} = \mathbf{v} \times \mathbf{u} + \mathbf{w} \times \mathbf{u}$.

(c) $\mathbf{v} \times (\mathbf{w} + \mathbf{u}) = \mathbf{v} \times \mathbf{w} + \mathbf{v} \times \mathbf{u}$.

(d) $c\mathbf{v} \times \mathbf{w} = c(\mathbf{v} \times \mathbf{w}) = \mathbf{v} \times c\mathbf{w}$.

(e) (Jacobi identity) $\mathbf{v} \times (\mathbf{w} \times \mathbf{u}) = (\mathbf{v} \times \mathbf{w}) \times \mathbf{u} + \mathbf{w} \times (\mathbf{v} \times \mathbf{u})$.

The Jacobi identity mirrors the product rule for differentiation. In other words, if we think of applying an operator $\mathbf{v} \times$ to the cross-product of two vectors the way we think of applying d/dx to the composition of two functions, we get the rule in Lemma B.11(e).

Consider $\mathbf{v}$ and $\mathbf{w}$ as arrows emanating from the same point in $\mathbb{R}^3$. Unless one is a scalar multiple of the other, the two vectors determine a plane in $\mathbb{R}^3$. It is an easy exercise to show that $\mathbf{v} \times \mathbf{w}$ is orthogonal to this plane. We can find its specific direction via the *right-hand rule*. Position the pinky-finger side of the right hand in the plane determined by $\mathbf{v}$ and $\mathbf{w}$, so that $\mathbf{v}$ extends from the wrist towards the end of your pinky. If necessary, twist your hand over so that you can curl your fingers towards $\mathbf{w}$. The thumb then points in the direction of $\mathbf{v} \times \mathbf{w}$.

We have seen the geometric interpretation of $\mathbf{v} \cdot \mathbf{w}$:

$$\mathbf{v} \cdot \mathbf{w} = \|\mathbf{v}\| \|\mathbf{w}\| \cos \varphi,$$

where φ is the angle formed by $\mathbf{v}$ and $\mathbf{w}$. What is the geometric interpretation of the cross-product?

Lemma B.12. If $\mathbf{v}$ and $\mathbf{w}$ belong to $\mathbb{R}^3$, then $\|\mathbf{v} \times \mathbf{w}\|$ is the area of the parallelogram determined by $\mathbf{v}$ and $\mathbf{w}$. Explicitly, $\|\mathbf{v} \times \mathbf{w}\| = \|\mathbf{v}\| \|\mathbf{w}\| \sin \varphi$, where φ is the angle in $(0, \pi)$ formed by $\mathbf{v}$ and $\mathbf{w}$.

Proof. Figure B.10 is an illustration to show that the quadrilateral determined by $\mathbf{v}$ and $\mathbf{w}$ has the same area as that of the rectangle in the same picture: $\|\mathbf{v}\| \|\mathbf{w}\| \sin \varphi$. To see that this is equal to $\|\mathbf{v} \times \mathbf{w}\|$, let $\mathbf{v} = (a_1, a_2, a_3)$ and $\mathbf{w} = (b_1, b_2, b_3)$. We have

(B.2) $$\|\mathbf{v} \times \mathbf{w}\|^2 = (a_2 b_3 - a_3 b_2)^2 + (-a_1 b_3 + a_3 b_1)^2 + (a_1 b_2 - a_2 b_1)^2.$$

On the other hand,

(B.3) $$(\|\mathbf{v}\| \|\mathbf{w}\| \sin \varphi)^2 = \|\mathbf{v}\|^2 \|\mathbf{w}\|^2 (1 - \cos^2 \varphi) = \|\mathbf{v}\|^2 \|\mathbf{w}\|^2 - (\mathbf{v} \cdot \mathbf{w})^2.$$

The proof follows once we expand and compare the expressions in (B.2) and (B.3). We leave the details to the reader. □

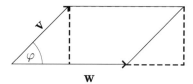

Figure B.10. The areas of the quadrilateral and the rectangle are identical.

When **v** and **w** belong to $\mathbb{R}^3$, the dot product of $\mathbf{v} \times \mathbf{w}$ with any other vector in $\mathbb{R}^3$ is called the *scalar triple product*.

A *parallelepiped* (pronounced with six syllables) is the region determined by three vectors in $\mathbb{R}^3$. You can picture it as a leaning box. (See Figure B.11.)

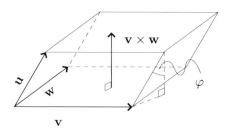

Figure B.11. If **v** and **w** determine the base of the parallelepiped given by $\mathbf{v}, \mathbf{w}, \mathbf{u}$, then the height of the parallelepiped is $\|\mathbf{u}\| |\cos \varphi|$, where φ is the angle between $\mathbf{v} \times \mathbf{w}$ and $\mathbf{u}$.

Lemma B.13. *If* $\mathbf{v}, \mathbf{w}, \mathbf{u}$ *all belong to* $\mathbb{R}^3$, *then* $|(\mathbf{v} \times \mathbf{w}) \cdot \mathbf{u}|$ *is the volume of the paralellpiped determined by* $\mathbf{v}, \mathbf{w}$, *and* $\mathbf{u}$. *Moreover, if* $\mathbf{v} = (a_1, a_2, a_3)$, $\mathbf{w} = (b_1, b_2, b_3)$, *and* $\mathbf{u} = (c_1, c_2, c_3)$, *then*

$$(\mathbf{v} \times \mathbf{w}) \cdot \mathbf{u} = \begin{vmatrix} a_1 & a_2 & a_3 \\ b_1 & b_2 & b_3 \\ c_1 & c_2 & c_3 \end{vmatrix}.$$

Proof. The volume of a parallelepiped is the product of its height and the area of its base. Consider the parallelepiped determined by $\mathbf{v}, \mathbf{w}, \mathbf{u}$ as in Figure B.11. We can take the base to be the region bounded by **v** and **w**. The height is then $\|\mathbf{u}\| |\cos \varphi|$, where φ is the angle between the perpendicular to the base. It follows that the volume is $|(\mathbf{v} \times \mathbf{w}) \cdot \mathbf{u}|$.

This proves the first statement of the lemma. Proof of the second statement is a calculation that we leave as an exercise. □

B.6. The Cross-Product

Example B.14. We can find the area of the parallelogram determined by the vectors $(1,2)$, $(-3,5)$ in $\mathbb{R}^2$ by finding the volume of the parallelepiped with base equal to the parallelogram and height 1 unit. We can find that by taking the scalar triple product of $\mathbf{v} = (1,2,0)$, $\mathbf{w} = (-3,5,0)$, and $\mathbf{u} = (0,0,1)$ in $\mathbb{R}^3$:

$$\begin{vmatrix} 1 & 2 & 0 \\ -3 & 5 & 0 \\ 0 & 0 & 1 \end{vmatrix} = 5 + 6 = 11.$$

The unit coordinate vectors $\{\mathbf{i}, \mathbf{j}, \mathbf{k}\}$, in that order, form a *right-handed coordinate system* for $\mathbb{R}^3$. Many applications in mathematics and physics can be approached efficiently through the construction of an appropriate right-handed coordinate system. To be able to construct a right-handed coordinate system for $\mathbb{R}^3$, we must specify what it is about the ordered set $\mathcal{E} = \{\mathbf{i}, \mathbf{j}, \mathbf{k}\}$ that makes it useful.

Definition B.15. An ordered set $\mathcal{B} = \{\mathbf{u}_1, \mathbf{u}_2, \mathbf{u}_3\}$ is a **right-handed coordinate system** for $\mathbb{R}^3$ provided

(a) $\mathbf{u}_i$ is a unit vector for $i = 1, 2, 3$;

(b) $\mathbf{u}_1 \times \mathbf{u}_2 = \mathbf{u}_3$, $\mathbf{u}_2 \times \mathbf{u}_3 = \mathbf{u}_1$, $\mathbf{u}_3 \times \mathbf{u}_1 = \mathbf{u}_2$; and

(c) $\mathbf{u}_i \times \mathbf{u}_j = -\mathbf{u}_j \times \mathbf{u}_i$.

Figure B.12 is a mnemonic for remembering the cross-products of elements in a right-handed coordinate system for $\mathbb{R}^3$. Proceeding around the circle counterclockwise, the cross-product of successive elements is equal to the following element in the circle.

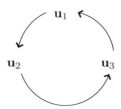

Figure B.12. Cross-products in a right-handed coordinate system for $\mathbb{R}^3$.

Usually when we need a right-handed coordinate system for $\mathbb{R}^3$, it must include two vectors lying in a particular plane through the origin. The third vector is then perpendicular to that plane. In this scenario, we can start with any nonzero vector $\mathbf{v}$ in our plane and then *normalize* $\mathbf{v}$ to get a unit vector in the same direction,

$$\mathbf{u}_1 = \frac{1}{\|\mathbf{v}\|}\mathbf{v}.$$

In a given plane, there will be a line perpendicular to the line determined by $\mathbf{u}_1$. We then have a choice of two unit vectors along that line: We take the one to the left of $\mathbf{u}_1$, that is, the one we come across first when we move counterclockwise from $\mathbf{u}_1$ in the plane. For the third vector, we take $\mathbf{u}_3 = \mathbf{u}_1 \times \mathbf{u}_2$.

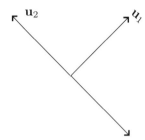

Figure B.13. To construct a right-handed coordinate system, start by choosing $\mathbf{u}_2$ to the left of $\mathbf{u}_1$.

Since $\|\mathbf{u}_1\| = \|\mathbf{u}_2\| = 1$ and the angle between $\mathbf{u}_1$ and $\mathbf{u}_2$ is $\pi/2$, we see that $\|\mathbf{u}_3\| = 1$. Putting all this together with the fact that $\mathbf{u}_1 \cdot \mathbf{u}_2 = 0$, we can churn through coordinates to determine that the ordered set $\{\mathbf{u}_1, \mathbf{u}_2, \mathbf{u}_3\}$ really is a right-handed coordinate system. For instance, if we take $\mathbf{u}_1 = (a_1, a_2, a_3)$ and $\mathbf{u}_2 = (b_1, b_2, b_3)$, then when we expand the coefficient of $\mathbf{i}$ in the expression

$$(\text{B.4}) \qquad \mathbf{u}_2 \times \mathbf{u}_3 = \begin{vmatrix} \mathbf{i} & \mathbf{j} & \mathbf{k} \\ b_1 & b_2 & b_3 \\ a_2 b_3 - a_3 b_2 & -a_1 b_3 + a_3 b_1 & a_1 b_2 - a_2 b_1 \end{vmatrix},$$

we get

$$a_1 b_2^2 - a_2 b_1 b_2 + a_1 b_3^2 - a_3 b_1 b_3 = a_1(b_2^2 + b_3^2) - b_1(a_2 b_2 + a_3 b_3)$$
$$= a_1(b_2^2 + b_3^2) - b_1(-a_1 b_1)$$
$$= a_1(b_1^2 + b_2^2 + b_3^2) = a_1.$$

If you prefer to think about it spatially, you can convince yourself that in the plane determined by $\mathbf{u}_2$ and $\mathbf{u}_3$, $\mathbf{u}_3$ is to the left of $\mathbf{u}_2$, and that $\mathbf{u}_2 \times \mathbf{u}_3$ must point in the direction of $\mathbf{u}_1$.

Example B.16. The set $\alpha = \{(x, y, z) \mid x - y + z = 0\}$ is a plane in $\mathbb{R}^3$ that passes through the origin. We would like to construct a right-handed coordinate system $\{\mathbf{u}_1, \mathbf{u}_2, \mathbf{u}_3\}$ such that $\mathbf{u}_1$ and $\mathbf{u}_2$ lie in α. We can start by normalizing any nonzero vector in α so we take $\mathbf{v} = (1, 1, 0)$ and then normalize to get

$$\mathbf{u}_1 = \left(1/\sqrt{2}, 1/\sqrt{2}, 0\right).$$

For $\mathbf{w}$ perpendicular to $\mathbf{v}$ in α we could take $\mathbf{w} = (-1, 1, 2)$ or $(1, -1, -2)$. The first choice is to the left of $\mathbf{u}_1$ so we take

$$\mathbf{u}_2 = \left(-1/\sqrt{6}, 1/\sqrt{6}, 2/\sqrt{6}\right).$$

Calculating $\mathbf{u}_1 \times \mathbf{u}_2$, we get

$$\mathbf{u}_3 = \left(1/\sqrt{3}, -1/\sqrt{3}, 1/\sqrt{3}\right).$$

It is easy enough to check here that $\mathbf{u}_2 \times \mathbf{u}_3 = \mathbf{u}_1$ and that $\mathbf{u}_3 \times \mathbf{u}_1 = \mathbf{u}_2$.

Right-handed coordinate systems come up when we study rotations.

Exercises B.6.

B.6.1. Prove Lemma B.11.

B.6.2. Is the cross-product bilinear?

B.6.3. Find vectors $\mathbf{u}, \mathbf{v}, \mathbf{w} \in \mathbb{R}^3$ so that $(\mathbf{u} \times \mathbf{v}) \times \mathbf{w} \neq \mathbf{u} \times (\mathbf{v} \times \mathbf{w})$.

B.6.4. Prove that $\mathbf{v} \times \mathbf{w}$ is orthogonal to both $\mathbf{v}$ and $\mathbf{w}$, for any $\mathbf{v}, \mathbf{w} \in \mathbb{R}^3$.

B.6.5. Prove that the expressions in (B.2) and (B.3) are identical.

B.6.6. Complete the proof of Lemma B.13.

B.6.7. Find the area of the parallelogram in Example B.14 without using $\mathbb{R}^3$ and the scalar triple product.

B.6.8. Verify that the ordered set $\mathcal{E} = \{\mathbf{i}, \mathbf{j}, \mathbf{k}\}$ is a right-handed coordinate system for $\mathbb{R}^3$.

B.6.9. Let $\mathcal{B} = \{\mathbf{u}_1, \mathbf{u}_2, \mathbf{u}_3\}$ be a right-handed coordinate system for $\mathbb{R}^3$. Show that $\{\mathbf{u}_2, \mathbf{u}_3, \mathbf{u}_1\}$ and $\{\mathbf{u}_3, \mathbf{u}_1, \mathbf{u}_2\}$ are also right-handed coordinate systems for $\mathbb{R}^3$. What about $\{\mathbf{u}_2, \mathbf{u}_1, \mathbf{u}_3\}$ and $\{\mathbf{u}_3, \mathbf{u}_2, \mathbf{u}_1\}$?

B.6.10. Calculate the coefficients of $\mathbf{j}$ and $\mathbf{k}$ in (B.4) to verify that they are, respectively, a_2 and a_3.

B.6.11. The set of points $\alpha = \{(x, y, z) \mid x + y + z = 0\}$ is a plane in $\mathbb{R}^3$ that passes through the origin. Construct a right-handed coordinate system for $\mathbb{R}^3$, $\{\mathbf{u}_1, \mathbf{u}_2, \mathbf{u}_3\}$, such that $\mathbf{u}_1$ and $\mathbf{u}_2$ belong to α. Check the cross-products to verify your calculations.

Appendix C

More Set Theory

The primary goals in this appendix are to explicate Zorn's Lemma and the notion of cardinality.

C.1. Partially Ordered Sets

There are occasions in any study of linear algebra in which questions about extremely large sets arise. In this section and the next, we develop tools that can help us understand these questions and how to approach answering some of them. Much of this involves slightly more advanced set theory.

Definition C.1. A **partial ordering** on a set A is a relation on A that is transitive and antisymmetric. A **partially ordered set** or **poset**, $(A, \leq)$, is a set A, with a partial ordering $\leq$. Elements a, b in a poset $(A, \leq)$ are **comparable** provided (a, b) or (b, a) is in $\leq$. When (a, b) is in $\leq$, we write $a \leq b$.

A **greatest element** in a poset $(A, \leq)$ is an element $m \in A$ where $a \leq m$ for all $a \in A$. A **maximal element** in A is $m \in A$ so that if, for some $a \in A$, $m \leq a$, then $m = a$.

A **total ordering** on a set A is a partial ordering $\leq$ on A such that any two elements $a, b \in A$ are comparable. A **totally ordered set** $(A, \leq)$ is a set A with a total ordering $\leq$.

A **chain** is a subset of a poset $(A, \leq)$ that is itself a totally ordered set under $\leq$. An **upper bound** for a chain C in a poset $(A, \leq)$ is an element $m \in A$ such that $c \leq m$ for all $c \in C$.

When $\leq$ is a partial ordering on a set, A, we frequently refer to A itself as a poset or a poset under $\leq$. It is common practice, when $a \leq b$, also to write $b \geq a$. Likewise, one may write $a < b$ or $b > a$, if $a \leq b$ and $a \neq b$.

The terms *upper bound*, *greatest element*, and *maximal element* all mean different things and they all have analogs in the other direction: lower bound, least element, minimal element. Notice for instance that a greatest element must be comparable to all other elements in a poset, whereas a maximal element only has to be greater than or equal to any other element with which it is comparable.

Upper and lower bounds for a chain, when they exist, need not be in the chain itself. Generally, they are in a larger poset from which the chain is drawn. Consider $\mathbb{R}$, for example, which is totally ordered under $\leq$ defined by

$$a \leq b \iff 0 \leq b - a.$$

The interval $(0,1) \subseteq \mathbb{R}$ has many upper bounds in $\mathbb{R}$, among them 1. The interval itself has neither a greatest element nor a maximal element. Likewise, $(0,1)$ has lower bound 0, but neither a least nor a minimal element. The half-open interval $(0,1]$ has the same upper bounds as $(0,1)$, but it does have a greatest element and a maximal element, viz., 1.

Example C.2. Let $A = \{\{1,2,3\},\{1,3,4\},\{1,3\},\{2\},\{3\},\emptyset\}$. Under set inclusion, $\subseteq$, A is a poset. Note for instance that $\{1,3\} \subseteq \{1,2,3\}$ while $\{1,2,3\}$ and $\{1,3,4\}$ are not comparable.

There is no element in A for which every other element in A is a subset so A does not have a greatest element. Since neither $\{1,2,3\}$ nor $\{1,3,4\}$ is a proper subset of any element in A, both of these are maximal elements in A. Since $\emptyset$ is a subset of each element in A, $\emptyset$ is both a least element and a minimal element in A.

Let $C = \{\{1,3\},\{3\},\emptyset\}$. Since any two elements in C are comparable, C is a chain in A. It has greatest and maximal element $\{1,3\}$, least and minimal element $\emptyset$.

The upper bounds for C in A are $\{1,3\}$, $\{1,2,3\}$, and $\{1,3,4\}$. It only has one lower bound, namely $\emptyset$.

Our first result is an easy application of the definitions.

Lemma C.3. Let $(A, \leq)$ be a poset. Suppose $S \subseteq A$. If we restrict $\leq$ to S, then $(S, \leq)$ is a poset. If $(A, \leq)$ is a totally ordered set, then $(S, \leq)$ is a totally ordered set.

Proof. Suppose $(A, \leq)$ and S are as hypothesized. If $a \in S$, then $a \in A$ so $a \leq a$. This shows that $\leq$ is reflexive on S. Suppose a, b, c are in S where $a \leq b$ and $b \leq c$. Since a, b, c are in A, $a \leq c$. This shows that $\leq$ is transitive on S. Suppose $a, b \in S$ and that $a \leq b$ and $b \leq a$. Since $a, b \in A$, $a = b$. This completes the proof that $(S, \leq)$ is a poset.

If $(A, \leq)$ is a totally ordered set and a, b are in S, then since a, b are in A, $a \leq b$ or $b \leq a$. This proves that any two elements in S are comparable, thus, that $(S, \leq)$ is a totally ordered set. □

Since $\mathbb{R}$ is a totally ordered set under $\leq$, so are its subsets. We can use induction to prove the following important fact about $\mathbb{Z}_+$.

Theorem C.4 (The Well-Ordering Principle). *Every nonempty set of positive integers has a least element.*

Proof. Let $A \subseteq \mathbb{Z}_+$ and assume that A does not have least element. Consider $B = \mathbb{Z}_+ \setminus A$. If 1 were in A, it would be the least element in A so 1 must belong to

C.2. Zorn's Lemma

B. Assume that $1, \ldots, k$ belong to B. If $k+1$ is in A, then it would be the least element in A, so $k+1$ must belong to B, as well. By the principle of induction, $\mathbb{Z}_+$ is contained in B, implying that A is empty. We conclude that any nonempty subset of $\mathbb{Z}_+$ must have a least element. $\square$

Definition C.5. The **power set** of a set A, $\mathcal{P}(A)$, is the set of all subsets of A.

We leave it as an easy exercise to show that there is no set A such that $\mathcal{P}(A)$ is empty.

If A is any set, then $\mathcal{P}(A)$ is a poset under set inclusion. We leave the verification of this as an exercise.

Example C.6. Consider $A = \{a, b, c\}$. We have
$$\mathcal{P}(A) = \{\emptyset, \{a\}, \{b\}, \{c\}, \{a,b\}, \{a,c\}, \{b,c\}, \{a,b,c\}\}.$$
Certainly $\emptyset$ is a least and minimal element in $\mathcal{P}(A)$, while $\{a, b, c\}$ is a greatest and maximal element.

Let $B = \mathcal{P}(A) \setminus \{\emptyset, \{a, b, c\}\}$. B is a poset under set inclusion and it has neither a least nor a greatest element. Notice that $\{a\}$, $\{b\}$, $\{c\}$ are all minimal elements of B and that $\{a, b\}$, $\{a, c\}$, $\{b, c\}$ are all maximal elements of B.

The following is a chain in $\mathcal{P}(A)$:
$$C = \{\emptyset, \{a\}, \{a,b\}, \{a,b,c\}\}.$$
In particular, $\emptyset \subseteq \{a\} \subseteq \{a, b\} \subseteq \{a, b, c\}$. We leave it as an exercise to find a few of the other chains in $\mathcal{P}(A)$.

Exercises C.1.

C.1.1. Show that a greatest element in a poset must also be a maximal element, as well as an upper bound.

C.1.2. Show that $\subseteq$ is a partial ordering on $\mathcal{P}(A)$, for any set A.

C.1.3. Prove that if $A \subseteq B$, then $\mathcal{P}(A) \subseteq \mathcal{P}(B)$. Use this fact to show that $\mathcal{P}(A)$ is nonempty for any set A.

C.1.4. Find $\mathcal{P}(\emptyset)$, $\mathcal{P}(\{1\})$, $\mathcal{P}(\{1,2\})$, and $\mathcal{P}(\{1,2,3\})$.

C.1.5. Find all four-element chains in $\mathcal{P}(A)$ from Example C.6.

C.2. Zorn's Lemma

We need Zorn's Lemma for our proofs that every vector space has a basis and that the cardinality of a basis is well-defined. The latter has its own appendix, Appendix D.

If I is an index set for a collection of nonempty sets, A_i, we denote the Cartesian product of the sets $A_i, i \in I$, by
$$\prod_{i \in I} A_i.$$
When $I = \{1, \ldots, n\}$ we may write
$$\prod_{i=1}^{n} A_i = A_1 \times \cdots \times A_n.$$

We know that
$$A_1 \times A_2 \times A_n = \{(a_1, \ldots, a_n) \mid a_i \in A_i\}.$$
What if I itself is more interesting? What if it is uncountable?

Definition C.7. Given an index set I for a collection of nonempty sets A_i, a **choice function** is a mapping
$$I \to \bigcup_{i \in I} A_i.$$
The **Cartesian product** $\prod_{i \in I} A_i$ is the set of all choice functions $I \to \bigcup_{i \in I} A_i$.

A familiar example helps us to understand a choice function. Consider the Cartesian product $\mathbb{R} \times \mathbb{R}$. We can take the index set to be $I = \{1, 2\}$. Any element $(a, b) \in \mathbb{R} \times \mathbb{R}$ is then the choice function $1 \mapsto a$, $2 \mapsto b$.

There is no proof that the Cartesian product of nonempty sets indexed by an uncountable set such as $\mathbb{R}$ exists. It is an axiom that we assume.

Axiom C.8 (Axiom of Choice). If I is an index set for a collection of nonempty sets, A_i, then the Cartesian product $\prod_{i \in I} A_i$ is nonempty.

The Axiom of Choice is one of the pillars of set theory that most mathematicians rely upon for their work, implicitly or explicitly.

Example C.9. Let $A_i = \{1\}$ for each $i \in \mathbb{R}$. The Cartesian product $\prod_{i \in \mathbb{R}} A_i$ is just the constant mapping, $f : \mathbb{R} \to \{1\}$.

Let $I = \mathcal{P}(\mathbb{R})$, the power set of the reals. For each $i \in I$, let $A_i = \mathbb{Z}$. A single element in $\prod_{i \in I} A_i$ assigns an integer to each member of $\mathcal{P}(\mathbb{R})$. The index set makes this difficult to think about. The Axiom of Choice is the assumption that there are choice functions $I \to \mathbb{Z}$.

The Axiom of Choice has several well-known logical equivalents. The one we rely on explicitly is *Zorn's Lemma*. As an equivalent to the Axiom of Choice, Zorn's Lemma holds if and only if the Axiom of Choice holds, under the axioms of set theory. While we present it as a lemma, we go forward treating Zorn's Lemma as an axiom. (See [9] for a proof that the Axiom of Choice implies Zorn's Lemma.)

Lemma C.10 (Zorn). If A is a partially ordered set for which every chain has an upper bound, then A contains a maximal element.

Let A be the set of open intervals in $\mathbb{R}$ that contain 0. Elements of A are subsets of $\mathbb{R}$. As a subset of $\mathcal{P}(\mathbb{R})$, A is partially ordered by set inclusion. The following subset of A is a chain: $\{(-1, 1), (-1/2, 1/2), (-1/4, 1/4), \ldots\}$. This chain has many upper bounds in A, among them $(-2, 1)$. The following is another chain in A: $\{(-1, 1), (-2, 2), (-3, 3), \ldots\}$. Since this chain does not have an upper bound in A, Zorn's Lemma does not apply to A.

Next is our first application of Zorn's Lemma. The proof we include here is based on [14] but the result can be taken on faith.

Theorem C.11. If A and B are nonempty sets, there is an injection from A to B or an injection from B to A.

Proof. Given nonempty sets A and B, let $\mathbf{F}$ be the collection of pairs (S, f) where $S \subseteq A$ and f is an injection from S into B. If we take $S = \{a\}$, for some $a \in A$, and if we choose any $b \in B$, then $f : S \to B$ given by $f(a) = b$ is an injection from S into B. In particular, $\mathbf{F}$ is nonempty. We define $\leq$ on $\mathbf{F}$ by $(S_1, f_1) \leq (S_2, f_2)$ provided $S_1 \subseteq S_2$ and $f_1 = f_2|_{S_1}$. We will show that $\leq$ is a partial ordering on $\mathbf{F}$.

It is certainly the case that for $(S, f) \in \mathbf{F}$, $S \subseteq S$ and $f = f|_S$. It follows that $(S, f) \leq (S, f)$. If $(S_1, f_1) \leq (S_2, f_2)$ and $(S_2, f_2) \leq (S_3, f_3)$, then $S_1 \subseteq S_2 \subseteq S_3$ so $S_1 \subseteq S_3$. Moreover, since $S_1 \subseteq S_2$, $f_3|_{S_1} = f_2|_{S_1} = f_1$. This proves that if $(S_1, f_1) \leq (S_2, f_2)$ and $(S_2, f_2) \leq (S_3, f_3)$, then $(S_1, f_1) \leq (S_3, f_3)$, which establishes that $\leq$ is transitive. If $(S_1, f_1) \leq (S_2, f_2)$ and $(S_2, f_2) \leq (S_1, f_1)$, then $S_1 \subseteq S_2 \subseteq S_1$, so $S_1 = S_2$. It follows then that $f_2 = f_2|_{S_1} = f_1$. This verifies that $\leq$ is a partial ordering on $\mathbf{F}$.

Let $C = \{(S_i, f_i) \mid i \in I\}$ be a chain in $\mathbf{F}$. C is then a set of pairs — each pair composed of a subset of A and an injection from that subset into B — such that any two pairs are comparable. Our next claim is that C has an upper bound in $\mathbf{F}$.

Let $S = \bigcup_{i \in I} S_i$. We then have $S \subseteq A$. Let $f : S \to B$ be given by $f(x) = f_i(x)$, where i is chosen so that $x \in S_i$. If $x \in S_i \cap S_j$, where $j \neq i$, then since (S_i, f_i), (S_j, f_j) are elements in a chain, either $S_i \subseteq S_j$ or $S_j \subseteq S_i$. We lose no generality in assuming $S_i \subseteq S_j$, in which case $f_j|_{S_i} = f_i$. For $x \in S_i \subseteq S_j$, then, $f_j(x) = f_i(x)$. This shows that f is well-defined.

We show next that f is an injection. This will prove that (S, f) is in $\mathbf{F}$.

Suppose there are $x, y \in S$ such that $f(x) = f(y)$. If $x \in S_i$ and $y \in S_j$ where $S_i \subseteq S_j$, then x and y are both in S_j. We then have $f(x) = f_j(x) = f(y) = f_j(y)$, which implies $x = y$, since f_j is an injection. This shows that f is an injection, as desired.

Given (S_i, f_i) in C, $S_i \subseteq S$ and $f|_{S_i} = f_i$. This is all we need to establish that $(S_i, f_i) \leq (S, f)$ for all (S_i, f_i) in C, thus, that (S, f) is an upper bound for C. This shows that $\mathbf{F}$ satisfies the hypothesis for Zorn's Lemma, thus, that it has a maximal element, (M, f_M). Note that $M \subseteq A$ and $f_M : M \to B$ is an injection.

If f_M is onto B, then $f_M^{-1} : B \to M \subseteq A$. Since f_M^{-1} is an injective mapping, then $f_M^{-1} : B \to A$ is the injective mapping, the existence of which we sought to prove.

If f_M is not onto B, we claim that $M = A$. Suppose by way of a contradiction that $M \neq A$ and that a is an element in $A \setminus M$. Since f_M is not onto B, we take $b \in B \setminus f_M[M]$. Extend f_M to a mapping $g : M \cup \{a\} \to B$, defining g by $g|_M = f_M$ and $g(a) = b$. Note that g is an injection and that $(M, f_M) \leq (M \cup \{a\}, g)$. This violates the maximality of (M, f_M). The contradiction proves our claim that if f_M is not onto B, then $M = A$. Since $f_M : A \to B$ is an injection, this completes the proof. $\square$

The next definition gives us vocabulary that helps us talk precisely about the relative sizes of large sets.

Definition C.12. When A and B are sets with a bijection $f : A \to B$, then A and B have the same **cardinality**, and we write $|A| = |B|$. If there is an injection $f : A \to B$, then the cardinality of A is less than or equal to the cardinality of B

and we write $|A| \leq |B|$. If $|A| \leq |B|$ and there is no bijection $f : A \to B$, then the cardinality of A is strictly less than the cardinality of B.

According to the definition, the cardinality of a set only makes sense in comparison to the cardinality of another set. We have some natural comparators though. If $|A| = |\{1, \ldots, n\}|$, then we say the cardinality of A is n.

Theorem C.11 says that the cardinalities of any two sets are comparable.

Proof of the next lemma is an exercise.

Lemma C.13. *The cardinality of a finite set is the number of elements in that set.*

If S is any set, then $|\mathcal{P}(S)| = 2^{|S|}$. That seems more difficult to prove than it is. When we have a subset of S in hand, each element in S is either in it or not. In choosing an element $A \in \mathcal{P}(S)$, we can then think of assigning a yes or a no to each element in S, according to whether it is or is not in A. Since there are two possibilities for each element in S whenever we choose $A \subseteq S$, there must be $2^{|S|}$ different subsets of S.

Cardinalities of infinite sets are labeled with *aleph numbers*.[1] The cardinality of $\mathbb{Z}$ is $\aleph_0$, read "aleph naught," for instance. The cardinality of the real numbers is $2^{\aleph_0}$: It is possible to define a bijection between the set of real numbers and the set of subsets in $\mathbb{Z}$. Any number that we can use to describe the cardinality of a set is called a **cardinal number**.

The next aleph number after $\aleph_0$ is $\aleph_1$. Georg Cantor, a founder of modern set theory, postulated in the 1870s that there was no set with cardinality strictly between $\aleph_0$ and $2^{\aleph_0}$, the cardinality of "the continuum." No one knows whether this is true. Cantor's idea that $\aleph_1 = 2^{\aleph_0}$ remains enshrined as the Continuum Hypothesis.

Lemma C.14. *If A and B are sets with $A \subseteq B$, then $|A| \leq |B|$.*

Proof. The mapping $\iota : A \to B$ given by $\iota(a) = a$ is an injection. □

We consider relationships among cardinalities of sets when we have mappings from one set to the other. The following is immediate by the last lemma.

Corollary C.15. *If A and B are sets and $f : A \to B$ is any mapping, then $|f[A]| \leq |B|$.*

Lemma C.16. *If A and B are sets and $f : A \to B$ is any mapping, then $|f[A]| \leq |A|$.*

Proof. If $f : A \to B$ is a mapping, then for each $b \in f[A]$, choose one $a_b \in A$ so that $f(a_b) = b$. This gives us a mapping $g : f[A] \to A$ defined by $g(b) = a_b$. If $g(b_1) = g(b_2) = a_b$, then since f is a mapping, $f(a_b) = b_1 = b_2$. This proves that g is injective, which proves the result. □

The previous two results give us another way to view a fundamental rule about mappings: A mapping may effect some collapsing in set size, but it cannot effect an expansion in set size.

---
[1] Aleph is the first letter of the Hebrew alphabet.

If $|A| = |\mathbb{Z}|$, we say A is **countable**. If A is an infinite set and it is not countable, we say that A is **uncountable**.

Infinite sets have surprising properties.

Consider $f : \mathbb{Z}_{\geq 0} \to \mathbb{Z}$ given by

(C.1) $\qquad\qquad 2k \mapsto k \quad \text{and} \quad 2k - 1 \mapsto -k, \quad \text{for } k \in \mathbb{Z}_{\geq 0}.$

Every element in $\mathbb{Z}_{\geq 0}$ can be written $2k$ or $2k - 1$ for some $k \in \mathbb{Z}_{\geq 0}$. For instance, since $0 = 2 \cdot 0$, we have $f(0) = 0$. Since $1 = 2 \cdot 1 - 1$, we have $f(1) = -1$, etc. We leave it as an exercise to verify that f is a bijection. It follows that $\mathbb{Z}_{\geq 0}$ is countable.

Now consider $g : \mathbb{Z}_+ \to \mathbb{Z}_{\geq 0}$ given by $k \mapsto k - 1$. It is easy to verify that g is a bijection. By Lemma A.32, the composition of bijections is a bijection. It follows that $\mathbb{Z}_+$ is also countable.

The example of $|\mathbb{Z}_{\geq 0}| = |\mathbb{Z}_+| = |\mathbb{Z}|$ underscores the fact that there can be a bijective correspondence between an infinite set A and a proper subset of A. This does not happen with finite sets!

It is important to understand that a countable set can always be identified with $\mathbb{Z}_+$.

Theorem C.17. If A is a countable set, every infinite subset of A is also countable.

Proof. Suppose A is countable. Let $f : A \to \mathbb{Z}_+$ be a bijection. Let $B \subseteq A$ be infinite. The Well-Ordering Principle guarantees that as a subset of $\mathbb{Z}_+$, $f[B]$ has a least element. Let that element be $f(b_1)$. Since f is injective, b_1 is well-defined. Define $g : B \to \mathbb{Z}_+$ inductively, starting with $g(b_1) = 1$. Assume that we have identified $b_1, \ldots, b_k$ in B so that $g(b_i) = i$. Notice that $f[B \setminus \{b_1, \ldots, b_n\}]$ has a least element, $f(b_{n+1})$, which gives us a well-defined element b_{n+1} in $B \setminus \{b_1, \ldots, b_n\}$. Define $g(b_{n+1}) = n + 1$. By the induction hypothesis, we can identify $b_k \in B$ and map $b_k \mapsto k$ under g, for all $k \in \mathbb{Z}_+$. Our choice of b_i guarantees that g is injective. Since B is infinite, g must also be onto, thus a bijection, which proves the theorem. $\square$

As the contrapositive of the last theorem, the next result is immediate.

Corollary C.18. A set with an uncountable subset is itself uncountable.

The next result underlies the proof that the rational numbers are countable. Our proof again follows [14].

Theorem C.19. If A is a countable set, then $A \times A$ is also countable.

Proof. Let A be countable and take $f : A \to \mathbb{Z}_+$ to be a bijection. Define

$$\mathbf{f} : A \times A \to \mathbb{Z}_+ \times \mathbb{Z}_+$$

by $\mathbf{f}(a, b) = (f(a), f(b))$. We leave it as an easy exercise to show that $\mathbf{f}$ is a bijection, which proves that $|A \times A| = |\mathbb{Z}_+ \times \mathbb{Z}_+|$. It is thus sufficient to show that $\mathbb{Z}_+ \times \mathbb{Z}_+$ is countable. To that end, consider the mapping $(m, n) \mapsto 2^m 3^n$. It is an easy exercise to show that this mapping is injective. Since its range is an infinite subset of $\mathbb{Z}_+$, Theorem C.17 guarantees that the range is countable. This is enough to establish that $\mathbb{Z}_+ \times \mathbb{Z}_+$ is countable, as desired. $\square$

The following is immediate by induction.

Corollary C.20. *If A is countable, then so is $A \times \cdots \times A$.*

Theorem C.19 generalizes: It is actually true that $|A| = |A \times \cdots \times A|$ for all infinite sets A. See [**14**] for details.

The next result is essential and does not require Zorn's Lemma or the Axiom of Choice in any other form.

Theorem C.21 (Schröder-Bernstein). *If A and B are sets with $|A| \leq |B|$ and $|B| \leq |A|$, then $|A| = |B|$.*

Sketch of proof. The Schröder-Bernstein Theorem is certainly true for finite sets and for countable sets so the content here is a statement about uncountable sets. Under the theorem's hypothesis, we have an injection $f : A \to B$ and an injection $g : B \to A$. The trick is to use f and g to construct a bijection from A to B, the obstacle being that we cannot assume that either mapping is onto. To deal with this, we partition A into subsets defined according to where those subsets are mapped into B under f. We can then find a bijection defined piecewise on A, viz., f on one piece, g^{-1} on the other. Details are in [**9**], [**14**], and [**20**], among other sources. □

The Schröder-Bernstein Theorem may seem simple but the proof is delicate. Interested readers should attempt to tease out the details for themselves or to dedicate some time to a close study of the approaches other mathematicians have used for the proof.

It is important to understand that the cardinality of $\mathbb{Z}$ and the cardinality of $\mathbb{R}$ are different. Since $\mathbb{Z} \subseteq \mathbb{R}$, this must mean $|\mathbb{Z}| < |\mathbb{R}|$.

Theorem C.22. $|\mathbb{Z}| < |\mathbb{R}|$.

Proof. We show that the cardinality of the open interval, $(0, 1) \subseteq \mathbb{R}$, is greater than the cardinality of $\mathbb{Z}_+$. The theorem then follows Corollary C.18.

Let S be the set of real numbers in the interval $(0, 1)$ that can be expressed as decimal expansions comprised of sequences of 0s and 1s. For example,

$$0.001, \quad 0.1001000110000\ldots, \quad 0.1010101\ldots$$

all belong to S. If we show S is uncountable, it follows that $(0, 1)$ and in turn $\mathbb{R}$ are uncountable. Since $\mathbb{R}$ includes numbers that can be expressed as nonterminating, nonrepeating decimals, S includes numbers that can be expressed as nonterminating, nonrepeating sequences of 0s and 1s.

The proof is by contradiction so we suppose S is countable with

$$S = \{a_1, a_2, a_3, \ldots\}.$$

Let $x = 0.b_1 b_2 b_3 b_4 \ldots$, where

$$b_i = \begin{cases} 0 \text{ if the } i\text{th decimal place of } a_i \text{ is } 1, \\ 1 \text{ if the } i\text{th decimal place of } a_i \text{ is } 0. \end{cases}$$

Evidently, x belongs to S. Notice, though, that x is different from every $a_i \in S$ because x and a_i differ in the ith decimal position. This is our contradiction. We conclude that S is uncountable, as claimed. □

The method applied in the argument to show that S is uncountable is called *Cantor diagonalization*.

Exercises C.2.

C.2.1. Let A be a finite set and let $B = A \cup \{x\}$, where $x \notin A$. Prove directly that $|\mathcal{P}(B)| = 2|\mathcal{P}(A)|$.

C.2.2. Prove Lemma C.13.

C.2.3. Find a bijection to establish that $|\mathbb{Z}_+| = |\mathbb{Z}|$.

C.2.4. We have talked about the nonnegative integers, $\mathbb{Z}_{\geq 0}$, and the positive integers, $\mathbb{Z}_+$. Give an example of another infinite proper subset $A \subsetneq \mathbb{Z}$ and find a bijection from A to some proper subset of A.

C.2.5. Let A be countable and take $f : A \to \mathbb{Z}_+$ to be a bijection. Define $\mathbf{f} : A \times A \to \mathbb{Z}_+ \times \mathbb{Z}_+$ by $\mathbf{f}(a, b) = (f(a), f(b))$. Show that $\mathbf{f}$ is a bijection.

C.2.6. Show that the mapping $f : \mathbb{Z}_+ \times \mathbb{Z}_+ \to \mathbb{Z}_+$ given by $f(m, n) = 2^m 3^n$ is injective.

Appendix D

Infinite Dimension

Here we show that when a vector space has an infinite basis, every basis for the space has the same cardinality. This allows us to define the dimension of an arbitrary vector space as the cardinality of a basis for the space.

This section requires knowledge of the material in Appendix C and is best approached after a review of linear transformations.

Theorem 1.35 says that in a finite-dimensional space, a linearly independent set can be no bigger than a spanning set. This gave us the key to determining that if one basis for a space had n elements, then every basis for the space had n elements. The central result in this section is that if we have two vector spaces over a field and a basis for each space, then there must be an injective linear mapping from one space to the other space that maps the one basis into the other basis. We can apply this to say that given two bases for one vector space, there must be an injective mapping from one of the bases into the other. From there we can prove that the two bases have the same cardinality.

The approach we take here, as well as the first two results and their proofs, are due to Miles Dillon Edwards, who generously supplied his analysis for our discussion here [7].

Lemma D.1. Let V and W be vector spaces over a field $\mathbb{F}$. Let $\mathcal{B}$ be a basis for V, and let $\mathcal{C}$ be a basis for W. Suppose there is an injective linear transformation $L: V \to W$. If $\mathbf{b} \in \mathcal{B}$ and $L(\mathbf{b}) \notin \mathcal{C}$, then there is an injective linear transformation, $T: V \to W$, for which $T(\mathbf{b}) \in \mathcal{C}$ and such that if $\mathbf{b}' \in \mathcal{B} \setminus \{\mathbf{b}\}$, then $T(\mathbf{b}') = L(\mathbf{b}')$.

Proof. Let V, W, $\mathbb{F}$, $\mathcal{B}$, $\mathcal{C}$, L, and $\mathbf{b}$ be as in the hypotheses of the lemma.

Since L is an injective linear transformation, $L[\mathcal{B}]$ is linearly independent in W. In particular, $L(\mathbf{b})$ cannot be in the span of $L[\mathcal{B} \setminus \{\mathbf{b}\}]$. Since $L[\mathcal{B} \setminus \{\mathbf{b}\}] \subseteq L[V]$ does not span $L[V]$, it cannot span W. Since $\mathcal{C}$ is a basis for W, there must then be $\mathbf{c} \in \mathcal{C}$ outside the span of $L[\mathcal{B} \setminus \{\mathbf{b}\}]$. Fix such an element $\mathbf{c} \in \mathcal{C}$.

Define $T: V \to W$ by setting

$$T(\mathbf{b}') := \begin{cases} L(\mathbf{b}') \text{ if } \mathbf{b}' \in \mathcal{B} \setminus \{\mathbf{b}\}, \\ \mathbf{c} \text{ if } \mathbf{b}' = \mathbf{b} \end{cases}$$

351

and then extending linearly. We will have proved the lemma once we show that T is injective.

Suppose that for $\{\mathbf{b}_1, \ldots, \mathbf{b}_k\} \subseteq \mathcal{B} \setminus \{\mathbf{b}\}$, $r, r_i \in \mathbb{F}$,
$$r\mathbf{b} + r_1\mathbf{b}_1 + \cdots + r_k\mathbf{b}_k$$
is in the kernel of T. Applying T to this expression, we have

(D.1) $\quad r\mathbf{c} + L(r_1\mathbf{b}_1 + \cdots + r_k\mathbf{b}_k) = r\mathbf{c} + r_1 L(\mathbf{b}_1) + \cdots + r_k L(\mathbf{b}_k) = \mathbf{0}_W.$

If $r \neq 0$ in (D.1), then $\mathbf{c}$ is in the span of $L[\mathcal{B} \setminus \{\mathbf{b}\}]$, a contradiction of our choice of $\mathbf{c}$, which forces us to conclude that $r = 0$. Since r must be zero, we have
$$r_1 L(\mathbf{b}_1) + \cdots + r_k L(\mathbf{b}_k) = \mathbf{0}_W.$$
As $\{\mathbf{b}_1, \ldots, \mathbf{b}_k\}$ is linearly independent and L is injective, $r_i = 0$ for all i. This establishes that $\operatorname{Ker} T$ is trivial, thus that T is injective, which completes the proof. $\square$

The lemma provides a critical step in the proof of the next theorem, an application of Zorn's Lemma.

Theorem D.2. Let V and W be vector spaces over a field $\mathbb{F}$. Let $\mathcal{B}$ be a basis for V and let $\mathcal{C}$ be a basis for W. If there is an injective linear transformation $V \to W$, then there is an injective linear transformation $T : V \to W$ such that $T[\mathcal{B}] \subseteq \mathcal{C}$.

Proof. Let V, W, $\mathbb{F}$, $\mathcal{B}$, and $\mathcal{C}$ be as in the hypotheses of the theorem. We construct a poset and then apply Zorn's Lemma.

Let $\mathbf{M}$ be the collection of ordered pairs $(\mathcal{B}', L)$, where $\mathcal{B}' \subseteq \mathcal{B}$ and $L : V \to W$ is an injective linear transformation for which $L[\mathcal{B}'] \subseteq \mathcal{C}$. $\mathbf{M}$ is nonempty: If L is the injective linear transformation from V to W that we are given and it does not map any subset of $\mathcal{B}$ into $\mathcal{C}$, then $(\emptyset, L) \in \mathbf{M}$, for instance.

Define $\leq$ on $\mathbf{M}$ by decreeing that $(\mathcal{B}_1, L_1) \leq (\mathcal{B}_2, L_2)$, provided
$$\mathcal{B}_1 \subseteq \mathcal{B}_2, \ L_1|_{\mathcal{B}_1} = L_2|_{\mathcal{B}_1}, \text{ and } L_1|_{\mathcal{B} \setminus \mathcal{B}_2} = L_2|_{\mathcal{B} \setminus \mathcal{B}_2}.$$
Note that when $(\mathcal{B}_1, L_1) \leq (\mathcal{B}_2, L_2)$, L_1 and L_2 agree on all of $\mathcal{B}$ except possibly on $\mathcal{B}_2 \setminus \mathcal{B}_1$. (See Figure D.1.)

(Informally, we can think of a pair in $\mathbf{M}$ as an approximation to an injective linear transformation that maps $\mathcal{B}$ to $\mathcal{C}$. If $(\mathcal{B}', L) \in \mathbf{M}$, then $L : V \to W$ is an injective linear transformation that maps $\mathcal{B}'$, a subset of $\mathcal{B}$, into $\mathcal{C}$. When $(\mathcal{B}_1, L_1) \leq (\mathcal{B}_2, L_2)$, $(\mathcal{B}_2, L_2)$ is an improvement on $(\mathcal{B}_1, L_1)$ in the sense that $\mathcal{B}_2$ is a larger subset of $\mathcal{B}$ than $\mathcal{B}_1$ and L_2 maps $\mathcal{B}_2$ into $\mathcal{C}$ while agreeing with L_1 on $\mathcal{B}_1$.)

We show that $\leq$ is transitive and antisymmetric which will prove that $\leq$ is a partial ordering on $\mathbf{M}$.

Suppose $(\mathcal{B}_1, L_1) \leq (\mathcal{B}_2, L_2)$ and that $(\mathcal{B}_2, L_2) \leq (\mathcal{B}_3, L_3)$. Since
$$\mathcal{B}_1 \subseteq \mathcal{B}_2 \subseteq \mathcal{B}_3,$$
it follows that
$$\mathcal{B} \setminus \mathcal{B}_3 \subseteq \mathcal{B} \setminus \mathcal{B}_2 \subseteq \mathcal{B} \setminus \mathcal{B}_1.$$
From there we see that L_1, L_2, and L_3 all agree on $\mathcal{B}_1$ and on $\mathcal{B} \setminus \mathcal{B}_3$. The implication is that $(\mathcal{B}_1, L_1) \leq (\mathcal{B}_3, L_3)$, which proves that $\leq$ is transitive.

D. Infinite Dimension

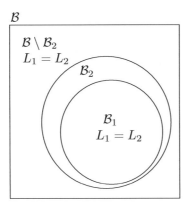

Figure D.1. $(\mathcal{B}_1, L_1) \leq (\mathcal{B}_2, L_2)$ means $L_1 = L_2$ on $\mathcal{B}$ except on $\mathcal{B}_2 \setminus \mathcal{B}_1$.

Next suppose $(\mathcal{B}_1, L_1) \leq (\mathcal{B}_2, L_2)$ and $(\mathcal{B}_2, L_2) \leq (\mathcal{B}_1, L_1)$. Since $\mathcal{B}_1 \subseteq \mathcal{B}_2$ and $\mathcal{B}_2 \subseteq \mathcal{B}_1$, $\mathcal{B}_1 = \mathcal{B}_2$. We then have that L_1 and L_2 agree on all of $\mathcal{B}$, thus on all of V, so $(\mathcal{B}_1, L_1) = (\mathcal{B}_2, L_2)$. This establishes that $\leq$ is antisymmetric, verifying that $(\mathbf{M}, \leq)$ is a poset.

Let $\mathbf{C} = \{(\mathcal{B}_i, L_i)\}_{i \in I} \subseteq \mathbf{M}$ be a chain. Let $\mathcal{B}' = \bigcup_{i \in I} \mathcal{B}_i$. Since $\mathcal{B}_i \subseteq \mathcal{B}$ for all $i \in I$, $\mathcal{B}' \subseteq \mathcal{B}$.

We define a linear transformation $L : V \to W$ with the property that $L[\mathcal{B}'] \subseteq \mathcal{C}$. We do this in two steps, by defining L first on $\mathcal{B}'$, then on $\mathcal{B} \setminus \mathcal{B}'$. With L defined on all of $\mathcal{B}$, we can then extend linearly.

If $\mathbf{b} \in \mathcal{B}'$, then $\mathbf{b} \in \mathcal{B}_i$ for some i so we let $L(\mathbf{b}) = L_i(\mathbf{b}) \in \mathcal{C}$. Suppose, though, that there is more than one $i \in I$ so that $\mathcal{B}_i$ contains $\mathbf{b}$. If $\mathbf{b}$ belongs both to $\mathcal{B}_i$ and $\mathcal{B}_j$, then since $(\mathcal{B}_i, L_i)$ and $(\mathcal{B}_j, L_j)$ are in a chain, we may assume that $(\mathcal{B}_i, L_i) \leq (\mathcal{B}_j, L_j)$. This means $\mathbf{b} \in \mathcal{B}_i \subseteq \mathcal{B}_j$ and $L_i|_{\mathcal{B}_i} = L_j|_{\mathcal{B}_i}$. It follows that $L_i(\mathbf{b}) = L_j(\mathbf{b})$. This proves that L is well-defined on $\mathcal{B}'$.

Now note that $\mathcal{B} \setminus \mathcal{B}' = \bigcap_{i \in I} \mathcal{B} \setminus \mathcal{B}_i$. Since $\mathcal{B} \setminus \mathcal{B}' \subseteq \mathcal{B} \setminus \mathcal{B}_i$ for every $i \in I$,
$$L_i|_{\mathcal{B} \setminus \mathcal{B}'} = L_j|_{\mathcal{B} \setminus \mathcal{B}'} \; \forall i, j \in I.$$
Given $\mathbf{b} \in \mathcal{B} \setminus \mathcal{B}'$, we can thus define $L(\mathbf{b}) = L_i(\mathbf{b})$ for any $i \in I$. Now we have L defined on all of $\mathcal{B}$. We refer to the linear extension of L also as $L : V \to W$.

We claim that $(\mathcal{B}', L)$ is an upper bound for $\mathbf{C}$ in $\mathbf{M}$. We already have $\mathcal{B}_i \subseteq \mathcal{B}' \subseteq \mathcal{B}$, $L|_{\mathcal{B}_i} = L_i|_{\mathcal{B}_i}$, and $L|_{\mathcal{B} \setminus \mathcal{B}'} = L_i|_{\mathcal{B} \setminus \mathcal{B}'}$, for all $i \in I$. The only thing left to show is that L is injective.

Let $\mathbf{x} \in V$ be in $\operatorname{Ker} L$. We can write
$$\mathbf{x} = c_1 \mathbf{b}'_1 + \cdots + c_k \mathbf{b}'_k + a_1 \mathbf{b}_1 + \cdots + a_r \mathbf{b}_r$$
for $\mathbf{b}'_i \in \mathcal{B}'$, $\mathbf{b}_i \in \mathcal{B} \setminus \mathcal{B}'$, and c_i, a_i in $\mathbb{F}$. For each $\mathbf{b}'_i$ we can fix $(\mathcal{B}_j, L_j) \in \mathbf{C}$ so that $\mathbf{b}'_i \in \mathcal{B}_j$. Since there are finitely many such $(\mathcal{B}_j, L_j)$s, and since $\mathbf{C}$ is a chain, there is a maximal $(\mathcal{B}_n, L_n)$ among the $(\mathcal{B}_j, L_j)$s. Note then that each $\mathbf{b}'_i$ belongs to $\mathcal{B}_n$, implying that $L(\mathbf{b}'_i) = L_n(\mathbf{b}'_i)$ for $i = 1, \ldots, k$. Since $L|_{\mathcal{B} \setminus \mathcal{B}'} = L_j|_{\mathcal{B} \setminus \mathcal{B}'}$ for all $j \in I$,

$L|_{\mathcal{B}\setminus\mathcal{B}'} = L_n|_{\mathcal{B}\setminus\mathcal{B}'}$, implying that $L(\mathbf{b}_i) = L_n(\mathbf{b}_i)$ for $i = 1, \ldots, r$. This gives us
$$L(\mathbf{x}) = L_n(\mathbf{x}) = \mathbf{0}_W.$$
Since L_n is injective, $\mathbf{x}$ must be zero, which establishes that L is injective, thus that $(\mathcal{B}', L)$ is an upper bound for $\mathbf{C}$ in $\mathbf{M}$.

Since $\mathbf{C}$ was arbitrary, we have shown that every chain in $(\mathbf{M}, \leq)$ has an upper bound, meaning that Zorn's Lemma applies to guarantee that $\mathbf{M}$ has a maximal element.

Let $(\widetilde{\mathcal{B}}, T)$ be a maximal element in $\mathbf{M}$. We have then that $\widetilde{\mathcal{B}} \subseteq \mathcal{B}$, that $T : V \to W$ is an injective linear transformation with $T[\widetilde{\mathcal{B}}] \subseteq \mathcal{C}$, and that if, for any other $(\mathcal{B}', L) \in \mathbf{M}$, $(\mathcal{B}', L) \geq (\widetilde{\mathcal{B}}, T)$, then $(\mathcal{B}', L) = (\widetilde{\mathcal{B}}, T)$.

We claim that $\widetilde{\mathcal{B}} = \mathcal{B}$.

Suppose there is $\mathbf{b} \in \mathcal{B} \setminus \widetilde{\mathcal{B}}$. Let $\mathcal{B}' = \widetilde{\mathcal{B}} \cup \mathbf{b}$. Lemma D.1 guarantees that there is an injective linear map $L : V \to W$ such that $L[\mathcal{B}'] \subseteq \mathcal{C}$, $L|_{\widetilde{\mathcal{B}}} = T|_{\widetilde{\mathcal{B}}}$, and
$$L|_{\mathcal{B}\setminus\mathcal{B}'} = T|_{\mathcal{B}\setminus\mathcal{B}'}.$$
This says, though, that $(\widetilde{\mathcal{B}}, T) \leq (\mathcal{B}', L)$, implying that $(\widetilde{\mathcal{B}}, T) = (\mathcal{B}', L)$, which is impossible since $\widetilde{\mathcal{B}}$ is a proper subset of $\mathcal{B}'$. By the contradiction, we conclude that $\widetilde{\mathcal{B}} = \mathcal{B}$ and that $T : V \to W$ is an injective linear transformation with $T[\mathcal{B}] \subseteq \mathcal{C}$, completing the proof of the theorem. $\square$

Theorem D.2 allows us to compare sizes of bases. For the next proof, we also require the services of the Schröder-Bernstein Theorem.

Theorem D.3. Let V and W be isomorphic vector spaces. If $\mathcal{B}$ is a basis for V and $\mathcal{C}$ is a basis for W, then there is an isomorphism $L : V \to W$ such that $L[\mathcal{B}] = \mathcal{C}$.

Proof. Let $V, W, \mathcal{B}$, and $\mathcal{C}$ be as in the hypotheses of the theorem. Since there must be an isomorphism $V \to W$, Theorem D.2 guarantees that there is an injective linear transformation $L : V \to W$ which satisfies $L[\mathcal{B}] \subseteq \mathcal{C}$. Since injections preserve cardinality, $|\mathcal{B}| = |L[\mathcal{B}]| \leq |\mathcal{C}|$.

There is also an isomorphism $W \to V$. Applying Theorem D.2 again, we know there must be an injective linear transformation $T : W \to V$ with $T[\mathcal{C}] \subseteq \mathcal{B}$. It follows that $|\mathcal{C}| = |T[\mathcal{C}]| \leq |\mathcal{B}|$. Since $|\mathcal{B}| \leq |\mathcal{C}|$ and $|\mathcal{C}| \leq |\mathcal{B}|$, the Schröder-Bernstein Theorem implies $|\mathcal{B}| = |\mathcal{C}|$.

Since $|\mathcal{B}| = |\mathcal{C}|$, there must be a bijection $f : \mathcal{B} \to \mathcal{C}$. Let L be the linear extension of f to all of V. Since $L[\mathcal{B}] = \mathcal{C}$, this completes the proof. $\square$

Since a vector space is isomorphic to itself, the following corollary is immediate.

Corollary D.4. If V is a vector space with bases $\mathcal{B}$ and $\mathcal{C}$, then $|\mathcal{B}| = |\mathcal{C}|$.

Corollary D.4 legitimizes Definition 1.42.

Corollary D.5. Let V and W be vector spaces over a field $\mathbb{F}$. V is isomorphic to W if and only if $\dim_\mathbb{F} V = \dim_\mathbb{F} W$.

D. Infinite Dimension

Proof. Let V, W, and $\mathbb{F}$ be as in the hypotheses. If $V \cong W$, Theorem D.3 implies that $\dim V = \dim W$. This proves the corollary in one direction.

Suppose $\dim V = \dim W$ so that a basis for V and a basis for W have the same cardinality. Let f be a bijection from a basis for V to a basis for W. Let $L : V \to W$ be the linear extension of f. Since L is an isomorphism, $V \cong W$. This proves the corollary in the other direction. $\square$

Since we arrive at Corollary D.5 after so much heavy lifting, it may seem like an afterthought. Nothing could be farther from the truth. This is the classification theorem for vector spaces over a given field, that is, it establishes a single measurement — dimension — that determines whether or not two vector spaces over a given field are isomorphic. This is a powerful type of result that one encounters rarely.

Bibliography

[1] Simon L. Altmann, *Rotations, quaternions, and double groups*, Dover Publications, Inc., Mineola, New York, 2005.

[2] E. Artin, *Geometric algebra*, Interscience Publishers, Inc., New York-London, 1957. MR0082463

[3] Michael Artin, *Algebra*, Prentice Hall, Inc., Englewood Cliffs, NJ, 1991. MR1129886

[4] Sheldon Axler, *Down with determinants!*, Amer. Math. Monthly **102** (1995), no. 2, 139–154, DOI 10.2307/2975348. MR1315593

[5] James Ward Brown and Ruel V. Churchill, *Complex variables and applications*, 7th ed., McGraw-Hill Higher Education, Boston, MA, 2003.

[6] M. D. Burrow, *The minimal polynomial of a linear transformation*, Amer. Math. Monthly **80** (1973), 1129–1131, DOI 10.2307/2318550. MR344271

[7] Miles Dillon Edwards, private communication.

[8] Paul R. Halmos, *Finite-dimensional vector spaces*, 2nd ed., Undergraduate Texts in Mathematics, Springer-Verlag, New York-Heidelberg, 1974. MR0409503

[9] Paul R. Halmos, *Naive set theory*, reprint of the 1960 edition, Undergraduate Texts in Mathematics, Springer-Verlag, New York-Heidelberg, 1974. MR0453532

[10] John Hannah, *A geometric approach to determinants*, Amer. Math. Monthly **103** (1996), no. 5, 401–409, DOI 10.2307/2974931. MR1400721

[11] Kenneth Hoffman and Ray Kunze, *Linear algebra*, 2nd ed., Prentice-Hall, Inc., Englewood Cliffs, N.J., 1971. MR0276251

[12] Victor J. Katz, *A history of mathematics: An introduction*, 3rd ed., Pearson Education, Inc., Boston, MA, 2009.

[13] Bernard Kolman, *Elementary linear algebra*, 6th ed., Prentice-Hall, Inc., Saddle River, NJ, 1996.

[14] Serge Lang, *Algebra*, 3rd ed., Graduate Texts in Mathematics, vol. 211, Springer-Verlag, New York, 2002, DOI 10.1007/978-1-4613-0041-0. MR1878556

[15] David C. Lay, *Linear algebra and its applications*, 5th ed., Pearson, New York, 1969.

[16] John Loustau and Meighan Dillon, *Linear geometry with computer graphics*, Monographs and Textbooks in Pure and Applied Mathematics, vol. 170, with 1 IBM-PC floppy disk (3.5 inch; DD), Marcel Dekker, Inc., New York, 1993. MR1200896

[17] Barbara D. MacCluer, *Elementary functional analysis*, Graduate Texts in Mathematics, vol. 253, Springer, New York, 2009, DOI 10.1007/978-0-387-85529-5. MR2462971

[18] John J. O'Connor and E. F. Robertson, "Matrices and determinants," *MacTutor History of Mathematics*, University of St. Andrews, Scotland, Feb. 1996. Web. Jan. 2019.

[19] Danny Otero, "Determining the determinant," Digital Commons, Ursinus College, 2018.

[20] "Schroeder-Bernstein Theorem," AoPS Online, https://artofproblemsolving.com/wiki/index.php/Schroeder-BernsteinTheorem

[21] Gilbert Strang, *The fundamental theorem of linear algebra*, Amer. Math. Monthly **100** (1993), no. 9, 848–855, DOI 10.2307/2324660. MR1247531

[22] Hermann Weyl, *The classical groups: Their invariants and representations*, 15th printing, Princeton Landmarks in Mathematics, Princeton Paperbacks, Princeton University Press, Princeton, NJ, 1997. MR1488158

[23] Thomas Yuster, *The Reduced Row Echelon Form of a Matrix is Unique: A Simple Proof*, Math. Mag. **57** (1984), no. 2, 93–94. MR1572501

Index

Altk(V), 137
$\mathcal{F}(A)$, 11
$\mathcal{F}(A,B)$, 11
$\mathbb{F}$-linear, 32, 34
$\mathbb{F}^n$, 9, 20, 47, 64, 66, 68, 70, 72–80, 83, 88, 91, 133, 149, 152, 223, 227, 229
    standard basis for, 18
    standard ordered basis for, 18
    usual basis for, 18
Klin(V), 135
k-linear form, 135
k-linear mapping, 134
    alternating, 136
    skew-symmetric, 137
    symmetric, 137
$\mathcal{L}(V,W)$, 69
L-invariant, 40, 41, 121, 124, 212, 243, 244, 247
    L-complement, 121, 243, 244, 246
    L-decomposition, 121, 124, 126, 232, 237, 240, 242, 243, 246, 247
    L-sum, 121, 125, 234
$M_{m,n}(\mathbb{F})$, 11
$M_n(\mathbb{F})$, 11
$\mathcal{S}_3$, 140, 147

adjoint of a linear operator, 187–189
adjoint/conjugate-transpose of a matrix, 189
affine geometry, 41, 324, 325, 330, 331
affine mapping, 43
    transformation, 43
affine set, 41, 44, 79

3-space, 330
    dimension, 41, 43
    parallel, 42
    plane, 324
aleph number, 346
algebra, $\mathbb{F}$-algebra, 26–28, 35, 71–73, 256, 289
    subalgebra, 27, 28
algebraic closure, 326
algebraic multiplicity, 239
alternating group, 145
analytic geometry, 324, 329
anticommutative multiplication, 266
antilinear, 131
antisymmetric relation, 312
arithmetic modulo p, 5, 7, 319, 320
associativity, 5, 319
automorphism
    algebra, 112, 202, 261
    group, 145, 261, 263
    vector space, 44, 59, 66, 73, 75, 83, 91, 126, 129, 225
Axiom of Choice, 344, 348

Babylonian clay tablet, 81
back-substitution, 98–100, 102
basic variable, 91
basis, 18–22
    algebraic/Hamel/linear, 22
    ordered, 65, 66
    standard ordered, 58, 84, 281
    standard ordered, for $M_{m,n}(\mathbb{F})$, 70
bidual space, 109, 110

Big Bang, 93
bijection, bijective, 22, 31, 34, 40, 41, 43, 44, 47, 139, 145, 147–149, 152, 187, 257, 260, 261, 265, 267, 270, 279, 282, 287, 316–318, 345–347, 349, 354, 355
bilinear, 332
bilinear extension, 116
bilinear form, 116, 120
   matrix representation, 116–121
bilinear mapping, 61, 116
binary operation, 3, 5, 8, 71, 256, 257, 265, 318–320
   associative, 318
   commutative, 318
   identity element, 318
   inverse element, 318
binomial, 25
binomial coefficient, 310
binomial expansion, 310
block matrix, 246–248, 252, 292, 295
   multiplication, 124

canonical isomorphism, 108, 111, 119, 136–138, 157, 159, 161
canonical mapping, 39, 45, 109, 114
Cantor diagonalization, 349
Cantor, Georg, 346
card games, 309, 310
cardinal number, definition, 346
cardinality, 13, 22, 40, 79, 270, 341, 343, 345, 346, 348, 351, 354, 355
Cartesian product, 49, 302, 343, 344
Cauchy-Schwarz inequality, 165
Cayley table, 5, 7, 117, 147, 149, 319
Cayley's Theorem, 257
Cayley-Hamilton Theorem, 213
center of a group, 274, 275
chain, 341–345, 353, 354
change of basis, 66, 68, 69
change of basis matrix, 66
characteristic equation, 212
characteristic polynomial, 212–214, 220, 224, 231, 234, 240, 242, 248, 251–253, 283
choice function, 344
circle group, 265
classical groups, 255, 290, 294
clearing a column, 87
closure under an operation, 318
codomain, 31, 316
coefficient matrix, 100

collinearity, 275–277, 279
column space, 58, 90, 92
column vector, 11
commutative diagram, 63
comparable, under an ordering, 341, 342, 345, 346
complement of a set, 304
complementary subspaces, 47, 52
complete quadrangle, 277–281
complete quadrilateral, 277, 280, 281
complex conjugation on functions, 30
complex conjugation on matrices, 30, 130
complex exponential, 178, 182, 328
complex function, 177
   conjugate of, 178
   continuous on real interval, 177
   real and imaginary parts, 177
complex number, 4, 12, 28, 201, 215, 216, 220, 221, 264–266, 285, 286, 321, 326, 327
   algebraic closure of $\mathbb{R}$, 201
   complex conjugation, 30, 326, 327
   complex plane, 264, 265, 285, 326
   exponential form, 265, 269, 282, 285, 327
   modulus, 327, 333
   real and imaginary parts, 30, 326
   unit complex numbers $\mathcal{U}(\mathbb{C})$, 264, 265
component, 9
componentwise, 9, 11
componentwise operation, 326
composition of mappings, 317
concurrence, 275–277
congruence, 312
congruent matrices, 118, 120
conjugate transpose/adjoint of a matrix, 130, 131, 286
conjugate-linear mapping, 131, 132, 159, 187
conjugation mapping $x \mapsto axa^{-1}$, 202, 257–259, 261–263, 269, 287, 288
continuous function, 31
continuum, 158
Continuum Hypothesis, 346
contradiction, proof by, 299
contrapositive, 16, 17, 297, 299
convergence
   in an inner product space, 165
   normwise or in the norm, 165
converse, 297, 298

Index

coordinate axes, 331
coordinate mapping, 62
coordinate planes, 82, 329, 331
coordinate system, 39, 154, 321
coordinate vector, 62, 245
countable, 24, 347–349
Cramer's determinant, 141
Cramer's Rule, 95
Cramer, Gabriel, 93
cross-product, 268, 269, 288, 334, 335, 337, 339
   Jacobi identity, 335
cycle
   length, 141

decimal expansion, 306
definite integral, 273
DeMorgan's Laws, 304
dependence relation, 15, 90
derangement, 96
determinant, 93, 95, 97, 149, 152, 212, 240, 268, 283, 335
   3×3 matrix, 97
   Cramer definition, modern version, 149, 151
   diagonal block matrix, 154
   linear operator, 154
   LU-factorization, 155
   multiplicative property of, 153, 155, 263, 264, 272
   of matrix inverse, 153
   orthogonal matrix, 155
   similar matrices, 154
   system of linear equations, 96
   triangular matrix, 152
   unitary matrix, 155
diagonal matrix, 129, 249, 274, 275, 280
diagonalizable, 216, 252, 283
diagonalizable operator, 216, 219
differentiable function, 33
differential equation, 65
differentiation, 122
dihedral group, 258
direct complement, 47, 52, 243, 244
direct product, 49, 50
direct sum, 122, 124
   external, 49, 50
   internal, 47, 48, 121
directed angle, 270
direction, 42, 43
disjoint union, 303
division algorithm, 200

division by zero, 306
domain, 31, 316
dot product, 118, 159, 255, 268, 281, 282, 291, 332–334, 336
   angle, 333
   distance, 157, 333
   length, 159
   on $\mathbb{C}^n$, 158, 159, 161
   on $\mathbb{R}^n$, 116, 120, 127, 157, 159, 332, 333
dot product, alternate names, 332
dual basis, 108, 111
dual space, 107
   algebraic, 107
   inner product space, 187
dual statement, 275, 276

Edwards, Miles Dillon, 351
eigenspace, 208, 242, 249, 250, 252, 283
eigenvalue, 206, 213, 242, 249, 251, 282, 283, 285
   algebraic multiplicity, 214, 240
   geometric multiplicity, 208
   zero, 208
eigenvector, 206, 222, 224, 240, 283–285
   nonzero, 208
elementary column operation, 85
elementary matrix, 84, 91, 99
   inverse, 100
   invertible, 85
   type-k, 84
elementary row operations, 83
empty set, 19, 241, 301
equilateral triangle, 259
equivalence class, 313
equivalence mod n, 313
equivalence relation, 312
Euclidean geometry, 41, 42, 127, 157, 158, 307, 323, 324, 328, 332
   on $\mathbb{F}^n$, 158
Euclidean inner product, 332
Euler's formula, 327, 328
Euler's identity, 182
Euler, Leonhard, 182
evaluation mapping, 32, 35, 45
even permutation, 144
existence and uniqueness, 298
existence statement, 298
exponential notation
   additive analog, 257
extended real numbers, 307

factorial notation, 309
Fano plane, 278–281
field, 3, 4, 9, 20, 22, 25, 32, 42, 45, 60,
    79, 107, 132, 200, 201, 227, 257,
    265, 267, 271, 326
  algebraic closure, 201
  algebraically closed, 200, 231
  characteristic, 5, 6, 17, 26, 120, 137,
    158, 252, 294
  distributive law, 3, 4
  exponential notation, 25
  extension, 204, 205, 216, 220
  nonzero elements, 155, 263
  notation na in, 7
  operations, 9
  ordered, 4, 158, 161
  subfield, 3, 7, 12
  underlying a vector space, 8
field with p elements, $\mathbb{F}_p$, 5, 6, 9, 11,
    201, 263
formal power series, 26, 50
formal sum, 24
forward-substitution, 98, 99, 102
four fundamental subspaces, 191
Fourier expansion, 181
free variable, 91
full rank, 92
function, 307, 316
function composition, 317, 319
function, image under, 316
Fundamental Theorem of Algebra, 201
Fundamental Theorem of Arithmetic, 7,
    300
Fundamental Theorem of Calculus, 177
Fundamental Theorem of Linear
    Algebra, 191

general linear group, $GL_n(\mathbb{F})$, 126, 127,
    129, 255–260, 268–270, 273, 274,
    280, 292, 294
Gram-Schmidt process, 171, 172, 291
greatest element, 341–343
group, 126, 129, 139, 256, 267
  abelian, 126, 146, 256, 257, 274
  alternating, 145
  coset, 146–149, 261–264, 273, 279
  cyclic, 257
  exponential notation, 140
  generator, 257
  normal subgroup, 259, 262–264, 270,
    272, 274
  order of an element, 257

quotient group, 262, 263, 268, 272,
    273, 275
subgroup, 126, 129, 256–263, 267,
    269, 272, 274, 280
symmetric group, 140

Halmos, Paul, 143, 199
Hamilton, William Rowan, 264, 267
Hermitian form, 159–161
Hermitian inner product (dot product)
    on $\mathbb{C}^n$, 158
Hermitian symmetric mapping, 159
homomorphism
  algebra, 71–73, 201, 202, 205
  group, 145, 148, 149, 259–264, 273,
    287, 289
hyperplane, 80, 227
hyperspace, 80, 109, 227

ideal, 205
identity mapping, 12, 32, 317
identity matrix, 59
identity, for binary operation, 3, 4, 8,
    27, 256
image, 31, 316
improper rotation, 284
incidence geometry, 275
incident, 41, 43
inclusive "or", 303
indeterminate, 24
indexed set, 302
induction, proof by, 19, 20, 60, 87, 88,
    105, 124, 125, 142, 150, 151, 154,
    219, 222, 233, 238–240, 243, 247,
    292, 298–300, 342, 343, 347, 348
initial conditions, 65
injective, 31, 316
inner product, 163
  Hermitian, 162
  matrix representation of, 163
  on $C([a, b])$, 164
  real, 163
  restriction to subspace, 164
  scalar multiples of, 163
  sums of, 163
inner product space, 163, 176, 221, 222,
    290, 291
  $C([0, 1])$, 174
  angle, 166
  complex, 178
  convergence, 165
  distance, 165

dual space, 187
length, 165
norm, 165
projection of a vector onto a subspace, 192
inner product, complex, 291
integers, $\mathbb{Z}$, 4
intersection of sets, 303
inverse, 256, 298
   additive, 4, 8
   multiplicative, 3, 4
invertible linear transformation, 75
invertible matrix, 59, 75
irrational numbers, 306
isomorphic vector spaces, 326
isomorphism, 44
   algebras, 71–73, 75, 199, 202, 205
   group, 145, 186, 260, 263, 273, 282
   inverse, 44
   vector space, 44–47, 50, 62–64, 68, 70, 72, 108, 111–113, 119, 125, 136, 137, 157, 159, 161, 164, 182, 183, 246, 247, 260, 268, 354, 355

Jacobi identity, 335
Jordan (canonical) form, 249–253
   theorem, 249
Jordan block, 248–252
juxtaposition, 9, 256

kernel, group homomorphism, 261–263
kernel, linear transformation, 33–35, 39, 40, 52, 53, 76, 78, 109, 112, 121, 125, 201, 207, 208, 212, 213, 217, 231, 236–243, 249–251, 253, 262, 352
kernel index, 241, 242, 249–251
Kronecker delta, 18

L'Hôpital's Rule, 307
Law of Cosines, 167, 324, 325
least element, 341–343, 347
least-squares solution, 192, 194
   formula, 194
Legendre polynomials, 180
limit, 306
line, 41
linear combination, 10
   coefficient, 10
   nontrivial, 10
   term, 10
   trivial, 10
   weight, 10
linear dependence, 15–17
linear equation, definition, 325
linear extension, 33
linear functional, 107, 111
linear independence, 14, 16, 18, 21, 29, 34
linear mapping, 32
linear operator, 32
   annihilator, 203
   characteristic polynomial, 214, 240
   diagonalizable, 216
   direct sum, 234, 239
   kernel index, 237
   minimal polynomial, 203–206, 210–215, 218–220
   nilpotence index, 235
   nilpotent, 235
   normal form, 242
   standard matrix representation, 221
linear projection, 51–53, 79, 121, 186
linear reflection, 186
linear space, 9
linear structure, 9
linear transformation, 31, 32, 43, 199, 260
   block matrix representation, 124
   codomain, 34
   dual, 110, 114
   injective, 35, 46
   kernel, 33, 34, 39
   matrix representation, 59, 62–64, 66–68, 70, 74, 75
   nontrivial, 32
   nullity, 46
   orthogonal, 127, 129, 183, 225, 282, 283, 285
   self-adjoint, 221
   standard matrix representation, 58, 61, 69, 114, 252
   surjective, 46
   symmetric, 221
logically equivalent, 298
lower bound, 341, 342
LU-factorization, 98, 100

Maclaurin series, 180
mapping, 316
matrix, 55
   arithmetic, 55
   block form, 123, 125
   column of row vectors, 57

congruence, 118, 120
conjugate transpose/adjoint of, $A^\dagger$, 130, 131, 286
diagonal, 98, 129
Hermitian, 131, 190
inverse, 59, 60, 100, 104
inverse, determinant of, 153
invertible, 59, 91
leading entry, 86
lower triangular, 98–100
multiplication, 55
  associativity, 56
  linear transformation, 57
  not commutative, 56, 60
nonsingular, 93, 125, 283
nullity, 58
orthogonal, 128, 129, 132, 155, 182, 223, 225, 281, 283, 285
postmultiplication, 58, 60
premultiplication, 58
rank, 58
row of column vectors, 57
self-adjoint, 131, 132, 161, 190, 223
similarity class, 75
skew-symmetric, 120, 121
subdiagonal, 246
super-diagonal, 246
symmetric, 120, 121, 131, 132, 160, 223, 280
trace, 111, 117
transpose, 89, 108, 112, 114
triangular, 245
unitary, 132
upper triangular, 98–100, 105
matrix equivalence, 73, 74
matrix representation
  bilinear or Hermitian form, 116–121
  inner product, 163
  linear operator, 252
  linear transformation, 59, 62–64, 66–68, 70, 74, 75, 124, 241, 245
  rotation, 226
  unitary operator, 185, 186
maximal element, 341–345, 353, 354
minimal distance, 193
minimal element, 341–343
minimal polynomial, 203–206, 210–214, 231, 236, 241, 248, 251–253
  roots, 210
monomial, 24
multilinear mapping, 133

negation, 298
nilpotence index, 242, 243, 245–248
nilpotent, 243, 245–248
Nine Chapters on the Mathematical Art, 81
nondegenerate form, 118–120, 162, 176
nonpivot column, 86, 91
norm, 168
  induced by inner product, 165
normal equations, 195
normal form, 242, 246, 247
normal matrix, 223, 283
normal vector, 174
normalize a vector, 337, 338
normwise convergence, 165
null space, 58

octant, 329
odd permutation, 144
onto, surjective, 31, 316
orbit, 141
order of a finite group, 139
ordered basis, 255, 257, 270, 271
ordered set, 301
ordering, 21
  antisymmetric, 21
  chain, 21, 23, 341
  maximal element, 21
  partial, 21
  poset, 21, 23
  total, 21
  transitivity, 21
  upper bound, 21
oriented parallelogram, 270
oriented parallelotope, 271, 272
orthogonal complement of a set, 174, 175
orthogonal equivalence, 223
orthogonal group, $O_n(\mathbb{F})$, 128, 129, 224, 228, 255, 281–284
orthogonal projection, 192, 284
orthogonal reflection, 282–284
orthogonal set of vectors, 169, 170, 333, 335
orthogonal transformation, 183
orthogonal, at right angles, 168
orthonormal basis, 225, 226, 228, 255, 272, 283, 285, 291
  algebraic, 170
  analytic, 179
orthonormal set, 170

paralellogram, 273
parallel, 42, 312, 330–332
parallel class, 36, 43, 97
parallelepiped, 271, 273, 336, 337
parallelogram, 270, 337, 339
parallelogram law, 168
Parallelogram Rule, 323
parametric equations, 43, 81, 82, 307, 330, 332
partial ordering, 341, 344, 345, 352
partially ordered set, 21
partition, 36, 261, 262, 273, 313–315, 348
Pauli spin matrices, 189, 289
permutation, 139, 247, 249, 251, 255, 257, 259
  cycle, 140, 141
  even and odd, 144
  orbit, 141
  product of disjoint cycles, 142
  product of transpositions, 142
  sign, 144
  stabilizer, 146, 147
  transposition, 140
permutation matrix, 259
perpendicular, 333
perpendicular lines, 67, 69, 120
perpendicular, at right angles, 168
pivot (verb), 86, 101, 102, 105
pivot column, 86, 90–92
pivot position, 86, 89, 90
plane, 41
point, 41
polar coordinates, 129, 273, 328
polynomial, 24
  nth coefficient, 24
  nth term, 24
  annihilator, 201–205
  constant, 24
  constant term, 24
  degree, 24, 200
  degree one factor, 200
  division, 200, 201
  factor, 200
  factor multiplicity, 200
  irreducible, 200, 283
  leading term, 25
  like terms, 24
  mapping, 25
  monic, 25, 27
  multiplication, 26
  nonzero term, 24
  reducible, 200
  root, 200, 213, 326
  root multiplicity, 200
  zero, 24, 200, 213
polynomial function, 24
poset, 341–344, 352, 353
positive definite, 162, 332
power set, 343, 344, 346, 349
preimage, 31
primary decomposition, 241, 242, 249
Primary Decomposition Theorem, 240
prime factorization, 300
principle of duality, 275–277
probability, 94, 310, 311
projection mapping, 51–53, 121, 202, 206, 212
  matrix representation, 79
projection, $\text{proj}_\mathbf{w}(\mathbf{v})$, 169, 333, 334
projective geometry, 275
projective group, $PGL_n(\mathbb{F})$, 275, 279–281
projective line, 277–281
projective plane, 275–281
projective plane over a field, 278
projective point, 278–280
projective triangle, 277, 279–281
proof by contradiction, 299
pure quaternion, 268, 269, 287–289
pure tensor, 134
Pythagorean Theorem, 157, 333

quadrant, 329
quadrilateral, 335, 336
quaternions, $\mathbb{H}$, 255, 264–269, 287, 288
  modulus, 267
  pure quaternion, 265
  quaternion conjugation, 266
  real part, 265
  unit, $\mathcal{U}(\mathbb{H})$, 267, 269, 287
  vector part, 268
quotient space, 39

range, 31, 316
rank
  linear transformation, 46, 63
  matrix, 58, 63, 90, 92
rank equivalence
  linear transformations, 74, 75
  matrices, 74, 75, 231
Rank Theorem, 109
  linear transformations, 46

matrices, 59, 80, 91
rational numbers, $\mathbb{Q}$, 4
real affine plane, 325
real numbers, $\mathbb{R}$, 4
reduced row echelon form, 231
reflection, 206, 209, 285
    orthogonal, 67, 68, 129, 130
reflexive relation, 312
regular polygon, 258
relation, 312, 341
reverse triangle inequality, 168
right-hand rule, 335
right-handed coordinate system, 225, 288, 337–339
rotation, 129, 187, 211, 223, 264, 268, 282–285, 287–290, 338
    improper vs proper, 284
rotation matrix, 224
rotoreflection, 284
row echelon form, 86
    reduced, 86, 92
row equivalence, 85, 89, 92, 231
Row Reduction Algorithm, 87
row space, 91, 92
row vector, 11
Russell's Paradox, 301

scalar, 8
    nonzero, 9
scalar mapping, 204
scalar matrix, 248, 274, 275, 279
scalar multiple, 9
scalar multiplication, 8
scalar product, 332
scalar triple product, 271, 336, 337, 339
scaling, 8, 9
Schröder-Bernstein Theorem, 348, 354
self-adjoint operator, 189, 190
    eigenvalues, 221
self-dual, 277
sequence, 24, 47
sesquilinear, 159
set, 301
set, complement of, 304
set, indexed, 302
set, ordered, 301
set, universal, 303
sets, intersection of, 303
sets, union of, 302
shear, 208, 272
signed area, 270
similar linear transformations, 74, 203

similar matrices, 74, 75, 231, 248
similar operators, 210, 213
similarity class, 245, 248
singleton, 301
singular event, 93
skew-field, 267
skew-symmetric k-linear mapping, 137
skew-symmetric form, 120, 290, 294
skew-symmetric matrix, 121, 291
slope, 94, 307, 324, 325
solution space, 65
span, 14, 17, 22
    spanning set, 14, 18, 22
special linear group, $SL_n(\mathbb{F})$, 263, 270–274, 281, 293
special orthogonal group, $SO_n(\mathbb{R})$, 224, 228, 282–285, 287
special unitary group, SU_n, 286, 287, 289
Spectral Theorem, Finite-Dimensional, 222, 283
spherical coordinates, 283
standard matrix representation, 242
    linear transformation, 58, 61, 69
Strang, Gilbert, 191
sub-1 matrix, 246–248, 252
subdiagonal, 246
submatrix, 86
subset, 301
sum, 10
sum of subspaces, 47
super-1 matrix, 246, 248, 249
super-diagonal, 246
surjective, onto, 31, 316
symmetric bilinear form, 119, 121, 160, 294
symmetric group on k letters, 139, 140, 270
symmetric matrix, 121, 294
symmetric operator, 190
symmetric relation, 312
symmetry, 270
symmetry group, 258
symplectic basis, 292, 294
symplectic form, 290, 291, 293, 294
symplectic form, matrix representation of, 291, 292
symplectic group, $Sp_{2n}(\mathbb{F})$, 290, 292–294
symplectic matrix, 292
symplectic space, 290–292, 294

Index

system of linear equations, 77, 330, 331
   a solution, 78
   associated system of normal equations, 195
   augmented matrix, 78
   coefficient matrix, 77
   consistent, 78, 91
   equivalent, 78, 83
   homogeneous, 77, 82
   inconsistent, 82, 88
   matrix equation, 78
   nonsingular, 93
   the solution, 78, 80
   the solution set, 91

total ordering, 341
totally ordered set, 341, 342
trace bilinear form, 117, 119, 160, 164
transitive relation, 312
translation, 43, 226
transposition, 140
triangle inequality, 158, 165
triangular matrix, 232
triangularizability, 232
trinomial, 25

uncountable, 344, 347–349
union of sets, 302
union, disjoint, 303
unit circle, 255, 265, 267, 285, 287
unit complex numbers, $\mathcal{U}(\mathbb{C})$, 269, 282, 283, 285, 286
unit coordinate vectors, 269, 334, 337
unit pure quaternions, 268, 269
unit quaternions, $\mathcal{U}(\mathbb{H})$, 269, 287, 288
unit sphere, S^n, 255, 267, 268, 287, 289
unit vector, 169–171, 180, 226, 289, 333, 337
unitary equivalence, 223
unitary group, U_n, 183, 285, 286
   on $\mathbb{C}^n$, 131
unitary matrix, 131, 155, 182, 185, 186, 285, 289
unitary transformation, 182–186, 285, 289
universal set, 303
upper bound, 341–345, 353, 354

Vandermonde matrix, 125
vector, 3, 8, 321–324, 330, 332–335
   addition, 8
   additive inverse, 8
   negative, 8
   nonzero, 9
   unit, 169–171, 180, 226, 289, 333, 337
vector space, 3, 8–12, 260, 264, 321, 323, 324, 326, 330, 332
   n-dimensional, 3
   ambient, 14
   complex, 12, 29, 130
   coset, 35–41, 78–80, 138, 146, 233
   dimension, 19, 20, 22, 23
   exterior power of, 137–139
   finite-dimensional, 18
   infinite-dimensional, 18
   nontrivial, 9
   polynomials, 24
   proper subspace, 9
   quotient space, 38–41, 45, 46, 53, 137, 261
   rational, 12
   real, 3, 12, 29
   subspace, 8, 13, 259
      four fundamental subspaces, 191
   tensor power, 134
   trivial, 9
vector, components, 333, 334
vector, coordinates, 322
vector, direction, 322
vector, magnitude, 322
vector, magnitude and direction, 321–323
Venn diagram, 303

well-definedness, 35, 38, 40, 273, 306, 307, 343
Well-Ordering Principle, 342, 347
Weyl, Hermann, 290
word, 309

zero, 3, 4, 7, 10, 144, 311, 326
zero divisor, 5, 6, 27
zero mapping, 12, 15, 28, 32, 49, 73, 109, 175, 202, 203, 235
zero matrix, 87, 117, 202
zero vector, 8–11, 14, 15, 38, 135, 169
zero, division by, 306
Zorn's Lemma, 21, 22, 341, 343–345, 348, 352, 354

Selected Published Titles in This Series

57 **Meighan I. Dillon,** Linear Algebra, 2023
55 **Joseph H. Silverman,** Abstract Algebra, 2022
54 **Rustum Choksi,** Partial Differential Equations, 2022
53 **Louis-Pierre Arguin,** A First Course in Stochastic Calculus, 2022
52 **Michael E. Taylor,** Introduction to Differential Equations, Second Edition, 2022
51 **James R. King,** Geometry Transformed, 2021
50 **James P. Keener,** Biology in Time and Space, 2021
49 **Carl G. Wagner,** A First Course in Enumerative Combinatorics, 2020
48 **Róbert Freud and Edit Gyarmati,** Number Theory, 2020
47 **Michael E. Taylor,** Introduction to Analysis in One Variable, 2020
46 **Michael E. Taylor,** Introduction to Analysis in Several Variables, 2020
45 **Michael E. Taylor,** Linear Algebra, 2020
44 **Alejandro Uribe A. and Daniel A. Visscher,** Explorations in Analysis, Topology, and Dynamics, 2020
43 **Allan Bickle,** Fundamentals of Graph Theory, 2020
42 **Steven H. Weintraub,** Linear Algebra for the Young Mathematician, 2019
41 **William J. Terrell,** A Passage to Modern Analysis, 2019
40 **Heiko Knospe,** A Course in Cryptography, 2019
39 **Andrew D. Hwang,** Sets, Groups, and Mappings, 2019
38 **Mark Bridger,** Real Analysis, 2019
37 **Mike Mesterton-Gibbons,** An Introduction to Game-Theoretic Modelling, Third Edition, 2019
36 **Cesar E. Silva,** Invitation to Real Analysis, 2019
35 **Álvaro Lozano-Robledo,** Number Theory and Geometry, 2019
34 **C. Herbert Clemens,** Two-Dimensional Geometries, 2019
33 **Brad G. Osgood,** Lectures on the Fourier Transform and Its Applications, 2019
32 **John M. Erdman,** A Problems Based Course in Advanced Calculus, 2018
31 **Benjamin Hutz,** An Experimental Introduction to Number Theory, 2018
30 **Steven J. Miller,** Mathematics of Optimization: How to do Things Faster, 2017
29 **Tom L. Lindstrøm,** Spaces, 2017
28 **Randall Pruim,** Foundations and Applications of Statistics: An Introduction Using R, Second Edition, 2018
27 **Shahriar Shahriari,** Algebra in Action, 2017
26 **Tamara J. Lakins,** The Tools of Mathematical Reasoning, 2016
25 **Hossein Hosseini Giv,** Mathematical Analysis and Its Inherent Nature, 2016
24 **Helene Shapiro,** Linear Algebra and Matrices, 2015
23 **Sergei Ovchinnikov,** Number Systems, 2015
22 **Hugh L. Montgomery,** Early Fourier Analysis, 2014
21 **John M. Lee,** Axiomatic Geometry, 2013
20 **Paul J. Sally, Jr.,** Fundamentals of Mathematical Analysis, 2013
19 **R. Clark Robinson,** An Introduction to Dynamical Systems: Continuous and Discrete, Second Edition, 2012
18 **Joseph L. Taylor,** Foundations of Analysis, 2012
17 **Peter Duren,** Invitation to Classical Analysis, 2012
16 **Joseph L. Taylor,** Complex Variables, 2011

For a complete list of titles in this series, visit the
AMS Bookstore at **www.ams.org/bookstore/amstextseries/**.